Springer Series in Statistics

More information about this series at http://www.springer.com/series/692

Anuj Srivastava • Eric P. Klassen

Functional and Shape Data Analysis

 Springer

Anuj Srivastava
Department of Statistics
Florida State University
Tallahassee, FL, USA

Eric P. Klassen
Department of Mathematics
Florida State University
Tallahassee, FL, USA

ISSN 0172-7397   ISSN 2197-568X   (electronic)
Springer Series in Statistics
ISBN 978-1-4939-8155-7   ISBN 978-1-4939-4020-2   (eBook)
DOI 10.1007/978-1-4939-4020-2

Printed on acid-free paper

This Springer imprint is published by Springer Nature
The registered company is Springer Science+Business Media LLC New York

1923–2016

In memory of Prof. Ulf Grenander, who led the way

Preface

Function and shape data analysis are old topics in statistics, studied off and on over the last several decades. However, the early years of the new millennium saw a renewed focus and energy in these areas. This focus was both exciting, because it sought new directions and resources, and productive, because it was application oriented and data driven. While the new interest was fueled by many factors, the most prominent amongst them was increasing availability of large datasets involving function and curve data, especially in the fields of computer vision and medical imaging. It was also propelled by increases in computational power and storage, a growing interest in Riemannian methods, and a favorable atmosphere for the confluence of ideas from geometry and statistics. Despite a long history of research and methods in function and shape analysis, several researchers took a fresh look at shape analysis during this period. As a result, they came up with novel approaches, based on mathematical tools that were new to this community, and made them practical using elegant computational solutions. This book started as a research monograph, as an exposition of these contemporary ideas in shape analysis of curves. It was heavily motivated by our desire to provide a self-contained treatment of this new generation of methods in shape analysis of curves, with a focus on statistical modeling and inference. However, like many other book projects, this project also grew beyond its original plan. Not only did it become a textbook with appendices, background material, and exercises, but also its scope grew to include a detailed treatment on function data analysis. Not surprisingly, it took much longer than expected to finish this manuscript. This delay allowed us to offer several graduate courses in statistics on the basis of this textbook.

The main topic area of this book is shape analysis of functions and curves—in one, two, and higher dimensions—both closed and open. What differentiates this material from past approaches is that it integrates the *registration problem* into shape analysis. Registration is concerned with matching of points across objects when their shapes are being compared and quantified. The past methods mostly treated registration as a pre-processing step, handled using an arbitrary off-the-shelf technique, followed by an unrelated metric for shape comparison. Instead, this textbook seeks a unified, comprehensive solution. It develops elegant Riemannian frameworks that provide both quantification of shape differences and registration of curves at the same time. Additionally, these methods are used for statistically summarizing given curve data, performing dimension reduction, and modeling the observed variability. This textbook investigates different mathematical representations and associated (invariant) Riemannian metrics that play a role in facilitating shape analysis. The focus is largely on certain square-root representations that *flatten* shape spaces and allow more traditional vector-space-based statistical analyses to become applicable.

Our main objective in writing this book is to familiarize graduate students with a broad array of tools that are relevant in developing computational solutions for shape and related analyses. These tools, gleaned from geometry, algebra, statistics, and computational science, are traditionally scattered across different courses and departments, which makes it difficult for graduate students to learn them in a reasonable time. While we wish to introduce readers to a diverse array of topics that are fast becoming important in current research, this book is not intended to be an elaborate mathematical treatment of those topics. Naturally, one cannot expect to become an expert in such multidisciplinary areas solely on basis of a single book. Besides, there are already many wonderful books that cover these items individually in great detail. Indeed, in this textbook, we have frequently been lax in presenting in-depth technical details, focusing instead on intuitive explanations and implementable solutions. Our hope is to provide a *working knowledge* of relevant material that is often present across different disciplines and better prepare graduate students for handling future scientific challenges.

This textbook is intended for graduate students who are interested in improving their skills in this broad interdisciplinary problem area. These students can be from statistics, engineering, applied mathematics, neuroscience, biology, bioinformatics, and other related areas. It seems necessary for a reader to have a background in calculus, linear algebra, and numerical analysis and computations. Of course, one can selectively read chapters of his/her interest, and a sequential reading of the entire chapter is not a necessity. This book covers a broad range of ideas, from introductory theory to algorithmic implementations and some statistical case studies. It is self-contained in terms of the background material needed for understanding methods described here. The appendices and background material are highly recommended for students without prior coursework in geometry and algebra.

The textbook starts with a motivation for development of knowledge in this area, by citing a number of current applications that are primarily dependent on analyzing shapes of functions and curves. In particular, it makes a case of using *elastic* functional representations of these objects, rather than sampling them with points in a pre-determined manner. Chapter 2 summarizes some current techniques in shape data analysis. These techniques mostly rely on point-based representations of objects and lack a natural system for registering points across objects. In order to understand the theory presented in this textbook, a reader needs a working knowledge of some relevant parts of algebra and geometry, and that is provided in Chapter 3. Chapter 4 deals with functional data analysis, starting with some standard tools from Hilbert space theory. Then, it raises the problem of pairwise registration of functions, highlights the shortcomings of what may be considered a natural solution (based on the $\mathbb{L}^2$ norm), and proposes a better solution using the SRSF (square-root slope function) representation of functions and extending the Fisher-Rao Riemannian metric to general function spaces. We take this opportunity to formally introduce the concepts of amplitude and relative phase of functional variables. This chapter ends with impositions of the Fisher-Rao metric on some spaces of interest, such as the sets of probability density functions and warping functions.

We present methods for shape analysis of planar curves in Chapter 5, with an emphasis on the elastic Riemannian metric and the square-root velocity function (SRVF) representation. This chapter describes the elastic metric and uses the SRVF mapping to convert the elastic metric into the standard $\mathbb{L}^2$ inner product. This allows for simultaneous registration of points and deformation of curves or *geodesics* using efficient numerical techniques. Chapter 6 focuses on *closed curves* by imposing an additional constraint of closure on the allowed curves. This nonlinear constraint results in representation spaces becoming nonlinear manifolds, and we introduce two general approaches—*shooting method* and *path-straightening method* for computing geodesic paths on such manifolds. In each case, we provide step-by-step algorithms and many examples to illustrate the ideas.

Chapter 7 serves as a general discussion on defining and computing summary statistics on sets that are not vector spaces. Here one uses ideas from differential geometry to specify the notions of sample means and covariances, and some associated parametric probability density functions on these sets. This is followed by tools for modeling functional data in Chapter 8, where an important idea is to separate functions into their amplitude and phase components, followed by modeling these components individually, but not independently. This chapter makes repeated using of functional principal component analysis (FPCA) for reaching finite-dimensional Euclidean representations of complex functional data. These representations are then used in analyzing, testing, classifying, and clustering given functions. Chapter 9 extends these ideas to modeling and analysis of planar curves using the representations developed in Chapters 5 and 6. It illustrates the use of geometries of shape spaces to locally linearize and to perform FCPA to reach tractable representations of planar shapes. These representations are then used in modeling and classifications of shapes using statistical shape analysis. Chapter 10 extends these ideas to curves in arbitrary Euclidean spaces and illustrates some of these ideas using examples from interesting application areas. The textbook ends with a collection of miscellaneous topics relating to shape analysis of curves.

This textbook can also serve as a textbook for an advanced topics course with interdisciplinary flavor. There are several possibilities in terms of teaching a course from this textbook. We list some of these ideas, but of course one can mix and match these ideas as well:

- A one-semester course in function data analysis using Appendices A and B, Chaps. 3, 4, 7, and 8, in conjunction with a more classical text on FDA.
- A one-semester course in shape data analysis using Appendices A and B, Chaps. 2, 3, 5, 6, 7, and 9.
- A two-semester course in function and shape data analysis using Appendices A and B, Chaps. 2–11, possibly in conjunction with a more classical text on FDA.

Tallahassee, FL, USA   Anuj Srivastava
                                                                            Eric P. Klassen

Acknowledgments

We were able to finish this textbook due to significant contributions from a large number of people. Indeed, this book uses a body of knowledge that was developed in collaboration with a number of wonderful colleagues. Without their contributions, this material would have never matured into this form. Therefore, we are deeply in debt to these people for their contributions and support.

Several of the ideas presented here were developed in collaboration with Prof. Ian H. Jermyn of Durham University, Durham UK. Indeed, he was the first one to introduce us to the literature on square-root representation of probability density functions and its connections with the Fisher-Rao metric. In some sense, this whole effort on elastic shape analysis of curves started for us at that moment in the summer of 2006. We thank him for all his shared wisdom and collaboration. We also acknowledge the efforts of Dr. Shantanu Joshi of UCLA, who was the first of many graduate students who helped transform conceptual ideas into implementations and demonstrated their impact on real data! We are thankful to all of our graduate students—Darshan Bryner, Wade Henning, David Kaziska, Sebastian Kurtek, Jose Laborde, Sayani Lahiri, Wei Liu, Sentibaleng Ncube, Dan Robinson, Michael Rosenthal, Chafik Samir, Jingyong Su, J. Derek Tucker, Linda Crystal White, Qian Xie, and Zhengwu Zhang—all of them chose to work on and develop topics covered in this book for their PhD dissertations. We also received continuous encouragement, comments, and feedbacks from many colleagues during the writing of this book. Of those, we point out Boulbaba Benamor, Rama Chellappa, Mohamed Daoudi, Zhaohua Ding, Hassen Drira, Hamid Laga, Steve Marron, Pavan Turaga, Wei Wu, and Jinfeng Zhang. We are very thankful to them for all their support.

This book exhibits, in its own small way, the beauty of general pattern theory pioneered by Prof. Ulf Grenander of Brown University. The most wonderful aspect of this theory is the rich set of tools it uses: algebra, geometry, statistics, and, of course, computational science. Our wanderings beyond classical multivariate and Euclidean statistics followed paths previously illuminated by Grenander's excursions. *"Discretize as late as possible"*—This was Prof. Grenander's advice to students in a class on pattern theory at Washington University in 1995 and became a source of inspiration for the material presented here. The first author is forever grateful to Prof. Grenander for his teachings, guidance, and mentorship. He also gratefully acknowledges training and guidance received from Prof. Michael Miller of Johns Hopkins University.

We also acknowledge support from the Army Research Office for our early work in shape analysis, especially with a focus on geometry and statistics. This effort was also supported in part by grants from the Office of Naval Research, the National Science Foundation, and the National Institutes of Health. The initialization and culmination of this writing project coincided with sabbatical leaves for AS during 2007–2008 and 2014–2015. He, therefore, acknowledges support from the University of Lille 1, Lille, France, and the Franco-American Fulbright commission in Paris, France, for supporting his long stays in France. AS also acknowledges support from the Statistics Division at the National Institute of Standards and Technology, Gaithersburg, MD, in completing this project.

We are thankful for a highly supportive environment for interdisciplinary research and creativity in our respective departments, Statistics and Mathematics, and the College of Arts and Sciences at FSU. We are thankful to our colleagues in these units for encouraging us to take up these challenges.

Finally, and most importantly, we thank our families for their support and encouragement in accomplishing this project. AS is very thankful of his family—wife Elise, parents Govind and Indraprabha, sons Alex and Neel, and parents-in-law André and Marie-Cécile—for their continuous support and cheer. EK thanks his wife Anna and daughter Rosemary because the happiness they bring to his life makes math so much more fun.

Contents

Chapter 1
Motivation for Function and Shape Analysis

This textbook is dedicated to the study of functional data analysis and shape analysis of curves in Euclidean spaces. In the first item, one develops tools for statistical analysis of real-valued functional data on fixed intervals. While function data analysis is a broad topic area, worthy of a textbook in itself, we will focus heavily on a specific aspect that deals with alignment or registration of functional data. In the second item, one studies shapes formed by curves in 2D, 3D, and higher dimensions, with a goal of performing statistical inferences. Since these curves are also functions, albeit vector valued, and the issue of registration is of prime importance in their shape analysis, we will cover these topics under a broad umbrella of *elastic* functional and shape data analysis!

There are several meanings of the word *shape* in the English language. According to the Oxford English Dictionary, when this word is used as a noun, it may mean:

1. *The external form or appearance of someone or something as produced by their outline.*
2. A piece of material, paper, etc., made or cut in a particular form.
3. A particular condition or state: the house was in poor shape.
4. A specific form or guise assumed by someone or something: a fiend in human shape.
5. Definite or orderly arrangement.

It is the first, and perhaps the most common, usage of the word shape that interests us in this book. The words *form, appearance, and outline* are all key to defining and understanding shapes. Shape is a basic, integral physical property of objects that plays a major role in analysis of their appearances. From developing a neuroscientific understanding of human vision to developing algorithms for automated image understanding, shape analysis is intimately involved in many processes. The importance of shape is highlighted by the fact that a growing child learns about the shapes and colors of objects before learning the alphabet or the numbers. It is no coincidence, thus, that many toys for toddlers involve matching simple shapes. For all human beings, observing a scene and identifying objects in that scene involves analyzing shapes and colors in a fundamental way, although many intricate details of that process are yet to be discovered and understood. A specific situation we are interested in arises when the actual scenes are replaced by their images. Since the scenes are imaged by cameras, or other similar sensors, the interpretations can naturally benefit from analysis of shapes of the objects

© Springer-Verlag New York 2016
A. Srivastava, E.P. Klassen, *Functional and Shape Data Analysis*,
Springer Series in Statistics, DOI 10.1007/978-1-4939-4020-2_1

contained in those scenes. Therefore, shape has emerged as one of the important ways to characterize objects seen in images. Although the notion of shape is used widely across the literature, it is seldom made precise in a mathematical sense. This may not be needed in all contexts, but if we wish to use computers to analyze images, then a formal, mathematically precise treatment of shapes is of profound importance. We start by motivating the use of shape analysis in a few applications.

1.1 Motivation

What is the motivation for studying functional and shape data analysis? Furthermore, why should shape analysis be performed using curve or functional data? We address these questions next.

1.1.1 Need for Function and Shape Data Analysis Tools

We start with the need for studying function data. The problem of statistical analysis in function spaces is important in a wide variety of applications, arising in nearly every branch of science, ranging from speech processing to geology, biology, and chemistry. One encounters many problems where the observations are real-valued functions on an interval, and the goal is to perform their statistical analysis. By statistical analysis we mean *to compare, align, average, and model* a collection of such random observations. These problems can, in principle, be addressed using tools from functional analysis, e.g., using the $\mathbb{L}^2$ Hilbert structure of the function spaces, where one can compute $\mathbb{L}^2$ distances, cross-sectional (i.e., point-wise) means and variances, and principal components of the observed functions. However, a serious challenge arises when functions are observed with flexibility or domain warping along the x axis. This warping may come either from an uncertainty in the measurement process or may simply denote an inherent variability in the underlying process itself that needs to be separated from the variability along the y axis (or the vertical axis). As another possibility, the warping may be introduced as a tool to horizontally align the observed functions, reduce their variance, and increase parsimony in the resulting model. Keep in mind that we allow only the x-axis (the domain) to be warped and the y-values to change only consequentially. In this situation, the mathematical framework needed for function data analysis is closely related to the notion of shape analysis of curves.

Moving on to the motivation for shape analysis of curves, we first consider traditional 2D image analysis where the goal is to try to recognize objects in images or videos. We consider this problem because images have become one of the largest sources of data in our digital society. The objects of interest are characterized by their appearances, and one looks for some distinguishing features that can be used to classify these appearances. There are two distinct (but not independent) features that can be used for such classifications: shapes and textures. The boundaries or silhouettes of objects in images characterize the objects themselves to a certain degree, and, thus, the shapes of these boundaries become an important feature. The other type of feature is the texture, or the pattern of pixel values, formed by the pixels falling on the object in the image. These pixels result from

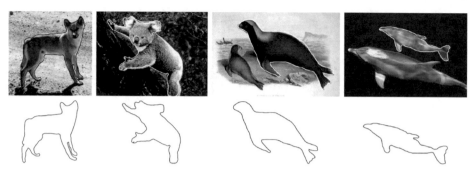

Fig. 1.1 Appearance of objects in images (*top*) can be partially characterized and analyzed using the shapes of their boundaries (*bottom*)

an electromagnetic process that involves a combination of the incident light, the surface reflectivity, the ambient light, and the observation angle and can be modeled using principles of physics. The patterns formed by pixels have already been utilized in a variety of ways by many researchers for detecting, classifying, and recognizing objects in images. It should be clarified that, eventually, in an image-based recognition system, we envision the use of shapes in conjunction with, rather than instead of, texture analysis.

Tools for studying shapes of objects in images have tremendous potential. The general area of computer vision that seeks to analyze and annotate elements of static and video images is full of situations where shape features play important roles. For example, the top row of Fig. 1.1 shows images of a few animals, whereas the bottom row displays only the silhouettes of these creatures. Even if one restricts to these contours as the only source of information, there is a strong possibility that we can successfully categorize these animals. We would like to develop automated procedures for this and related tasks. Of course, the task of extracting contours or silhouettes from real images is nontrivial in itself. However, to focus on the main task, i.e., shape analysis, we can assume that the contours in images are readily available. Then, one would like to develop a computer program, founded on sound mathematical principles, that can classify contours on the basis of their shapes, in a completely automated fashion.

1.1.2 Why Continuous Shapes?

In all the motivating examples provided so far and later on, we have targeted *functions* or *continuous curves* for shape analysis. We have identified continuous objects despite the fact that the actual observations of functions and shapes are often discrete, a finite number of points on a certain parameterization domain. An alternative representation of objects that uses a finite, sampled set of points is quite common in the literature. So a reader may ask the question: What is the need to study continuous objects? Analysis of functions and curves, instead of more common vectors, brings new challenges since they are elements of infinite-dimensional spaces. Are we unnecessarily complicating our problem by focusing on these infinite-dimensional representations of shapes? In our opinion the answer to the last question is no! Here is why:

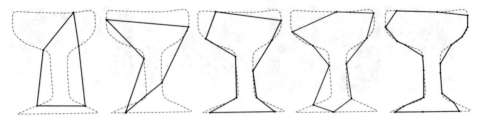

Fig. 1.2 Illustration of variability in shapes of polygons resulting from different samplings of the same continuous curve

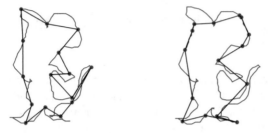

Fig. 1.3 Representation of shapes by point sets fixes their pairing or matching across objects. A better approach is to consider underlying contours and to solve the registration problem at that level

1. Firstly, we argue that underlying the shapes we observe—natural, man-made or biological—are continuous objects with precise contours. Thus, it is fundamentally superior to study/analyze these continuous shapes even though their observations are often obtained by "sampling" points along those shapes. Since the process of sampling, in general, is often random, one should not completely rely on the sampled points for analyzing the underlying shapes. As an example, consider the continuous curve shown in Fig. 1.2. Here we see five different examples of samplings of this curve using a finite number of points. In these examples the number and the placements of the sampled points are different, and, naturally, result in sampled curves, or polygons, that are quite different in shape from each other. If we restrict to the sampled points only, it will be difficult to infer that all these polygons are observations of the same shape. Instead, our approach mimics the construction in Fig. 1.2. As a first step, we shall develop statistical models for shapes of continuous functions and curves. Then, we will develop models for sampling these continuous shapes and for generating finite observations. Thus, a fully statistical framework will allow us to analyze shapes of finite point sets by relating them to infinite-dimensional shape models in a principled way.
2. Secondly, one of the most important problems in shape analysis is registration of points. Registration is concerned with setting up a correspondence between points on one curve and points on a second curve. In case one restricts to discrete representations of functions and curves, resulting in finite vectors, the pairing of points across vectors is already determined. The first point is matched with first, second with second, and so on. However, that may be far from optimal in general situations. Consider Fig. 1.3 where a curve is sampled twice using the same number of points. If one restricts to ordered lists of the sampled points for each sampling, the pairing of points seems quite arbitrary. In fact, it is

hard to find a linear registration that will suffice for the two samplings shown here. A better approach is to analyze underlying contours and to solve for the registration problem at that level.

1.2 Important Application Areas

What are the important application areas for function and shape analysis?

Biological Growth Curves Function data appears fundamentally in biological systems that measure some aspect of growth. Consider, for example, the height evolution of subjects in the famous Berkeley growth data.[1] Figure 1.4 shows height functions (left) and their time derivatives (right), for female and male subjects, to highlight periods of faster growth. Although the growth rates associated with different individuals are different, it is of great interest to discover broad common patterns underlying the growth data, particularly after aligning functions using time warping. This requires advanced techniques, beyond just a standard Hilbert space structure, to discover underlying patterns and to make inferences.

Mass Spectrometry Data Analysis The use of mass spectrometry data to profile metabolites present in a specimen is important in biomarker discovery, enzyme substrate assignment, drug development, and similar applications. The use of liquid chromatography-mass spectrometry (LC-MS) is quite common, and it provides data for retention times of different metabolites identified by peaks in observed chromatograms at the corresponding retention times (x axis). However, an analysis is faced with the challenge of random nonlinear shifts in the peaks, due to variability in retention times across measurements and equipment. Figure 1.5 show an example of proteomics data collected for patients having therapeutic treatments for acute myeloid leukemia (AML). These two functions represent the same chemical specimen but display a nonlinear misalignment in their peaks. Thus, an important challenge in LC-MS data analysis is to align peaks in a principled way using nonlinear time warping.

Biosignals Computer intervention in medicine provides many advantages ranging from automatic and quantitative analysis to removal of subjective variation in diagnosis across physicians. In many medical applications, the quantities used for disease diagnosis and monitoring are functions of time such as blood pressure, electrocardiogram signals, gait measurements, and so on. But, because of the

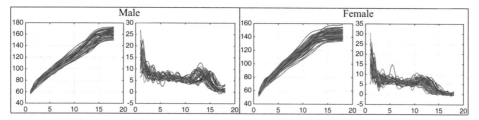

Fig. 1.4 Berkeley growth data for female and male subjects. For each gender we show the height functions and their time derivatives

[1] http://www.psych.mcgill.ca/faculty/ramsay/datasets.html.

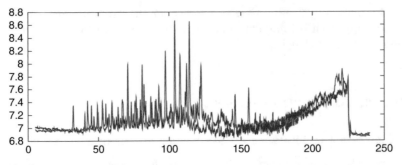

Fig. 1.5 Example of LC-MS data samples from a proteomics study

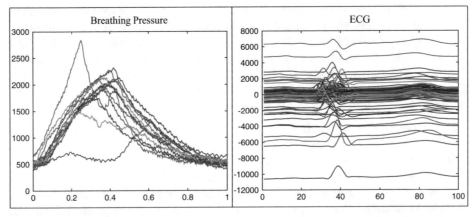

Fig. 1.6 Examples of biosignals used in medical diagnostics: Left side shows breathing pressure profiles and the right side shows some ECG signal recordings

large subject variability in the studies, the observed random temporal variability has to be correctly accounted for. Figure 1.6 shows some examples of biosignals, coming from breathing pressure measurements and ECG recordings, that require alignment techniques for statistical modeling. Another example of a biosignal is given by neuronal spike trains used to communicate information from the brain to different parts of the body. Although the original data is discrete, researchers frequently convolve these spike trains with Gaussian kernels to convert them into smooth data, ready for function data analysis, as shown in Fig. 1.7.

Biometric Security A really important, and fascinating, problem is the characterization of individual humans using easily measurable traits, for applications ranging from security to social media to welfare services. In cooperative situations, where the subjects are willing or interested in providing data, one can use more intimate biometrics such as fingerprints, palm prints, ear shapes, nose profiles, iris, knuckle outlines, etc. In the noncooperative environments, such as surveillance situations, the imaging sensors dominate. A large variety of sensors—visible-spectrum cameras, infrared or night vision imagers, hyperspectral sensors, satellite and aerial cameras, surveillance video monitors—are all capable of generating high-throughput volume data that requires automated processing. Since the shape has a potential for characterizing objects in images, it becomes important for security applications. Consider the problem of recognizing human beings when observed from a distance in a noncooperative environment using a remote camera.

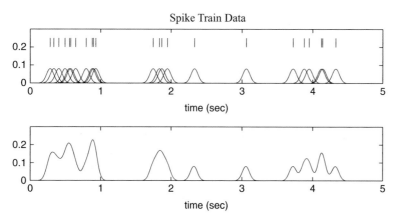

Fig. 1.7 An example of a neuronal spike train converted into function data

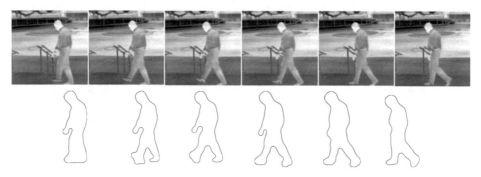

Fig. 1.8 Examples of temporal evolution of human silhouettes (*bottom*) in an infrared video sequence (*top*)

One possibility is to study the gait, i.e., the style of walking, of a person to recognize them from a distance. In a typical scenario for gait-based human recognition, one obtains a sequence of silhouettes of a person, and the goal is to study this temporal evolution of shapes and to compare it across people. Shown in Fig. 1.8 is an example of this case where the top row shows a sequence of surveillance-quality infrared images of a person walking, while the bottom row shows the corresponding sequence of silhouettes obtained from images using certain image processing techniques. An emerging and related area is activity recognition, where one tries to characterize activities of human subjects using video data. An activity is made up of a sequence of actions performed in a certain order and at a certain execution rate. One of the central features being used currently in detecting actions and categorizing activities is the shape.

Anatomical Shape Analysis One of the most important areas for applying shape analysis is in performing medical diagnosis using medical images. Noninvasive imaging has become the centerpiece of modern diagnostics and treatments. Imaging modalities here include sensors like ultrasound, X-rays, MRI, fMRI, CT-scans, etc. These images typically are low resolution and low contrast, relative to even a cheap (visible-spectrum) digital camera. Therefore, the methods based on texture analysis are not as reliable, and shape analysis becomes even more important in this situation. If one considers analysis of medical images by human experts, i.e., physicians, this analysis currently relies on a coarse quantification of shapes using distances and sizes. For example, when quantifying growths and decays of tumors, the temporal changes are routinely monitored by manually drawing polygons around the tumors and by measuring simple quantities such as perimeter, diameter, and thickness. Shape changes in organs are important indicators of the progression of diseases and general health. In particular, ultrasound imaging is used in cardiology (echocardiography), gynecology, obstetrics, and endocrinology to study the shapes, sizes, locations, and orientations of relevant anatomical parts. It seems natural to use a full analysis of shapes, rather than restrict to some coarser representations, to perform medical diagnosis. Shape estimation and analysis is also important in locating organs for planning radiation therapy. Although there are a large number of applications of shape analysis in medical diagnostics, we provide only a few examples.

In current medical image analysis, a prominent source of data is magnetic resonance imaging (MRI), which provides 3D images of anatomical parts. In particular, diffusion-tensor MRI (DT-MRI) has become increasingly popular to study the "connectivity" of subparts of the brain and other organs (for instance, refer to human connectome project `http://humanconnectomeproject.org`). This connectivity is conceptualized through the tracts formed by bundles of neurons, termed *fiber tracts*, that form "pipelines" for transmitting signals and thus information from one region to another. An ongoing effort in the medical image community is to study the shapes of these fiber tracts to associate them with functionality of the subparts and the health of the patient. Shown in Fig. 1.9 is a graphical rendering of the bundles of fiber tracts in a human brain; these fiber tracts were estimated from DT-MRI scan of a human brain.

Bioinformatics Bioinformatics is the science of using computational and statistical approaches to the field of molecular biology. This is a fast growing area of

Fig. 1.9 A graphical rendering of fiber tracts in a human brain, generated from DT-MRI data (Data courtesy of Wikipedia Commons)

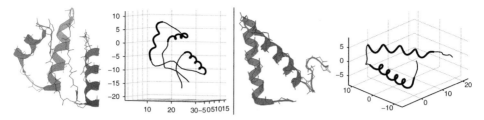

Fig. 1.10 Structural analysis of proteins includes shapes analysis of their backbones. The figure shows backbones of two simple proteins: 1CTF on the left and 2JVD on the right

research and application where a large suite of techniques from statistical pattern recognition is needed. In particular, there is an important need to associate patterns of shapes of biochemical structures, such as proteins, with their functionality. Proteins are linear polymers formed by concatenating amino acids. The primary structure in a protein is a linked chain of carbon, nitrogen, and oxygen atoms known as the backbone. Additionally, a protein also has side chains that are connected to the backbone, and it is the combined effect of the backbone and the side chains that ultimately determines the three-dimensional structure of a protein. There are two types of protein structure comparison problems, comparison of backbone structures (structure alignment) and comparison of the binding or active sites of proteins (surface matching). The backbones contain certain distinct geometrical pieces and one prominent type is the so-called α-helix. Figure 1.10 shows examples of backbones of two simple proteins—1CTF and 2JVD—that contain three and two α-helices, respectively. In analyzing shapes of backbones, it seems important to match not only their global geometries but also the local features (such as α-helices) that appear along these curves.

Forensic Applications Researchers for long have been interested in developing systems for automated reading and understanding of human handwriting. One application of this tool can be in automated sorting of mail by the postal service. Another use can be in authentication of human signatures in banks, credit card transactions, wills, and so on. The structure of an alphabet, a word, or a sentence can vary greatly depending on the writer, and that makes automated reading a challenging problem. Since signatures are forged by either other humans or machines, there is a possibility of these writers introducing their own characteristics in the process. The top part of Fig. 1.11 shows examples of genuine signatures and forgeries. The bottom part displays a representation of signatures based on the magnitude of the second derivatives along these curves. A systematic comparisons of shapes of letters in signatures can lead to automated techniques for signature authentication.

General Morphometrics Beyond the applications involving image data, there are several branches of sciences—archeology, fisheries, cartography, botany, biochemistry, etc.—that involve the study of shapes of objects. In some of these applications, data is available in the form of images, although that is not always the case. In Fig. 1.12, we show four examples of the pubis bone of a dinosaur from the Hadrosauridae family. Here, one is interested in studying shapes of such bones for inferring phylogeny, i.e., relatedness among various groups of dinosaurs during their evolution. Figure 1.13 shows an example from morphometrics in fisheries

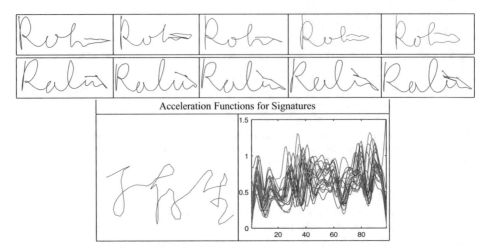

Fig. 1.11 Examples of signatures for a person. The top row shows genuine signatures and the middle shows forgeries. The bottom row shows a sample signature on the left and the acceleration functions associated with all genuine signatures of that class (Data courtesy Signature Verification Competition 2004)

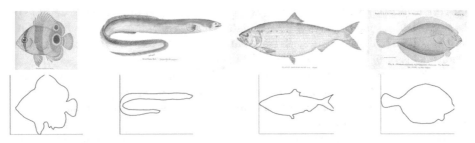

Fig. 1.12 Images of pubis bones of a Hadrosaurid dinosaur (Data courtesy Mr. Albert Prieto-Marquez of Florida State University)

Fig. 1.13 Images of some fishes (*top*) and their boundaries (*bottom*) (Data courtesy the Surrey fish database [80])

where the objective is to cluster and classify aquatic animals using their shapes, commonly observed as images. The Surrey database [80] contains silhouettes of around 1100 marine animals, a rich source for characterizing species according to the shapes of their specimens.

One potentially important application of shape analysis in images can be the recognition of symbols (alphabets) from gestures in a sign language. Shown in Fig. 1.14 are several pictures of hand gestures forming English alphabets in American Sign Language (ASL). Although the full pictures are important for precise distinction between symbols, a shape analysis of hand contours may provide an acceptable level of distinction between the signs. Note that, in the context of a

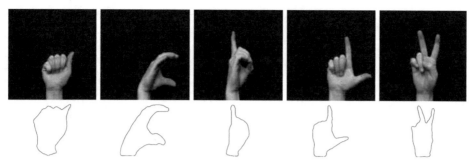

Fig. 1.14 Examples of gestures from the American sign language alphabet (Data courtesy ASL alphabets)

language, one does not need to have each and every alphabet decoded correctly; instead, the neighboring symbols in a sentence can provide additional knowledge for classification and language understanding.

In summary, there are a large number of applications, spread across the vast landscape of science and technology, where tools for shape analysis can play a central role in solving problems.

1.3 Specific Technical Goals

Functional and shape analysis is a broad subject with a variety of problem formulations. What does one want to accomplish in shape analysis? Focusing on typical scenarios, we enumerate a set of tasks that one wants to perform in shape analysis of functions and curves:

1. **Registration of Points Across Functions and Curves**:
   Since we have chosen to represent objects of our interests using functions and curves, we need to solve the registration problem mentioned earlier. This problem is concerned with finding pairings of points across objects that are useful in their comparisons. Figure 1.15 shows some simple illustrations of this problem. The left panel shows two functions, and we want to find appropriate pairings of points across their domains. This registration is established by (nonlinearly) time warping one of the functions so that its peaks and valleys are well aligned with those of the other (second panel). Similarly, in the context of curves, the registration implies finding optimal pairings of ordered points across objects, as shown in the next two panels, with optimality defined using a chosen objective function.

2. **Quantification of Shape Similarities/Dissimilarities**:
   Perhaps the most frequently asked question about functions and shapes is: How different are their shapes? One would like to quantify the similarities and dissimilarities between given functional objects. It is important to note that these quantifications should not depend on the rotation, placement, and other transformations that can change curves but do not change their shapes. A more formal question is: How can we define a metric space of shapes and compute distances between shapes as elements of this space? There exist many techniques

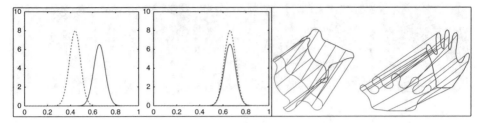

Fig. 1.15 Illustration of registration of points across objects. The left two panels show registration of function using time warping. The right two panels show the matched points for two examples

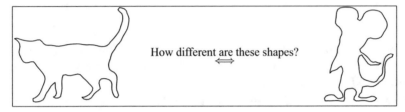

How different are these shapes?
$\Longleftrightarrow$

Fig. 1.16 A central aim in shape analysis is to quantify dissimilarities of given shapes

for comparing shapes and generating some kind of "matching scores," but we want a distance function that satisfies the three properties of a distance, including the triangle inequality. Specifically, if s_1, s_2, and s_3 are three shapes and $d(\cdot, \cdot)$ is a distance for comparing them, then this distance should satisfy the inequality: $d(s_1, s_2) + d(s_2, s_3) \geq d(s_1, s_3)$.

   Shown in Fig. 1.16 are two examples of 2D shapes: a cat and Mickey mouse. We seek a framework which can automatically quantify the differences in those two and other shapes.

Of course, it is desirable to have computerized methods that mimic human responses (based on visual inspections) to such questions; however, one should not insist on a full agreement at the outset. A human visual system is a complex mechanism that is invariably influenced by many other factors such as the situation, the presentation of data, and the prior experiences. Instead of seeking to build such a complex system, we will present several shape metrics and algorithms, all within a principled mathematical framework, and will try to obtain their physical interpretations. Such interpretations and explanations will help justify their usage in certain situations and rule them out in others.

3. **Representative Function or Shape:**

Just as we can study the average height of trees in a forest, the average value of homes in a county, or the average score of students in class, we would like to know the average shape of a collection of objects. Sometimes one needs a shape that is a good representative or a "summary" of the shapes present in a collection. In more formal language, one can ask the following question: What is a statistical mean or a mode for a sample taken from a population of shapes? The role of such a representative is important in many applications. It can be used in a compression algorithm to replace a large number of observations by a single shape. It can be useful in an indexing and retrieval algorithm to search for shapes from a large database; instead of trying to compare a query with all the

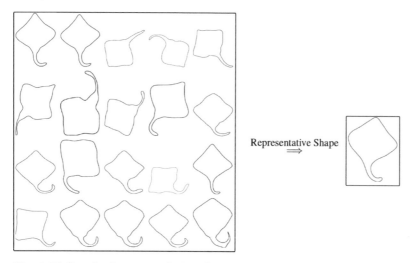

Fig. 1.17 Samples from a population of stingrays from the Surrey fish database on the left and a representative shape for this sample on the right

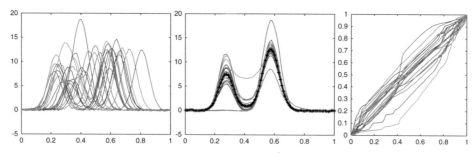

Fig. 1.18 The problem of alignment of given functions and computation of their mean under an elastic metric. Left shows given functions, middle shows their alignment and mean, and the right shows the time-warping functions used in alignment

shapes, one can try to match the query with only the representative shape(s) for each class and gain efficiency. It can also be used in shape classification and hypothesis selection. Shown in the left panel of Fig. 1.17 are some observations from a population of sting rays, and the task here is to find a representative shape, such as the one shown in the right, to represent this sample. This idea is especially useful in medical image analysis, where one can build a template for the shape of a healthy organ or an anatomical unit, using averages, and compare individual patients with that template to help in the diagnosis.

A related problem in analysis of function data is the computation of the sample mean of a set of given functions while simultaneously registering points along them. This framework, termed summarization under *elastic metrics*, is illustrated in Fig. 1.18 where the functions shown in the left panel are aligned, as shown in the middle panel, using nonlinear warping functions shown in the right panel. These warping functions are diffeomorphic deformations of the domain, $[0, 1]$ in this example, such that the original functions composed with these warpings are optimally aligned. Overlaid on the aligned functions is the mean function, or representative, of this given set of functions.

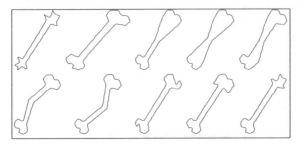

Fig. 1.19 Observed variability in a population of shapes of bones

4. **Modeling Shape Variability in a Class**:

   In most applications of interest, shapes exhibit variation even within the same homogeneous population. Analogous to statistical modeling of real-valued random variables, one would like to develop probability models—parametric or nonparametric—that capture this variability. As an example, Fig. 1.19 shows the variability of shapes in a small sample of bones taken from the Kimia database. The challenge is to develop a probabilistic framework to capture this variability, so that one can analyze it, sample from it, and use it for shape classification. Our approach will be to define and estimate probability densities on appropriate spaces of shapes; the spaces will differ based upon the application of interest. In case these densities take a parametric form, both the estimation and the analysis of resulting shape models becomes more efficient. This is because the estimation of a probabilistic shape model is then equivalent to an estimation of parameters which can be performed using maximum likelihood estimation, for instance. In probabilistic modeling, synthesis of new shapes is akin to generating random samples from the shape density, and classification of shapes is the same as the problem of hypothesis testing using likelihood functions under class-specific densities. Probability models of shapes are central to any comprehensive framework for statistical analysis of shapes.

5. **Shape Classification**:

   One basic task in a shape-based application is to classify shapes into predetermined categories. However, the nature of the problem depends on the form in which the data is presented. For instance, the data can be available directly in the form of parameterized curves, or it comes in the form of cluttered point clouds. The first case is shown in Fig. 1.20, and the main question is the following: Given a few samples from each of the two distinct populations, classify an independent sample, called a *test shape*, into one of the two populations and provide a confidence level for this classification. In case the random variables of interest are $\mathbb{R}^n$-valued, there is an abundance of classifiers in the classical pattern theory to handle this problem. However, the analysis of shape is more complex, and it involves modifying the classical theory before applying it to shape analysis. We briefly mention two possible approaches to shape classification. In case there is a way to quantify pairwise shape differences, one can use a *nearest-neighbor classifier*. In this scheme, one computes the "distance" between the test shape and all the given samples from the two populations and declares the test class to be that of the nearest shape to the test shape. A more comprehensive, and often efficient, classifier is the binary hypothesis testing using the likelihood ratio test. Using results from the previous item, i.e., estimated probability densities for each of the possible populations, one can compute likelihoods of the test shape under different classes and assign it

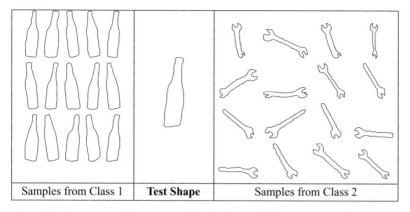

| Samples from Class 1 | **Test Shape** | Samples from Class 2 |

Fig. 1.20 Example of binary shape classification: Which shape class does the test shape belong to: left or right?

Fig. 1.21 *Top*: Examples of point clouds extracted from images for use in shape classification. *Bottom*: estimation of shapes in unordered, cluttered point clouds

to the class with the highest likelihood. Furthermore, one can also provide a measure of confidence in that assignment.

The second possibility, where the data is in the form of unordered (and cluttered) point clouds, is shown in Fig. 1.21. In this case one has to estimate as well as classify the shape from the point cloud data. One type of shape estimation involves selecting relevant subset of points, impose a certain ordering, and connect them to form polygonal shapes. Some examples are shown in the bottom row of Fig. 1.21.

6. **Symmetry Analysis**:

Symmetry of objects plays an important role in several applications, including object design, packaging, medical diagnosis, and surgery. In this case one asks questions of the type: Is a given object symmetric? What is the level (amount) of asymmetry in an object? What is the nearest symmetric object for a given asymmetric object? What are the planes(s) of symmetry of a given symmetric object? Take a look at the sixteen planar shapes shown in Fig. 1.22 and answer the following questions: Which shape is the most symmetric and which one is the least symmetric? Is the bird contour more symmetric than the cat contour? How can we deform the wine glass to make it perfectly symmetric?

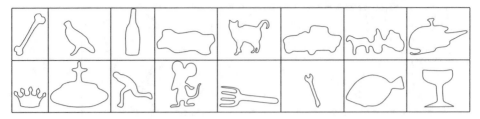

Fig. 1.22 Symmetry analysis: What is the amount of asymmetry in these shapes?

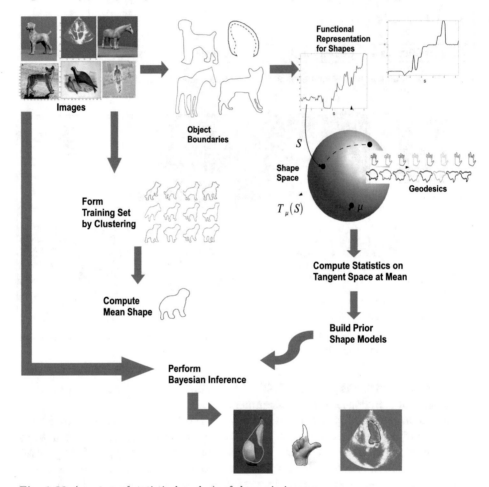

Fig. 1.23 A system of statistical analysis of shapes in images

We will develop a formal approach for analyzing symmetry of objects using deformations; here the objects are deformed until they become symmetric and the amount of deformation measures the level of asymmetry.

In summary, our goal is to develop a fully statistical framework in which we can treat shapes of functions and curves as random variables taking values in well-defined shape spaces and governed by underlying probability densities. This will enable us to derive shape-based inferences using well-established statistical techniques, albeit modified to account for the geometries of shape space. As an example, Fig. 1.23 outlines a system for generating shape-based inferences using image data. In this system, the input images are processed to extract contours of interest, either manually or automatically or both. These contours are mathematically represented as points on certain infinite-dimensional shape spaces. The dissimilarities between shapes of two contours are quantified using lengths of geodesic paths between the corresponding points on the shape space (Task 2). This process naturally includes registration of points along the curves (Task 1). Additionally, tools for statistical analysis of shapes, such as computations of moments or probability models on shape space, are derived. In particular, the concept of an average shape is developed on the shape space (Task 3). Probability models, estimated from the training shapes in a shape class (Task 4), are used for future Bayesian inferences on image data. A contour estimated in this Bayesian framework can then be used for classifying objects in images (Task 5). Geodesic paths between a shape and its reflection are useful in symmetry analysis (Task 6).

1.4 Issues and Challenges

What makes statistical analysis of functions and shapes difficult? It is quite easy for us, as human beings, to observe and to analyze shapes and to perform many of the aforementioned tasks without much difficulty. But doing so on computers using mathematical representations is a completely different proposition. First and foremost, one has to formulate functions and curves as numerical quantities. As described in Chaps. 4, 5, and 6, there are a variety of ways of representing functions and curves mathematically, each with its own pros and cons. However, all these representations entail certain common challenges from the perspective of function and shape analysis. In this section, we discuss some of those challenges.

1. **Invariance**:
   One important challenge in shape analysis comes from the fact that our notion of shape is *invariant to certain transformations*, such as translations, rotations, and rescaling. Abstractly, Kendall [49] described shape as a property that remains unchanged under these transformations, and this aspect must be included in our mathematical formulations. As an example, Fig. 1.24 shows sixteen different closed curves but they all have the same shape. Again, we have to develop

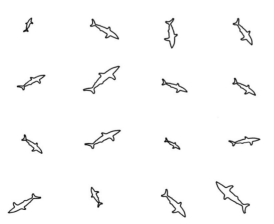

Fig. 1.24 All these closed curves have the same shape

mathematical representations and metrics in such a way that the pairwise distances between shapes of all these curves are zero. There are at least two ways of obtaining such invariance:

- One is to come up with a mathematical representation such that all the curves with the same shape map to the same point in the representation space. For example, if we represent a curve based on the velocity function along that curve, then that representation is automatically invariant to rigid translation. No additional work is needed.
- The other is to define an equivalence relation such that different curves with the same shape are deemed equivalent; i.e., they belong to the same equivalence class. The set of all such equivalence classes then becomes the desired representation space. Any ensuing analysis of shapes will have to respect this equivalence.

Another aspect of invariance of functions and curves, although not readily visible through pictures, is the issue of parameterization. A curve can be parameterized in many ways; a re-parameterization does not change its shape. Since this text is primarily concerned with shapes of functions and curves, it becomes critical to have the invariance with respect to re-parameterizations also. In the case of curves, there exists a canonical parameterization, namely, the arc-length parameterization, so that one can restrict each of the curves under study to this parameterization. In other words, if a curve is not parameterized by arc length, then re-parameterize it to make it arc length and then perform its shape analysis. This will potentially remove the parameterization variability. However, as will be demonstrated in later chapters, it is often useful to allow curves to have variable parameterizations, especially to improve matching of points across curves and, thus, enlarge the representation space. In those cases, one has to perform additional mathematical operations to ensure that the results remain invariant to re-parameterization of the curves under study.

2. **Nonlinearity**:
   Although there are a variety of mathematical representations for studying shapes, many of them share the following important property. The spaces formed by these representations are nonlinear. That is, they are not vector spaces and one cannot use the classical vector calculus to perform operations on shapes. Operations such as addition, multiplication, and subtraction are not valid on these spaces. How can we add two shapes like we add two real-valued variables? We cannot! Therefore, one has to devise a special framework for performing "calculus of shapes," a framework that will enable us to define probability densities on shapes, to perform differentiation and integration on shape spaces, to solve optimization problems, to take averages of shapes, and to perform other statistical inferences.

3. **Infinite Dimensionality**:
   Lastly, the study of shapes of functions and curves introduces an additional challenge of high, theoretically infinite, dimensionality. This is because curves are formally represented by functions, and spaces of functions are usually infinite dimensional. Although computer experiments will ultimately involve discretization of these functions into finite sets of points, the basic theory shall be developed assuming continuous, or functional, representations.

In summary, the nonlinearity and the infinite dimensionality of shapes, and the need for certain invariances, make it difficult to perform shape analysis of curves. However, we show in this textbook that with proper mathematical tools and concepts, this analysis can potentially be as powerful as the statistical analysis of real-valued random variables. Indeed we will develop a rich set of tools for performing statistical inferences on shape spaces and for handling the variability and uncertainties associated with corrupted observations of shapes.

1.5 Organization of this Textbook

This textbook is organized as follows. In Chap. 2, we briefly summarize some of the commonly used techniques in shape analysis, along with discussions on their strengths and limitations. In Chap. 3 and Appendix A, we provide a concise summary of mathematical concepts that are relevant to understanding our approach. Specifically, they cover basic material from differential geometry, algebra, and functional analysis. We have tried to highlight definitions and examples that are particularly important in shape analysis. We start the development of our approach by considering function data analysis, especially the problem of pairwise registration of functions, in Chap. 4. This chapter introduces the Fisher-Rao Riemannian metric and its extensions to general functions. This discussion is extended to study shapes of planar curves in Chap. 5. It presents different representations of parameterized curves, impositions of Riemannian structures, and constructions of optimal shape deformations. Chapter 6 imposes an additional closure constraint, to restrict to only *closed* planar curves, and provides their shape analysis. Changing focus to statistical modeling, we introduce some basic techniques in modeling manifold-valued random variables in Chap. 7. In particular, we discuss the computation of sample means and covariances from random samples on these manifolds. These ideas are first applied to function data in Chap. 8 and then to planar curves in Chap. 9. They introduce a framework for computing sample statistics and defining statistical models for functions and planar curves. Chapter 10 extends this shape analysis of curves to general $\mathbb{R}^n$. We finish the textbook with a number of shape-related topics in Chap. 11.

Chapter 2
Previous Techniques in Shape Analysis

A recent search on Google using the query: *shape analysis* results returned more than 24 million hits, approximately 5 million for *statistical shape analysis* and 71 million for *functional data analysis*! Of course, not all of them are relevant to our use of this phrase, but a large fraction of them partly or completely share the same goals as this textbook. This shows the scope and involvement of these topics in all areas of science and engineering and indeed life in general. This also makes it difficult for us to provide a complete picture of the previous efforts in shape analysis. We choose to focus on those works where statistical tools, based on precise mathematical representations, have been used to address shape-related issues.

Perhaps the earliest known efforts in formalizing shape analysis came from D'Arcy Thompson who tried to relate the shapes of functionally similar objects. He explored the possibility of making shapes visually similar by applying simple transformations, making them closer than they originally appeared. His treatment of shapes appears in the form of a 1917 book titled On Growth and Form. Figure 2.1 shows two examples of his work on using nonrigid transformations for matching two seemingly different but functionally similar objects. The top example shows three crocodilian skulls and a way of comparing their shapes. D'Arcy Thompson transformed the coordinate systems of the two skulls on the right so as to look as close to the left skull as possible. Notice that the transformation is applied to the coordinate system in which the object is represented and not to the object itself. The appearance of the object changes accordingly. As another example, he compared the shapes of two fishes—*Argyropelecus olfersi* and *Sternoptyx diaphana*—using a similar transformation of the coordinates, shown in the bottom row.

An interesting aspect of shape analysis is that there are multitudes of techniques that are available. We can try to distinguish them using the way they represent shape themselves. Consider the problem of analyzing shapes formed by closed, planar curves. While it seems natural to study their shapes by treating them as parameterized curves, there are some other possibilities. Figures 2.2, 2.3 illustrate some of these representations. The top row, second panel, in these figures show points sampled along these curves to form an ordered set of points in $\mathbb{R}^2$ (or a polygon). This representation is used in two methods—*landmark-based* shape analysis and *active shape analysis*. A related possibility is to discard the ordering of points and simply treat them as a set of points, as shown in the top row third panel. While we lose some information when we discard the ordering of points,

© Springer-Verlag New York 2016
A. Srivastava, E.P. Klassen, *Functional and Shape Data Analysis*,
Springer Series in Statistics, DOI 10.1007/978-1-4939-4020-2_2

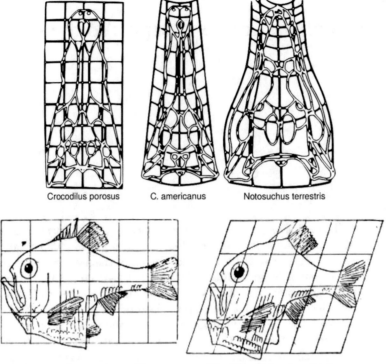

Crocodilus porosus   C. americanus   Notosuchus terrestris

Fig. 517. *Argyropelecus Olfersi.*   Fig. 518. *Sternoptyx diaphana.*

Fig. 2.1 Examples from D'Arcy Thompson's work on measuring differences in shapes of related objects by means of simple mathematical transformations. The top example studies variations in shapes of crocodilian skulls, while the bottom example compares the shape of an *Argyropelecus olfersi* with that of a *Sternoptyx diaphana*. (Data courtesy of Wikipedia Commons)

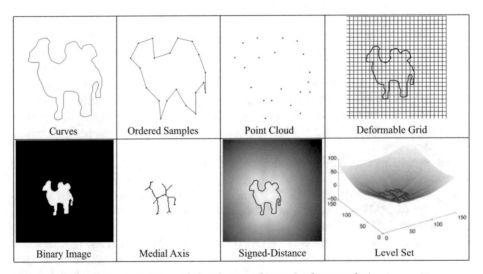

| Curves | Ordered Samples | Point Cloud | Deformable Grid |
| Binary Image | Medial Axis | Signed-Distance | Level Set |

Fig. 2.2 Different representations of closed curves for use in shape analysis

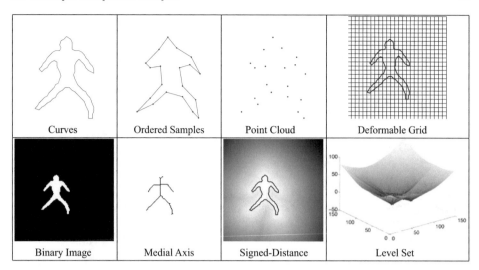

Fig. 2.3 Same as Fig. 2.2

we can now use set-theoretic methods for shape analysis. In contrast to the point-based methods, there are also methods involving the full two-dimensional domains that contain these shapes.

2.1 Principal Component Analysis

We start the discussion by introducing a standard technique for linear dimensional reduction that is commonly used in multivariate statistical analysis. The basic idea is to find an orthogonal matrix that can project the observed data to a lower dimensional vector space such that the structure of the original data is maximally preserved.

Now we make this precise. Let $\mathbf{x} \in \mathbb{R}^n$ denote a vector of random variables, with mean assumed to be to zero. The goal is to find a d-dimensional subspace of $\mathbb{R}^n$, for $d << n$, such that the orthogonal linear projection of $\mathbf{x}$ on that subspace is optimal. The optimality is measured using the expected value of the squared error due to projection. If $U \in \mathbb{R}^{n \times d}$ is an orthogonal matrix, i.e., $U^T U = I_d$, then $U^T \mathbf{x}$ is an orthogonal projection of $\mathbf{x}$ and $U U^T \mathbf{x}$ is a reconstruction of $\mathbf{x}$. In PCA the goal is to find a $\hat{U} \in \mathbb{R}^{n \times d}$ such that $\hat{U}^T \hat{U} = I_d$ and

$$\hat{U} = \underset{\{U \in \mathbb{R}^{n \times d} | U^T U = I_d\}}{\mathrm{argmin}} (E[\|\mathbf{x} - U U^T \mathbf{x}\|^2]), \tag{2.1}$$

where E denotes the expectation over the joint probability density function of $\mathbf{x}$. Simplifying this expression,

$$\hat{U} = \underset{\{U \in \mathbb{R}^{n \times d} | U^T U = I_d\}}{\mathrm{argmin}} E[(\mathbf{x} - U U^T \mathbf{x})^T (\mathbf{x} - U U^T \mathbf{x})] = \underset{\{U \in \mathbb{R}^{n \times d} | U^T U = I_d\}}{\mathrm{argmax}} E[\mathrm{tr}(U^T \mathbf{x} \mathbf{x}^T U)]$$

$$= \underset{\{U \in \mathbb{R}^{n \times d} | U^T U = I_d\}}{\mathrm{argmax}} \mathrm{tr}(U^T E[\mathbf{x} \mathbf{x}^T] U) = \underset{\{U \in \mathbb{R}^{n \times d} | U^T U = I_d\}}{\mathrm{argmax}} \mathrm{tr}(U^T (V \Lambda V^T) U) \,,$$

where $E[\mathbf{x}\mathbf{x}^T] \in \mathbb{R}^{n \times n}$ is the covariance matrix of $\mathbf{x}$ and $V \Lambda V^T$ is its singular value decomposition (SVD). Hence, the solution $\hat{U}$ is given by the first d columns of V provided that the entries of diagonal matrix Λ are non-increasing from top to bottom.

In practical situations, where the covariance K is not known, and only some observations of X are available, one uses the sample covariance matrix, $\hat{K} \in \mathbb{R}^{n \times n}$:

$$\hat{\Sigma} = \frac{1}{M-1} \sum_{m=1}^{M} (x_i - \bar{x})(x_i - \bar{x})^T \ , \quad \text{where} \quad \bar{x} = \frac{1}{M} \sum_{m=1}^{M} x_i \ .$$

This calculation of $\hat{U}$ is called *principal component analysis* (PCA) and is the most commonly used technique for statistical dimensional reduction.

Once we have the principal components of a random vector $\mathbf{x}$, we can model it as a linear combination: $\hat{\mathbf{x}} \approx \bar{\mathbf{x}} + \sum_{i=1}^{d} z_i \hat{U}_i$, where $\bar{\mathbf{x}}$ is the sample mean, $\hat{U}_i$ denotes the ith column of the matrix $\hat{U}$, and z_is are independent and identically distributed standard normal random variables.

2.2 Point-Based Shape-Analysis Methods

In this section we briefly introduce shape analysis methods that represent the underlying objects using points sampled along their boundaries. Let $\mathbf{x} = \{x_1, x_2, \ldots, x_k\} \in \mathbb{R}^{2k}$ and $\mathbf{y} = \{y_1, y_2, \ldots, y_k\} \in \mathbb{R}^{2k}$ denote the sampled points for any two shapes, respectively. Depending on the application, these points are either ordered or unordered. We start with the more general case where the points are unordered, and later we will assume a specific ordering.

2.2.1 ICP: Point Cloud Analysis

In case the points are unordered, we call these sets *point clouds*. One of the most common methods for aligning and comparing point clouds is the iterative closest point (ICP) algorithm. It uses the root-mean squared distance (RMSD) as objective function to pose an optimization problem as follows. (While the algorithm is applicable to points in any Euclidean space, we will state it here for planar sets.) Let $\varsigma : \{1, 2, \ldots, k\} \to \{1, 2, \ldots, k\}$ be a mapping that associates an element of $\mathbf{y}$ to each element of $\mathbf{x}$; ς is also called the *registration* of $\mathbf{y}$ to $\mathbf{x}$. Let Σ denotes the set of all such mappings. Then, we solve for the alignment of two point sets as the joint optimization problem:

$$\text{RMSD} = \min_{O \in SO(2), \rho \in \mathbb{R}_+, T \in \mathbb{R}^2, \varsigma \in \Sigma} \sum_{i=1}^{k} \|(T + \rho O x_i) - y_{\varsigma(i)}\|^2 \ . \tag{2.2}$$

The joint optimization problem over the space $SO(2) \times \mathbb{R}_+ \times \mathbb{R}^2 \times \Sigma$ is a complex one. The variables O, ρ, and T are called the *transformation variables* since they rotate, scale, and translate the set $\mathbf{x}$, while ς is called the *registration variable*.

It turns out that given transformation variables, it is relatively easy to solve for the registration variable and vice-versa.

1. **Given Registration Variable**: If ς is kept fixed, we can solve for the transformation variables as follows. For the current ς, we utilize only the set $\mathbf{y}_\varsigma = \{y_{\varsigma(i)}, \ i = 1, \ldots, k\}$ and one-by-one we will solve for the three transformation variables, keeping others fixed.

   - **Translation**: In this step we solve for the optimal translation while keeping the rotation and the scaling of $\mathbf{x}$ fixed. In fact, let us assume that $\mathbf{x}_t$ denotes the configuration of points in $\mathbf{x}$ at the current orientation and scale. Then,

$$T^* = \underset{T \in \mathbb{R}^2}{\operatorname{argmin}} \sum_{i=1}^{k} \|y_{\varsigma(i)} - x_{i,t} - T\|^2 = \underset{T \in \mathbb{R}^2}{\operatorname{argmin}} \sum_{i=1}^{k} \langle y_i - x_{i,t} - T, y_{\varsigma(i)} - x_{i,t} - T \rangle.$$

   Taking the gradient of this cost function with respect to T and setting it to zero, we get

$$T^* = \frac{1}{k} \sum_{i=1}^{k} y_{\varsigma(i)} - \frac{1}{k} \sum_{i=1}^{k} x_{i,t} \ . \tag{2.3}$$

   - **Rotation**: In this step we solve for the optimal rotation while keeping the translation and scaling fixed. Once again assume that $\mathbf{x}_t$ denotes the configuration of points in $\mathbf{x}$ at the current position and scale. Then, we define:

$$O^* = \underset{O \in SO(2)}{\operatorname{argmin}} \sum_{i=1}^{k} \|y_{\varsigma(i)} - O x_{i,t}\|^2 = \underset{O \in SO(2)}{\operatorname{argmin}} \sum_{i=1}^{k} \langle y_{\varsigma(i)} - O x_{i,t}, y_i - O x_{i,t} \rangle$$

$$= \underset{O \in SO(2)}{\operatorname{argmin}} \sum_{i=1}^{k} \operatorname{trace}(y_{\varsigma(i)} y_{\varsigma(i)}^T + x_{i,t} x_{i,t}^T - 2 O y_{\varsigma(i)} x_{i,t}^T)$$

$$= \underset{O \in SO(2)}{\operatorname{argmax}} \operatorname{trace}(OA), \quad A = \sum_{i=1}^{k} y_{\varsigma(i)} x_{i,t}^T \ .$$

   Let $A = U\Sigma V^T$, the SVD of A. Then, the optimal rotation is given by:

$$O^* = \begin{cases} UV^T & \text{if } \det(A) > 0 \\ U \begin{bmatrix} 1 & 0 \\ 0 & -1 \end{bmatrix} V^T & \text{otherwise.} \end{cases} \tag{2.4}$$

   - **Scale**: Now we denote by $\mathbf{x}_t$ the configuration points in $\mathbf{x}$ at the current position and orientation, and solve for the optimal scale:

$$\rho^* = \underset{\rho \in \mathbb{R}_+}{\operatorname{argmin}} \sum_{i=1}^{k} \|y_{\varsigma(i)} - \rho x_{i,t}\|^2 = \underset{\rho \in \mathbb{R}_+}{\operatorname{argmin}} \sum_{i=1}^{k} \langle y_{\varsigma(i)} - \rho x_{i,t}, y_{\varsigma(i)} - \rho x_{i,t} \rangle \ .$$

   Taking the derivative of the right side with respect to ρ and setting it equal to zero, we obtain:

$$\rho^* = \frac{\sum_{i=1}^{k} \langle y_{\varsigma(i)}, x_{i,t} \rangle}{\sum_{i=1}^{k} \langle x_{i,t}, x_{i,t} \rangle} \ . \tag{2.5}$$

2. **Given Transformation Variables**: Now we fix the transformation variables, and treat ς as the variable of interest. Now we need to solve for:

$$\varsigma^* = \operatorname*{argmin}_{\varsigma \in \Sigma} \sum_{i=1}^{k} \|y_{\varsigma(i)} - x_{i,t}\|^2 \, ,$$

where $\mathbf{x}_t$ denotes a transformed version of $\mathbf{x}$. The solution to this problem depends on what further assumptions we make about ς. If we constrain ς to be one-to-one, then this problem becomes the so-called *assignment problem*. The solution to the assignment problem comes from a famous algorithm called the *Hungarian algorithm*. However, if this additional constraint is not imposed, i.e., ς is allowed to take the same values for different arguments, then the solution comes from the nearest-neighbor search. That is,

$$\varsigma^*(i) = \operatorname*{argmin}_{j=1,2,...,k} \|x_i - y_j\| \, . \tag{2.6}$$

Since the individual solutions are well defined, a reasonable approach is to iterate between the two solutions until convergence. This is often called the *ICP algorithm*. We state this algorithm next.

Algorithm 1 (ICP Algorithm). *Inputs:* $\mathbf{x}$, $\mathbf{y}$, *initial values of* O, ρ, T; *stopping criterion* ϵ.

1. *Find the transformed set* $x_{i,t} = \rho O x_i + T$.
2. *For each* i, *solve for* $\varsigma^*(i)$ *using either the nearest-neighbor search (Eq. 2.6) or the Hungarian algorithm.*
3. *Compute the objective function* $\sum_{i=1}^{k} \|x_{i,t} - y_{\varsigma(i)}\|^2$. *If the change in its value from the previous iteration is less than* ϵ, *then stop. Else, continue.*
4. *For the fixed* ς^*, *solve for the transformation variables using Eqs. 2.3, 2.4, and 2.5. Return to Step 1.*

Figure 2.4 shows three examples of matching point clouds using ICP algorithm. In each row, the first two panels show the two clouds $\mathbf{y}$ and $\mathbf{x}$, the third panel shows the optimal transformation of $\mathbf{x}$ and the resulting registration ς^*, and the final panel shows the evolution of the cost function versus the iteration index.

There are certain pros and cons associated with the ICP algorithm. Since this method is based on matching points, it is generally very fast and can be implemented in real time. One of the main strengths of the ICP algorithm is that it estimates the registration ς, rather than assuming that it is known. This is certainly important in situations where data consists of unordered points. However, it has limited utility in situations where ordering of points is available. Another important limitation of ICP is that the resulting matching function is not a proper distance. In fact, it is not even symmetric. We illustrate an example in Fig. 2.5 where the RMSD values for two point sets are computed twice, once for the original $\mathbf{x}$ and $\mathbf{y}$ and then for the roles switched. The RMSD values are found to be 3.374 and 1.5356, respectively. This asymmetry in RMSD values arises due to the nature of the nearest neighbor registration in Eq. 2.6. This process, by definition, takes into account all the elements of $\mathbf{x}$ but not necessarily all the elements of $\mathbf{y}$.

In some situations one can force the symmetry of RMSD by replacing the nearest neighbor registration with a more symmetric assignment problem. As mentioned

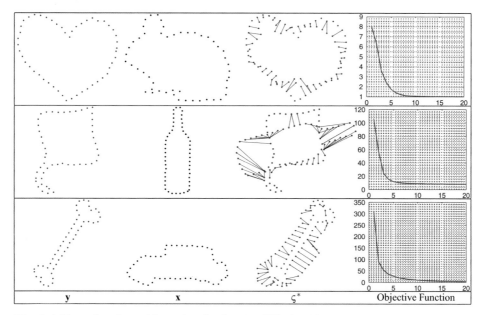

Fig. 2.4 Examples of matching point clouds using ICP algorithm

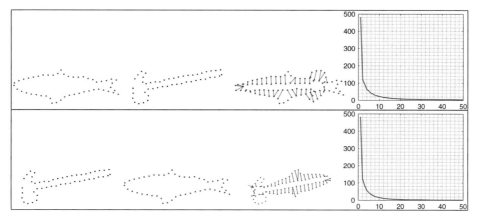

Fig. 2.5 Asymmetry of RMSD. The RMSD for the top case is 3.374 and for the bottom case is 1.5356. The only difference between them is the order in which shapes are considered

earlier, here the two configurations are assumed to have same number of points, and the mapping $\varsigma : \{1, 2, \ldots, k\} \rightarrow \{1, 2, \ldots, k\}$ is assumed to be one-to-one and onto. In other words, one finds for each element of $\mathbf{x}$ a *unique* element of $\mathbf{y}$ (using the Hungarian algorithm) while minimizing the objective function in Eq. 2.2. In Fig. 2.6, the two RMSD values for the nearest-neighbor registration are 1.7149 and 0.9181, while the RMSD value for the Hungarian algorithm is 5.2618.

Another limitation in the ICP based matching is the locality of the solution. The final transformation and registration variables are highly dependent on their initial values. Different initializations can lead to different results, as demonstrated in Fig. 2.7. In this example, we take a set $\mathbf{x}$ and rotate it arbitrarily to generate the point set $\mathbf{y}$. Then, we try to match them using ICP algorithm. Since $\mathbf{x}$ and $\mathbf{y}$ differ only in rigid motion, the resulting RMSD between them should be zero,

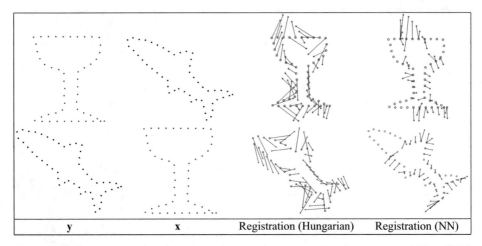

| y | x | Registration (Hungarian) | Registration (NN) |

Fig. 2.6 Different optimal registrations of two points sets using the nearest-neighbor (NN) algorithm and the Hungarian algorithm

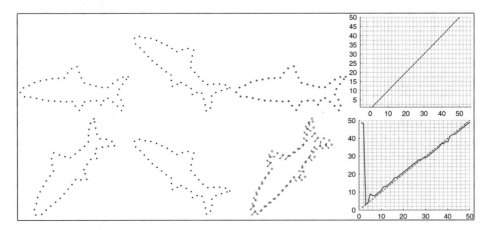

Fig. 2.7 Dependence of ICP on initial condition

which is the case in the top row. However, in the solution displayed in the bottom row, the two sets do not align completely and the resulting RMSD is not zero.

Another feature of ICP algorithm is the instability in matching resulting from the Nearest-neighbor registration. A small transformation of **x** can result in a completely different registration and, thus, a different RMSD value. This is illustrated in Fig. 2.8 where only a slight initial transformation of **x** results in two vastly different RMSD values.

In terms of the goals laid out in Sect. 1.3, this method can at best provide the first item: quantification of shape similarities/dissimilarities, but not any one of the remaining items. For instance, in its original form, it cannot help provide a notion of a statistical mean of shapes. Even in that first item, not all the invariances are satisfied.

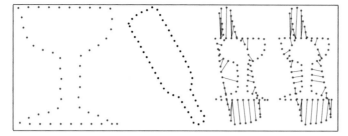

Fig. 2.8 Two seemingly similar matches between two shapes results in RMSD values of 5.6508 and 0.0053, respectively

2.2.2 Active Shape Models

In case one assumes that the point sets are sampled from curves and there are known orderings associated with them, the task of comparing them becomes considerably easier. Since the two sets $\mathbf{x}$ and $\mathbf{y}$ are ordered, it is natural to use $\varsigma(i) = i$ as the registration. (Here we are assuming that the number of points on $\mathbf{x}$ and $\mathbf{y}$ are same.) Now that the registration is taken care off, one only needs to solve for the transformation variables. This is accomplished as follows.

We start with the simple change in notation. It is often convenient to identify points in $\mathbb{R}^2$ with elements of $\mathbb{C}$, i.e., $x_i \equiv z_i = (x_{i,1} + j x_{i,2})$, where $j = \sqrt{-1}$. Thus, in this complex representation, a configuration of k points $\mathbf{x}$ is now $\mathbf{z} \in \mathbb{C}^k$. Before analyzing the shape of $\mathbf{z}$, it is "standardized" by moving its centroid to the origin (of the coordinate system):

$$z_i \mapsto \left(z_i - \frac{1}{k}\sum_{i=1}^{k} z_i\right) .$$

To remove the scale variability, z is rescaled to have norm one, i.e., $\mathbf{z} \mapsto \mathbf{z}/\|\mathbf{z}\|$. (These transformations effectively project a configuration in the orthogonal section of its orbit under the nuisance group—translation and scaling—as explained in Sect. 3.14). Then, one uses tools from standard multivariate statistics to analyze and model them. So far, the translation and the scale variability of a configuration are removed but the rotation remains. That is, two configurations, $\mathbf{z}$ and a rotation of $\mathbf{z}$, will have a nonzero distance between them even when they have the same shape. Using an additional step of rotational alignment solves the problem, as follows:

$$
\begin{aligned}
\phi^* &= \operatorname*{argmin}_{\phi \in \mathbb{S}^1} \|\mathbf{z}_1 - e^{j\phi}\mathbf{z}_2\|^2 \\
&= \operatorname*{argmin}_{\phi \in \mathbb{S}^1} (\|\mathbf{z}_1\|^2 + \|\mathbf{z}_2\|^2 - 2\Re(\langle \mathbf{z}_1, e^{j\phi}\mathbf{z}_2 \rangle)) \\
&= \operatorname*{argmax}_{\phi \in \mathbb{S}^1} (\Re(e^{-j\phi} \langle \mathbf{z}_1, \mathbf{z}_2 \rangle)) = \theta, \quad \text{where} \quad \langle \mathbf{z}_1, \mathbf{z}_2 \rangle = r e^{j\theta} .
\end{aligned}
\tag{2.7}
$$

Here $\Re$ denotes the real part of a complex number. The last inner product is the Hermitian inner product between $\mathbf{z}_1$ and $\mathbf{z}_2$. The distance between the two configurations is then $\|\mathbf{z}_1 - e^{j\phi^*}\mathbf{z}_2\| = \sqrt{2(1-r)}$. The corresponding optimal

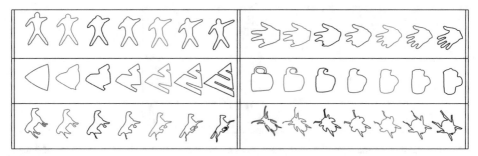

Fig. 2.9 Examples of geodesic paths between same shapes using ASM

deformation from one shape to another is simply a straight line between $\mathbf{z}_1$ and $e^{j\phi^*}\mathbf{z}_2$, i.e., $\alpha(\tau) = (1-\tau)\mathbf{z}_1 + \tau e^{j\phi^*}\mathbf{z}_2$ for $\tau \in [0,1]$.

One remaining issue in this analysis is the selection of a starting point, z_1 or the *origin*, on a closed curve. If there are k points sampled on a curve, then there are k candidates for the origin. The solution is to select the best seed during a pairwise comparison of configurations. That is, select any point on the first configuration as the origin for the first shape and try all k points in the second configuration as the origin for the second shape. Of those, select the one that results in the smallest distance from the first configuration. Figure 2.9 shows several examples of these deformations: one between a pair of human silhouettes, one between a pair of hands, and so on. These deformations have been computed using $k = 100$ points on each configuration so that the resulting polygons look like smooth curves.

While this approach is computationally very efficient, it has some limitations in terms of shape analysis. As discussed later in Sect. 2.2.4, a major limitation of this technique is the selection and registration of points across curves. In particular, the fact that it assumes a linear registration of points leads to unnatural shape comparisons and analysis. Also, this method does not completely take care of the scale variability of objects in analyzing their shapes. Although it starts by rescaling the complex vectors to be of unit length, it ignores this constraint in the remaining analysis. This particular problem is addressed by the next method that carefully restricts its analysis to only the unit-length vectors.

2.2.3 Kendall's Landmark-Based Shape Analysis

David Kendall was among the earliest researchers to identify a certain space, whose elements represent shapes of interest, as a Riemannian manifold. This pioneering idea has become a guiding principle for many subsequent ideas in shape analysis, including those presented in this text. Furthermore, most of the terminology used in this text, for instance, the use of pre-shape and shape spaces, is also borrowed from that landmark-based shape analysis literature. In view of the historical importance and relevance of this approach to our framework, we present it in some detail.

Pre-shape Space Similar to ASM, an object is also represented here by an ordered set of points, called **landmarks** and for the purpose of shape analysis, the objects are replaced by the corresponding landmarks. No other information is retained from the objects. Since shapes are treated as being invariant with respect to rotations, translations, and scales, one develops an algebraic representation that incorporates this invariance. In this discussion, we will again identify points in $\mathbb{R}^2$ with elements of $\mathbb{C}$, i.e., $x_i \equiv z_i = (x_{i,1} + j x_{i,2})$, where $j = \sqrt{-1}$. A configuration of k points is thus denoted by a $\mathbf{z} \in \mathbb{C}^k$. Before analyzing the shape of $\mathbf{z}$, it is "standardized" by moving its centroid to the origin (of the coordinate system): $z_i \mapsto (z_i - \frac{1}{k}\sum_{i=1}^{k} z_i)$ and by rescaling to have norm one, i.e., $\mathbf{z} \mapsto \mathbf{z}/\|\mathbf{z}\|$. This results in a set:

$$\mathscr{C} = \{\mathbf{z} \in \mathbb{C}^n | \frac{1}{k}\sum_{i=1}^{k} z_i = 0, \|\mathbf{z}\| = 1\} \ .$$

$\mathscr{C}$ is essentially an orthogonal section of the translation and scaling groups on $\mathbb{C}^n$ (see Example 3.14) is not a vector space because if a_1, $a_2 \in \mathbb{R}$ and $\mathbf{z}_1$, $\mathbf{z}_2 \in \mathscr{C}$, then $a_1\mathbf{z}_1 + a_2\mathbf{z}_2$ is typically not in $\mathscr{C}$, due to the unit norm constraint. However, $\mathscr{C}$ is a unit sphere and one can utilize the geometry of a sphere to analyze points on it. It is called a *pre-shape space* because there are many vectors that have the same shape but different values in $\mathscr{C}$. This is because the rotation has not yet been removed from this representation. Under the Euclidean metric, the shortest path between any two elements $\mathbf{z}_1$, $\mathbf{z}_2 \in \mathscr{C}$, also called a *geodesic*, is given by the great circle: $\alpha_{ksa} : [0,1] \to \mathscr{C}$, where

$$\alpha_{ksa}(\tau) = \frac{1}{\sin(\vartheta)} \left[\sin(\vartheta(1-\tau))\mathbf{z}_1 + \sin(\tau\vartheta)\mathbf{z}_2 \right], \quad \text{and} \quad \vartheta = \cos^{-1}(\Re(\langle \mathbf{z}_1, \mathbf{z}_2 \rangle)) \ .$$

$$(2.8)$$

The geodesic distance between $\mathbf{z}_1$ and $\mathbf{z}_2$ is given by $d_c(\mathbf{z}_1, \mathbf{z}_2) = \vartheta$.

Shape Space In order to compare the shapes represented by $\mathbf{z}_1$ and $\mathbf{z}_2$, we need to align them rotationally, as was done earlier in ASM, but the shape space is defined more formally this time. Let $[\mathbf{z}]$ be the set of all rotations of a configuration z according to:

$$[\mathbf{z}] = \{e^{j\phi}\mathbf{z} | \phi \in \mathbb{S}^1\} \quad \subset \mathscr{C} \ .$$

One defines an equivalence relation on $\mathscr{C}$ as follow: $z_1 \sim z_2$ if there exists an angle $\phi \in \mathbb{S}^1$ such that $z_1 = e^{j\phi}z_2$. Equivalence classes under this relation are exactly the sets $[\mathbf{z}]$ for different $\mathbf{z} \in \mathscr{C}$. The set of all such equivalence classes is the quotient space $\mathscr{C}/U(1)$ (see Definition 3.15), where $U(1) = SO(2) = \mathbb{S}^1$ is the set of all rotations in $\mathbb{R}^2$. This space is called the *complex projective space* and is denoted by $\mathbb{CP}^{k-1}$. It is quickly checked that the action of $\mathbb{S}^1$ on $\mathscr{C}$, defined as:

$$\mathbb{S}^1 \times \mathscr{C} \to \mathscr{C}, \quad (\phi, \mathbf{z}) = e^{j\phi}\mathbf{z} \ ,$$

is by isometries under the standard Euclidean metric $\| \cdot \|$. That is, for any $\mathbf{z}_1, \mathbf{z}_2 \in \mathscr{C}$, we have $\|\mathbf{z}_1 - \mathbf{z}_2\| = \|e^{j\phi}\mathbf{z}_1 - e^{j\phi}\mathbf{z}_2\|$ for any $\phi \in \mathbb{S}^1$. Therefore, using Theorem 3.17, the Euclidean distance descends to the quotient space $\mathbb{CP}^{n-1}$, and we can define:

$$d_s([\mathbf{z}_1], [\mathbf{z}_2]) = \min_{\phi_1} \ d_c(e^{j\phi_1}\mathbf{z}_1, \mathbf{z}_2) = \min_{\phi_2 \in \mathbb{S}^1} \ d_c(\mathbf{z}_1, e^{j\phi_2}\mathbf{z}_2) \ .$$

The last equality comes from the isometric nature of the group action. A geodesic between two elements $\mathbf{z}_1, \mathbf{z}_2 \in \mathbb{CP}^{n-1}$ is given by computing α_{ksa} between $\mathbf{z}_1$ and $e^{\phi^*}\mathbf{z}_2$, where ϕ^* is the optimal rotational alignment of $\mathbf{z}_2$ to $\mathbf{z}_1$. As shown in the

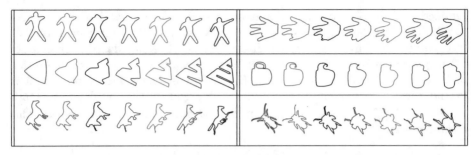

Fig. 2.10 Examples of geodesic paths between same shapes using Kendall's shape analysis

previous item, this optimal rotation is given by $\phi^* = \theta$, where $\langle \mathbf{z}_1, \mathbf{z}_2 \rangle = re^{j\theta}$. With this rotation, we get $\langle \mathbf{z}_1, \mathbf{z}_2^* \rangle = r$, and the geodesic path between two equivalence classes, represented by $\mathbf{z}_1$ and $\mathbf{z}_2$, is given by:

$$\alpha_{ksa}(\tau) = \frac{1}{\sin(\vartheta)} \left[\sin(\vartheta(1 - \tau))\mathbf{z}_1 + \sin(\tau\vartheta)\mathbf{z}_2^* \right] , \qquad (2.9)$$

where $\vartheta = \cos^{-1}(\Re(\langle \mathbf{z}_1, \mathbf{z}_2^* \rangle)) = \cos^{-1}(r)$. The length of the geodesic is given by ϑ and that quantifies the difference in shapes of the boundaries represented by $\mathbf{z}_1$ and $\mathbf{z}_2$. Figure 2.10 shows several examples of geodesic paths between the same shapes as for the ASM examples.

One of the most important contributions of this approach has been the advanced statistical framework that has been developed using it. For a collection of shapes $[\mathbf{z}_1], [\mathbf{z}_2], \ldots, [\mathbf{z}_n]$ in $\mathbb{CP}^{k-1}$, define the sample mean to be the quantity:

$$\mu_{ksa} = \operatorname*{argmin}_{[\mathbf{z}] \in \mathbb{CP}^{k-1}} \sum_{i=1}^{n} d_s([\mathbf{z}], [\mathbf{z}_i])^2 , \qquad (2.10)$$

where d_s is the geodesic distance as defined earlier. This mean is actually computed using an iterative algorithm that uses the gradient of the cost function given in Eq. 2.10 to iteratively update the estimate. (This algorithm is presented for computing means on general Riemannian manifolds later in Chap. 7.) Once the mean μ_{ksa} is computed, one can project all the observed shapes into the tangent space of $\mathbb{CP}^{k-1}$ at μ_{ksa}. Since this tangent space is a finite-dimensional vector space, one can use the standard multivariate calculus for posing and solving statistical problems. Dryden and Mardia [28] provide an excellent treatment on this subject with lots of illustrative examples. In the methodology presented in the text, we will borrow several ideas, notations, and definitions from this approach. In that sense, this landmark-based analysis of shapes can be considered a precursor to the approach outlined in this text.

2.2.4 Issue of Landmark Selection

Although Kendall's approach succeeds in preserving the unit-length constraints on the landmark configurations, it does not address a very important practical issue: How to systematically select points on objects, say curves, to form representative

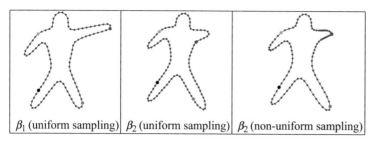

| β_1 (uniform sampling) | β_2 (uniform sampling) | β_2 (non-uniform sampling) |

Fig. 2.11 Selection of landmarks using the uniform and a convenient nonuniform sampling. Nonuniform sampling allows a better matching of features between β_1 and β_2

point sets? This process is difficult to standardize and different selections can lead to drastically differing solutions. This issue is present in any point-based approach, including the ASM method discussed above. While it may be tempting to sample a curve uniformly along its length, i.e., parameterize a unit-length curve β using arc length and sample $\{\beta(t_i)|i = 0, 1, 2, \ldots, n\}$ where $t_i = i/n$, the results are not always good since this forces a particular registration of points. The point $\beta_1(t_i)$ on the first curve is matched to the point $\beta_2(t_i)$ on the second curve, irrespective of the shapes involved. Figure 2.11 illustrates this point using an example. Shown in the left two panels are two curves: β_1 and β_2, sampled uniformly along their lengths. For $t_i = i/4$, $i = 1, 2, 3, 4$ the corresponding four points on each curve $\{\beta_1(t_i)\}$ and $\{\beta_2(t_i)\}$ are shown in the same color. While two of the four pairs seem to match well, the pairs shown in red and green fall on different parts of the body, resulting in a mismatch of features. This example shows the pitfall of using uniform sampling of curves. In fact, any predetermined sampling and preregistration of points will, in general, be problematic. A more natural solution is to treat the boundaries of objects as *continuous* curves, rather than discretize them in point sets at the outset, and find an optimal (perhaps nonuniform) sampling, such as the one shown in the rightmost panel, that better matches features across curves. This way one can develop a more comprehensive solution, including a theory and algorithms, assuming continuous objects and one discretizes them only at the implementation stage.

2.3 Domain-Based Shape Representations

The next set of representations can be described as domain-based in the sense that the whole space in which the shape is present is taken into account. In the case of planar shapes, each shape is treated as a subset of $\mathbb{R}^2$, say $[0, 1]^2$, and one represents shapes as some predetermined family of functions on that domain. Then, shapes are compared and analyzed by comparing the corresponding functions. We briefly describe some of these methods next.

2.3.1 Level-Set Methods

One prominent idea has been to use level sets to study dynamics of closed contours [85]. In the planar case, the contours are embedded in a subset of $\mathbb{R}^2$ in an arbitrary

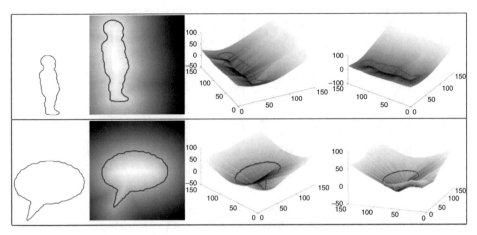

Fig. 2.12 Level-set representation of closed planar curves (*leftmost panel*) using signed-distance functions (*remaining three panels*)

way, say in $[0,1]^2$, and are represented by certain functions on $[0,1]^2$ such that the level curves of those functions are the original contours. A common choice of function is the signed-distance function. Let β denotes a simple closed curve in $\mathbb{R}^2$.

Definition 2.1. A signed-distance function $\phi : \mathbb{R}^2 \to \mathbb{R}_{\geq 0}$ is a function whose value at a point is the Euclidean distance in $\mathbb{R}^2$ to the nearest point on the curve. Additionally, the function values for points inside the contour are negative:

$$\phi(x) = \begin{cases} -d(x,\beta), & x \text{ is inside the curve} \\ d(x,\beta), & \text{otherwise} \end{cases}, \quad d(x,\beta) = \inf_{y \in \beta} \|x - y\| . \quad (2.11)$$

It is easy to see that the set $\{x \in \mathbb{R}^2 | \phi(x) = 0\}$ is exactly the set of points on the curve β. In other words, β is the zero level set of ϕ. Figure 2.12 shows two examples of this representation: the left curve in each row is represented by its signed-distance function shown in three different ways in the remaining panels. The second column of this figure shows the signed-distance functions as gray scale images, while the last two columns show them as height functions from different views.

For closed contours β_1 and β_2 with given embeddings in $[0,1]^2$, let ϕ_1 and ϕ_2 denote the corresponding signed-distance functions on $[0,1]^2$. Then, the shapes of β_1 and β_2 can be compared by using the $\mathbb{L}^2$ metric:

$$d(\beta_1, \beta_2) = \|\phi_1 - \phi_2\| = \sqrt{\int_{[0,1]^2} (\phi_1(x) - \phi_2(x))^2 \, dx} .$$

There are several issues with this distance function. Firstly, any rigid motion of the βs will change the distance between the corresponding level functions. One can handle two of the transformations: translation and scaling, by pre-processing the curves β_1 and β_2 as in the previous sections. For example, we can translate them so that their mean is at the origin and scale them so that their norm is one:

$$\beta(t) \to \beta(t) - \bar{\beta}(t), \quad \bar{\beta}(t) = \int \beta(t) dt ,$$

and

$$\beta(t) \to \frac{\beta(t)}{\sqrt{\int \|\beta(t)\|^2 dt}} \ .$$

The removal of the rotation requires solving an optimization problem, as was the case in the earlier frameworks.

While there are distinct advantages of using this representation in modeling dynamical evolution of curves, especially when the curves are allowed to split and merge and to change topologies, their use in shape analysis is more difficult. The main issues come in imposing invariances with respect to shape-preserving transformations. Secondly, it is difficult to construct an optimal deformation of one shape into another using this framework. In the earlier two methods, ASM and KSA, we gave equations of geodesic paths whose lengths achieve the distance used to compare shapes. Since the distance between any two level-set functions ϕ_1 and ϕ_2 is $\mathbb{L}^2$, it is reasonable to connect using the straightline:

$$\psi(\tau) = (1 - \tau)\phi_1 + \tau\phi_2 \ .$$

Each element along this path, $\psi(\tau)$, is a function from $[0, 1]^2$ to $\mathbb{R}$, but is it a valid signed-distance function? An important property of a signed-distance function $\phi : [0,]1^2 \to \mathbb{R}$ is that $\|\nabla\phi(x)\| = 1$ for all $x \in [0, 1]^2$. It is easy to see that, for a $\tau \in (0, 1)$,

$$\|\nabla\psi(\tau)(x)\| = \|(1 - \tau)\nabla\phi_1(x) + \tau\nabla\phi_2(x)\|$$

is not necessarily equal to one. Hence, an intermediate point in the path ψ is not a proper signed-distance function, and it becomes difficult to talk about the level sets of this function as a shape.

Although this framework is not ideally suited for shape analysis of objects, it has been used extensively for extracting boundaries of contours in images. Its strength in dealing with changes in topology is very helpful in finding multiple objects in images.

2.3.2 Deformation-Based Shape Analysis

Another fundamental idea for shape analysis comes from Grenander's theory of deformable templates. In this approach, the shapes are considered as points on an infinite dimensional, differentiable manifold, and variations between shapes are modeled by actions of Lie groups on this manifold. Low-dimensional groups, such as rotation, translation, and scaling, change the object instances keeping the shape fixed, while high-dimensional groups, such as the diffeomorphism group, smoothly change the object shapes. This representation forms the mathematical basis of *deformable templates.*

A brief introduction to this approach follows: Let $\mathscr{I}$ be the space of all gray scale images defined on a domain D; so $\mathscr{I} = \mathbb{R}^D$. Here, D is usually a unit square in $\mathbb{R}^2$ for two-dimensional images or a unit cube in $\mathbb{R}^3$ for three-dimensional images. Changes in shapes of objects contained in images are obtained by deforming the domains. Let Γ be the space of diffeomorphisms from D to itself. Then, the elements of Γ can be used to deform the elements of $\mathscr{I}$ according to the map:

$$\gamma \cdot I(x) = I(\gamma(x)), \ \gamma \in \Gamma, \ \ x \in D \ , \ I \in \mathscr{I} \ .$$

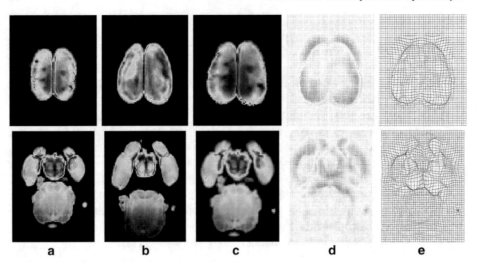

a   b   c   d   e

Fig. 2.13 (a) Image I_1, (b) Image I_2, (c) Deformed image is $I_1(\gamma)$, (d) deformation vector field generating γ, (e) deformed grid $\gamma(D)$

Associated with an image I, there is an equivalence class of images that can be reached by deforming that I.

$$[I] = \{\gamma \cdot I : \gamma \in \Gamma\} \ .$$

Similar to the previous section, this is defined to be the equivalence class of shapes associated with I.

A transformation between two images I_1 and I_2 can be modeled by diffeomorphically aligning I_1 to I_2. The resulting registration problem on the image space is given by:

$$E(I_1, I_2) = \min_{\gamma} \left(\lambda_1 \|I_2 - \gamma \cdot I_1\|^2 + \lambda_2 \mathscr{R}(\gamma)\right) \ , \tag{2.12}$$

where $\|\cdot\|$ is an appropriate distance on the image space $\mathscr{I}$ and $\mathscr{R}$ is a roughness penalty on Γ. Shown in Fig. 2.13 are two examples of this idea using the MRI images of mouse brains; in each case the two images are taken a few weeks apart to measure the anatomical growth. The leftmost image is I_1, the second image is I_2, and the third image is the deformed $I_1 \circ \gamma$, with γ being the minimizer of the cost function in Eq. 2.12. The right two panels show this optimal γ in a couple of ways: using the displacement vector field $(\gamma(x) - x)$ and using a deformed grid $(\gamma(D))$.

Although this method is quite effective in studying variability of objects as they appear in images, taking into account both the boundaries and the interiors, the invariance to translation, rotation, and scaling are difficult to apply here. Also, it is computationally expensive to deform the full image domain D when our interest is only in some curves or surfaces contained in D.

2.4 Exercises

Theoretical and computational exercises associated with techniques presented in this chapter are covered in the later chapters.

2.5 Bibliographic Notes

D'arcy Thompson's work appears in the second edition of his book *On Growth and Form* [114].

ICP is a widely used algorithm and some of the earlier references for ICP are [15, 9, 25]. Cootes and Taylor introduced the active shape model framework in a series of papers including [26]. Osher and Fedkiw [84] provide a recent, comprehensive review of the techniques used in level-set analysis. A good reference for principle component analysis is [41]. The first formal mathematical/statistical theory of shapes is due to the pioneering efforts of David Kendall [50, 49]. A remarkable body of work on this type of shape analysis exists due to works of Bookstein [18], Dryden and Mardia [28], Small [102], Le [49, 66], and several others. Diffeomorphism-based image matching and shape analysis is a major focus in *Computational Anatomy* [115, 33, 32, 7]. Interested readers are encouraged to read [34] for further information.

Chapter 3
Background: Relevant Tools from Geometry

Developing mathematical and/or statistical analysis of shapes involves two important steps: (i) formulating mathematical representations for shape and (ii) devising a framework for performing calculations under those representations. In a general shape analysis system, one would like to accomplish the tasks laid out in Sect. 1.3, namely, quantify similarities and differences between shapes, calculate integrals (or probabilities) on spaces of shapes, optimize cost functions on shape spaces, and design a representative shape or a template for a class of shapes. Three peculiar aspects of shape analysis make it necessary to involve some advanced mathematical tools:

- **Nonlinearity $\rightarrow$ Differential Geometry**:
  There are two important reasons for using differential geometry in the study of shapes. First, the shapes themselves are often curved, or nonlinear, objects. Therefore, the commonly used tools for vector-space calculus—addition, multiplication, differentiation, and integration—do not suffice for the study of these shapes. Also, irrespective of how shapes are mathematically represented, the resulting *spaces* of shapes are also nonlinear, i.e., they are not vector spaces. Thus, the study of shapes, as well as spaces of shapes, both require ideas from differential geometry, the part of mathematics that deals with calculus on non-linear manifolds. Since shape spaces are studied as nonlinear manifolds and individual shapes as points on these manifolds, differential geometry plays a very important role in shape analysis.
- **Invariance $\rightarrow$ Algebra**:
  Another aspect of analyzing shapes (of objects) is their invariance to certain transformations, such as translations, rotations, and scaling. Abstractly, shape is described as a property that remains unchanged under these transformations, but how should one incorporate this phenomenon into mathematical formulations? In the framework used in this book, originally proposed by Grenander [32] and Kendall [50], these transformations are conveniently represented as actions of groups on shape spaces. Equivalence classes of shapes are then defined as orbits under these group actions. Hence, algebra, the study of groups and group actions, becomes important in the discussion of invariant shape analysis.
- **Infinite Dimensionality $\rightarrow$ Functional Analysis**:
  Lastly, the study of shapes of (continuous) curves introduces an additional challenge of high, often infinite, dimensionality. This is because curves are formally represented as functions, and spaces of functions are usually infinite

dimensional. To analyze these spaces of functions, we require some tools from functional analysis. Although computer experiments will ultimately involve discretizing these functions into finite sets of points, the basic theory shall be developed assuming continuous, or functional, representations.

In summary, the three main mathematical tools for the shape analysis presented in this book are differential geometry, algebra, and functional analysis. Along with the appendix, this chapter serves as a basic and a brief refresher for some pertinent parts of those topics. If the readers wish to gain a deeper understanding of these areas, they are encouraged to refer to textbooks on these subjects. We recommend Boothby [19], Lang [64], and Do Carmo [23] for differential geometry; Munkres [83] and Milnor [78] for topology; Lang [65] for algebra; and Rudin [96] for functional analysis.

3.1 Equivalence Relations

In order to focus on the analysis of shapes and their representations as mathematical quantities, and especially to understand invariance of shapes to different transformations (see item 1 in Sect. 1.4), we need to use the concept of equivalence relations. Since a shape will typically be represented by many elements of a representation space, we will define equivalence relations to unify these elements. This will result in a consolidated space in which each shape is represented by a unique class. In this section, we give the definition of an equivalence relation and present some examples.

Definition 3.1 (Equivalence Relation). A relation $\sim$ on a set X is called an **equivalence relation** if, for all $x, y, z \in X$, we have the following properties:

- reflexivity, i.e., $x \sim x$,
- symmetry, i.e., $x \sim y \implies y \sim x$, and
- transitivity, i.e., $x \sim y, y \sim z \implies x \sim z$

The **equivalence class** of $x \in X$, denoted by $[x]$, is the set of all $y \in X$ such that $y \sim x$. For a set X, the **quotient space** of X under the equivalence relation $\sim$ is the set of all equivalence classes in X and is denoted by $X/\sim$. An equivalence relation partitions the set X into disjoint sets, each of which is an equivalence class. In other words, any two equivalence classes $[x]$ and $[y]$ are either disjoint or they are identical; they cannot have partial overlaps. Here are some examples.

Example 3.1. 1. For the set X=$\{1, 2, 3, 4\}$, define an equivalence relation $\sim$ such that $1 \sim 3$ and $2 \sim 4$. For this relation, we have the quotient space $X/\sim=$ $\{\{1, 3\}, \{2, 4\}\}$.

2. For $X = \mathbb{R}^n \backslash \{0\}$ and any $x, y \in X$, define $x \sim y$ if there exists a nonzero scalar $a \in \mathbb{R}$, such that $y = ax$. For this relation, an equivalence class is $[x] = \{ax | a \in \mathbb{R}, a \neq 0\}$, the straight line passing through the origin that contains x (not including the origin). The quotient space $X/\sim$ can be thought of as the set of all lines passing through the origin. If we represent each line by its intersection with one hemisphere of the unit sphere, we have a way of visualizing this quotient space as this hemisphere. Figure 3.1 shows an example for $n = 2$ in the left panel. The case of points lying on the equator is exceptional. For those

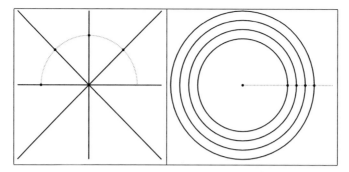

Fig. 3.1 Pictorial illustrations of the equivalence classes presented in Example 3.1, item 2 (*left*) and item 3 (*right*). Equivalence classes are shown in *dark lines* and their representatives in *broken lines*

points, we need to identify each point with its antipodal point since they both lie on the same line through the origin. This space is called $(n-1)$-dimensional *projective space*.

3. The equivalence relation complementary to the previous case is for $x, y \in \mathbb{R}^n$, $x \sim y$ when $|x| = |y|$ (under the two norm). In this case, all the points lying on a sphere (centered at the origin) are equivalent and the set of all spheres in $\mathbb{R}^n$ centered at the origin forms $\mathbb{R}^n/\sim$. Each sphere can simply be represented by its radius, so $\mathbb{R}^n/\sim$ in this case is identical to $\mathbb{R}_{\geq 0}$, the set of non-negative real numbers. Figure 3.1 shows an example for $n = 2$ in the right panel.

Later on, when we introduce mathematical representations of shapes of interest, we will apply the concept of equivalence relations to unify shape representation and to facilitate invariance.

Next, we shift attention to some basic tools from differential geometry for use in shape representation and analysis.

3.2 Riemannian Structure and Geodesics

To refresh some basic concepts in differential geometry, we refer the reader to Appendix A, where we introduce the concept of a differentiable manifold, its tangent spaces, and submanifolds. Here we introduce the relatively advanced concept of a Riemannian manifold. One of the most fundamental tasks in analyzing shapes as points on a manifold is to calculate distances between them. Distances on shape manifolds can, for instance, be used to quantify dissimilarities between shapes. These distances will be calculated by constructing shortest paths between shapes and by measuring the lengths of these paths. However, both the definition of a shortest path and the measurement of its length requires the additional concept of a Riemannian metric. In this section, we introduce this concept and illustrate it with a few examples.

This terminology needs some additional clarification. A Riemannian metric, often written simply as a metric, should be distinguished from a distance function on a space. As we shall see, the metric is defined infinitesimally using elements of the tangent space at a point, whereas the distance function is defined between any two elements of a space. We will also see that a metric can be used to define a distance function, while not every distance function may be derived from a Riemannian metric.

Informally, a Riemannian metric is an inner product defined on the tangent spaces of the manifold. Since tangent spaces are vector spaces, we begin by defining a bilinear form on a vector space.

Definition 3.2. A bilinear form on a vector space V over $\mathbb{R}$ is a map $\Phi : V \times V \to \mathbb{R}$, that satisfies the following requirements:

1. $\Phi(\alpha v_1 + \beta v_2, w) = \alpha \Phi(v_1, w) + \beta \Phi(v_2, w)$
2. $\Phi(v, \alpha w_1 + \beta w_2) = \alpha \Phi(v, w_1) + \beta \Phi(v, w_2)$

where $\alpha, \beta \in \mathbb{R}$ and $v, w, v_1, v_2, w_1, w_2 \in V$.

A bilinear form is symmetric if $\Phi(v, w) = \Phi(w, v)$ and skew symmetric if $\Phi(v, w) = -\Phi(w, v)$. A symmetric form becomes an inner product if it is positive definite, i.e., $\Phi(v, v) \geq 0$ and $\Phi(v, v) = 0 \iff v = 0$.

Definition 3.3. A Riemannian metric on a differentiable manifold M is a map Φ that smoothly associates to each point $p \in M$ a symmetric, bilinear, positive definite form on the tangent space $T_p(M)$.

A differentiable manifold with a Riemannian metric on it is called a **Riemannian manifold**. Very often we will reduce notation by writing $\Phi(v_1, v_2)$ simply as $\langle v_1, v_2 \rangle$. So, $\langle \cdot, \cdot \rangle$ will not always imply the Euclidean inner product; the actual definition will come from the Riemannian metric being used.

Here are a few examples.

In Example 3.2, it is assumed that the reader is familiar with the material in Appendix A.1.

Example 3.2. 1. As stated in Example A.1, the tangent space at any point in $\mathbb{R}^n$ is $\mathbb{R}^n$ itself. It becomes a Riemannian manifold with the Riemannian metric $\Phi(v_1, v_2) = v_1^T v_2$, the standard Euclidean product. Here, v_1 and v_2 are assumed to be column vectors so that $v_1^T v_2$ is their scalar product.

2. The upper half-plane $\mathbb{R}_+^2 = \{p = (p_1, p_2) \in \mathbb{R}^2 | p_2 > 0\}$ is a Riemannian manifold with the hyperbolic inner product, defined as follows. For any point $p \in \mathbb{R}_+^2$, the tangent space $T_p(\mathbb{R}_+^2)$ is $\mathbb{R}^2$. For any two vectors $v_1, v_2 \in T_p(\mathbb{R}_+^2)$, we define the hyperbolic inner product as:

$$\Phi(v_1, v_2) = \frac{1}{p_2^2} v_1^T v_2. \tag{3.1}$$

3. We examined the manifold $O(n)$ in Appendix A.1, Example A.6 and calculated its tangent space: $T_A O(n) = \{AX : X \text{ is skew symmetric}\}$. Define the inner product for any $Y, Z \in T_A O(n)$ by $\Phi(Y, Z) = trace(YZ^T)$, where *trace* denotes the sum of diagonal elements. With this metric, $O(n)$ becomes a Riemannian manifold. It will be important later to note that this metric is preserved under left and right multiplication by elements of $O(n)$:

$$\Phi(AY, AZ) = trace(AY(AZ)^T) = trace(AYZ^T A^T) = trace(AYZ^T A^{-1})$$
$$= trace(YZ^T) = \Phi(Y, Z).$$

This shows the metric is preserved by left translations; for right translations, the calculation is similar.

4. In case of the unit circle, the tangent space at any point p is given by $T_p(\mathbb{S}^1) = \{\alpha(-p_2, p_1) | \alpha \in \mathbb{R}\}$. Let us impose the standard Euclidean inner product on the tangent space as the Riemannian metric. Then, for any two vectors $\alpha(-p_2, p_1)$ and $\beta(-p_2, p_1)$ in $T_p(\mathbb{S}^1)$, we get $\Phi(\alpha(-p_2, p_1), \beta(-p_2, p_1)) = \alpha\beta$. (Note that this is the same inner product we would obtain by considering both these vectors as tangent vectors to $\mathbb{R}^2$ and then evaluating the ordinary Euclidean inner product, as in item (1) above.) This Riemannian metric makes $\mathbb{S}^1$ a Riemannian manifold.

5. Similarly, for the unit sphere $\mathbb{S}^n$ and a point $p \in \mathbb{S}^n$, the Euclidean inner product on the tangent vectors make $\mathbb{S}^n$ a Riemannian manifold. That is, for any $v_1, v_2 \in T_p(\mathbb{S}^n)$, we used the Riemannian metric $\Phi(v_1, v_2) = v_1^T v_2$.

Definition 3.4 (Isometry). A diffeomorphism $f : M \to N$, where M, N are Riemannian manifolds, is an **isometry** if $\langle u, v \rangle_p = \langle df_p(u), df_p(v) \rangle_{f(p)}$, for all $p \in M$ and all $u, v \in T_p(M)$. The left inner product is computed using the Riemannian metric in M, while the right one is computed using the metric in N.

Example 3.3. Consider the mapping from $\mathbb{R}^{n+1}$ to itself, given by $x \mapsto Ox$, for a fixed $O \in SO(n+1)$. Restricting it to the unit sphere $\mathbb{S}^n$, endowed with the Euclidean metric, we obtain a mapping $f : \mathbb{S}^n \to \mathbb{S}^n$ that is an isometry.

Using the Riemannian structure, it becomes possible to define lengths of paths on a manifold. Let $\alpha : [0, 1] \mapsto M$ be a parameterized path on a Riemannian manifold M, such that α is differentiable everywhere on $[0, 1]$. Then $\frac{d\alpha}{dt}$, the velocity vector at t, is an element of the tangent space $T_{\alpha(t)}(M)$ (as an equivalence class of curves, this tangent vector can simply be identified with $[\alpha]$) and its length is defined to be $\sqrt{\Phi(\frac{d\alpha}{dt}, \frac{d\alpha}{dt})}$. The length of the path α is then given by:

$$L[\alpha] = \int_0^1 \sqrt{\left(\Phi \left(\frac{d\alpha(t)}{dt}, \frac{d\alpha(t)}{dt} \right) \right)} dt . \tag{3.2}$$

This is the integral of the lengths of the velocity vectors along α and, hence, is the length of the whole path α. For any two points $p, q \in M$, one can define the distance between them as the infimum of the lengths of all smooth paths on M that start at p and end at q:

$$d(p, q) = \inf_{\{\alpha: [0,1] \mapsto M | \alpha(0) = p, \alpha(1) = q\}} L[\alpha] . \tag{3.3}$$

Definition 3.5. If there exists a path $\hat{\alpha}$ that achieves the above minimum, then it is called a **geodesic** between p and q on M.

We will give a more formal definition of geodesic in Chap. 6 and only observe here that it is heavily dependent upon the Riemannian metric Φ on M.

It can be seen that the minimization problem given in Eq. 3.3 is degenerate. Let α^* be a path that minimizes $L[\alpha]$. Since there are infinitely many ways to re-parameterize α^*, and each re-parameterization has the same length, there are infinitely many solutions to this problem. The problem of re-parameterization variability is handled by changing the objective functional from length to an energy functional given by:

$$E[\alpha] = \int_0^1 \Phi \left(\frac{d\alpha(t)}{dt}, \frac{d\alpha(t)}{dt} \right) dt \ . \tag{3.4}$$

The only difference here from the path length in Eq. 3.2 is that the square root has been removed. It can be shown that a critical point of this functional restricted to paths that start at p and end at q is a geodesic path on M. Furthermore, of all the re-parameterizations of a geodesic path, the one with constant speed has the minimum energy.

We should also note that, while distance-minimizing paths are always geodesics, there may be other geodesics between p and q that do not minimize distance. As a simple example, consider two points on a unit circle that are not antipodal. There are two arcs, or geodesics, connecting them: the major arc and the minor arc. Only the latter one minimizes the distance between them globally. However, it is always true that geodesics are "locally distance minimizing": i.e., if two points on a geodesic are sufficiently close to each other, then the geodesic will be a distance-minimizing path between them. In fact, a path in a Riemannian manifold is a geodesic if and only if it is locally distance minimizing. So, for now, we can consider that property as the definition of the word "geodesic," though it is not the usual definition given by differential geometers. We should also note that, depending on the Riemannian manifold and on the points p and q, there may not exist any geodesics at all between p and q. Here is a rather trivial example of this phenomenon: denote by M the manifold $\mathbb{R}^2$ with the origin $(0,0)$ removed, and with the standard Euclidean metric. Then, for any point $(x, y) \in M$, there is no distance-minimizing path, and in fact no geodesic, joining (x, y) to $(-x, -y)$.

Example 3.4. 1. Geodesics on $\mathbb{R}^n$ with the Euclidean metric are straight lines: $\alpha(t) = tq + (1 - t)p$ is a geodesic between p and q. The length of this geodesic is $\int_0^1 |q - p|dt = |q - p|$. So the distance between points in $\mathbb{R}^n$, as we already know, is simply the (Euclidean) length of these straight lines.

2. In case of the upper half-plane with the hyperbolic metric, as given in Eq. 3.1, the geodesic between any two points p, q in $\mathbb{R}_+^2$ has two forms:

   - If the two points p, q have the same first coordinate, i.e., $p_1 = q_1$, then the geodesic between them is the vertical line connecting them.
   - Otherwise, the geodesic between them is an arc of a half-circle whose center lies on the x axis. The center is located at $(x_1, 0)$ and radius is r where:

$$x_1 = \frac{|p|^2 - |q|^2}{2(p_1 - q_1)} \quad \text{and} \quad r = |p - (x_1, 0)| \ .$$

   Shown in Fig. 3.2 are some examples of geodesics between points in $\mathbb{R}_+^2$ under this metric. For comparison, we have also drawn the Euclidean geodesics (straight lines) between those points.

3. Geodesics on $\mathbb{S}^2$ under the Euclidean metric are given by great circles. We can prove this using polar coordinates (θ, ϕ), where θ is the latitude and ϕ is the longitude angle. Without loss of generality, we will assume that the points lie on the same longitudinal line. Let one of the points be represented by $p_1 \equiv (\theta_1, \phi_1)$ and the other by $p_2 \equiv (\theta_2, \phi_1)$ on $\mathbb{S}^2$. A continuous path connecting them is

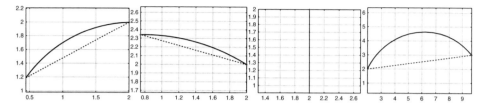

Fig. 3.2 Geodesics between points in a plane under two different Riemannian metrics: Euclidean (*broken lines*) and hyperbolic (*solid lines*)

given by $(\theta(t), \phi(t))$, $t \in [0,1]$ where $\theta(0) = \theta_1$, $\phi(0) = \phi_1$, $\theta(1) = \theta_2$ and $\phi(1) = \phi_1$. The length of this path, using Eq. 3.2, is given by:

$$L = \int_0^1 \sqrt{(\dot\theta(t)^2 + \dot\phi(t)^2 \sin^2(\theta(t)))} dt \geq \int_0^1 |\dot\theta(t)| dt = |\theta_2 - \theta_1|$$

The last quantity does not depend on the path and is thus a global minimum. The equality is achieved when $\dot\phi(t) = 0$ for all $t \in [0,1]$. Thus the meridian line, characterized by $\phi(t) = $ constant is a geodesic path between p_1 and p_2. For arbitrary two points on $\mathbb{S}^2$, we can rewrite this expression using Cartesian coordinates, as described next.

4. Similar to the case above, geodesics on a general unit sphere $\mathbb{S}^n$ are great circles. The distance-minimizing geodesic between any two points is the shorter of the two arcs of a great circle joining them between them. If p and q are points on the unit sphere (with $p \neq \pm q$), then the path:

$$\alpha(t) = \frac{1}{\sin(\vartheta)} (\sin(\vartheta(1-t))p + \sin(\vartheta t)q) \qquad (3.5)$$

gives a constant-speed parameterization of the unique shortest geodesic (i.e., great circle arc) from p to q, where ϑ is determined by $\cos(\vartheta) = \langle p, q \rangle$ and $0 < \vartheta < \pi$.

Let us prove this result. Clearly there is a unique vector w such that $\{p, w\}$ forms an orthonormal pair, and $q = (\cos(\vartheta))p + (\sin(\vartheta))w$. We then have the standard parameterization of the great circle through p and w, given by $\alpha(t) = (\cos(t))p + (\sin(t))w$. To express this path in terms of p and q, we first solve for w in terms of p and q, obtaining:

$$w = -\left(\frac{\cos(\vartheta)}{\sin(\vartheta)}\right)p + \left(\frac{1}{\sin(\vartheta)}\right)q,$$

and then substitute this into our formula for $\alpha(t)$. A little algebra and trigonometry then yields $\alpha(t) = \frac{1}{\sin(\vartheta)}(\sin(\vartheta - t)p + \sin(t)q)$. In order to arrange that $\alpha(0) = p$ and $\alpha(1) = q$, we replace t by ϑt, which yields the desired formula.

5. To define geodesics on $O(n)$, we introduce the notion of matrix exponential. For a matrix $A \in M(n)$, define its matrix exponential $\exp(A)$ by:

$$\exp(A) = I + \frac{A}{1!} + \frac{A^2}{2!} + \frac{A^3}{3!} + \ldots \qquad (3.6)$$

The matrix exponential is a mapping from $M(n)$ to $GL(n)$ and is defined for all elements of $M(n)$. The inverse of this mapping is called the matrix logarithm: $\log : GL(n) \to M(n)$. The logarithm (or log) does not exist for all invertible matrices and, when it does exist, it is not necessarily unique. It is important to note that the familiar property $\exp(A + B) = \exp(A)\exp(B)$ does not always hold for matrices A and B, but it does hold if A and B satisfy $AB = BA$. This is quite easy to see, since, if A and B commute, we can reorganize the power series for $\exp(A+B)$ just as we do in the case of the scalar exponential. It is also immediate from the power series definition that for any matrices $A \in GL(n)$ and $B \in M(n)$, $\exp(ABA^{-1}) = A\exp(B)A^{-1}$ and $\exp(X^T) = (\exp(X))^T$. We also observe that given any skew-symmetric matrix X, $\exp(X) \in O(n)$. To see this, assume X is skew symmetric and compute:

$$\exp(X)\exp(X)^T = \exp(X)\exp(X^T) = \exp(X + X^T) = \exp(0) = I \ .$$

Note that, for the second equality, we used the fact that X commutes with X^T, since $X^T = -X$. Using the matrix exponential, one can define geodesics on $O(n)$ (with respect to the Riemannian metric defined earlier) as follows: for any $A \in O(n)$ and any skew-symmetric matrix X:

$$\alpha(t) \equiv A\exp(tX) \ ,$$

is the unique geodesic in $O(n)$ passing through A with velocity vector AX at $t = 0$.

As we have mentioned earlier, the tangent space $T_p(M)$ is a vector space even though the manifold M may not be. One can view $T_p(M)$ as a locally flat approximation of M. It is local since this approximation is valid only in some neighborhood of p. One use of this approximation is in imposing probability models on M by first imposing a density on $T_p(M)$ and then transferring it to M. Defining a density first on $T_p(M)$ is an attractive option because it is a vector space and traditional multivariate densities (parametric or nonparametric) can be used. To exploit such a linear approximation, one needs a mapping to transfer points back and forth between M and $T_p(M)$. Next, we describe one such mapping in the form of the exponential map (a generalization of the matrix exponential).

Theorem 3.1. *Let M be a Riemannian manifold. Given a point $p \in M$ and a tangent vector $v \in T_p(M)$, there exists a unique constant-speed parameterized **geodesic** $\alpha_v : (-\epsilon, \epsilon) \to M$, for some $\epsilon > 0$, such that $\alpha_v(0) = p$ and $\dot{\alpha}_v(0) = v$.*

It is important to note that the domain $(-\epsilon, \epsilon)$ of α_v depends on the point p and the tangent vector v. However, if M is complete, then this domain is always $(-\infty, \infty)$. For a proof of this theorem, please refer to [36] pp. 30.

Definition 3.6 (Exponential Map). If M is a Riemannian manifold and $p \in M$, the **exponential map** $\exp_p : U \subset T_p(M) \to M$ is defined by $\exp_p(v) = \alpha_v(1)$ where α_v is as defined above and the domain U contains an open neighborhood of 0 in $T_P(M)$.

The exponential mapping exp maps a vector $v \in T_p(M)$ to a point of M. In words, to reach the point $\exp_p(v)$, one starts at p and then moves for time 1 along the unique constant speed geodesic whose velocity vector at p is v. Note that if $v = 0$, the corresponding geodesic is just the constant path at p, so $\exp_p(0) = p$.

The inverse of an exponential map takes a point on the manifold M and maps it to an element (or multiple elements) of the tangent space $T_p(M)$. A vector v is said to be the inverse exponential of $q \in M$ at p if $\exp_p(v) = q$. It is denoted by $v = \exp_p^{-1}(q)$ and is often not a unique point. That is, the inverse may be set-valued.

Example 3.5. 1. For $\mathbb{R}^n$, under the Euclidean metric, since geodesics are given by straight lines, the exponential map is a simple addition: $\exp_p(v) = p + v$, for $p, v \in \mathbb{R}^n$. The inverse exponential is simply $\exp_p^{-1}(q) = q - p$. Similarly for $M(n)$, the space of $n \times n$ matrices, the exponential map is given by a simple addition of matrices, i.e., $\exp_A(X) = A + X$, and the inverse exponential is $\exp_A^{-1}(B) = B - A$.

2. The geodesics on a sphere $\mathbb{S}^n$ under the Euclidean metric can also be parameterized in terms of a direction v in $T_p(\mathbb{S}^n)$:

$$\alpha_t(v) = \cos(t|v|)p + \sin(t|v|)\frac{v}{|v|}. \tag{3.7}$$

As a result, the exponential map, $\exp_p : T_p(\mathbb{S}^n) \mapsto \mathbb{S}^n$, has a simple expression:

$$\exp_p(v) = \cos(|v|)p + \sin(|v|)\frac{v}{|v|}. \tag{3.8}$$

The exponential map is a bijection if we restrict $|v|$ so that $|v| \in [0, \pi)$. For a point $q \in \mathbb{S}^n$ ($q \neq p$), the inverse exponential map $\exp_p^{-1}(q)$ is given by u, where:

$$u = \frac{\theta}{\sin(\theta)}(q - \cos(\theta)p), \quad \text{where} \quad \theta = \cos^{-1}(\langle p, q \rangle).$$

3. The exponential map for $O(n)$ at any point $O \in O(n)$ is given by $\exp_O : T_O(O(n)) \to O(n)$, $\exp_O(OX) = O\exp(X)$, where X is skew symmetric and the last term is the matrix exponential (Fig. 3.3). Note that this map is not one-to-one unless we restrict to a small enough neighborhood of O. The inverse exponential map on $O(n)$ is defined as follows: For $O_1, O_2 \in O(n)$, define $\exp_{O_1}^{-1}(O_2) = O_1 \log(O_1^T O_2)$.

Fig. 3.3 Exponential map of a tangent vector X_p

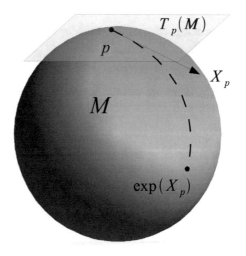

The tangent spaces at different points on a manifold are together referred to as the tangent bundle of the manifold.

Definition 3.7. The **tangent bundle** of M, denoted by TM, is defined as the disjoint union of the tangent spaces at all points of M. It is specified as $TM = \cup_{p \in M} T_p(M)$.

The tangent bundle itself is a smooth manifold whose dimension is twice that of M. To see that TM is a smooth manifold, we define charts on it as follows. Assume M has dimension n. Given any point $p \in M$, let (U, ϕ) be a chart for M, with $p \in U$. Denote by $TU \subset TM$ the tangent bundle over U and define $\theta : TU \to \mathbb{R}^{2n}$ by $\theta(v) = (\phi(p), d\phi_p(v))$, where $v \in T_p(M)$. θ is a bijection from TU to $\phi(U) \times \mathbb{R}^n \subset \mathbb{R}^{2n}$. Charts defined in this way clearly cover TM, and their overlapping maps are smooth, so they provide TM with the structure of a smooth manifold.

3.3 Geodesics in Spaces of Curves on Manifolds

Let M be a Riemannian manifold and denote by I the unit interval $[0, 1]$. Also, for some $L > 0$, denote by $\mathcal{M}$ the space of measurable functions $[0, L] \to M$. $\mathcal{M}$ is also a manifold (see, e.g., [87]); its tangent space is given as follows. If $\alpha \in \mathcal{M}$, then:

$$T_\alpha(\mathcal{M}) = \{w : [0, L] \to TM | \ \forall \tau \in [0, L], w(\tau) \in T_{\alpha(\tau)}(M)$$
$$\text{and } \int_0^L \langle w(\tau), w(\tau) \rangle \, d\tau < \infty \} \ .$$

In other words, this is just the set of first-order deformations of $\alpha \in \mathcal{M}$. We now make $\mathcal{M}$ into a Riemannian manifold. If $w_1, w_2 \in T_\alpha(\mathcal{M})$, define:

$$\langle w_1, w_2 \rangle = \int_0^L \langle w_1(\tau), w_2(\tau) \rangle_{\alpha(\tau)} \, d\tau$$

where the inner product inside the integral uses the Riemannian metric on M.

Theorem 3.2. *Suppose we are given a path in $\mathcal{M}$ represented as $\alpha : [0, L] \times I \to M$. For each $\tau \in [0, L]$, define $\alpha_\tau : I \to M$ by $\alpha_\tau(t) = \alpha(\tau, t)$. Then α is a geodesic in $\mathcal{M}$ if $\forall \tau \in [0, L]$, α_τ is a geodesic in M.*

Proof. Here we will use the fact that a path on a Riemannian manifold is a geodesic if and only if the gradient of the energy given in Eq. 3.4, with respect to the path, is zero. In other words, that path is a critical point of the energy function E.

Now suppose that $\tilde{\alpha} : [0, L] \times I \times (-\epsilon, \epsilon) \to M$ is an arbitrary variation of α in the space of curves in $\mathcal{M}$, i.e., we are assuming that for all $\tau \in [0, L]$ and $t \in I$, $\tilde{\alpha}(\tau, t, 0) = \alpha(\tau, t)$ and for all $\tau \in [0, L]$ and $h \in (-\epsilon, \epsilon)$, $\tilde{\alpha}(\tau, 0, h) = \alpha(\tau, 0)$ and

$\tilde{\alpha}(\tau, 1, h) = \alpha(\tau, 1)$. For each value of $h \in (-\epsilon, \epsilon)$, we calculate the energy of the path $\tilde{\alpha}(., ., h)$ in $\mathcal{M}$ as follows:

$$E(\tilde{\alpha}(., ., h)) = \frac{1}{2} \int_0^1 \int_0^L \left\langle \frac{\partial \tilde{\alpha}}{\partial t}(\tau, t, h), \frac{\partial \tilde{\alpha}}{\partial t}(\tau, t, h) \right\rangle_{\tilde{\alpha}(\tau, t, h)} d\tau \ dt$$

$$= \frac{1}{2} \int_0^L \int_0^1 \left\langle \frac{\partial \tilde{\alpha}}{\partial t}(\tau, t, h), \frac{\partial \tilde{\alpha}}{\partial t}(\tau, t, h) \right\rangle_{\tilde{\alpha}(\tau, t, h)} dt \ d\tau$$

Differentiating with respect to h at h = 0 gives:

$$\frac{d}{dh}|_{h=0} E(\tilde{\alpha}(., ., h)) = \frac{1}{2} \int_0^L \frac{d}{dh}|_{h=0} \left(\int_0^1 \left\langle \frac{\partial \tilde{\alpha}}{\partial \tilde{t}}(\tau, t, h), \frac{\partial \tilde{\alpha}}{\partial \tilde{t}}(\tau, t, h) \right\rangle_{\tilde{\alpha}(\tau, t, h)} dt \right) d\tau$$

If we assume that α_τ is a geodesic in M for every $\tau \in [0, L]$, then it follows immediately that the function we are integrating over $[0, L]$ in the right-hand side of the above expression is 0 for every τ, proving that α is a geodesic in $\mathcal{M}$. □

Example 3.6. 1. Let $M = \mathbb{R}^n$ with the Euclidean metric and $\mathcal{M}$ be the set of continuous maps from $[0, L]$ to $\mathbb{R}^n$. For any $\alpha \in \mathcal{M}$, we will have the Riemannian metric: for any $w_1, w_2 \in T_\alpha(\mathcal{M})$,

$$\langle w_1, w_2 \rangle = \int_0^L \langle w_1(\tau), w_2(\tau) \rangle \, dt \ .$$

The corresponding distance between any two paths α_1 and α_2 is simply $\int_0^L |\alpha_1(\tau) - \alpha_2(\tau)| d\tau$.

2. This time let $M = \mathbb{S}^2$ with the Euclidean metric and $\mathcal{M}$ be the set of continuous maps from $[0, L]$ to $\mathbb{S}^2$. For any $\alpha \in \mathcal{M}$, we will have the Riemannian metric: for any $w_1, w_2 \in T_\alpha(\mathcal{M})$: $\langle w_1, w_2 \rangle = \int_0^L \langle w_1(\tau), w_2(\tau) \rangle \, d\tau$. The corresponding distance between any two paths α_1 and α_2 is simply $\int_0^L \theta(\tau) d\tau$, where $\theta(\tau) = \cos^{-1}(\langle \alpha_{(\tau)}, \alpha_2(\tau) \rangle)$. This assumes that $\alpha_1(\tau) \neq -\alpha_2(\tau)$ for all τ. A pictorial depiction of the geodesic path between two curves α_1 and α_2 on $\mathbb{S}^2$ is shown in Fig. 3.4.

Fig. **3.4** A depiction of geodesic path between two curves, α_1 and α_2 on $\mathbb{S}^2$, under the Euclidean metric, with *dotted lines* denoting great circles

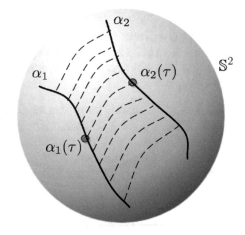

3.4 Parallel Transport of Vectors

An important concept in differential geometry is the *parallel transport* of tangent vectors along paths in manifolds. Here is the basic setup: Suppose M is a Riemannian manifold, $p \in M$, $v \in T_p(M)$, and let $\alpha : [0, 1] \to M$ be a smooth path in M such that $\alpha(0) = p$. We wish to find the most "natural" way to transport v along α. What we mean by this is that, for each $t \in [0, 1]$, we wish to produce a tangent vector $w_v(t) \in T_{\alpha(t)}(M)$ that results from transporting v along α with as little "twisting" as possible. Consider the special case in which $M = \mathbb{R}^n$; make M into a Riemannian manifold by using the standard Euclidean inner product on each tangent space $T_x(\mathbb{R}^n) \cong \mathbb{R}^n$. Then the obvious solution is just to let $w_v(t) = v$ for each t. However, for a general Riemannian manifold, each tangent space is a different vector space, so it is not obvious how to transport vectors from one point to another.

For general Riemannian manifolds, the precise definition of "parallel transport" requires the concept of covariant derivative, which we have not introduced yet. So we will consider the following special case:

Definition 3.8. Suppose M is a submanifold of $\mathbb{R}^n$, with Riemannian metric defined by restricting the Euclidean inner product on $\mathbb{R}^n$ to each tangent space of M. Let $p \in M$, $v \in T_p(M)$ and suppose $\alpha : [0, 1] \to M$ is a smooth path with $\alpha(0) = p$. Then the *parallel transport* of v along α is defined to be the unique vector field $w_v : [0, 1] \to \mathbb{R}^n$ satisfying:

1. For all $t \in [0, 1]$, $w_v(t) \in T_{\alpha(t)}(M)$.
2. $w_v(0) = v$.
3. For all $t \in [0, 1]$, $\dot{w}_v(t) \perp T_{\alpha(t)}(M)$.

The "dot" in item 3 refers to the ordinary derivative with respect to t. It is a fact that the parallel transport $w_v(t)$ always exists and is unique. It follows that, for each t, parallel transport induces a mapping $\Psi : T_{\alpha(0)}(M) \to T_{\alpha(t)}(M)$ defined by $\Psi(v) = w_v(t)$. This mapping has the following important properties:

1. It is a linear isomorphism.
2. It is an isometry between these two tangent spaces, i.e., $\langle v_1, v_2 \rangle = \langle \Psi(v_1), \Psi(v_2) \rangle$ for each v_1, v_2 in $T_{\alpha(0)}(M)$.

If p and q are nearby points in M, we sometimes refer simply to "parallel transport from $T_p(M)$ to $T_q(M)$," without specifying the path α. In this case, α is taken to be the shortest geodesic from p to q. We illustrate this idea with some simple examples.

Example 3.7. 1. **Euclidean Space**: Since the tangent spaces $T_p(\mathbb{R}^n) = \mathbb{R}^n$ for all $p \in \mathbb{R}^n$, and a geodesic path under the Euclidean metric is simply a straight line, a transport of a vector is rather simple in this case. For any vector $v \in T_p(\mathbb{R}^n)$, we can set a constant vector field $w(t) = v$ and since $\dot{w}(t) = 0$, we automatically have $\dot{w}(t) \perp T_{\alpha(t)}(\mathbb{R}^n)$. As a result, $w(1) = v$ is simply a copy of the original vector. This fact holds more generally for any path $\alpha : [0, 1] \to \mathbb{R}^n$. If $v \in \mathbb{R}^n$ is a tangent vector at $\alpha(0)$, then the parallel transport of v along α is simply the constant vector field $w_v(t) = v$.

2. **Unit Sphere**: Consider the unit sphere $\mathbb{S}^{n-1} \subset \mathbb{R}^n$, with the standard Riemannian metric inherited from $\mathbb{R}^n$. In this case, there is a straightforward formula for parallel transport of tangent vectors.

Theorem 3.3. *Let x_1 and x_2 be two points in $\mathbb{S}^{n-1}$, with $x_1 \neq \pm x_2$. Let Ψ denote the parallel transport map from $T_{x_1}(\mathbb{S}^{n-1}) \to T_{x_2}(\mathbb{S}^{n-1})$, along the shortest geodesic from x_1 to x_2. Then Ψ is given by the formula:*

$$\Psi(v) = v - (2(v \cdot x_2)/|x_1 + x_2|^2)(x_1 + x_2). \tag{3.9}$$

Proof. First, consider the case $\mathbb{S}^1 \subset \mathbb{R}^2$. Let $\{e_1, e_2\}$ be any orthonormal basis of $\mathbb{R}^2$. There is only one geodesic in $\mathbb{S}^1$; its unit-speed parameterization is given by $\alpha(t) = \cos(t)e_1 + \sin(t)e_2$. An arbitrary element of the tangent space $T_{e_1}(\mathbb{S}^1)$ will be of the form $v = re_2$, where r is any real number. It is easy to verify that the parallel transport of v along the above geodesic is given by $w_v(t) = -r\sin(t)e_1 + r\cos(t)e_2$. This exercise has been included in the list at the end of this chapter. Now consider the given elements x_1 and x_2 of $\mathbb{S}^1$. Without loss of generality, we may assume that $x_1 = e_1$ and $x_2 = \cos(\theta)e_1 + \sin(\theta)e_2$. Then the proof for $\mathbb{S}^1$ is completed by observing that, if we plug these expressions for x_1, x_2, and v (in terms of e_1, e_2, and θ) into the formula for $\Psi(v)$, we obtain $w_v(\theta)$.

To prove the Theorem for general $\mathbb{S}^{n-1}$, one simply observes that the formula given for Ψ is the identity on the component of v perpendicular to the plane spanned by x_1 and x_2. Since the derivative of this component will be 0, the three requirements are still satisfied.

This theorem covers only the case in which $x_1 \neq \pm x_2$. How do we parallel transport if $x_1 = \pm x_2$? If $x_1 = x_2$, the answer is easy: the shortest geodesic is just the constant path, and parallel transport is clearly just given by the identity. (In fact, the formula in the Theorem still works in this case, if you substitute $x_2 = x_1$.)

If $x_1 = -x_2$, things are a little trickier, because there are an infinite number of equally short geodesics from x_1 to x_2. To visualize this situation, consider the 2-sphere (surface of the earth) with x_1 and x_2 equal to the north and south poles. Then all of the longitude lines are shortest geodesics from x_1 to x_2. By parallel transporting along different geodesics, one obtains different parallel transport maps. In fact, each unit vector in $T_{x_1}(\mathbb{S}^{n-1})$ is tangent to one of these geodesics. If $w \in T_{x_1}(\mathbb{S}^{n-1})$ is a unit vector, and we wish to parallel transport from $T_{x_1}(\mathbb{S}^{n-1})$ to $T_{-x_1}(\mathbb{S}^{n-1})$ along the geodesic tangent to w, the formula is $\Psi(v) = v - 2(v \cdot w)w$. Three pictorial examples of parallel transport along geodesic paths in $\mathbb{S}^2$ are presented in Fig. 3.5. In each case, the thick line shows a geodesic path α on $\mathbb{S}^2$ and vectors show the parallel transport of the vector at the point $\alpha(0)$.

Now we consider parallel translation of vectors along a path on $\mathbb{S}^2$. Let $\alpha : [0, 1] \to \mathbb{S}^2$ be a path in $\mathbb{S}^2$ and let $v \in T_{\alpha(0)}(\mathbb{S}^2)$ be a tangent vector. There is no simple formula for the parallel transport of v along α in this case. However, in the special case that α is a great circle (which is a geodesic on $\mathbb{S}^2$), one may parallel transport tangent vectors along α using the equation developed above:

$$R_\alpha(v) = v - (2(v \cdot \alpha(1))/(|\alpha(0) + \alpha(1)|^2))(\alpha(0) + \alpha(1)). \tag{3.10}$$

For the general case in which α is not a great circle, one may proceed as follows. Choose a large integer N. For $i = 1$ to N, define α_i to be the small arc of a great

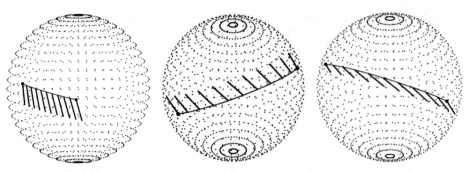

Fig. 3.5 Three examples of forward parallel transport of tangents along paths on $\mathbb{S}^2$

circle from $\alpha(\frac{i-1}{N})$ to $\alpha(\frac{i}{N})$. The original curve α is approximated by the union of the small arcs α_i. Hence, we approximate parallel transport of a tangent vector from $T_{\alpha(0)}(\mathbb{S}^2)$ to $T_{\alpha(1)}(\mathbb{S}^2)$ by the composition $R_\alpha \approx R_{\alpha_N} \circ R_{\alpha_{N-1}} \circ \cdots \circ R_{\alpha_1}$, where each of these maps R_{α_i} is simply a rotation, since each α_i is a great circle. As $N \to \infty$, this approximation approaches an exact formula. $\quad\square$

We pause here to discuss the concept of "flatness," which is closely related to parallel transport of tangent vectors. As seen in the first example above, if we are given a tangent vector $w \in T_{p_1}(\mathbb{R}^n)$ and wish to parallel transport it to $T_{p_2}(\mathbb{R}^n)$, the result will not depend on which path α from p_1 to p_2 we use. This is because $\mathbb{R}^n$ (with the standard Riemannian metric) is a flat manifold. On the other hand, $\mathbb{S}^2$ is not flat with respect to the standard metric. (In fact, it can be proven that $\mathbb{S}^2$ is not flat with respect to *any* Riemannian metric.) The fact that $\mathbb{S}^2$ is not flat means that parallel transport of a tangent vector from $T_{p_1}(\mathbb{S}^2)$ to $T_{p_2}(\mathbb{S}^2)$ will depend upon which path α from p_1 to p_2 we use. Here is a definition of flatness:

Definition 3.9 (Flat Riemannian Manifold). A Riemannian manifold M is flat if and only if it can be covered by a collection of open sets $\{U_i\}$, where each U_i has the property that given any two points $x, y \in U_i$; the parallel transport map from $T_x(M)$ to $T_y(M)$ does not depend on which path from x to y we use (as long as the path lies entirely in the set U_i).

3. **Special Orthogonal Group**: We state the following result without proof. For any two elements O_1 and O_2 in $SO(n)$, and a tangent vector $W \in T_{O_1}(SO(n))$, the tangent vector at O_2 that is the parallel transport of W along the shortest geodesic from O_1 to O_2 is:

$$W' = O_1(O_1^T O_2)^{1/2}(O_1^T W)(O_1^T O_2)^{1/2}.$$

Note that $SO(n)$ is not a flat manifold.

3.5 Lie Group Actions on Manifolds

Central to our formulation of shape analysis is the use of transformations to model the variations of objects, especially within the same shape classes. These transformations are mathematically formulated as actions of certain groups on objects, as described next.

So far we have discussed definitions and examples of manifolds and groups. In some cases, the sets have both the structures—manifold and group; these sets are called Lie groups. In view of their dual structures, they are sets of great importance in algebra and pattern theory.

Definition 3.10 (Lie Groups). A group G is a **Lie group**, if (i) it is a smooth manifold and (ii) the group operations $G \times G \to G$ defined by $(g, h) \to gh$ and $G \to G$ defined by $g \to g^{-1}$ are both smooth mappings.

Conveniently, most of the groups we have considered earlier are also Lie groups.

Example 3.8. 1. The general linear group $GL(n)$, by the virtue of being an open subset of $\mathbb{R}^{n \times n}$, is also a differentiable manifold. It is, therefore, a Lie group. Any subgroup of $GL(n)$, if it is also a submanifold of $GL(n)$, will be a Lie group. An important example in this category is the rotation group $SO(n)$.
2. The translation group $\mathbb{R}^n$ is both a manifold and a group (with addition operation). Therefore, it is a Lie group.
3. The scaling group, $R^\times$ with multiplication operation, is a Lie group.
4. The unit circle $\mathbb{S}^1$ is a group although it is not straightforward to see why from its definition. Using the mapping:

$$\mathbb{S}^1 \to SO(2), \quad \text{given by} \quad (x_1, x_2) \mapsto \begin{bmatrix} x_1 & -x_2 \\ x_2 & x_1 \end{bmatrix},$$

we can identify $\mathbb{S}^1$ to $SO(2)$. (In fact, this mapping provides a diffeomorphism between the two manifolds.) The latter is a group with matrix multiplication as the group operation. Through this identification, $\mathbb{S}^1$ also inherits a group structure. Since it is also a differentiable manifold, $\mathbb{S}^1$ becomes a Lie group. However, the two-dimensional sphere $\mathbb{S}^2$ is not a Lie group as it does not have a group structure.

3.5.1 Actions of Single Groups

Now we take a manifold M and study how the points change on M when operated on by some kind of a transformation group. Our motivation, of course, is shape analysis where M will be a manifold formed by curves and we want to study variations of their shapes under different transformations - rotations, translations, and scalings. Mathematically, this is managed using group actions on manifolds.

Definition 3.11 (Group Action). Given a manifold M and a Lie group G, a left **group action** of G on M is a map $G \times M \to M$, written as $(g, p) \mapsto g * p$, such that:

1. $g_1 * (g_2 * p) = (g_1 \cdot g_2) * p$, $\forall g_1, g_2 \in G$ and $p \in M$.
2. $e * p = p$, $\forall p \in M$.

In item 1, the symbol $\cdot$ denotes the group operation in G.

Another way to phrase this relation is to say that G *acts on* M. We say that G acts **smoothly** on M if the map $G \times M \to M$ is a smooth map. In the definition above, the group element g is applied from the left, so it is also called the *left* group action. Instead, if it is applied from the right, i.e., $(p, g) \mapsto p * g$, then it is called the *right* group action.

In case M has a Riemannian structure and, therefore, we can define distances between points on M, we can study the effect of the group action on these distances.

Definition 3.12. A group action of G on a Riemannian manifold M is called **isometric** if it preserves the Riemannian metric on M. In other words, for all $g \in G$, the map $M \to M$ given by $p \mapsto g * p$ is an isometry. For the same situation, we sometimes say that G acts on M by isometries.

It then also follows that for all $g \in G$ and $x, y \in M$, $d(x, y) = d(g * x, g * y)$, where $d(x, y)$ is the distance function on M resulting from the Riemannian metric. Compare this with Definition 3.4 that specifies an isometry between Riemannian manifolds. Clearly, this is a specific case of that definition; here the mapping is between the same manifold and it is defined using a group action.

There is some additional notation associated with group actions that will be useful later in shape analysis.

Definition 3.13 (Orbit). Assume that a group G acts on a manifold M. For any $p \in M$, the **orbit** of p under the action of G is defined as the set $G \cdot p = \{g \cdot p : g \in G\}$. We will also denote it by $[p]$.

If the orbit of any $p \in M$ is the whole of M, then the group action is said to be **transitive**.

The orbit of a point in M refers to all possible points one can reach in M using the action of G on that point. The orbit of a point can vary in size from a single point to the entire manifold M (which is the case if the action is transitive). Let us consider some examples to understand this point.

Example 3.9. In this example, we study the actions of some familiar transformation groups on $\mathbb{R}^n$.

1. **Translation Group**: The translation group $\mathbb{R}^n$ acts on the vector space $\mathbb{R}^n$ by the action $x * y = x + y$ for any $x, y \in \mathbb{R}^n$. This group action is transitive since the orbit of any vector v is the whole of $\mathbb{R}^n$, i.e., $[y] = \mathbb{R}^n$. Also, this group action is isometric using the Euclidean structure on $\mathbb{R}^n$, since $|y_1 - y_2| = |(x + y_1) - (x + y_2)|$ for all $x, v_1, v_2 \in \mathbb{R}^n$.

2. **Scaling Group**: The scale group $\mathbb{R}^\times$ acts on $\mathbb{R}^n$ by $a * x = ax$ for any $a \in \mathbb{R}^\times$ and $x \in \mathbb{R}^n$. This group action is not transitive on $\mathbb{R}^n$. Under the action, the orbit of the zero vector consists of the zero vector alone. The orbit of a nonzero vector is a straight line spanned by positive scalings of this vector. If we impose the Euclidean metric on $\mathbb{R}^n$ to make it a Riemannian manifold, then the action of $\mathbb{R}^\times$ on $\mathbb{R}^n$ is not isometric because $|x_1 - x_2| \neq |ax_1 - ax_2|$ for any $a \neq 1$.

3. **Rotation Group**: The rotation group $SO(n)$ acts on $\mathbb{R}^n$ by the action $O * x = Ox$, the matrix-vector multiplication, for all $O \in SO(n)$ and $x \in \mathbb{R}^n$. The orbit of any vector x is simply a sphere centered at zero and radius $|x|$, i.e., $[x] = \{|x|u | u \in \mathbb{S}^{n-1}\}$. Therefore, this group action is not transitive but is isometric, under the Euclidean structure on $\mathbb{R}^n$, because $|x_1 - x_2| = |Ox_1 - Ox_2|$, for all $O \in SO(n)$, $x_1, x_2 \in \mathbb{R}^n$.

Our introduction of the group action is for an important reason. We shall use membership of orbits to define equivalence relations.

Lemma 3.1. *If we define a relation $\sim$ by saying that $p \sim q$ if and only if p is an element of the orbit $[q]$, then $\sim$ is an equivalence relation.*

The proof is left as an exercise to the reader. These equivalence relations, in turn, define quotient spaces that are natural domains for statistical shape analysis. We cover this topic in more detail in Sect. 3.6.

3.5.2 Actions of Product Groups

In case there are multiple groups acting on a manifold, what is the nature of their joint action? There are at least two ways of categorizing these joint actions.

3.5.2.1 Direct Product Action

If G and H are groups, the easiest way to endow the product $G \times H$ with a group structure is simply to define $(g_1, h_1) \cdot (g_2, h_2) = (g_1 \cdot g_2, h_1 \cdot h_2)$. With this definition, $G \times H$ is called the **direct product** of G and H. Suppose G and H both act on the manifold M. We say these actions **commute** if $g * (h * p) = h * (g * p)$ for all $g \in G$, $h \in H$, and $p \in M$. In the case that these actions commute, they combine to give us an action of the product group $G \times H$ on M defined by $(g, h) * p = g * (h * p)$.

Example 3.10. **Scale and Rotate**: Consider the actions of the rotation group $SO(n)$ and the scaling group $\mathbb{R}^{\times}$ on $\mathbb{R}^n$. We can form the product group $SO(n) \times \mathbb{R}^{\times}$ with the group operation: $(O_1, a_1) \cdot (O_2, a_2) = (O_1 \cdot O_2, a_1 \cdot a_2)$. We check if the actions of these two groups on $\mathbb{R}^n$ commute. For $x \in \mathbb{R}^n$, $O \in SO(n)$, and $a \in \mathbb{R}^{\times}$, we have:

$$a(Ox) = O(ax) .$$

These actions do commute, and therefore, we can define the combined action of $SO(n) \times \mathbb{R}^{\times}$ on $\mathbb{R}^n$ as:

$$(SO(n) \times \mathbb{R}^{\times}) \times \mathbb{R}^n \mapsto \mathbb{R}^n \quad \text{by} \quad (O, a) * x = aOx .$$

3.5.2.2 Semi-direct Product Action

If the actions of G, H on M do not commute, then we do not obtain an action of the product group $G \times H$ on M. In some cases, however, we can define an action of a semi-direct product as follows. If G and H are groups, we say that H **acts on G by isomorphisms** if we have a group action of H on G (denoted by $h * g$) and if, for every $h \in H$, the function $G \to G$ given by $g \mapsto h * g$ is a group isomorphism from G to itself. Given an action of H on G by isomorphisms, we can define a corresponding group operation on $G \times H$ as follows: $(g_1, h_1) \cdot (g_2, h_2) = (g_1 \cdot (h_1 * g_2), h_1 \cdot h_2)$. When endowed with this group operation, $G \times H$ is called the **semi-direct product** of G and H and is denoted by $G \rtimes H$. To reiterate an important point, let us emphasize that we cannot form the semi-direct product $G \rtimes H$ without first being provided with an action of H on G by isomorphisms.

Example 3.11. **Scale, Rotate, and Translate**: As an example, recall that the three groups $\mathbb{R}^\times$, $SO(n)$, and $\mathbb{R}^n$ all act on $\mathbb{R}^n$ (the first by scalar multiplication, the second by matrix multiplication, and the third by vector addition). We can see that the actions of $SO(n)$ and $\mathbb{R}^\times$ do not commute with the action of $\mathbb{R}^n$. That is, $Ox + v \neq O(x + v)$ and $a(x + v) \neq (ax + v)$, in general, for $a \in \mathbb{R}^\times$, $v \in \mathbb{R}^n$, $O \in SO(n)$, and $x \in \mathbb{R}^n$. Thus, we have an action of $\mathbb{R}^\times \times SO(n)$ on $\mathbb{R}^n$, and an action of $\mathbb{R}^n$ on $\mathbb{R}^n$, but not an action of $\mathbb{R}^n \times (\mathbb{R}^\times \times SO(n))$ on $\mathbb{R}^n$. However, $\mathbb{R}^\times \times SO(n)$ acts on the translation group $\mathbb{R}^n$ in the same way it acts on the vector space $\mathbb{R}^n$. Hence, we can form the semi-direct product $\mathbb{R}^n \rtimes (\mathbb{R}^\times \times SO(n))$ with the group operation being:

$$(v_1, a_1, O_1) * (v_2, a_2, O_2) = (v_1 + a_1 O_1 v_2, a_1 a_2, O_1 O_2) \ .$$

The reader can then check that we then have a well-defined action of $\mathbb{R}^n \rtimes (\mathbb{R}^\times \times SO(n))$ on $\mathbb{R}^n$ defined by $(v, a, O) * \mathbf{x} = aO\mathbf{x} + v$. The action of an element of the semi-direct product $\mathbb{R}^n \rtimes SO(n)$, also called $SE(n)$, on a solid object is called a rigid motion because this action is by isometries.

Extending this idea to a certain shape representation, consider the shapes represented by an ordered k-tuple in $\mathbb{R}^n$. Each such configuration is represented by $X \in \mathbb{R}^{n \times k}$ where each column denotes a point in $\mathbb{R}^n$. However, we require that the space spanned by the columns of X be all of $\mathbb{R}^n$.

Definition 3.14 (Landmark Space). The landmark space, denoted by $\mathcal{L}_{n,k}$ (with $k \geq n$), is defined as the subset of all $X \in \mathbb{R}^{n \times k}$ such that $\dim(\text{span}(X)) = n$.

The tangent space at $X \in \mathcal{L}_{n,k}$ is given by $T_X(\mathcal{L}_{n,k}) = \mathbb{R}^{n \times k}$. We will generally assume the standard Euclidean metric on $\mathcal{L}_{n,k}$, with the resulting distance between any two points $X_1, X_2 \in \mathcal{L}_{n,k}$ given by $|X_1 - X_2|$ and where $|\cdot|$ denotes the Frobenius norm $\left(|X| = \sqrt{\sum_{i,j} X_{i,j}^2} \right)$.

The action of $\mathbb{R}^n \rtimes (\mathbb{R}^\times \times SO(n))$ on the landmark space $\mathcal{L}_{n,k}$ is given by:

$$(v, a, O) * X = aOX + v\mathbf{1}_k^T \ ,$$

and the orbit of X is:

$$[X] = \{aOX + v\mathbf{1}_k^T | a \in \mathbb{R}^\times, O \in SO(n), v \in \mathbb{R}^n\} \ .$$

Here, $\mathbf{1}_k$ is a column vector of length k containing all ones so that $v\mathbf{1}_k^T$ is an $n \times k$ matrix with identical columns.

3.6 Quotient Spaces of Riemannian Manifolds

Just as cosets are used to define equivalence relations, as explained after Definition A.11, the orbits of G can also be used to define an equivalence relation in M. This leads us to the notion of quotient space.

Definition 3.15. Let M be a finite-dimensional manifold and G be a Lie group that acts of M. Then, the quotient space M/G is defined to be the set of all orbits of G in M:

$$M/G = \{[p] | p \in M\} \ . \tag{3.11}$$

We will use group actions and their orbits to incorporate the role of shape-preserving transformations in shape analysis. Let us motivate this idea using a simple example.

Example 3.12. 1. Let $X \in \mathcal{L}_{n,k} \subset \mathbb{R}^{n \times k}$, a matrix whose k columns denote k-ordered points in $\mathbb{R}^n$. The action of $SO(n)$ on $\mathcal{L}_{n,k}$ is defined in Example 3.9, except that the same rotation is applied to all the columns, and this defines the orbit of X:

$$[X] = \{OX | O \in SO(n)\} \ .$$

This set consists of all $n \times k$ matrices whose columns can be obtained by simultaneously rotating the columns of X using the same rotation. Treating elements of $[X]$ as an equivalence class implies that all rotated versions of X are equivalent. The resulting quotient space is given by $\mathcal{L}_{n,k}/SO(n) = \{[X] | X \in \mathcal{L}_{n,k}\}$.

2. One can go beyond rotations and include a larger set of transformations in defining an equivalence relation. Consider the action of $\mathbb{R}^n \rtimes (\mathbb{R}^\times \times SO(n))$ on $\mathcal{L}_{n,k}$ given by $(v, a, O) * X = aOX + v\mathbf{1}_k^T$, and the equivalence relation $X \sim Y$ if $Y \in [X]$. The group $\mathbb{R}^n \rtimes (\mathbb{R}^\times \times SO(n))$ represents translation, scaling, and rotation (all shape-preserving transformations) of elements of $\mathcal{L}_{n,k}$. Thus, each such equivalence class represents a unique shape and individual shapes can be identified as elements of the quotient space $\mathcal{L}_{n,k}/(\mathbb{R}^n \rtimes (\mathbb{R}^\times \times SO(n)))$. This quotient space is termed *landmark shape space*, the shape space of configurations of k-ordered points in $\mathbb{R}^n$ and is a manifold.

One can immediately notice the importance of quotient spaces in shape analysis, where one wants the analysis and metrics to be invariant to certain (shape-preserving) transformations. An obvious way to accomplish this invariance is to impose equivalence relations between objects whose shapes are deemed equivalent. If the chosen metric respects this equivalence relation, then the ensuing analysis will have the desired invariance. This is exactly what is frequently done in shape analysis. Since the so-called shape-preserving transformations can be realized as actions of certain Lie groups, shape spaces are treated as quotient spaces of certain manifolds, called pre-shape spaces, under the actions of these groups. This discussion underscores the need to better understand and to develop tools for analyzing elements of quotient spaces.

The first item of interest is: Given the Riemannian structure on M, what Riemannian structure can be imposed on the quotient space M/G? Suppose a group G acts by *isometries* on a Riemannian manifold M. This implies that the group action preserves distances between points in M. Depending upon the action, this quotient space M/G may or may not be a smooth manifold. If it is a smooth manifold, it inherits a Riemannian metric from the Riemannian metric on M as follows. Let $T_p(M)$ and $T_p([p])$ denote the tangent spaces at p to the manifold M and the orbit $[p]$, respectively. Clearly, $T_p([p]) \subset T_p(M)$. Let $N_p(M)$ be the orthogonal complement of $T_p([p])$ in $T_p(M)$, i.e., it is the set of vectors that are perpendicular to $T_p([p])$ in $T_p(M)$. Elements of $T_p([p])$ are tangent to the orbit $[p]$ at p and the elements of $N_p(M)$ are perpendicular to the orbit at p. Figure 3.6 shows an illustration of this setup. The direct sum:

$$T_p([p]) \oplus N_p(M) = T_p(M) \ ,$$

is the full tangent space at p. We can identify the perpendicular space $N_p(M)$ with the tangent space on the quotient set $T_{[p]}(M/G)$. That is, for every element

Fig. 3.6 Illustration of
vector spaces tangent and
normal to an orbit $[p]$ at $p \in M$, and the identification
of tangent space $T_{[p]}(M/G)$
with the normal space
$N_p(M)$

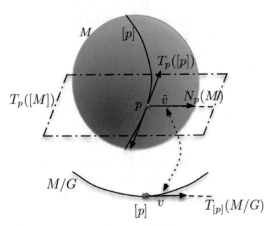

$v \in T_{[p]}(M/G)$, there is a corresponding element $\tilde{v} \in N_p(M)$, and vice versa. We will use this identification to specify tangent vectors in the quotient space. Once we have made this identification, we simply define the inner product of two vectors in this orthogonal complement to be their inner product using the Riemannian metric on M.

Definition 3.16 (Inherited Riemannian Metric). Let M be a Riemannian manifold and let G act on M by isometries. The inherited Riemannian metric on the quotient space M/G is given by, for any $v_1, v_2 \in T_{[p]}(M/G)$, define:

$$\langle\langle v_1, v_2 \rangle\rangle = \langle \tilde{v}_1, \tilde{v}_2 \rangle_p \ ,$$

where $\tilde{v}_1, \tilde{v}_2$ are considered as elements of $N_p(M) \subset T_p(M)$ on the right side.

Due to the isometry of the group action, it does not matter which $p \in [p]$ is selected for this definition. We can then define arc-length, geodesic, and a distance function on the manifold M/G using this inherited Riemannian metric.

If M/G is not a smooth manifold, then it does not make sense to talk about a Riemannian metric, but the space M/G still inherits a distance function from the distance function on M. We can define the distance function on the quotient space as follows.

Definition 3.17. If a Lie group G acts on a Riemannian manifold M by isometries and the orbits under G are closed, then define a distance function d_q on the quotient space M/G, using the original distance function d_m on the manifold M, as follows:

$$d_q([p_1], [p_2]) = \min_{g \in G} d_m((g, p_1), p_2) = \min_{g \in G} d_m(p_1, (g, p_2)) \ . \qquad (3.12)$$

Theorem 3.4. *The distance d_q defined in Eq. 3.12 forms a proper distance.*

Proof. There are three properties that d_q should satisfy—symmetry, positive definiteness, and triangle inequality. We will consider them in that order:

(1) Since the action of G is by isometries, it is easy to show that d_q is symmetric.
(2) For positive definiteness, we need to show that $d_q([p_1], [p_2]) = 0 \Rightarrow [p_1] = [p_2]$. Suppose that $d_q([p_1], [p_2]) = 0$. By definition, it follows immediately that, for all $\epsilon > 0$, there exists a $g \in G$ such that $d_m(p_1, (g, p_2)) < \epsilon$. From this, it

follows that p_1 is in the closure of the orbit p_2. Since the orbits are assumed to be closed, it follows that $p_1 \in [p_2]$, so $[p_1] = [p_2]$.

(3) To establish the triangle inequality, we need to show that $d_q([p_1], [p_3]) \leq d_q([p_1], [p_2]) + d_q([p_2], [p_3])$, for any $p_1, p_2, p_3 \in M$. Seeking contradiction, suppose that $d_q([p_1], [p_3]) > d_q([p_1], [p_2]) + d_q([p_2], [p_3])$. Let $\epsilon = \frac{1}{3}(d_q([p_1], [p_3]) - d_q([p_1], [p_2]) - d_q([p_2], [p_3])])$; by our supposition, $\epsilon > 0$. From the definition of ϵ, it follows that $d_q([p_1], [p_3]) = d_q([p_1], [p_2]) + d_q([p_2], [p_3]) + 3\epsilon$.

By the definition of d_q, we can choose $g_1, g_2 \in G$, such that $d_m((g_1, p_1), p_2) \leq d_q([p_1], [p_2]) + \epsilon$ and $d_m(p_2, (g_2, p_3)) \leq d_q([p_2], [p_3]) + \epsilon$. Now by the triangle inequality for d_m, we know that:

$$d_m((g_1, p_1), (g_2, p_3)) \leq d_m((g_1, p_1), p_2) + d_m(p_2, (g_2, p_3))$$
$$\leq d_q([p_1], [p_2]) + d_q([p_2], [p_3]) + 2\epsilon .$$

By definition of d_q, it follows that $d_m([p_1], [p_3]) \leq d_q([p_1, [p_2]) + d_q([p_2], [p_3]) + 2\epsilon$. But this contradicts that fact that $d_q([p_1], [p_3]) = d_q([p_1], [p_2]) + d_q([p_2], [p_3]) + 3\epsilon$. Hence our supposition that $d_q(p_1], [p_3]) > d_q([p_1], [p_2]) + d_q([p_2], [p_3])$ must be false. The triangle inequality follows. $\square$

(In case M/G is a smooth manifold, then this distance is the same that results from the inherited Riemannian metric given in Definition 3.16.) Assume that d_m is a geodesic distance on M, i.e., for any $p_1, p_2 \in M$, there exists a geodesic path between p_1 and p_2 such that the length of that geodesic is $d_m(p_1, p_2)$. In that case the following question becomes interesting: What is the corresponding geodesic path in M/G, connecting $[p_1]$ and $[p_2]$, whose length achieves the quotient distance $d_q([p_1], [p_2])$? The following path provides the answer. Let $\alpha : [0, 1] \to M$ be the geodesic between the points p_1 and $p_2^* = (g^*, p_2)$ in the original manifold M, where we are assuming that $g^* \in G$ can chosen in order to minimize the distance between p_1 and (g^*, p_2). Then, $[\alpha(\tau)]$, indexed by $\tau \in [0, 1]$, forms the desired geodesic path between $[p_1]$ and $[p_2]$ in M/G. Here, $[\alpha(\tau)]$ denotes the orbit of the point $\alpha(\tau)$ in M under the action of G. This geodesic can be shown to be perpendicular to each orbit it meets. In other words, the velocity vector $\frac{d\alpha(\tau)}{d\tau}$ is in the normal space $N_{\alpha(\tau)}(M)$. Specifically, the shooting vector for the geodesic $v = \frac{d\alpha(\tau)}{d\tau}|_{\tau=0}$ is in $N_{p_1}(M)$, or equivalently in $T_{[p_1]}(M/G)$.

Example 3.13. **Rotation Group:** Once again, let $X \in \mathcal{L}_{n,k}$ be an ordered k-tuple of points in $\mathbb{R}^n$ with the Euclidean metric and consider the action of $SO(n)$ on $\mathcal{L}_{n,k}$ given by $(O, X) = OX$. Since $SO(n)$ acts on $\mathcal{L}_{n,k}$ by isometries, we can inherit a distance on the quotient space $\mathcal{L}_{n,k}/SO(n)$ as follows: for any two $X_1, X_2 \in \mathcal{L}_{n,k}$

$$d_q([X_1], [X_2]) = \operatorname{argmin}_{O \in SO(n)} \sqrt{\langle X_1 - OX_2, X_1 - OX_2 \rangle} = \max_{O \in SO(n)} \sqrt{\operatorname{trace}(X_1 X_2^T O^T)}$$

$$= \max_{O \in SO(n)} \sqrt{\operatorname{trace}(U \Sigma V^T O^T)}, \quad U \Sigma V^T = \operatorname{SVD}(A), \quad A \equiv X_1 X_2^T$$

$$= \max_{O \in SO(n)} \sqrt{\operatorname{trace}(\Sigma W)} \tag{3.13}$$

where $W = V^T O^T U$ is an element of $O(n)$. In computing the SVD, we assume that the singular values have been arranged in a descending order from top-left to bottom-right in Σ. Since $W \in O(n)$, we know that $W_{ij} \leq 1$ for all i, j. Now there are two cases:

1. If $\det(A) > 0$, then $\text{trace}(\Sigma W) \leq \text{trace}(\Sigma)$ and the equality is achieved when $W = I_n$, which implies that $O^* = UV^T \in SO(n)$.
2. If $\det(A) < 0$, then UV^T has the determinant -1. This can be fixed by inserting a matrix in the product as follows:

$$O^* = U \begin{bmatrix} 1 & 0 & \dots & 0 \\ 0 & 1 & \dots & 0 \\ & \vdots & & \\ 0 & 0 & \dots & -1 \end{bmatrix} V^T \ .$$

The resulting distance on the quotient space is given by $d_q([X_1], [X_2]) = |X_1 - O^* X_2|$, where $|\cdot|$ denotes the Frobenius norm of a matrix. The geodesic between $[X_1]$ and $[X_2]$ in $\mathcal{L}_{n,k}/SO(n)$ is computed as follows. Let $\alpha(\tau) = (1-\tau)X_1 + \tau O^* X_2$. Then, the desired geodesic in the quotient space is given by $[\alpha(\tau)]$.

In the later chapters on shapes of curves, we will repeatedly use isometric actions of groups on Riemannian manifolds to define distances between orbits. This is a very important concept in our analysis of shapes and the key idea here is the *isometry* of the group action. If the action of G is not by isometries, then this construction of distances on M/G does not work.

3.7 Quotient Spaces as Orthogonal Sections

We now discuss a method for realizing the quotient space M/G in a simple way, as a submanifold of M. This method is not always available, but when it is, it can be very useful! The basic idea is to identify a subset S of M, termed a global orthogonal section, with the set M/G and, if possible, using an isometric map. In this way, one can perform all relevant operations on S instead of M/G and solve inference problems on M/G indirectly. In the following, we develop this idea starting with a general global section and then particularizing it to the situation of interest.

Definition 3.18 (Section). Define a *section* of the action of G on M to be a submanifold S of M that intersects each orbit of the action in at most one point.

The left panel of Fig. 3.7 shows an illustration of this idea. If S intersects every orbit, we say S is a *global section*. In the case of a global section, one can use the set S as a representative of the quotient space M/G to perform certain computations. For instance, if we want to compare any two orbits $[p]$ and $[q]$, we can represent these orbits by their intersections p_1 and q_1 with S and use a certain distance between p_1 and q_1 in S as the desired distance. Note that the choice of metric on S is arbitrary at this stage and need not necessarily correspond to the Riemannian structure of M. However, for comparing elements of M/G, it seems natural to use the metric inherited from M. In that case, to use a section, one needs some additional conditions.

Definition 3.19 (Orthogonal Section). Define an *orthogonal section* of the action of G on M to be a submanifold S of M with the following three properties:

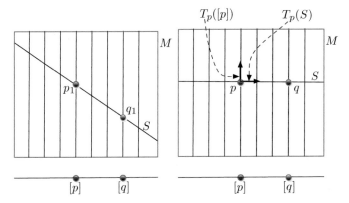

Fig. 3.7 *Left*: A section of M under the group G. *Right*: An orthogonal section of M under the action of group G

1. S intersects each orbit of the action in at most one point.
2. For each $p \in S$, $T_p([p]) \perp T_p(S)$ (perpendicularity is defined using the Riemannian metric on M).
3. For each $p \in S$, $T_p([p]) \oplus T_p(S) = T_p(M)$.

If S intersects every orbit, we say S is a *global orthogonal section*. Otherwise, we say S is a *local orthogonal section*.

The right panel of Fig. 3.7 shows a pictorial illustration of this setup.

Example 3.14. We continue with the example of group actions on the Euclidean manifold $\mathcal{L}_{n,k}$. We will use X_j to denote the jth column of X.

1. **Translation Group**: Let the group $G = \mathbb{R}^n$ act on the manifold $\mathcal{L}_{n,k}$ according to the action: for $x \in \mathbb{R}^n$ and $X \in \mathcal{L}_{n,k}$, $(x, X) = X + x\mathbf{1}_k^T$, where $\mathbf{1}$ is column vector of k ones. The orbits of $\mathcal{L}_{n,k}$ under $\mathbb{R}^n$ are given by $[X] = \{X + x\mathbf{1}_k^T | x \in \mathbb{R}^n\}$. Define a subset of $\mathcal{L}_{n,k}$:

$$S_t = \{X \in \mathcal{L}_{n,k} | \sum_{j=1}^{k} X_j = 0\} , \tag{3.14}$$

We will now show that S_t is a global, orthogonal section of $\mathcal{L}_{n,k}$. For an arbitrary $X \in \mathcal{L}_{n,k}$, let the column sum be a nonzero vector $x \in \mathbb{R}^n$, i.e., $\sum_{i=1}^{k} X_i = x$. Then, $X - (x/n)\mathbf{1}_k^T$ is the only element of the orbit $[X]$ that is in S. For any $X \in S_t$, the space tangent to the orbit is:

$$T_X([X]) = \{ \begin{bmatrix} v_1 & v_1 & \dots & v_1 \\ v_2 & v_2 & \dots & v_2 \\ \dots & & & \\ v_n & v_n & \dots & v_n \end{bmatrix} | (v_1, v_2, \dots, v_n) \in \mathbb{R}^n\},$$

and the space tangent to S_t is:

$$T_X(S_t) = \{ \begin{bmatrix} w_{1,1} & w_{1,2} & \cdots & w_{1,n} \\ w_{2,1} & w_{2,2} & \cdots & w_{2,n} \\ \cdots & & & \\ w_{n,1} & w_{n,2} & \cdots & w_{n,n} \end{bmatrix} \mid \sum_{j=1}^{k} w_{i,j} = 0, \quad \text{for all } i \},$$

It is easy to see that $T_X([X])$ and $T_X(S_t)$ are perpendicular to each other under the Euclidean metric. Furthermore, we can easily establish that $T_X([X]) \oplus T_X(S_t) = T_X(\mathcal{L}_{n,k}) = \mathbb{R}^{n \times k}$. Thus, the set S defined in Eq. 3.14 is an orthogonal section of $\mathbb{R}^{n \times k}$.

2. **Scaling Group**: Let the group $G = \mathbb{R}^{\times}$ act on manifold $\mathcal{L}_{n,k}$ according to the action: for $a \in \mathbb{R}^{\times}$ and $X \in \mathcal{L}_{n,k}$, $(a, X) = aX$. The orbits of $\mathcal{L}_{n,k}$ under $\mathbb{R}^{\times}$ are given by radial lines $[X] = \{aX | a \in \mathbb{R}^{\times}\}$. Define a subset of $\mathcal{L}_{n,k}$:

$$S_s = \{ X \in \mathcal{L}_{n,k} | \sum_{j=1}^{k} \sum_{i=1}^{n} X_{i,j}^2 = 1 \} . \tag{3.15}$$

We now show that this S_s is an orthogonal section of $\mathcal{L}_{n,k}$. If for an X, we have that $(\sum_{j=1}^{k} \sum_{i=1}^{n} X_{i,j}^2) = b \neq 1$, then $\frac{1}{\sqrt{b}} X$ is the only element of the orbit $[X]$ in S_s. The relevant tangent spaces are $T_x([X]) = \{aX | a \in \mathbb{R}^{\times}\}$ and $T_X(S_s) = \{Y \in \mathbb{R}^{n \times k} | \langle Y, X \rangle = 0\}$, and it is easy to verify that $T_x([X]) \perp T_X(S_s)$ under the Euclidean metric. Additionally, we have $T_X([X]) + T_X(S_s) = T_X(\mathcal{L}_{n,k}) = \mathbb{R}^{n \times k}$. Thus, S_s is an orthogonal section of $\mathcal{L}_{n,k}$.

3. **Rotation Group**: Let the action of $SO(n)$ on $\mathcal{L}_{n,k}$ be given by $(O, X) = OX$. Does this action admit an orthogonal section of $\mathcal{L}_{n,k}$, even locally? The answer is no! We illustrate it with an example for $n = 2$ and $k = 2$. We further restrict to the set:

$$\mathbb{S}^3 = \{ X \in \mathbb{R}^{2 \times 2} | \sum_{i=1}^{2} \sum_{j=1}^{2} X_{i,j}^2 = 1 \} .$$

We restrict to $\mathbb{S}^3$ since this set is closed under the action of $SO(2)$. It is well known that this action of $\mathbb{S}^1$ on $\mathbb{S}^3$ does not admit any orthogonal section, even a local one! We will not give a proof of this fact, but it is a straightforward consequence of the famous Frobenius theorem, a fundamental result in the theory of differential manifolds (the Frobenius theorem is presented in Boothby [19], p. 155). Since the smaller set $\mathbb{S}^3$ does not admit an orthogonal section, the same holds for the larger set $\mathbb{R}^{2 \times 2}$.

In case we know the orthogonal sections of a manifold M under the individual actions of several groups, then sometimes we can determine the orthogonal section under the joint actions by taking simple intersections.

Example 3.15. **Joint Scaling and the Translation Group**: Consider the action of the semi-direct product group $\mathbb{R}^n \rtimes \mathbb{R}^{\times}$ on the Euclidean space $\mathcal{L}_{n,k}$, given by:

$$((y, a), X) = aX + y\mathbf{1}_k^T .$$

The orbit of the joint action associated with $X \in \mathcal{L}_{n,k}$ is given by:

$$[X] = \{aX + y\mathbf{1}_k^T | a \in \mathbb{R}^{\times}, y \in \mathbb{R}^n \} .$$

The orthogonal section for scaling is $S_s = \{X \in \mathcal{L}_{n,k} | \sum_{j=1}^{k} \sum_{i=1}^{n} X_{i,j}^2 = 1\}$ and orthogonal section for translation is $S_t = \{X \in \mathcal{L}_{n,k} | \sum_{j=1}^{k} X_j = 0\}$. Then, it can be shown (see the Exercise list at the end of the chapter) that their intersection:

$$S_{st} = \{X \in \mathcal{L}_{n,k} | \sum_{j=1}^{k} \sum_{i=1}^{n} X_{i,j}^2 = 1, \ \sum_{j=1}^{k} X_j = 0\} \ . \tag{3.16}$$

forms an orthogonal section of the joint scaling-translation group $\mathbb{R}^n \rtimes \mathbb{R}^\times$.

Suppose M is a Riemannian manifold, G is a Lie group acting on M by isometries, and S is an orthogonal section for this action. Since S is a submanifold of M, we can define a Riemannian metric on S by restricting the Riemannian metric on M to S. There is an obvious map $S \to M/G$ defined to be the composition $S \to M \to M/G$ where the first map is the inclusion of S into M and the second is the quotient map from M to M/G (taking $p \to [p]$). It can be shown that, with respect to the Riemannian metrics on S and M/G, this map is an isometry of S onto an open submanifold of M/G. If S is a global section, then this map is simply an isometry from S to M/G. In this case, if we wish to make distance computations or find geodesics in M/G, we may make them in S instead! So, for the purposes of statistically analyzing points on M/G, one can directly work on the corresponding points on S, and this greatly simplifies the analysis. Let us take a few examples.

Example 3.16. Once again, we will consider the actions of different groups on the manifold $\mathcal{L}_{n,k}$, although sometimes under a metric that is not the standard Euclidean metric.

1. **Translation**: We assume the standard Euclidean metric on $\mathcal{L}_{n,k}$, i.e., for any u, $v \in \mathbb{R}^{n \times k}$, the Riemannian metric is the standard Euclidean product $\langle u, v \rangle$. We have already verified in Example 3.9 that the action of $\mathbb{R}^n$ on Euclidean spaces is by isometries. Therefore, the section S_t given in Eq. 3.14 is isometric to the quotient space $\mathcal{L}_{n,k}/\mathbb{R}^n$.

    Now that we have identified the Riemannian structures of S_t and the quotient space $\mathcal{L}_{n,k}/\mathbb{R}^n$, we can compute geodesics between elements of $\mathcal{L}_{n,k}/\mathbb{R}^n$ as follows. Let X_1, $X_2 \in \mathcal{L}_{n,k}$ and we want to compute a geodesic between their equivalence classes $[X_1], [X_2] \in \mathcal{L}_{n,k}/\mathbb{R}^n$. Let $\sum_{i=1}^{k} X_{1,i} = x_1$ and $\sum_{i=1}^{k} X_{2,i} = x_2$ be the two column sums. Then, define $\tilde{X}_1 = X_1 - x_1 \mathbf{1}_k^T$ and $\tilde{X}_2 = X_2 - x_2 \mathbf{1}_k^T$ as the representatives of the two equivalence classes in S_t. The geodesic between $\tilde{X}_1$ and $\tilde{X}_2$ in S_t is given by a straight line:

    $$\alpha(\tau) = (1 - \tau)\tilde{X}_1 + \tau \tilde{X}_2 \ ,$$

    and the length of this geodesic is $|\tilde{X}_1 - \tilde{X}_2| = (\sum_{i,j}(\tilde{X}_{1,i,j} - \tilde{X}_{2,i,j})^2)^{1/2}$. Therefore, the geodesic path between the two equivalence classes $[X_1]$ and $[X_2]$ in the quotient space $\mathcal{L}_{n,k}/\mathbb{R}^n$ is $[\alpha(\tau)]$ and the distance between them is $d([X_1], [X_2]) = |\tilde{X}_1 - \tilde{X}_2|$. This $\alpha(\tau)$ is a constant-speed geodesic with the speed given by $|\tilde{X}_1 - \tilde{X}_2|$. Since S_t is a vector space with standard metric, the exponential map and its inverse can be expressed using a simple formula (Example 3.5).

2. **Scaling Group**: This time let $\mathbb{R}^\times$ act on $\mathcal{L}_{n,k}$ by scalar multiplication. We already verified in Example 3.9 that this action is not by isometries with respect to the usual inner product. As a result, the quotient space M/G does not naturally inherit a Riemannian metric.

We remedy this by defining a new Riemannian metric on $\mathcal{L}_{n,k}$.

Definition 3.20 (Scaled-Euclidean metric). If $U, V \in T_X(\mathcal{L}_{n,k})$, define the scaled-Euclidean metric as:

$$\langle\langle U, V \rangle\rangle_X = \frac{\langle U, V \rangle}{\langle X, X \rangle} \tag{3.17}$$

where $\langle \cdot, \cdot \rangle$ denotes the standard inner product in $\mathbb{R}^{n \times k}$.

We now establish the desired isometry condition.

Lemma 3.2. *With respect to the scaled-Euclidean metric on $\mathcal{L}_{n,k}$, the action of $\mathbb{R}^\times$ is by isometries.*

Proof. Fix $a \in \mathbb{R}^\times$ and define $L_a : \mathcal{L}_{n,k} \to \mathcal{L}_{n,k}$ by $L_a(X) = aX$. Since this action is a linear transformation on the underlying vector space $\mathbb{R}^{n \times k}$, it follows that its derivative is given by the same formula, i.e., given $U \in T_X(\mathcal{L}_{n,k})$, $d(L_a)_X(U) = aU$. We then compute:

$$\langle\langle d(L_a)_X(U), d(L_a)_X(V) \rangle\rangle_{L_a(X)} = \frac{\langle aU, aV \rangle}{\langle aX, aX \rangle} = \frac{\langle U, V \rangle}{\langle X, X \rangle} = \langle\langle U, V \rangle\rangle_X \; ,$$

which shows that L_a is an isometry for all $a \in R^\times$. $\square$

The next question is: Can we find a submanifold of $\mathcal{L}_{n,k}$ that is a section of this action? The answer is yes: in fact, the set S_s defined in Eq. 3.15 is also a section of this action. How do we see this? Properties (1) and (3) have already been proven in the previous example (they don't depend on which Riemannian metric we are using). Property (2) involves orthogonality and has also been proven earlier under the Euclidean metric. It follows that it holds under the scaled-Euclidean metric also, because for an $X \in S_s$, the scaled-Euclidean metric reduces to the standard Euclidean metric. Thus, the set S_s is an orthogonal section of the action of $SO(n)$ on $\mathcal{L}_{n,k}$. Note that S_s is a unit sphere in $\mathbb{R}^{n \times k}$, and under the Euclidean metric, it is rather straightforward to analyze points in S_s.

Now consider the quotient space $\mathcal{L}_{n,k}/\mathbb{R}^\times$ and for any two X_1 and X_2 in $\mathcal{L}_{n,k}$; we want to find a geodesic between their equivalence classes in the quotient space using the scaled-Euclidean metric. We find the intersection points of these orbits with S_s, which are $\tilde{X}_1 = X_1/|X_1|$ and $\tilde{X}_2 = X_2/|X_2|$. We then parameterize the arc of the great circle in S_s joining $\tilde{X}_1$ to $\tilde{X}_2$:

$$\alpha(\tau) = \frac{1}{\sin(\theta)}(\sin((1-\tau)\theta)\tilde{X}_1 + \sin(\tau\theta)\tilde{X}_2), \quad \theta = \cos^{-1}(\langle \tilde{X}_1, \tilde{X}_2 \rangle) \; .$$

Then, the path of orbits $[\alpha(t)]$ gives the geodesic in $\mathcal{L}_{n,k}/\mathbb{R}^\times$ from $[X_1]$ to $[X_2]$. Likewise, the distance between the two orbits is given by:

$$d([X_1], [X_2]) = \cos^{-1}(\langle \tilde{X}_1, \tilde{X}_2 \rangle) \; .$$

The exponential map and its inverse can be expressed analytically using the spherical geometry of S_s (see Example 3.5).

One can easily encounter a situation where a group G does not act on M by isometries but the group action has an orthogonal section S. We would want to study elements of the quotient space M/G using a Riemannian analysis on the orthogonal section S. However, since the action is not by isometries, the inherited metric is not available on the quotient space M/G. In the absence of a metric on this space, we cannot talk about the isometry of the map from S to M/G, although the map still exists. One possible solution is to use this natural map $S \to M \to M/G$ to induce a metric on M/G. By construction, this map will be isometric and we can return to analyzing elements of M/G using the corresponding elements of S.

Let us understand this using an example. As described in Example 3.15, the semi-direct product $\mathbb{R}^n \rtimes \mathbb{R}^\times$ acts on $\mathcal{L}_{n,k}$ and this action has an orthogonal section S_{st} defined earlier (Eq. 3.16). We want to utilize the Riemannian structure of S_{st}, obtained as a restriction of the Riemannian metric on $\mathcal{L}_{n,k}$, to analyze elements of the quotient space $\mathcal{L}_{n,k}/(\mathbb{R}^n \rtimes \mathbb{R}^\times)$. As mentioned already, the action of $\mathbb{R}^n \rtimes \mathbb{R}^\times$ on $\mathcal{L}_{n,k}$ is not by isometries and the inherited metric is not available on the quotient space. So, we are going to identify S_{st} with the quotient space and transfer the Euclidean metric on S_{st} to the quotient space. For any X_1, $X_2 \in \mathcal{L}_{n,k}$, the intersections of their orbits with S_{st} is given by:

$$
\tilde{X}_1 = \bar{X}_1/|\bar{X}_1|, \ \bar{X}_1 = X_1 - x_1 \mathbf{1}_k^T
$$
$$
\tilde{X}_2 = \bar{X}_2/|\bar{X}_2|, \ \bar{X}_2 = X_2 - x_2 \mathbf{1}_k^T \ , \tag{3.18}
$$

where x_1 and x_2 are the column sums of X_1 and X_2, respectively. With these representatives, the geodesic and the geodesic distances are same as in the previous item:

$$
\alpha(\tau) = \frac{1}{\sin(\theta)}(\sin((1-\tau)\theta)\tilde{X}_1 + \sin(\tau\theta)\tilde{X}_2), \ \ \theta = \cos^{-1}(\langle \tilde{X}_1, \tilde{X}_2 \rangle) \ .
$$

and $d([X_1],[X_2]) = \cos^{-1}(\langle \tilde{X}_1, \tilde{X}_2 \rangle)$. Since S_{st} is a unit sphere, albeit smaller in dimension than the orthogonal section of $\mathbb{R}^n$, one can easily write down the exponential map and its inverse using the spherical geometry of S_{st} (see Example 3.5).

3.8 General Quotient Spaces

As we have seen, it becomes much easier to analyze a quotient space if it can be identified with an orthogonal section, isometrically or otherwise. What about the cases where no orthogonal section exists? In such cases, one has to develop the analysis starting from the first principles.

According to Definition 3.17, a geodesic in M/G is realized using a corresponding geodesic in M that is the shortest between a fixed point on one orbit and all the elements of the second orbit. Let $p_1, p_2 \in M$; we want to compute a geodesic between their orbits $[p_1]$ and $[p_2]$ in M/G. Let $\tilde{p}_2 \in M$ be the point in $[p_2]$ nearest to p_1 under the distance d_m, a distance that results from the chosen Riemannian metric on M. Then, let $\alpha(\tau)$ be a shortest geodesic path between p_1 and $\tilde{p}_2$ in M under this metric. The quotient path $[\alpha(\tau)]$, i.e., the quotient of each point along the path, indexed by the time τ, is the desired geodesic in M/G.

Example 3.17. Consider the action of the group $G = \mathbb{R}^n \rtimes (\mathbb{R}^\times \times SO(n))$ on $\mathcal{L}_{n,k}$ and the problem of computing geodesics and geodesic distances on the quotient space $\mathcal{L}_{n,k}/G$. Denoting the subgroup $(\mathbb{R}^n \rtimes \mathbb{R}^\times)$ by G_1, we will use two different approaches for the two components of G. For G_1, we will use the orthogonal section of its action on $\mathcal{L}_{n,k}$, while for $SO(n)$ we will use the approach of finding nearest points. Since $\mathcal{L}_{n,k}/G_1$ can be identified with S_{st}, we can rewrite the quotient space $\mathcal{L}_{n,k}/G$ as $S_{st}/SO(n)$.

For any X_1, $X_2 \in \mathcal{L}_{n,k}$, we can compute the geodesics between them as follows. Find their representatives in S_{st} using Eq. 3.18. Then, solve for:

$$O^* = \underset{O \in SO(n)}{\operatorname{argmin}} \cos^{-1}(\langle \tilde{X}_1, O\tilde{X}_2 \rangle) = \underset{O \in SO(n)}{\operatorname{argmax}} \langle \tilde{X}_1, O\tilde{X}_2 \rangle \ .$$

Then, the desired geodesic in the quotient space $\mathcal{L}_{n,k}/G$ is given by $[\alpha(\tau)]$ where:

$$\alpha(\tau) = \frac{1}{\sin(\theta)}(\sin((1-\tau)\theta)\tilde{X}_1 + \sin(\tau\theta)\tilde{X}_2),$$

with $\theta = \cos^{-1}(\langle \tilde{X}_1, O^*\tilde{X}_2 \rangle)$.

The exponential map in M/G can be realized using a unit-speed geodesic in the given direction. For a point $[p] \in M/G$, and a tangent vector $v \in T_{[p]}(M/G)$ (i.e., in $N_p(M)$), let us construct a unit-speed geodesic $\alpha(\tau)$ from p in the direction of v (existence of ψ is shown in Theorem 3.1). Then, since $\dot{\alpha}(0) \perp T_p([p])$, we have $\dot{\alpha}(\tau) \perp T_{\alpha(\tau)}([\alpha(\tau)])$ for all τ, by isometry, and the exponential map $\exp : T_{[p]}(M/G) \to M/G$ is given by $\exp_{[p]}(v) = [\alpha(1)]$. The inverse of an exponential map takes a point on the quotient space M/G and maps it to an element (or multiple elements) of the tangent space $T_{[p]}(M/G)$. A vector $v \in T_{[p]}(M/G)$ is said to be the inverse exponential of $[q] \in M/G$ at p if $\exp_{[p]}(v) = [q]$. It is denoted by $v = \exp_{[p]}^{-1}([q])$ and is often not a unique point. That is, the inverse may be set-valued.

Example 3.18. Continuing with the quotient space, $S_{st}/SO(n)$. We want to compute geodesics in this space and define exponential map and its inverse. For any $\tilde{X}_1, \tilde{X}_2 \in S_{st}$, with the rotation orbits $[\tilde{X}_1]$ and $[\tilde{X}_2]$, let $O^* \in SO(n)$ be the optimal rotation on $\tilde{X}_2$ that minimizes the distance between $\tilde{X}_1$ and $\tilde{X}_2$, as described in the previous example. The shooting vector (initial velocity) of the geodesic path given there is $\dot{\alpha}(0) = O^*\tilde{X}_2 - \tilde{X}_1$. Hence, the inverse exponential map is given by:

$$\exp_{[\tilde{X}_1]}^{-1}([\tilde{X}_2]) = [\frac{\theta}{\sin(\theta)}(O^*\tilde{X}_2 - \cos(\theta)\tilde{X}_1)]_o, \quad \text{where} \ \ [V]_o = \{OV | O \in SO(n)\} \ ,$$

and $\theta = \cos^{-1}(\langle \tilde{X}_1, \tilde{X}_2 \rangle)$. Similarly, for any $\tilde{W} \in \mathbb{R}^{n \times k}$ such that $W \perp [W]_o$, with $[\tilde{W}]_o$ as defined above, the exponential map is given by:

$$\exp_{[\tilde{X}]}(\tilde{V}) = \cos(|\tilde{V}|)\tilde{X} + \sin(|\tilde{V}|)\frac{\tilde{V}}{|\tilde{V}|} \ .$$

Table 3.1 Options for inducing distances in quotient spaces

	Action is by isometries	Action is not by isometries
Admits an orthogonal section	**(1) = (2)**, or **(3)**	**(2)** or **(3)**
Admits a section that is not orthogonal	**(1)** or **(2)** or **(3)**	**(2)** or **(3)**
Does not admit a section	**(1)**	None

3.9 Distances in Quotient Spaces: A Summary

Consider the general set up where a group G acts on a Riemannian manifold M and, for simplicity, let us assume that the orbits of G are closed sets in M. In the previous sections, we have described several ways of computing distances in quotient space M/G, depending on the nature of the group action. In the following, we list these different possibilities for defining and computing distances in a quotient space.

In general, we have the following methods available:

1. **Method (1)**: Use Definition 3.17 to impose the distance:

$$d_{M/G}([p_1], [p_2]) = \min_{g \in G} d_M(p_1, (p_2, g)) \ .$$

2. **Method (2)**: Restrict the metric from M to S and compute geodesic distances on the section S. Assign those distances to the corresponding orbits. If d_S is the distance in S, then set $d_{M/G}([p_1], [p_2]) = d_S(p_1^*, p_2^*)$ where $p_1^* = [p_1] \cap S$ and $p_2^* = [p_2] \cap S$.

3. **Method (3)**: Ignore the original metric on M and choose a different metric on the section S. Define the distance between any two orbits as the distance between the corresponding elements of S on S. That is, if d_S is the distance in S, then set $d_{M/G}([p_1], [p_2]) = d_S(p_1^*, p_2^*)$ where $p_1^* = [p_1] \cap S$ and $p_2^* = [p_2] \cap S$. This, of course, is a less interesting option since the Riemannian structure of M is completely ignored here.

We remark that in cases of a group action by isometries, and where the action admits an orthogonal section, the results from method **(1)** and **(2)** will be identical. It may however be computationally cheaper to use method **(2)** in this case since one does not need to solve the optimization over G. In the case where the group action admits a section but it is not orthogonal, and the action is not by isometries, one can choose either **(2)** or **(3)**, but the former is preferred because it uses the original metric of M rather than choosing an arbitrary metric on the section (Table 3.1).

3.10 Center of an Orbit

As described in the previous sections, the action of a group G on a Riemannian manifold M generates orbits whose membership defines an equivalence relation on M. While technically all the elements of an equivalence class are equally good for analysis, it is sometimes more efficient or meaningful to select a particular element of this set to represent its class. This element termed the *center* of the orbit is defined with respect to a set of points, called the *reference set*, in M. To define the center precisely, let the action of G be by isometries with respect to the metric of M. We will also need this action to be *free*, which means the following: if for any $p \in M$, the fact $(g, p) = p$ implies that $g = e$ (the identity element), then the action is called *free*. Assume further that there is a metric structure on G itself that enables one to compute the *mean* (e.g., Karcher mean) of a finite number of points in G.

The center of the orbit of a point $p \in M$ is defined as follows.

Definition 3.21 (Center of An Orbit). Let a group G act freely on a Riemannian manifold M such that the action of G is by isometries. For any point p, let $[p]$ denote the orbit of p and let $R = \{p_1, p_2, \ldots, p_k\}$ for a reference set. Then, the center of the orbit of p with respect to the set R is given by a point q if it satisfies the following properties:

1. $q \in [p]$.
2. Let $g_i^* = \operatorname{argmin}_g d_m(g \cdot q, p_i)^2$, for $i = 1, 2, \ldots, k$, where d_m is the geodesic distance on M. Then, the mean of the g_i^*s is the identity in G.

We provide a simple algorithm for computing the center of the orbit of a given point p, with respect to a reference set R. This algorithm assumes that the natural action of G on itself is by isometries under the chosen metric of G. Now, for any set $g_1, g_2, \ldots, g_k$, let their mean be given by:

$$\bar{g} = \operatorname*{argmin}_{g \in G} d_g(g, g_i)^2 \ .$$

For the shifted problem, where each g_i is replaced by $g_i \cdot \bar{g}^{-1}$, the mean is given by:

$$\bar{g}_0 = \operatorname*{argmin}_{g \in G} d_g(g, g_i \cdot \bar{g}^{-1})^2 = \operatorname*{argmin}_{g \in G} d_g(g \cdot \bar{g}, g_i)^2 \ ,$$

This mean $\bar{g}_0$ satisfies $\bar{g} = \bar{g}_0 \cdot \bar{g}$. Hence, $\bar{g}_0 = g_{id}$, the identity element of G. To summarize this result, if we replace each g_i by $g_i \cdot \bar{g}^{-1}$, then the mean changes from $\bar{g}$ to g_{id}. This idea is depicted in Fig. 3.8 and it leads to the following algorithm for computing the center of an orbit.

Algorithm 2 (Computation of Center of an Orbit).

1. Let $r \in [p]$ be any arbitrary point.
2. Compute $g_i = \operatorname{argmin}_{g \in G} d_m(g \cdot q, p_i)$, for all $i = 1, 2, \ldots, k$.
3. Compute the mean $\bar{g}$ of the set $\{g_1, g_2, \ldots, g_k\}$ using the distance d_g on G.
4. Update r by $q = \bar{g}^{-1} \cdot r$.

Example 3.19. Consider the action of $G = SO(n)$ on the space $M = \mathcal{L}_{n,k}$ under the Euclidean metric. We know that this action is by isometries and is free. The latter implies that whenever we have $OX = X$, for an $X \in \mathcal{L}_{n,k}$, then $O = I_n$. Let G have a metric structure given by the Frobenius norm, i.e., $d_g(O_1, O_2) = \sqrt{\sum_{l,j}(O_1(l,j) - O_2(l,j))^2}$. The mean of a set of rotation matrices $O_1, O_2, \ldots, O_k$ is given by forming the element-wise average $A = \frac{1}{k}\sum_i O_i$ and using its SVD decomposition $A = U\Sigma V^T$. If the determinant of A is positive, then $\bar{O} = UV^T$,

Fig. 3.8 The point $q = \bar{g}^{-1} \cdot r$ form the center of the orbit $[p]$ with respect to the reference set $R = \{p_1, p_2, \ldots, p_n\}$

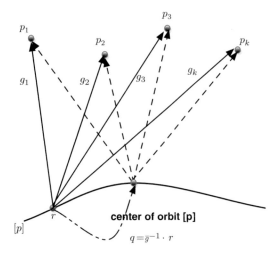

else the last column of V changes sign (this assumes that the entries in Σ are non-increasing from top-left to bottom-right).

Given a point $X \in \mathcal{L}_{n,k}$ and a reference set $R = \{X_1, X_2, \ldots, X_k\}$, let:

$$O_i = \operatorname*{argmin}_{O \in SO(3)} |OX - X_i|^2 , \quad i = 1, 2, \ldots, k .$$

Then, set $Y = \bar{O}^T X$ where $\bar{O}$ is the average of the rotations $\{O_1, O_2, \ldots, O_k\}$. The point $Y \in \mathcal{L}_{n,k}$ becomes the center of $[X]$ with respect to the reference set R.

3.11 Exercises

3.11.1 Theoretical Exercises

1. Is "$\leq$" an equivalence relation on the set of real numbers?
2. In $\mathbb{R}^2$, define a relation $x \sim y$ if x and y lie on the same straight line passing through the origin. Is this an equivalence relation? What if this relation is defined for the set $\mathbb{R}^2 \setminus \{(0,0)\}$?
3. In the set $\mathbb{R}^n \setminus \{0\}$, define the relation $x \sim y$ if there exists a nonzero scalar a such that $y = ax$. Verify that this an equivalence relation. (Example 3.1, item 2).
4. Verify that, in $\mathbb{R}^n$, the relation $x \sim y$ if $|x| = |y|$ forms an equivalence relation. (Example 3.1, item 3).
5. Verify that the following maps are symmetric, positive definite, and bilinear:

   - $\Phi : \mathbb{R}^{n \times n} \times \mathbb{R}^{n \times n} \to \mathbb{R}$ given by: $\Phi(A, B) = \operatorname{trace}(AB^T)$.
   - For any $p \in \mathbb{S}^n$, let $\Phi : T_p(\mathbb{S}^n) \times T_p(\mathbb{S}^n) \to \mathbb{R}$ given by $\Phi(v, w) = v^T w$.
   - For any $x \in \mathbb{R}^2_+ = \{(x_1, x_2) | x_1 \in \mathbb{R}, x_2 \geq 0\}$, let $\Phi_x : \mathbb{R}^2 \times \mathbb{R}^2 \to \mathbb{R}$, given by $\Phi_x(v, w) = \frac{v^T w}{x_2^2}$.

6. Show that $\Phi : \mathbb{R}^3 \times \mathbb{R}^3 \to \mathbb{R}$ given by

$$\Phi(x, y) = x_1 y_1 + x_2 y_2 - x_3 y_3 \ ,$$

   is not a Riemannian metric on $\mathbb{R}^3$.

7. Consider the upper half-plane $H = \{(x_1, x_2) \in \mathbb{R}^2 | x_2 > 0\}$. Show that the inner product given by $\langle v, w \rangle_x = \frac{v^T w}{x_2^2}$ defines a Riemannian metric on H. (This metric is called the *hyperbolic metric*.)

8. Show that the mapping of $\mathbb{R}^n$ to itself, given by $x \mapsto Ox$, for a fixed $O \in SO(n)$, is an isometry under the Euclidean metric (Example 3.3). If we replace O by an $A \notin SO(n)$, is the resulting mapping an isometry, under the Euclidean metric?

9. Is the mapping of $\mathbb{R}^n$ to itself, given by $x \mapsto ax$, for a fixed $a \in \mathbb{R}_+$, an isometry under the Euclidean metric? What about under the metric $\langle v, w \rangle_x = (v^T w)/(x^T x)$?

10. Show that the geodesics in the following cases have these known structures:

    - Geodesics in $\mathbb{R}^n$ under the Euclidean metric are straight lines.
    - Geodesics on a unit sphere under the Euclidean metric are great circles.
    - Geodesics between points in the 2D upper half-plane (H define above in Problem 7), under the hyperbolic metric, are as follows. If the two points have the same y coordinate, then the geodesic is a straight line passing through those points. Otherwise, the geodesic is a part of the circle that goes through these points and has its center on the x axis.

11. Let $M = \mathbb{R}^n$ with the Euclidean Riemannian metric and let $\mathcal{M}$ be the space of absolutely continuous functions from $[0, 1]$ to M. Using Theorem 3.2, derive an expression for the geodesic between any two elements of $\mathcal{M}$. Repeat this problem for $M = \mathbb{S}^2$ with the Euclidean metric.

12. Here we verify the expression for parallel transport of a vector along a geodesic path in $\mathbb{S}^1$, that was used in the proof of Theorem 3.3. Let e_1, e_2 be an orthonormal basis of $\mathbb{R}^2$ and let $v = re_2$ be an arbitrary element of $T_{e_1}(\mathbb{S}^1)$. Verify that the parallel transport of v along the geodesic $\alpha(t) = \cos(t)e_1 + \sin(t)e_2$ is given by $-r\sin(t) + r\cos(t)e_2$. In other words, show that this expression satisfy the three properties given in Eq. 3.8.

13. Show that the mapping $v \mapsto \Psi(v)$, given in Eq. 3.9, forms a rotation of $\mathbb{S}^{n-1}$.

14. Verify that the following three maps define group actions on $\mathbb{R}^n$:

    - **Translation**: $\mathbb{R}^n \times \mathbb{R}^n \to \mathbb{R}^n$ given by $(x, y) = x + y$.
    - **Rotation**: $SO(n) \times \mathbb{R}^n \to \mathbb{R}^n$ given by $(O, x) = Ox$.
    - **Scaling**: $\mathbb{R}^\times \times \mathbb{R}^n \setminus \{0\} \to \mathbb{R}^n \setminus \{0\}$ given by $(a, x) = ax$.

15. First, verify that the semi-direct product $\mathbb{R}^n \rtimes (\mathbb{R}^\times \times SO(n))$ forms a group. Then, verify that it forms a group action on $\mathbb{R}^n \setminus \{0\}$ according to the map: $((y, a, O), x) = aOx + y$.

16. Show that if a group G acts on a manifold M, we can define an equivalence relation as follows: for any two $p, q \in M$, we have $p \sim q$ if $p \in [q]$. Show that this forms an equivalence relation between elements of M. In other words, the orbits of G form the equivalence classes under this relation.

17. Consider the action of $SO(2)$ on $\mathbb{R}^2$ given by $(O, x) = Ox$ with the Euclidean metric on $\mathbb{R}^2$. Write down the quotient space $\mathbb{R}^2/SO(2)$ and derive an expression for distance inherited on the quotient space using Definition 3.17.

18. Study of sections of landmark spaces:

   a. For an $X \in \mathcal{L}_{n,k}$, let $[X] = \{X + x\mathbf{1}_k^T | x \in \mathbb{R}^n\}$ and S_t, as defined in Eq. 3.14, show that the vector spaces $T_X([X])$ and $T_X(S_t)$ are orthogonal.
   b. For an $X \in \mathcal{L}_{n,k}$, let $[X] = \{aX | a \in \mathbb{R}^\times\}$ and S_s, as defined in Eq. 3.15, show that the vector spaces $T_X([X])$ and $T_X(S_s)$ are orthogonal.
   c. Show that the set S_{st} in Eq. 3.16 is an orthogonal section of the action of $(\mathbb{R}^n \rtimes \mathbb{R}^\times)$ on $\mathcal{L}_{n,k}$.

19. For a group G acting on a Riemannian manifold M, let (a) denote the condition that the group action is by isometries and (b) denote that there is an orthogonal section for this action. For $M = \mathcal{L}_{n,k}$, find groups that provide an example of each of the three situations:

   a. (a) is true and (b) is not.
   b. (b) is true and (a) is not.
   c. Both (a) and (b) are true.

20. Consider the action of the translation group $G = \mathbb{R}^n$ on $M = \mathcal{L}_{n,k}$. Here G acts on M by isometries and the group action admits an orthogonal section. As described in Sect. 3.9, both methods **(1)** and **(2)** can be used in this instance. Show that the distances resulting from the two methods are identical.

21. **Affine Group Action**: Here we are going to consider the action of group $G = GL(2) \rtimes \mathbb{R}^2$ on $\mathcal{L}_{2,k}$.

   a. First, verify that the standard actions of $GL(2)$ and $\mathbb{R}^2$ on $\mathbb{R}^2$ do not commute but they form a semi-direct product group using the operation: $(A_1, b_1) \cdot (A_2, b_2) = (A_1 A_2, A_2 b_1 + b_2)$. $G = GL(2) \rtimes \mathbb{R}^2$ is called the *affine group*.
   b. Show that this semi-direct group $G = GL(2) \rtimes \mathbb{R}^2$ acts on $\mathbb{R}^2$ according to the map:

   $$((A, b), x) = Ax + b, \quad A \in GL(2), \quad b \in \mathbb{R}^2, \quad x \in \mathbb{R}^2 .$$

   c. Furthermore, show that the action of G on $\mathbb{R}^2$, under the Euclidean metric, is not by isometries.
   d. Let $S \subset \mathcal{L}_{2,k}$ be a set of k-tuples in R^2 that satisfy:

   $$\frac{1}{k} \sum_{i=1}^{k} x_i = 0, \quad \text{and} \quad \frac{1}{k} \sum_{i=1}^{k} x_i x_i^T = I_2 ,$$

   where I_2 is the 2×2 identity matrix. Show that S intersects every orbit of the action of G on $R^{2 \times k}$ in at least one point. Is it a section of the action? The mapping of a configuration $x \in \mathcal{L}_{2,k}$ into the section S is called a *standardization* of x. For efficient algorithms on standardization of planar configurations, please refer to [21].

22. Consider the action of $\mathbb{R}$ on $\mathbb{R}^2$ given by *vertical translations*: for any $a \in \mathbb{R}$ and $x \in \mathbb{R}^2$ the action is given by $(a, x) = x + \begin{pmatrix} 0 \\ a \end{pmatrix}$. For a point $x \in \mathbb{R}^2$ and a reference set $R = \{x_1, x_2, \ldots, x_k\} \subset \mathbb{R}^2$, find the center of the orbit $[x]$ with respect to R.

23. Similarly, now consider the action of the scaling group $\mathbb{R}^\times$ on $\mathbb{R}^n - \{0\}$ given by for any $a \in \mathbb{R}^\times$ and $x \in \mathbb{R}^n - \{0\}$ the action is given by $(a, x) = ax$. For a point $x \in \mathbb{R}^n - \{0\}$ and a reference set $R = \{x_1, x_2, \ldots, x_k\} \subset \mathbb{R}^d$, find the center of the orbit $[x]$ with respect to R.

3.11.2 Computational Exercises

1. Write a program to compute the shortest, unit-speed geodesic path $\alpha : [0, 1] \to \mathbb{S}^2$ between any two non-antipodal points. The program should compute and display geodesic points at times $t = 0, \frac{1}{T}, \frac{2}{T}, \ldots, 1$ for a given positive integer $T = 10$.

2. Write subroutines to compute the exponential map, inverse exponential map, and parallel transport equation along a geodesic path for the unit sphere $\mathbb{S}^k$.

3. Write a program to compute and display geodesic paths (not necessarily constant speed) between arbitrary points in the upper half-plane under the hyperbolic metric. (A description of these geodesics is provided in Example 3.4.)

4. **Landmark Shape Geodesics**: Write a program to compute and display geodesic paths and geodesic distances between elements of the following quotient spaces: (1) $\mathcal{L}_{n,k}$, (2) $\mathcal{L}_{n,k}/SO(n)$, (3) $\mathcal{L}_{n,k}/\mathbb{R}^n$, (4) $\mathcal{L}_{n,k}/\mathbb{R}^\times$, (5) $\mathcal{L}_{n,k}/(\mathbb{R}^n \rtimes SO(n))$, (5) $\mathcal{L}_{n,k}/(\mathbb{R}^n \rtimes \mathbb{R}^\times)$ and (6) $\mathcal{L}_{n,k}/(\mathbb{R}^n \rtimes (\mathbb{R}^\times \times SO(n)))$. Demonstrate this program using some triangular shapes for $n = 2$ and $k = 3$.

5. Implement the formulas given in Example 3.17 to compute (discrete-time) geodesic paths, exponential map, and its inverse on the quotient space $\mathcal{L}_{n,k}/(\mathbb{R}^n \rtimes (\mathbb{R}^\times \times SO(n)))$.

6. Take two landmark configurations, i.e., elements of $\mathcal{L}_{n,k}$; repeatedly apply random rotations, translations, and scales to each of them; and compute the geodesic distance between them in the quotient space: $\mathcal{L}_{n,k}/(\mathbb{R}^n \rtimes (\mathbb{R}^\times \times SO(n)))$. Study the variability associated with the resulting distances.

3.12 Bibliographic Notes

This chapter provides an overview of some relevant tools from geometry and algebra. For more detailed discourse on these topics, the reader is referred to full texts such as Boothby [19], Lang's textbooks on algebra and geometry [64], Helgason [36], and Warner [120]. The tools for landmark-based shape analysis have been developed by many researchers, leading to full textbook-level treatments in Dryden and Mardia [28], Small [102], and Kendall et al. [49]. The concept of the center of an orbit was discussed in [108] but only for the specific warping group introduced in the next few chapters.

Chapter 4
Functional Data and Elastic Registration

Functional data analysis (FDA) is a branch of statistics where one observes, models, and analyzes quantities that are functions on certain intervals. This kind of data naturally arises in nearly every branch of science, ranging from engineering to geology, biology, medicine, and chemistry. Similar to more classical statistics, where one summarizes, models, regresses, estimates, and tests data involving Euclidean vectors, in FDA, one performs these tasks while working with functional data. The functions are treated as elements of *function spaces* and one performs calculus on these spaces in a manner similar to the Euclidean calculus. In traditional FDA, statistical tasks have commonly been performed using the Hilbert structure of the function spaces resulting from the $\mathbb{L}^2$ norm. This framework leads to simple algorithms for computing $\mathbb{L}^2$ distances, cross-sectional (i.e., point-wise) means and variances, and functional principal component analysis (FPCA) of given function data. Note that even though computer implementations require us ultimately to discretize functional data, the development of methodologies and statistical solutions in FDA regards them as functions and exploits additional structure present in functional data. We refer the reader to Appendix A.3 where we present some basic elements of functional analysis, including definitions of Banach and Hilbert manifolds and submanifolds. (We also provide some background material on FPCA later, in Sect. 4.3.1.)

What is the fundamental difference between multivariate statistics and FDA? This question is quite relevant considering the fact that data is generally collected and stored in the form of discrete point sets. Even if the data is available in a function form, most computational techniques discretize such functions and perform computations on the resulting finite vectors, much like multivariate statistics. So, what makes FDA different from multivariate analysis? We highlight a fundamental difference using Fig. 4.1, where the left panel shows a set of points $\{(t_i, y_i)\}$ sampled from a function on the domain $[0, 1]$. In multivariate statistics, one works with the vector $\{y_i\}$ using vector calculus and analyzes it using statistical methods. In FDA, one keeps the association of values $\{y_i\}$ with time indices $\{t_i\}$, seeks the underlying function itself, and estimates it using some interpolation or estimation method (see the middle panel). Having obtained this function, one can resample it at arbitrary points on the domain for performing functional calculus (differentiation, integration, etc). This resampling aspect distinguishes FDA from multivariate analysis. Keep in mind that function spaces are infinite dimensional and that necessitates some additional considerations beyond Euclidean calculus.

© Springer-Verlag New York 2016

A. Srivastava, E.P. Klassen, *Functional and Shape Data Analysis*,
Springer Series in Statistics, DOI 10.1007/978-1-4939-4020-2_4

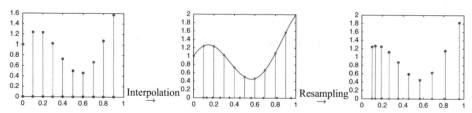

Fig. 4.1 FDA allows us to interpolate between given data points (*middle panel*) and sample the fitted curve at arbitrary points in the domain (*right panel*). This distinguishes FDA from multivariate statistics where the sample points are fixed as given

FDA also plays an important role in shape analysis of curves since curves are naturally represented as functions, more precisely parameterized curves, rather than as Euclidean vectors.

An interesting and challenging situation arises when the given data is not registered. Generally speaking, this implies that the peaks and valleys of given functions are not well aligned. This situation is the main focus of this chapter where we discuss how to handle the problem of registering two or more functions. We describe some specific function spaces of interest and study their Riemannian geometries under a specific Riemannian metric that is especially relevant for functional registration and shape analysis. This Riemannian metric, called the *Fisher-Rao metric*, is closely related to the elastic metric for shape analysis of curves, introduced later in Chap. 5. In fact, that elastic metric for curves can be considered as a natural extension of the Fisher-Rao metric to $\mathbb{R}^n$-valued functions.

4.1 Goals and Challenges

Functional data analysis or FDA is the analysis of random variables that take values in certain function spaces. In FDA, a serious challenge arises when functions are observed with flexibility or domain warping along the x axis (or the horizontal axis). This warping may come either from an uncertainty in the measurement process or may simply denote an inherent variability in the underlying process itself that needs to be separated from the variability along the y axis (or the vertical axis). As another possibility, the warping may be introduced as a tool to horizontally align the observed functions, reduce their variance, and increase parsimony in the resulting model. This process is analogous to using polar decomposition of Euclidean data: take an $x \in \mathbb{R}^n$ and form the pair $(|x|, x/|x|) \in (\mathbb{R}_+ \times \mathbb{S}^{n-1})$. This decomposition maps x into its amplitude $|x|$ and direction $x/|x|$, and it often is useful in some situations. For instance, in some cases, we may not be interested in one of these two components in our analysis. Or, it may be more natural to model these two separately.

Consider the two functions, f_1 and f_2, shown in the top-left panel of Fig. 4.2. Each of these functions has two peaks. However, because the peaks are centered at different points, their cross-sectional mean (a function formed by taking the mean of heights at each t), shown in the next panel, has three peaks. If we warp the domain of f_1, such that its peaks/valleys are now aligned with those of f_2, as shown in bottom-left panel, then the cross-sectional mean preserves that bimodal structure (bottom-middle panel). The warping function γ used here is shown in

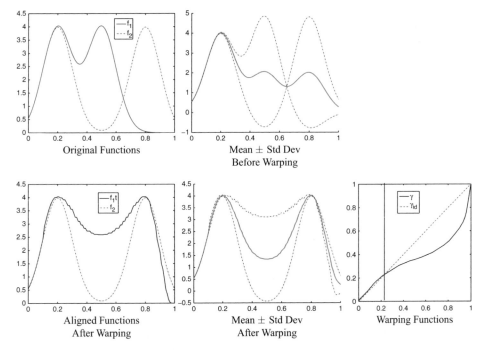

Fig. 4.2 Averaging of two functions, f_1 and f_2, before and after warping of f_1

the far right panel. This picture illustrates the benefits of decomposing the x-variability from the y-variability in the observed functions before doing statistical analysis. After warping of the domain of f_1, the two resulting functions $f_1 \circ \gamma$ and f_2 are vertically aligned and represent the y-variability in the data. The actual warping function used in this alignment, thus, represents the x-variability in the data. In addition to better preserving the structure of the observed data, a separate modeling of y and x variability can be more natural, parsimonious, and efficient.

The variability associated with warping of functions is called the *phase* variability and the remaining variability along the vertical axis is called the *amplitude* variability. We will formally introduce the concept of phase and amplitude components of functional data in Sect. 4.7 and will continue studying them in Chap. 8. We will also study a related problem of registration between any two functions, i.e., which point on one function corresponds to which point on the second function. We will use an extension of the classical Fisher-Rao metric, albeit in its nonparametric form, to tackle this problem.

We start, however, with a very practical problem of estimating full functions from discrete observations.

4.2 Estimating Function Variables from Discrete Data

We are interested in analyzing functional data, but most applications typically result in discrete data of the type $\{(t_i, y_i) \in ([0,1] \times \mathbb{R}) | i = 1, 2, \ldots, m\}$. So, the very first problem one faces is the estimation of a function f from its discrete observations. The main goal of function estimation, at least from the perspective

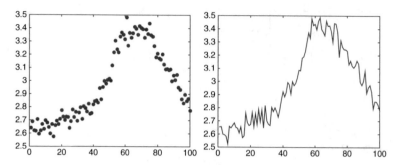

Fig. 4.3 If we connect the given data points (*left*) using straight lines, we get a noisy function (*right*)

of shape and functional data analysis, is to be able to provide values of f at arbitrary points in the domain $[0, 1]$. While there are numerous solutions to this problem, depending on the nature of the data and the assumptions about f, we briefly discuss a couple of standard solutions that will be useful throughout this textbook. The most straightforward way, of course, is to form a piecewise linear function by simply connecting the given data points, as shown in Fig. 4.3. This naive solution is prone to noise and it is hard to control the level of smoothness of the solution. Therefore, one seeks other ideas for estimating f.

Least-Squares Estimation Using a Basis:
One idea is to formulate a least-squares problem by using a truncated basis of an appropriate function space. We will assume that the desired function belongs to a Hilbert space $\mathscr{F}$ with a given inner product structure, denoted by $\langle \cdot, \cdot \rangle$. Let $\mathscr{B} = \{b_1, b_2, \dots\}$ denote a complete basis of $\mathscr{F}$, not necessarily orthogonal. Then, for any $f \in \mathscr{F}$, we can represent it using the expansion: $f(t) = \sum_{k=1}^{\infty} c_k b_k(t)$. Truncating the series to first K terms, a least-squares estimate of f can be derived as:

$$\hat{f} = \underset{c \in \mathbb{R}^K}{\operatorname{argmin}} \sum_{i=1}^{m} \left(y_i - \sum_{k=1}^{K} c_k b_k(t_i) \right)^2 \tag{4.1}$$

Using a matrix notation where $B_{ik} = b_k(t_i)$, $c = [c_1, c_2, \dots, c_K]$, and $y = [y(t_1), y(t_2), \dots, y(t_K)]$, we can rewrite this problem as:

$$\hat{c} = \underset{c \in \mathbb{R}^K}{\operatorname{argmin}} \left((y - Bc)^T (y - Bc) \right) = (B^T B)^{-1} B^T y \,, \text{ assuming } B^T B \text{ is non-singular.}$$

The vector of fitted values is then $\hat{y} = B(B^T B)^{-1} B^T y$. In case B forms an *orthonormal* set, then $B^T B$ is simply the identity.

We illustrate this idea using a simple example. Assuming that f is a square-integrable function on the desired interval $[0, T]$, we choose the following basis set:

$$b_1(t) = 1, \quad b_{2j}(t) = \sin(2\pi j \frac{t}{T}), \quad b_{2j+1}(t) = \cos(2\pi j \frac{t}{T}), \; j = 1, \dots, J \,,$$

with $K = 2J + 1$ total basis elements. The top-left panel of Fig. 4.4 shows a collection of ordered points that form the set (t_i, y_i) with $T = 100$. Our goal is to estimate the underlying function f using a least-squares criterion. The top row shows estimates of f for values $J = 1, 3$, and 20. As the value of K increases, the

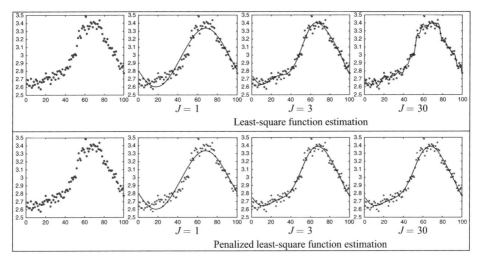

Fig. 4.4 Example of fitting a curve to noisy, discrete data using Fourier harmonics of increasing order

estimate increasingly fits better to the given data but, at the same time, becomes noisier also. One can control the smoothness of the solution using the parameter K.

An important question here is: How to choose the basis size K? Some insight into this choice is obtained by the following decomposition:

$$Var(\hat{f}) = E[\|\hat{f} - E[\hat{f}]\|^2] = E[\|\hat{f} - f + f - E[\hat{f}]\|^2]$$
$$= E[\|\hat{f} - f\|^2] + E[\|f - E[\hat{f}]\|^2] + 2E[\langle\hat{f} - f, f - E[\hat{f}]\rangle]$$
$$= E[\|\hat{f} - f\|^2] - [\|f - E[\hat{f}]\|^2].$$

Since $E[\|\hat{f} - f\|^2]$ is the mean-squared error of $\hat{f}$, or $MSE(\hat{f})$, and $(E[\hat{f}] - f)$ is the bias of $\hat{f}$, with its norm denoted by $Bias(\hat{f})$, we get:

$$MSE(\hat{f}) = Bias(\hat{f})^2 + Var(\hat{f}).$$

This decomposition represents the trade-off that governs the selection of J ($= 2K + 1$) in the definition of $\hat{f}$. As J increases, the $Var(\hat{f})$ increases and $Bias(\hat{f})$ decreases. This is depicted in Fig. 4.5 where the $MSE(\hat{f})$ is shown in the solid line. It reaches the minimum value at $J = 7$ and increases in either direction. One can choose the minimum of MSE as a guiding principal for selecting K, but that requires the knowledge of f itself (to compute the $Bias$ and the MSE). Thus, while this principal provides an important insight into the choice being made, it does not actually help in making a decision.

Penalized Least Squares : Another way to control the bias-variance trade-off is by adding an extra penalty term in Eq. 4.1:

$$\hat{f} = \underset{c \in \mathbb{R}^K}{\text{argmin}} \left(\sum_{i=1}^{m}\left(y_i - \sum_{k=1}^{K} c_k b_k(t_i)\right)^2 + \lambda\mathscr{R}(f)\right), \tag{4.2}$$

where $\lambda > 0$ is a constant and $\mathscr{R}(f)$ is a measure of roughness of f. A commonly used term for capturing roughness of a function is $\mathscr{R}(f) = \int_0^1 (\ddot{f}(t))^2\, dt$. Continuing

Fig. 4.5 Illustration of
bias-variance trade-off in
choice of K in estimating
f using the least-squares
criterion. The *solid line* is
MSE, the *broken line* is
$Var(\hat{f})$, and the *marked
line* is $Bias(\hat{f})$

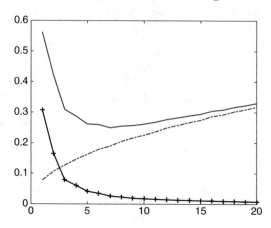

with the basis representation of f, we can write $\ddot{f}(t) = \sum_k c_k \ddot{b}_k(t)$. Truncating the
basis, as earlier, at K terms, let us write the penalty term in the discrete form as
the matrix:

$$\mathscr{R}(f) = \int_0^1 \ddot{f}(t)^2 \, dt = \int_0^1 (\sum_k c_k \ddot{b}_k(t))(\sum_j c_j \ddot{b}_j(t)) \, dt$$

$$= \sum_k \sum_j \left(c_j c_k \int_0^1 \ddot{b}_k(t) \ddot{b}_j(t) \, dt \right) = c^T R c \, ,$$

where R is a $K \times K$ symmetric matrix with elements $R_{jk} = \int_0^1 \ddot{b}_k(t) \ddot{b}_j(t) \, dt$. Thus,
we reach the discrete optimization problem:

$$\hat{c} = \underset{c \in \mathbb{R}^K}{\operatorname{argmin}} \left((y - Bc)^T (y - Bc) + \lambda c^T R c \right) \, . \tag{4.3}$$

The solution is given by $\hat{c} = (B^T B + \lambda R)^{-1} B^T y$ with the estimated function
being $\hat{f}_\lambda(t) = \sum_{k=1}^K \hat{c}_k b_k(t)$. The bottom row of Fig. 4.4 shows examples of fitted
functions for $\lambda = 0.001$ for $J = 1, 3$, and 30.

4.3 Geometries of Some Function Spaces

In this section, we discuss basic geometries of some function spaces of interest.

4.3.1 Geometry of Hilbert Spaces

Consider the set of square integrable, real-valued functions, on the unit interval
$[0, 1]$, denoted by $\mathbb{L}^2([0, 1], \mathbb{R})$ or simply $\mathbb{L}^2$. $\mathbb{L}^2$, is a vector space and, as described
in Appendix A.3, becomes a Hilbert space using the standard $\mathbb{L}^2$ inner product:

$$\langle f, g \rangle = \int_0^1 f(t)g(t)dt \, , \quad \text{for } f, g \in \mathbb{L}^2 \, . \tag{4.4}$$

The resulting $\mathbb{L}^2$ norm is given by $\|f\| = \sqrt{\int_0^1 f(t)^2 dt}$. Since $\mathbb{L}^2$ is a Hilbert space, it is also a Hilbert manifold, whose tangent space is given by $T_f(\mathbb{L}^2) = \mathbb{L}^2$ for all $f \in \mathbb{L}^2$. Therefore, one can define geodesic paths, exponential maps, etc., very easily, as follows:

1. **Shortest Geodesic**: For any $f_1, f_2 \in \mathbb{L}^2$, the shortest geodesic path between them is given by a "straight line":

$$\alpha : [0, 1] \to \mathbb{L}^2, \quad \alpha(\tau)(t) = (1 - \tau)f_1(t) + \tau f_2(t) . \qquad (4.5)$$

   This is a constant-speed geodesic with the speed given by $\|\dot{\alpha}(\tau)\| = \|f_2 - f_1\|$. Its shooting vector (initial velocity $\dot{\alpha}(0)$) is given by point-to-point difference $f_2 - f_1$.

2. **Exponential Map**: Given a starting point $f \in \mathbb{L}^2$ and a shooting direction $v \in \mathbb{L}^2$, the unique constant-speed geodesic with those initial values is given by the vector translation $\alpha(\tau)(t) = f(t) + \tau v(t)$. Thus, using Definition 3.6, the exponential map $\exp_f : \mathbb{L}^2 \to \mathbb{L}^2$ is given by a simple sum (translation): $\exp_f(v) = f + v$.

3. **Inverse Exponential Map**: Since the exponential map is given by a sum, its inverse is given by a difference (also a translation). For any $f_1, f_2 \in \mathbb{L}^2$: $\exp_{f_1}^{-1}(f_2) = f_2 - f_1$.

In view of these simple formulas, a basic statistical analysis on $\mathbb{L}^2$, such as computing sample means under the $\mathbb{L}^2$ norm, is straightforward. Given a set of functions $f_1, f_2, \ldots, f_n \in \mathbb{L}^2$, their mean is given by:

$$\bar{f} = \underset{f \in \mathbb{L}^2}{\arg\min} \left(\sum_{i=1}^n \|f_i - f\|^2 \right) = \underset{f \in \mathbb{L}^2}{\arg\min} \left(\sum_{i=1}^n \int_0^1 (f_i(t) - f(t))^2 \, dt \right)$$

Taking the summation inside, and since the cost function is additive over t, one can solve for each t separately, leading to the solution:

$$\bar{f}(t) = \underset{f(t) \in \mathbb{L}^2}{\arg\min} \left(\sum_{i=1}^n (f_i(t) - f(t))^2 \, dt \right) = \frac{1}{n} \sum_{i=1}^n f_i(t) . \qquad (4.6)$$

This is called the *cross-sectional* sample mean.

From a practical point of view, the only difficulty in analyzing elements of $\mathbb{L}^2$ is its dimensionality: $\mathbb{L}^2$ is infinite dimensional. This challenge is usually handled using a dimension-reduction tool, such as the principal component analysis (PCA), so that the analysis is restricted to a finite-dimensional subspace of $\mathbb{L}^2$. We describe this idea next.

PCA in Hilbert Spaces The application of PCA to function spaces is naturally called functional PCA or *FPCA*. One way to motivate FPCA is to consider the model:

$$f_i(t) = \mu_f(t) + \sum_{j=1}^\infty c_{i,j} b_j(t) + \epsilon_i(t) , \qquad (4.7)$$

where:

- $\mu_f(t)$ is the expected value of $f_i(t)$.
- $\{b_j\}$ forms an orthonormal basis of $\mathbb{L}^2$.

- $\epsilon_i \in \mathbb{L}^2$ is considered the noise process, typically chosen to be a white Gaussian process with zero mean.
- $c_{i,j} \in \mathbb{R}$ are coefficients of f_i with respect to $\{b_j\}$. In order to ensure that μ_f is the mean of f_i, we impose the condition that the sample mean of $\{c_{.,j}\}$ is zero.

Given a set of observed functions $\{f_i\}$, the estimation of these model parameters is performed using the minimization:

$$(\hat{\mu}_f, \hat{b}) = \underset{\mu_f, \{b_j\}, \{c_{i,j}\}}{\text{argmin}} \left(\sum_{i=1}^{n} \| f_i - \mu_f - \sum_{j=1}^{J} \langle f_i, b_j \rangle b_j \|^2 \right) , \qquad (4.8)$$

and set $\hat{c}_{i,j} = \left\langle f_i, \hat{b}_j \right\rangle$.

This minimization is achieved as follows. For the mean function, we can use the same mean estimate $\hat{\mu}_f(t) = \frac{1}{n} \sum_{i=1}^{n} f_i(t)$. For estimating the basis set, define the sample covariance function $\hat{C} : [0,1] \times [0,1] \to \mathbb{R}$ according to:

$$\hat{C}(s,t) = \frac{1}{n-1} \sum_{i=1}^{n} (f_i(s) - \hat{\mu}_f(s))(f_i(t) - \hat{\mu}_f(t)) ,$$

with $\hat{\mu}_f$ as given above. The function $\hat{C}$ is by definition symmetric and positive semidefinite, where the latter means that for any $g \in \mathbb{L}^2$, we have $\int_0^1 \int_0^1 \hat{C}(s,t)g(s)g(t) \, ds \, dt \geq 0$. This covariance function $\hat{C}$ defines a linear operator on $\mathbb{L}^2$ using the formula:

$$A : \mathbb{L}^2 \to \mathbb{L}^2, \quad Af(t) = \int_0^1 \hat{C}(s,t)f(s) \, ds .$$

Since $\hat{C}$ is positive semidefinite, we have $\langle Af, f \rangle \geq 0$ for all $f \in \mathbb{L}^2$. According to the Karhunen-Loeve expansion theorem [72], the eigenfunctions of A provide the principal components of the function data. A function $b \in \mathbb{L}^2$ is an eigenfunction of $\hat{C}$ if $Ab(t) = \int_0^1 \hat{C}(s,t)b(s)ds = \lambda b(t)$, for $t \in [0,1]$ and $\lambda \in \mathbb{R}$ is a constant. Let $\hat{b}_1, \hat{b}_2, \ldots, \hat{b}_J$ be the eigenfunctions of A, i.e., $A\hat{b}_j(t) = \lambda_j \hat{b}_j(t)$, so that the corresponding eigenvalues $\{\lambda_j\}$ satisfy $|\lambda_1| \geq |\lambda_2| \geq \ldots$. Then, $\{\hat{b}_j, \ j = 1, 2, \ldots, J\}$ solves the optimization problem in Eq. 4.8; they are termed the first J *principal directions* of variations in the given data. The space spanned by them is called the *principal subspace*.

In practice, a sampling of functions on a finite partition of $[0,1]$ is used to perform computations. For convenience, consider the uniform partition $\{0, \frac{1}{T}, \frac{2}{T}, \ldots, 1\}$ and, with a slight abuse of notation, let $f_i \in \mathbb{R}^{T+1}$ denote a vector of values of the original functions sampled at these points. Since these vectors are finite dimensional, the standard multivariate statistical tools apply. The sample covariance matrix of this dataset, also denoted by $\hat{C}$, is a $(T+1) \times (T+1)$ symmetric, positive semidefinite matrix. (This matrix is often singular because the number of observations n is usually less than the partition size T.) Consequently, the SVD of this covariance matrix leads to the principle directions and dominant modes of variability in the given data. One can use interpolation to obtain values of functions at arbitrary points and, thus, to recreate full functions. Figure 4.6 shows an example of FPCA. The leftmost panel shows 10 unimodal functions with synchronized peaks and the only difference is in the heights of their peaks.

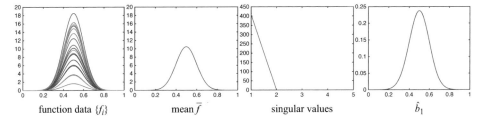

Fig. 4.6 FPCA of function well-aligned data

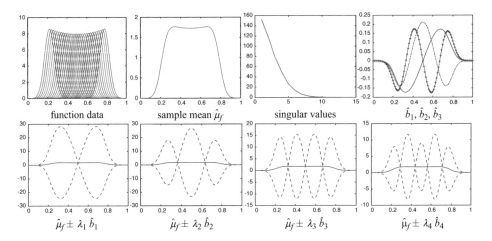

Fig. 4.7 FPCA of the function data shown in top left. We caution the reader that y axes are scaled differently in different plots

The next panel shows their cross-sectional mean $\hat{\mu}_f(t)$. The singular values $\{\lambda_i\}$ are plotted in the third panel. In this example, there is only one nonzero singular value and the corresponding eigenfunction $\hat{b}_1$ is shown in the last panel. This makes intuitive sense since there is only one mode of variation in the given functions, namely, the height and, thus, only one singular value is nonzero.

We consider another example in Fig. 4.7 where the functions are still unimodal but now they differ in both the locations and heights of their peaks (top-left panel). In fact, the differences in the locations of peaks overshadow the differences in their heights. Their cross-sectional mean μ is shown in the next panel. The vector of singular values of the covariance matrix is shown in the third panel, and the top three eigenfunctions $\hat{b}_1$, $\hat{b}_2$, and $\hat{b}_3$ are shown in the last panel. In the bottom row, we see the top four directions of variation in the data, using the function band $\hat{\mu}_f \pm \lambda_j \hat{b}_j$ around $\hat{\mu}_f$, for $j = 1, 2, 3$, and 4. This example illustrates an important limitation of FPCA, if applied directly to these functions. The sample mean $\hat{\mu}_f$ and the principle modes fail to capture the main source of variability in the original functions, namely, the location of the modes. This example motivates the need for registration of functions, i.e., alignment of peaks and valleys in f_is, before or during FPCA. This alignment is performed by warping the domains of f_is and introduces a certain amount of "elasticity" in the given functions. The problem of pairwise alignment of functions is considered in Sect. 4.4, while the problem of joint alignment of multiple functions is studied in Chap. 8.

Next we will look at some submanifolds of $\mathbb{L}^2$ that are of particular interest in our approach to shape analysis.

4.3.2 Unit Hilbert Sphere

A submanifold of $\mathbb{L}^2$ that is both interesting and useful in shape analysis is the unit sphere $\mathbb{S}_\infty \subset \mathbb{L}^2$:

$$S_\infty = \{f \in \mathbb{L}^2 | \|f\| = 1\} .$$

As mentioned in Appendix A.3, S_∞ is a submanifold of $\mathbb{L}^2$ and becomes a Hilbert manifold with the $\mathbb{L}^2$ Riemannian metric (Eq. 4.4). Although $\mathbb{S}_\infty$ is no longer a vector space, its geometry is still relatively simple since the formulas presented for finite-dimensional spheres also apply to $\mathbb{S}_\infty$. For an $f \in \mathbb{S}_\infty$, the tangent space $T_f(\mathbb{S}_\infty) = \{v \in \mathbb{L}^2 | \langle v, f \rangle = 0\}$ (the inner product is given in Eq. 4.4) and other basic items are as follows:

1. **Geodesic:** Similar to Eq. 3.5, the (arc-length parameterized) geodesic path between any two points $f_1, f_2 \in \mathbb{S}_\infty$ ($f_1 \neq -f_2$) is given by:

$$\alpha(\tau) = \frac{1}{\sin(\theta)} \left[\sin(\theta(1-\tau))f_1 + \sin(\tau\theta)f_2 \right] \in S_\infty , \qquad (4.9)$$

   where $\theta = \cos^{-1}(\langle f_1, f_2 \rangle)$. For any $\tau \in [0, 1]$, $\alpha(\tau)$ is a function given by:

$$\alpha(\tau)(t) = \frac{1}{\sin(\theta)} \left[\sin(\theta(1-\tau))f_1(t) + \sin(\tau\theta)f_2(t) \right] .$$

   Figure 4.8 shows two examples of geodesics between unimodal and bimodal functions when represented as elements in S_∞. In each case, the left panel shows the functions f_1 and f_2, and the right panel shows equally spaced functions between them on a constant-speed geodesic connecting them, i.e., plots of $\alpha(\tau)(t)$, $\tau = 0, 1/5, 2/5, \ldots, 1$, versus t. It is interesting to note that in the first case where the modes of f_1 and f_2 are distant, the geodesic path has the effect of flattening a bump and creating a new bump. In the second case, where the peaks are aligned, the two bumps simply change their shapes along the geodesic.

2. **Exponential Map:** The geodesic on a unit sphere $\mathbb{S}_\infty$ can also be expressed in terms of a starting point $f \in \mathbb{L}^2$ and a shooting vector v in $T_f(\mathbb{S}_\infty)$:

$$\alpha(\tau)(t) = \cos(\tau\|v\|)f(t) + \sin(\tau\|v\|)\frac{v(t)}{\|v\|}, \quad t \in [0, 1] . \qquad (4.10)$$

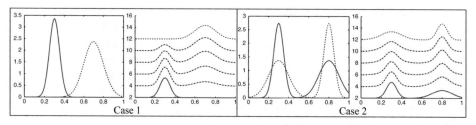

Fig. 4.8 Examples of geodesic paths between functions in $\mathbb{S}_\infty$. The functions along the paths have been stacked to improve display

Therefore, the exponential map $\exp_f : T_f(\mathbb{S}_\infty) \to \mathbb{S}_\infty$, once again using Definition 3.6, is given by:

$$\exp_f(v)(t) = \cos(\|v\|)f(t) + \sin(\|v\|)\frac{v(t)}{\|v\|}, \quad t \in [0,1] . \qquad (4.11)$$

Similarly to the case of $\mathbb{S}^2$, this exponential map is many to one. If we restrict the domain of $\exp_f$ by requiring $\|v\| < \pi$, then the exponential map becomes one-to-one and its range is missing only the "antipodal point," $-f$.

3. **Inverse Exponential Map**: Likewise, the inverse of this exponential map has a simple analytical form. For $f_1, f_2 \in \mathbb{S}_\infty$, $\exp_{f_1}^{-1} : \mathbb{S}_\infty/\{-f_1\} \to T_{f_1}(\mathbb{S}_\infty)$ is:

$$\exp_{f_1}^{-1}(f_2)(t) = \frac{\theta}{\sin(\theta)}(f_2(t) - f_1(t)\cos(\theta)), \quad \theta = \cos^{-1}(\langle f_1, f_2 \rangle) \text{ and } t \in [0,1] .$$
$$(4.12)$$

This expression simply provides the shooting vector for a uniform-speed geodesic from f_1 to f_2 on S_∞. One can easily verify this formula using $(\frac{d}{d\tau}\alpha(\tau))|_{\tau=0}$ for the path α given in Eq. 4.9.

These formulas are useful for computing summary statistics of functional data, as described later in Chap. 7.

4.3.3 Group of Warping Functions

Yet another set of functions that is of great interest in this textbook is the set of *warpings*. These maps deform a function by warping its domain in a certain constrained way. This set plays an important role in parameterizations of curves and alignments of functions. We introduce the set of warping functions here, but its geometry is presented a little later in this chapter.

Consider a function $\gamma : [0,1] \to [0,1]$ that satisfies the following properties: $\gamma(0) = 0$, $\gamma(1) = 1$, γ is invertible, and both γ and γ^{-1} are smooth. Such a γ is called a *boundary-preserving diffeomorphism* of $[0,1]$; let Γ_I denote the set of all such functions (the subscript I implies that underlying domain is an interval, as opposed to being a circle that occurs later in this text). For details on the geometry of this set, we encourage the reader to consult [45]. The set Γ_I is interesting for a variety of reasons. Firstly, it is a Lie group with the group operation given by composition: for any $\gamma_1, \gamma_2 \in \Gamma_I$, the group operation is $(\gamma_1 \circ \gamma_2)(t) = \gamma_1(\gamma_2(t))$. The identity element of Γ_I is the identity map $\gamma_{id}(t) = t$, and for any $\gamma \in \Gamma_I$, its inverse γ^{-1} is well defined such that $\gamma \circ \gamma^{-1} = \gamma^{-1} \circ \gamma = \gamma_{id}$. Secondly, it acts on a function space by right composition. Let $\mathscr{F}$ denote the space of all functions on $[0,1]$, i.e., $\mathscr{F} = \{f : [0,1] \to \mathbb{R}\}$. Then, a right action of Γ_I on $\mathscr{F}$ is given by:

$$\mathscr{F} \times \Gamma_I \to \mathscr{F}, \quad (f, \gamma) = f \circ \gamma . \qquad (4.13)$$

The proof that this mapping forms a group action is left as an exercise. (This composition is also a group action if $\mathscr{F}$ is replaced by certain subspaces, such as the set of absolutely continuous functions, or the square-integrable functions.) Shown in Fig. 4.9 are two examples of this action on a function f. As can be seen

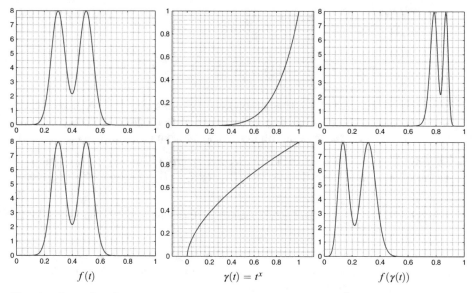

$f(t)$ $\gamma(t) = t^x$ $f(\gamma(t))$

Fig. 4.9 Examples of warping a function. The top row is for $\gamma(t) = t^5$ and the bottom row for $\gamma(t) = t^{0.6}$

in these examples, the effect of warping a function f is to horizontally displace its values but no new y values are introduced in the process. Another way to say this is that the phase of f is changed but its amplitude remains the same. The concepts of phase and amplitude are defined formally later in Chap. 8.

While Γ_I is a group, it is not a vector space; a linear combination of elements of Γ_I may not be in Γ_I. (However, Γ_I is closed under the operation of taking *convex* combinations, i.e., linear combinations where the coefficients add up to 1 and are all non-negative.) It is interesting to note that one can impose a nice manifold structure on Γ_I. However, the model space (the space for defining local charts) is not a Banach space, but rather a more general *Fréchet space*. The tangent structure of Γ_I at γ_{id} can be derived from the definition of tangent spaces (see Sect. A.1.1). Let $\alpha(\tau, \cdot)$ be a differentiable path in Γ_I passing through γ_{id} at 0, i.e., $\alpha(0, t) = \gamma_{id}(t)$. Since $\alpha(\tau, \cdot) \in \Gamma_I$, we have $t \mapsto \alpha(\tau, t)$ which is a diffeomorphism. The derivative of $\alpha(\tau, t)$ with respect to τ, evaluated at $\tau = 0$, given by $v(t) \equiv \frac{d\alpha}{d\tau}(0, t)$, is a scalar quantity. At the boundaries $\frac{d\alpha}{d\tau}(\tau, 0) = \frac{d\alpha}{d\tau}(\tau, 1) = 0$ since $\alpha(\tau, 0) = 0$ and $\alpha(\tau, 1) = 1$. Also, because α has smooth derivatives, the function v is a smooth function on $[0, 1]$. Thus, the tangent space $T_{\gamma_{id}}(\Gamma_I)$ is the set of smooth functions that are zero at the boundaries:

$$T_{\gamma_{id}}(\Gamma_I) = \{v : [0, 1] \mapsto \mathbb{R} | v(0) = 0, v(1) = 0, \quad v \text{ is smooth}\} .$$

This tangent space is a vector space but unfortunately not a Hilbert space because it is not complete. There are really two reasons why it is not complete: it is quite easy to produce a sequence $\{f_i\}$, with each $f_i \in T_{\gamma_{id}}(\Gamma_I)$ such that $f_i \to g$ with respect to the $\mathbb{L}^2$ norm, but g fails to be smooth, or g fails to satisfy the boundary conditions $g(0) = g(1) = 0$, or it fails both of these!

We are interested in both the geometry of Γ_I and its role in warping functions on $[0, 1]$. Specifically, we would like to impose a Riemannian structure on Γ_I and compute elements such as geodesics, exponential map, and the inverse exponential map under that metric. We postpone this discussion until Sect. 4.10.2.

4.4 Function Registration Problem

An important problem in statistical analysis of function data (and by extension to shape analysis of curves) is the alignment of function data using domain warping. The broad goal of an alignment process is to warp the time (or parameter) axis in such a way that their peaks and valleys are better aligned. This alignment problem has also been referred to as the separation of *phase* and *amplitude* of functions in the given data, or the *registration* of data, or the correspondence problem. We describe a comprehensive solution to this problem that separates phases and amplitudes of the given functions and develops their statistical models separately. One reason for taking this separation approach is that it better preserves the modal structure of the given functions and leads to efficient statistical models for capturing the main source of data variability. In this section, we will use $\mathscr{F} \subset \mathbb{L}^2$ to denote a certain function space of interest. The actual definition of $\mathscr{F}$ comes later when we present more details. The group of warping functions Γ_I continues to be the same as in the last section. There are two kinds of registration problems:

1. **Pairwise Alignment Problem**: Given any two functions f_1 and f_2 in $\mathscr{F}$, we define their pairwise alignment or registration to be the problem of finding a warping function γ such that a certain energy term $E[f_1, f_2 \circ \gamma]$ is minimized. That is, we solve for:

$$\gamma^* = \operatorname*{argmin}_{\gamma \in \Gamma_I} \quad E[f_1, f_2 \circ \gamma] \ .$$

   Then, for any $t \in [0,1]$, the value $f_1(t)$ is said to be registered to $f_2(\gamma^*(t))$. An example of this alignment is shown in Fig. 4.10.

2. **Groupwise or Multiple Alignment**: Given a set of functions $\{f_i \in \mathscr{F} | i = 1, 2, \ldots, n\}$, we call the problem of finding a set of warping functions $\{\gamma_i | i = 1, 2, \ldots, n\}$ such that, for any $t \in [0,1]$, the values $f_i(\gamma_i(t))$ are said to be registered with each other, the problem of joint or multiple alignment. Figure 4.11 shows an example of this idea where the given functions $\{f_i\}$ (left panel) are aligned (middle panel) using the warping functions shown in the right panel.

In this chapter, we address the first problem, namely, the pairwise registration, and postpone the solution of the multiple registration problem until later in Chap. 8.

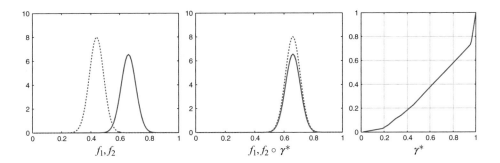

Fig. 4.10 Illustration of pairwise alignment

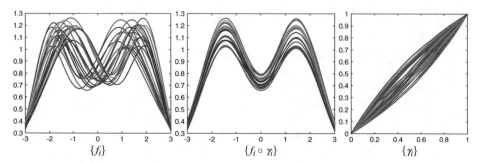

Fig. 4.11 Illustration of multiple function alignment

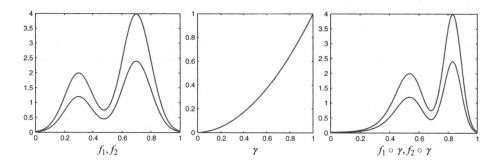

Fig. 4.12 Identical warping of f_1, f_2 by γ preserves their registration

The main question in solving the pairwise registration problem is: What should be the optimization criterion E? In other words, what is a mathematical definition of a *good registration*? Visually one can evaluate an alignment by comparing the locations of peaks and valleys, but how should one do it in a formal, quantifiable, and, most importantly, automated way? Before we establish a criterion for evaluating alignments, we should first focus on the basic properties that such a criterion should satisfy. Then, we can propose a formal criterion and demonstrate its effectiveness using theory and experiments. Below, we list three important properties that E should satisfy. Actually, only the first one is a fundamental property; the remaining two are simple consequences of the first one (and some additional structure). Still, we list all three of them to highlight different aspects of the registration problem.

1. **Invariance to Simultaneous Warping**: We start by noting that an identical warping of any two functions preserves their registration. That is, for any $\gamma \in \Gamma_I$ and $f_1, f_2 \in \mathscr{F}$, the function pair (f_1, f_2) has the same registration as the pair $(f_1 \circ \gamma, f_2 \circ \gamma)$. What we mean by that is the application of γ has not disturbed their point-to-point correspondence. This is easy to see since γ is a diffeomorphism. The two height values across the functions that were matched, say $f_1(t_0)$ and $f_2(t_0)$, for a parameter value t_0, remain matched. They are now labeled $f_1(\gamma(t_0))$ and $f_2(\gamma(t_0))$, and the parameter value has changed to $\gamma(t_0)$, but they still have the same parameter and, thus, are still matched to each other. This is illustrated using an example in Fig. 4.12 where the left panel

shows a pair f_1 and f_2. Note that the two functions are nicely registered since their peaks and valleys are perfectly aligned. If we apply the warping function shown in the middle panel to both their domains, we get the functions shown in the right. The peaks and valleys, and in fact all the points, have the same matching as before.

This motivates the following invariance property of E. Since E is expected to be a measure of registration of two functions, it should remain unchanged if the two functions are warped identically. That is:

Invariance Property : $E[f_1, f_2] = E[f_1 \circ \gamma, f_2 \circ \gamma]$, for all $\gamma \in \Gamma_I$.   (4.14)

2. **Effect of Random Warpings**: Suppose we have found the optimal warping function $\gamma^* \in \Gamma_I$, defined by:

$$\gamma^* = \underset{\gamma \in \Gamma_I}{\operatorname{argmin}} E[f_1, f_2 \circ \gamma] \ .$$

Once this γ^* is found, the resulting matched height values are $f_1(t)$ and $f_2(\gamma^*(t))$, for all t. Now, let us suppose that we replace f_1 and f_2 by random warpings of these functions, say $f_1 \circ \gamma_1$ and $f_2 \circ \gamma_2$. What is the optimal correspondence between these functions? It will be given by:

$$\tilde{\gamma}^* = \underset{\gamma \in \Gamma_I}{\operatorname{argmin}} E[f_1 \circ \gamma_1, (f_2 \circ \gamma_2) \circ \gamma] = \underset{\gamma \in \Gamma_I}{\operatorname{argmin}} E[f_1, f_2 \circ (\gamma_2 \circ \gamma \circ \gamma_1^{-1})],$$

where the last equality follows from the above *invariance property*. The last two equations immediately imply that $\gamma^* = \gamma_2 \circ \tilde{\gamma}^* \circ \gamma_1^{-1}$, in other words $\tilde{\gamma}^* = \gamma_2^{-1} \circ \gamma^* \circ \gamma_1$. More interestingly, the optimal registration of functions and the minimum value of E remains unchanged despite the presence of random γ_1 and γ_2, i.e.:

$$\underset{\gamma \in \Gamma_I}{\min} E[f_1, f_2 \circ \gamma] = \underset{\gamma \in \Gamma_I}{\min} E[f_1 \circ \gamma_1, (f_2 \circ \gamma_2) \circ \gamma] \ .$$

This important equality intimately depends on the invariance property and the group structure of Γ_I. Without the invariance property, one can expect the results of registration to be highly dependent on γ_1 and γ_2. For instance, this undesirable situation will occur if we use the $\mathbb{L}^2$ metric between the functions to define E.

3. **Inverse Symmetry**: We know that registration is a symmetric property. That is, if f_1 is registered to f_2, then f_2 is also registered to f_1. Similarly, if f_1 is optimally registered to $f_2 \circ \gamma$, then f_2 is optimally registered to $f_1 \circ \gamma^{-1}$. Therefore, the choice of E should be such that this symmetry is preserved. That is:

$$\gamma^* = \underset{\gamma \in \Gamma_I}{\operatorname{argmin}} E[f_1, f_2 \circ \gamma] \Rightarrow \gamma^{*-1} = \underset{\gamma \in \Gamma_I}{\operatorname{argmin}} E[f_1 \circ \gamma, f_2] \ .$$

This symmetry property has also been termed as *inverse consistency*. If the invariance property holds, then this inverse symmetry follows immediately.

In addition to registering any two functions, it is often important to compare them and to quantify their differences. For this, we need a proper distance function to be able to compare f_1 and $f_2 \circ \gamma^*$, where γ^* is the optimal warping of f_2 that registers it with f_1. One can always choose an independent distance function on the space, e.g., the $\mathbb{L}^2$ norm, that measures differences in the registered functions f_1 and $f_2 \circ \gamma^*$. However, this makes the process of registration isolated from the mechanism for comparison. Ideally, we would like to jointly solve these two problems. Therefore, it will be useful if the quantity $\inf_{\gamma \in \Gamma_I} E[f_1, f_2 \circ \gamma]$ is also a proper distance in some sense. That is, in addition to symmetry, it also satisfies non-negativity and the triangle inequality. The sense in which we want it to be a distance is that the result does not change if we randomly warp the individual functions in different ways.

We will focus on deriving a cost function E that satisfies all of these nice properties and we will do so by extending a famous Riemannian metric called the (nonparametric) Fisher-Rao metric. However, we start by taking a very popular approach based on the $\mathbb{L}^2$ metric and highlight its limitations.

4.5 Use of $\mathbb{L}^2$-Norm and Its Limitations

Although the $\mathbb{L}^2$ norm, or its variants, have been used widely in registration of functional data, it has many shortcomings. We check the desired properties listed in the last section for cost functions based on this metric.

1. **Lack of Isometry Under Warping**: Let us check the isometry property for the $\mathbb{L}^2$ norm under identical warping. For the example shown in Fig. 4.12, the $\mathbb{L}^2$ norms of differences before and after warping are 0.7566 and 0.6263, respectively. For any general $f_1, f_2 \in \mathbb{L}^2$ and $\gamma \in \Gamma_I$, we have:

$$\|f_1 \circ \gamma - f_2 \circ \gamma\|^2 = \int_0^1 (f_1(\gamma(t)) - f_2(\gamma(t)))^2 dt$$

$$= \int_0^1 (f_1(s) - f_2(s))^2 \frac{1}{\dot{\gamma}(\gamma^{-1}(s))} ds, \quad s = \gamma(t) .$$

Since, in general $\dot{\gamma}(\gamma^{-1}(s)) \neq 1$, we can see that generally:

$$\|f_1 \circ \gamma - f_2 \circ \gamma\| \neq \|f_1 - f_2\| . \tag{4.15}$$

Despite identical warping and preservation of matches between the two functions, the $\mathbb{L}^2$ norm of their difference changes. In other words, the action of Γ_I on $\mathscr{F}$ under the $\mathbb{L}^2$ metric is *not* by isometries. This rules out the use of $E[f_1, f_2] = \|f_1 - f_2\|$ directly in function registration!

2. **Pinching Effect**: A related problem in using $\mathbb{L}^2$ to perform matching is called the *pinching effect*. The basic idea here is that in matching of two functions, say f_1 and f_2, using $\inf_{\gamma \in \Gamma_I} \|f_1 - f_2 \circ \gamma\|$, one can squeeze or pinch a large part of f_2 and make this cost function arbitrarily close to zero. An illustration of this problem is presented in Fig. 4.13 using a simple example. Here a part of f_2 is identical to f_1 over $[0, 0.6]$ and is completely different over the remaining domain $[0.6, 1]$. Since f_1 is essentially zero and f_2 is strictly positive in $[0.6, 1]$, there

is no warping that can match f_1 with f_2 over that subinterval. The optimal solution is, therefore, to decimate that part of f_2 by using the following γ^*: it coincides with γ_{id} over $[0, 0.6]$, climbs rapidly to $1 - \epsilon$ around 0.6, and then goes slowly from $1 - \epsilon$ to 1 over the interval $[0.6, 1]$. The precise expression for a small $\epsilon > 0$ is given by:

$$\gamma_\epsilon(t) = \begin{cases} t, & \text{for } 0 \le t \le 0.6 \\ (\frac{0.4 - \epsilon}{\epsilon})(t - 0.6) + 0.6, & \text{for } 0.6 \le t \le 0.6 + \epsilon \\ (\frac{\epsilon}{0.4 - \epsilon})(t - 0.6 - \epsilon) + 1 - \epsilon, & \text{for } 0.6 + \epsilon \le t \le 1 \end{cases}$$

It is easy to check that in the limit:

$$\lim_{\epsilon \to 0} \|f_1 - f_2 \circ \gamma_\epsilon\| = 0 \ .$$

The top panel of Fig. 4.13 shows the limiting case where f_1 and $f_2 \circ \gamma_0$ are identical. In the bottom row, we provide results from a numerical procedure for this optimization problem. Sometimes, in practice, we restrict γ to have positive, bounded slope and do not obtain the same result as the theoretical limit. However, one can still see the pinching effect in the second panel of this row where the second part of f_2 is being squeezed into a small domain since it is mismatched with f_1. The last columns of this figure show the $\dot{\gamma}^*$ in the two cases—the picture in the bottom row shows that the slope of γ in the numerical implementation was bounded by 12.

To avoid the pinching problem, a common approach is to impose an additional term in the optimization that constrains the roughness of γ. This term, also called a *regularization term*, results in the registration problem of the type:

$$\inf_{\gamma \in \Gamma_I} \left(\|f_1 - f_2 \circ \gamma\| + \lambda \mathscr{R}(\gamma) \right) \ , \tag{4.16}$$

where $\mathscr{R}(\gamma)$ is the regularization term, e.g., $\mathscr{R}(\gamma) = \int \ddot{\gamma}^2(t) dt$, and $\lambda > 0$ is a constant. This solution has several problems including the fact that it does not

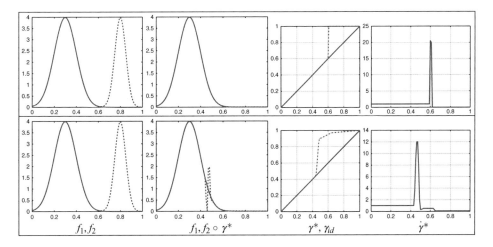

Fig. 4.13 Illustration of the pinching effect. The top row shows a degenerate analytical solution to matching under $\mathbb{L}^2$ norm for functions. The bottom row shows a similar problem arising in a numerical procedure for the same matching

satisfy inverse symmetry. The asymmetry comes from first term that has already been shown to be asymmetric earlier. A more comprehensive and satisfactory solution to the registration problem is developed in the next few sections.

3. **Inverse Inconsistency**: For registration under the $\mathbb{L}^2$ distances between functions, the results are inconsistent between inverse registrations. That is, the registration of f_1 to f_2 may lead to a completely different result than that of f_2 to f_1. We illustrate this using an example.

Example 4.1. As a simple example, let $f_1(t) = t$ and $f_2(t) = 1 + (t - 0.5)^2$. In this case, for the minimization problem $\gamma_{12} = \min_{\gamma \in \Gamma_I} \|f_2 - f_1 \circ \gamma\|^2$, the optimal solution is as follows. Define a warping function that climbs quickly (linearly) from 0 to $1 - \epsilon$ on the interval $[0, \epsilon]$ and then climbs slowly (also linearly) from $1 - \epsilon$ to 1 in the remaining interval $[\epsilon, 1]$. The limiting function, when $\epsilon \to 0$, results in the optimal γ for this case.

For the inverse problem, $\gamma_{21} = \min_{\gamma \in \Gamma_I} \|f_1 - f_2 \circ \gamma\|^2$, the optimal warping is the following. Define a warping function that rises quickly from 0 to $0.5 - \epsilon$ in the interval $[0, \epsilon]$, climbs slowly from $0.5 - \epsilon$ to $0.5 + \epsilon$ in the interval $[\epsilon, 1 - \epsilon]$, and finally climbs quickly from $0.5 + \epsilon$ to 1 in the interval $[1 - \epsilon, 1]$. The optimal solution is obtained when $\epsilon \to 0$.

A computer implementation of these two solutions, based on the dynamic programming algorithm, are shown in Fig. 4.14. Since this implementation allows only a small number of possible slopes for optimal γ, the results are not as accurate as the analytical solution. Still, it is easy to see a large difference in the solutions for the two cases.

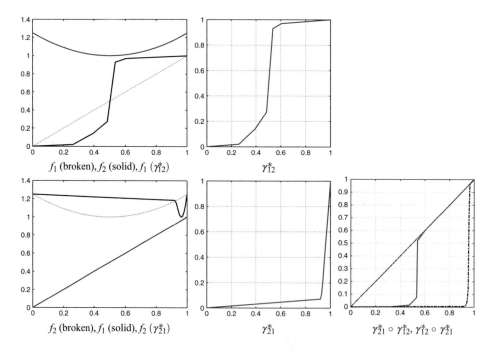

Fig. 4.14 Example of asymmetry in registration using $\mathbb{L}^2$ norm. The top row show solution of registering f_1 to f_2, while the bottom row shows the reverse case. Bottom right emphasizes that the two γ functions are not inverses of each other

4.6 Square-Root Slope Function Representation

In order to define a cost function E that avoids the pinching effect, satisfies the *invariance property* and, as a consequence, allows inverse symmetry and invariance of E to random warping, we introduce a new mathematical representation for functions. Starting fresh, this time we are going to restrict to those f that are absolutely continuous on $[0, 1]$; let $\mathscr{F}$ denote the set of all such functions. For the reader's convenience, we will give a definition of *absolutely continuous*. Though it is not the one most commonly given in real analysis textbooks, it is logically equivalent to that one and is more useful for our purposes.

Definition 4.1 (Absolutely Continuous). A function $f : [0, 1] \to \mathbb{R}$ is *absolutely continuous* if it satisfies the following two conditions:

1. f is differentiable almost everywhere on $[0, 1]$; denote its derivative by $\dot{f}$.
2. $f(t) = f(0) + \int_0^t \dot{f}(u)du$ for all $t \in [0, 1]$.

The absolutely continuity is stronger than continuity, but weaker than C^1 (and much weaker than being smooth!). It is easy to show that $\mathscr{F}$ is a vector space. This is left as an exercise for the reader.

The new representation of functions is based on the following transformation. Define a mapping: $Q : \mathbb{R} \to \mathbb{R}$ according to:

$$Q(x) \equiv \text{sign}(x)\sqrt{|x|} \ . \tag{4.17}$$

Note that Q is a continuous map. For the purpose of studying the function f, we will represent it using the SRSF defined as follows.

Definition 4.2 (SRSF Representation of Functions). Assume that $f : I \to R$ is absolutely continuous. Define the square-root slope function of f to be the function $q : [0, 1] \to \mathbb{R}$, where:

$$q(t) \equiv Q(\dot{f}(t)) = \text{sign}(\dot{f}(t))\sqrt{|\dot{f}(t)|} \ .$$

This representation includes those functions whose parameterization can become singular in the analysis. In other words, if $\dot{f}(t) = 0$ at some point, it does not cause any problem in the definition of $q(t)$. It can be shown that, if the function f is absolutely continuous, then the resulting SRSF is square integrable. This is left as an exercise to the reader. Thus, we will define $\mathbb{L}^2([0, 1], \mathbb{R})$ (or simply $\mathbb{L}^2$) to be the set of all SRSFs. For every $q \in \mathbb{L}^2$, there exists a function f (unique up to a constant) such that the given q is the SRSF of that f. In fact, this function can be obtained precisely using the equation: $f(t) = f(0) + \int_0^t q(s)|q(s)|ds$. Thus, the representation $f \Leftrightarrow (f(0), q)$ is invertible. Some examples of SRSFs are shown in Fig. 4.15. Note that the locations of peaks and valleys of f correspond to the zero crossings of q.

The next question is: If a function is warped, then how does its SRSF change? For an $f \in \mathscr{F}$ and $\gamma \in \Gamma_I$, let q be the SRSF of f. Then, what is the SRSF of $f \circ \gamma$? This can simply be derived as:

$$\tilde{q}(t) = Q\left(\frac{d}{dt}\left(f \circ \gamma\right)(t)\right) = \text{sign}\left(\frac{d}{dt}\left(f \circ \gamma\right)(t)\right)\sqrt{\left|\frac{d}{dt}(f \circ \gamma)(t)\right|} = q(\gamma(t))\sqrt{\dot{\gamma}(t)} \ .$$

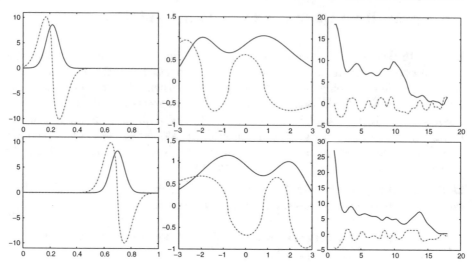

Fig. 4.15 Examples of functions $f(t)$ in *solid lines* and their SRSFs $q(t)$ in *broken lines*

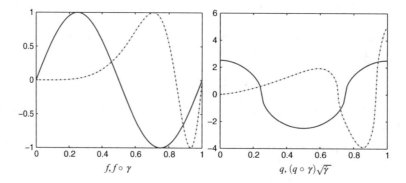

$f, f \circ \gamma$ $q, (q \circ \gamma)\sqrt{\dot{\gamma}}$

Fig. 4.16 *Left panel*: two functions $f(t)$ (*solid line*) and its warped version $f(\gamma(t))$ (*broken line*). *Right*: their SRSFs $q(t)$ (*solid line*) and $q(\gamma(t))\sqrt{\dot{\gamma}(t)}$ (*broken line*)

In fact, this denotes an action of group Γ_I on $\mathbb{L}^2$ from the right side: $\mathbb{L}^2 \times \Gamma_I \to \mathbb{L}^2$, given by $(q, \gamma) = (q \circ \gamma)\sqrt{\dot{\gamma}}$. The fact that this mapping is a group action is left as an exercise. One can show that this action of Γ_I on $\mathbb{L}^2$ is *compatible* with its action on $\mathscr{F}$ given earlier, in the following sense:

$$
\begin{array}{ccc}
 & f \xrightarrow{SRSF} q & \\
\text{action on } \mathscr{F} \quad \downarrow & & \downarrow \quad \text{action on } \mathbb{L}^2 \\
 & f \circ \gamma \xrightarrow{SRSF} (q \circ \gamma)\sqrt{\dot{\gamma}} &
\end{array}
$$

We can apply the group action and compute SRSF in any order, and the result remains same. Two examples of the changes in SRSFs, when the underlying function is warped, are shown in Fig. 4.16. It can be seen that, while the vertical coordinates of the original functions remain unchanged, the vertical coordinates of SRSFs are changed by this group action.

The most important advantage of using SRSFs in functional data analysis comes from the following result. Recall that the $\mathbb{L}^2$ inner product is given by $\langle v_1, v_2 \rangle = \int_0^1 v_1(t) v_2(t)\, dt$.

Lemma 4.1. *The mapping $\mathbb{L}^2 \times \Gamma_I \to \mathbb{L}^2$ given by $(q, \gamma) = (q \circ \gamma)\sqrt{\dot\gamma}$ forms an action of Γ_I on $\mathbb{L}^2$ by isometries.*

Proof. That the mapping is a group action has been mentioned earlier. The proof of isometry is an easy application of integration by substitution. For any $v_1, v_2 \in \mathbb{L}^2$:

$$\langle (v_1, \gamma), (v_2, \gamma) \rangle = \int_0^1 v_1(\gamma(t)) \sqrt{\dot\gamma(t)} v_2(\gamma(t)) \sqrt{\dot\gamma(t)} dt$$
$$= \int_0^1 v_1(\gamma(t)) v_2(\gamma(t)) \dot\gamma(t) dt = \int_0^1 v_1(s) v_2(s) ds = \langle v_1, v_2 \rangle \ . \square$$

We already know that the geodesics in $\mathbb{L}^2$, under the standard metric, are straight lines and the geodesic distance between any two elements $q_1, q_2 \in \mathbb{L}^2$ is given by $\|q_1 - q_2\|$ (Sect. 4.3.1). Since the action of Γ_I on $\mathbb{L}^2$ is by isometries, the following result is automatic. Still, for the sake of completeness, we provide a short proof.

Lemma 4.2. *For any two SRSFs $q_1, q_2 \in \mathbb{L}^2$ and $\gamma \in \Gamma_I$, we have that $\|(q_1, \gamma) - (q_2, \gamma)\| = \|q_1 - q_2\|$.*

Proof. For an arbitrary element $\gamma \in \Gamma_I$, and q_1, $q_2 \in \mathbb{L}^2$, we have:

$$\|(q_1, \gamma) - (q_2, \gamma)\|^2 = \int_0^1 (q_1(\gamma(t)) \sqrt{\dot\gamma(t)} - q_2(\gamma(t)) \sqrt{\dot\gamma(t)})^2 dt$$
$$= \int_0^1 (q_1(\gamma(t)) - q_2(\gamma(t)))^2 \dot\gamma(t) dt = \|q_1 - q_2\|^2 \ . \square$$

An interesting corollary of this lemma is the following.

Corollary 4.1. *For any $q \in \mathbb{L}^2$ and $\gamma \in \Gamma_I$, we have $\|q\| = \|(q, \gamma)\|$.*

This implies that the action of Γ_I on $\mathbb{L}^2$ is actually a norm-preserving transformation. Conceptually, it can be equated with the rotation of vectors in Euclidean spaces. Later on, we will introduce another kind of transformation that relates to warping of functions and will organize them in different groups.

4.7 Definition of Phase and Amplitude Components

As mentioned earlier, the problem of registration of functions is also called the problem of phase-amplitude registration. In this section, we will introduce a formal specification of these terms, but before we do that, we allude to the usage of phase and amplitude of waves in classical physics. In physics, a modulated sine wave is given by:

$$f(t) = a(t) \sin(2\pi\omega t + \phi(t)) \ ,$$

where $a(t)$ is called the instantaneous amplitude and $(\omega + \dot\phi(t))$ the instantaneous frequency. The function $\phi(t)$ determines the instantaneous phase of the function.

One can characterize a wave using $a(t)$, ω, and $\phi(t)$, as amplitude, frequency, and phase functions, respectively. The process of encoding information for transmitting signals by instantaneous modifications of these components is termed *modulation*. The three types of modulations are amplitude, phase, and frequency modulation, although the last two are closely related. Motivated by this basic terminology, we wish to develop formal notions of phase and amplitude for arbitrary function data.

4.7.1 Amplitude of a Function

Recall that $\mathscr{F}$ is the set of real-valued, absolutely continuous functions on $[0, 1]$. For any $f \in \mathscr{F}$ and an arbitrary warping function γ, consider the composition or warping $f \circ \gamma$. We stipulate that *amplitude is a property of a function that remains unchanged under warping*; warping of a function only changes its phase. (This is similar to the definition of shape as a property that is preserved under certain transformations (rigid motions, etc.), as described in later chapters.) Thus, f and $f \circ \gamma$ have the same amplitude for all $\gamma \in \Gamma_I$. Figure 4.17 shows an example where a number of functions are generated by warping the same function f_0, i.e., $f_i = f_0 \circ \gamma_i$. One can verify that all these functions go through the same set of heights (or amplitudes) and in the same order, just the rate of traversal is different. According to our definition, all these functions have the same amplitude.

It is natural to use the notion of orbits, under the action of a warping group, to specify amplitudes of functions as equivalence classes. However, since the set Γ_I is not closed, we actually use a larger set $\widetilde{\Gamma_I}$ in order to get a precise definition of orbits.

Definition 4.3. Let $\widetilde{\Gamma_I}$ denote the set of all functions $\gamma : [0, 1] \to [0, 1]$ that satisfy the following three properties:

1. $\gamma(0) = 0$ and $\gamma(1) = 1$.
2. γ is absolutely continuous.
3. For all $t_1, t_2 \in [0, 1]$, if $t_1 < t_2$, then $\gamma(t_1) \leq \gamma(t_2)$ (i.e., γ is "weakly increasing").

It is easy to show that $\widetilde{\Gamma_I}$ satisfies all the properties of a group except that not every element has an inverse. Such an object is called a *monoid*. Note that $\tilde{\Gamma}_I$ includes elements γ with $\dot{\gamma}(t) = 0$ on some subset of $[0, 1]$ with nonzero measure. Figure 4.18 shows an example of such a γ and its action on a bimodal function f.

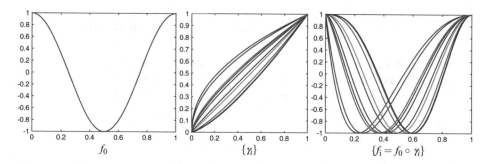

Fig. 4.17 Different functions generated by warping the same function are said to have the same amplitude

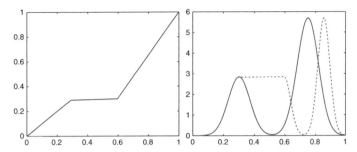

Fig. 4.18 An example of an element of $\widetilde{\Gamma}_I$ whose derivative equals zero over an interval, and its warping action on a function

An interesting and helpful fact in this setup is this: If $\dot{f} = 0$, only on a set of measure zero, then the closure of the set $\{f \circ \gamma | \gamma \in \Gamma_I\}$ equals the orbit of f under $\tilde{\Gamma}_I$. (The closure is with respect to the $\mathbb{L}^2$ metric on the SRSF space of the function.) The assumption on $\dot{f}$ is not very limiting since, according to Lemma 4.3, one can re-parameterize any nontrivial function to satisfy this property. Another useful fact is that the group Γ_I is a dense subset of $\tilde{\Gamma}_I$, with respect to the Fisher-Rao metric on Γ_I (covered later in Sect. 4.10.2).

Now we can specify amplitudes as orbits under the action of $\tilde{\Gamma}_I$. Instead of using the original function space $\mathscr{F}$, we will do so on the SRSF space since it allows the use of a simpler metric, i.e., the $\mathbb{L}^2$ metric. Let $f_1 : [0, 1] \to \mathbb{R}$ and $f_2 : [0, 1] \to \mathbb{R}$ be two absolutely continuous functions, with SRSFs q_1 and q_2, respectively. Define a relation $\sim$ on $\mathbb{L}^2$ by defining $q_1 \sim q_2$ if and only if q_1 is an element of the $\mathbb{L}^2$ closure of $[q_2]$. This relation is easily shown to be an equivalence relation. Let's see what it means for $q_1 \sim q_2$, in terms of the original functions f_1 and f_2. We can use this monoid to define constant-speed parameterization of a function.

Lemma 4.3. *If $f : [0, 1] \to \mathbb{R}$ is absolutely continuous, then there exists a unique absolutely continuous function $h : [0, 1] \to \mathbb{R}$ with $|\dot{h}(t)| = $ constant almost everywhere in $[0, 1]$, and a unique $\gamma \in \tilde{\Gamma}_I$ such that $f = h \circ \gamma$. This function h is called the constant-speed parameterization of f.*

The proof of this Lemma is standard.

Theorem 4.1. *Let $f_1 : [0, 1] \to \mathbb{R}$ and $f_2 : [0, 1] \to \mathbb{R}$ be two absolutely continuous functions, with SRSFs q_1 and q_2, respectively. Then $q_1 \sim q_2$ if and only if f_1 and f_2 have the same constant-speed parameterization h.*

The proof of this theorem is involved and is skipped for brevity. We refer the reader to the paper [62] for details. This theorem shows us that when we form the quotient $\mathbb{L}^2/\sim$, we are identifying two functions precisely if they have identical constant-speed parameterizations, i.e., if they traverse the same points of $\mathbb{R}$ in the same order. Note that either or both of these functions are allowed (but not required) to "stop and rest" at various times as they traverse these points.

Now we are ready to define the amplitude of a function.

Definition 4.4 (Amplitude). Let f be an element of $\mathscr{F}$ such that $\dot{f} \neq 0$ almost everywhere. The amplitude of f, with associated SRSF q, is defined as its orbit $[q]$ under the monoid $\tilde{\Gamma}_I$: $[q] = \{(q \circ \gamma)\sqrt{\dot{\gamma}} | \gamma \in \tilde{\Gamma}_I\}$.

We will use $\mathscr{A}$ to denote the space of all amplitudes or orbits.

4.7.2 Relative Phase Between Functions

The next step is to specify the phase component of function. Before we do that, we make a few remarks about what properties such a specification should satisfy.

1. While the notion of amplitude is absolute, we can only define phase of one function *w.r.t.* the other.
2. So far we have not used any metric or statistical model (to introduce the notion of amplitude). In contrast, we will need either a metric or a statistical model to define relative phase.
3. The phase of a function, relative to another, is changed by a warping of either function but their amplitudes remain unchanged by warpings.
4. We expect the relative phase of f_1 *w.r.t* f_2 to be exactly the inverse of the relative phase of f_2 *w.r.t* f_1. This is analogous to the fact that if the rotation from one element of $\mathbb{R}^n$ to another is $O \in SO(n)$, then the rotation from the second to the first is $O^{-1} = O^T$.
5. Also, similar to rotation, if we change the relative phase of a function twice—using one warping after another—then the total change should be a composition of the two warpings. That is, changing phase is an action of a monoid or a group!
6. Multiplying a function by a positive scalar, or adding a constant to a function, should not change its relative phase with respect to any other function.
7. The relative phase is not well defined if either of the functions involved is constant.

The SRSF of a function $f \in \mathscr{F}$ is given by $q(t) = \text{sign}(\dot{f}(t))\sqrt{|\dot{f}(t)|}$ and the SRSF of $f \circ \gamma$, for any $\gamma \in \tilde{\Gamma}_I$, is given by $(q, \gamma)(t) = q(\gamma(t))\sqrt{\dot{\gamma}(t)}$. Also, we remind the reader that $\|(q, \gamma)\| = \|q\|$ for all $q \in \mathbb{L}^2$ and $\gamma \in \tilde{\Gamma}_I$ (See Lemma 4.1).

Definition 4.5 (Relative Phase). The relative phase of a function f_1 *w.r.t* a function f_2 is defined by the pair:

$$(\gamma_1^*, \gamma_2^*) = \underset{\gamma_1, \gamma_2 \in \tilde{\Gamma}_I}{\text{argmin}} \left(\|(q_1, \gamma_1) - (q_2, \gamma_2)\| \right) \in \tilde{\Gamma}_I \times \tilde{\Gamma}_I, \tag{4.18}$$

where q_1 and q_2 are the SRSFs of f_1 and f_2, respectively.

One can show that this relative phase exists if f_1 or f_2 is a piecewise linear function. However, it is not necessarily unique. The same holds if f_1 and f_2 are C^1-functions. For more general results on existence of optimal warpings, please refer to the paper [20]. The two components of the relative phase are needed to bring q_1 and q_2 to the closest points in their respective orbits. Note that the relative phase is defined only up to a $\gamma \in \tilde{\Gamma}_I$. That is, if (γ_1^*, γ_2^*) is a solution to Eq. 4.18, then so is $(\gamma_1^* \circ \gamma, \gamma_2^* \circ \gamma)$, for any $\gamma \in \tilde{\Gamma}_I$.

The optimization problem in Eq. 4.18 is precisely the pairwise alignment solution for f_1 and f_2 proposed in Chap. 4. Figure 4.19 (left) illustrates these ideas using a sketch. The two orbits have been drawn parallel to each other because $\|q_1 - q_2\| = \|(q_1, \gamma) - (q_2, \gamma)\|$ for all γ. One moves along an orbit using time warpings or phase changes and moves orthogonal to the orbits by changing amplitudes. For the relative phase as defined above, the following statements hold.

1. The relative phase of f_1 *w.r.t* f_2 is exactly the reverse of the relative phase of f_2 *w.r.t* f_1, i.e., the relative phase of f_2 *w.r.t* f_1 is given by (γ_2^*, γ_1^*).

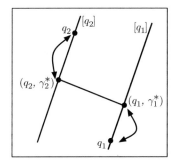 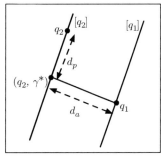

Fig. 4.19 *Left:* The definition of phase and amplitudes of function: amplitudes are defined by the orbits and phases are defined by the pair (γ_1^*, γ_2^*) that finds the nearest elements of the orbits. *Right:* Approximation of phase using a single warping γ^* and the phase, amplitude distances between functions

2. Two functions are said to have zero phase between them if their relative phase is $(\gamma_{id}, \gamma_{id})$. In Fig. 4.19(left), this situation will occur if q_1, q_2 happen to be the nearest points across orbits. For example, this is the case when $f_1 = f_2$.
3. The relative phase of f_1 *w.r.t* $f_2 \circ \gamma$, for any $\gamma \in \Gamma_I$, is $(\gamma_1^*, \gamma^{-1} \circ \gamma_2^*)$. Similarly, the relative phase of $f_1 \circ \gamma_1$ *w.r.t* f_2 is $(\gamma^{-1} \circ \gamma_1^*, \gamma_2^*)$. Combining these results, the relative phase of $f_1 \circ \gamma_a$ *w.r.t* $f_2 \circ \gamma_b$ is $(\gamma_a^{-1} \circ \gamma_1^*, \gamma_b^{-1} \circ \gamma_2^*)$.
4. As a corollary to the last property, the relative phase of $f_1 \circ \gamma$ *w.r.t* $f_2 \circ \gamma$, for any $\gamma \in \Gamma_I$, is $(\gamma^{-1} \circ \gamma_1^*, \gamma^{-1} \circ \gamma_2^*)$.
5. The relative phase between any two functions is unchanged by vertical translations and scalar multiplications of those functions. That is, the relative phase of $(c_1 f_1 + e_1)$ relative to $(c_2 f_2 + e_2)$ is also (γ_1^*, γ_2^*), where c_1, $c_2 \in \mathbb{R}_+$ and e_1, $e_2 \in \mathbb{R}$. This is a consequence of the fact that SRSF is independent of vertical translation and Lemma 4.5.

The proofs of these properties have been left as exercises to the reader.

4.7.3 A Convenient Approximation

Although the definitions laid out so far have nice theoretical properties, the ensuing computations are relatively complicated. For instance, the precise solution of Eq. 4.18 is not easy to reach in general cases (in case the functions are piecewise linear, an algorithm is presented in [62]). However, a simple approximation leads to convenient computational solutions that allows us to use existing algorithms. Using the fact that Γ_I is dense in $\tilde{\Gamma}_I$, we get the equality:

$$\inf_{\gamma_1, \gamma_2 \in \tilde{\Gamma}_I} \|(q_1, \gamma_1) - (q_2, \gamma_2)\| = \inf_{\gamma_1, \gamma_2 \in \Gamma_I} \|(q_1, \gamma_1) - (q_2, \gamma_2)\| = \inf_{\gamma \in \Gamma_I} \|q_1 - (q_2, \gamma)\| .$$
(4.19)

Thus, one can approximate the amplitude distance by fixing q_1 and minimizing only on Γ_I orbit of q_2. This minimization can be performed in several ways, one of which is the DPA described in Appendix B. In the same spirit, one can use a single function:

$$\gamma_{12}^* = \operatorname*{arginf}_{\gamma \in \Gamma_I} \|q_1 - (q_2, \gamma)\|$$
(4.20)

as an *approximation* for the relative phase for f_1 *w.r.t* f_2. Note that this γ_{12}^* replaces the pair (γ_1^*, γ_2^*) in studying the relative phase. From here onward, we will use this single function to define the relative phase between functions, instead of the pair. Also, note that the optimization used here is exactly the same as defined as the registration problem in Eq. 4.7. This approximation is depicted pictorially in the right panel of Fig. 4.19.

4.8 SRSF-Based Registration

Using the tools introduced in this section, we can formulate a new registration approach as follows.

4.8.1 Registration Problem

Definition 4.6 (Registration Energy). For any two functions f_1 and f_2, let q_1 and q_2 denote their SRSFs, respectively. We define the cost function for registering f_1 and f_2 to be $E[f_1, f_2] = \|q_1 - q_2\|$.

Let us check if this choice of E satisfies the three properties listed earlier.

1. **Invariance to Simultaneous Warping**: This important property is already established in Lemma 4.2, which states that $E[f_1, f_2] = E[f_1 \circ \gamma, f_2 \circ \gamma]$ for all $f_1, f_2 \in \mathscr{F}$ and all $\gamma \in \Gamma_I$.

2. **Invariance to Random Warpings**: For any $f_1, f_2 \in \Gamma_I$ and $\gamma, \gamma_1, \gamma_2 \in \Gamma_I$, we have

$$
\begin{aligned}
\inf_{\gamma \in \Gamma_I} E[f_1 \circ \gamma_1, f_2 \circ \gamma_2 \circ \gamma] &= \inf_{\gamma \in \Gamma_I} \|(q_1, \gamma_1) - ((q_2, \gamma_2), \gamma)\| \\
&= \inf_{\gamma \in \Gamma_I} \|(q_1, \gamma_1) - (q_2, \gamma_2 \circ \gamma)\| \\
&= \inf_{\gamma \in \Gamma_I} \|q_1 - ((q_2, \gamma_2 \circ \gamma), \gamma_1^{-1})\| \\
&= \inf_{\gamma \in \Gamma_I} \|q_1 - (q_2, \gamma_2 \circ \gamma \circ \gamma_1^{-1})\| = \inf_{\gamma \in \Gamma_I} \|q_1 - (q_2, \gamma)\| \\
&= \inf_{\gamma \in \Gamma_I} E[f_1, f_2]
\end{aligned}
$$

The three equalities used above are based on the group structure of Γ_I and on Lemma 4.2. Thus, the cost function is invariant to simultaneous warpings of input functions and the resulting registration problem is unaffected by random warpings of the input data.

3. **Inverse Symmetry**: This framework ensures inverse symmetry in registration, as shown in the following lemma.

Lemma 4.4. *Let $\gamma^* \in \Gamma_I$ be a minimizer of $\|q_1 - (q_2, \gamma)\|$. In general, this may not exists. But if it does then we have that:*

$$
\gamma^{*-1} \in \operatorname*{argmin}_{\gamma \in \Gamma_I} \|(q_1, \gamma) - q_2\| . \tag{4.21}
$$

Proof. Since $\|q_1 - (q_2, \gamma^*)\| \leq \|q_1 - (q_2, \gamma)\|$, this implies that $\|(q_1, \gamma^{*-1}) - q_2\| \leq \|(q_1, \gamma^{-1}) - q_2\|$ for all $\gamma \in \Gamma_I$. Equation 4.21 directly follows from this condition.
□

In other words, this choice of E ensures that interchanging of the arguments does not change the registration result, and the optimal registration of (f_1, f_2) is the same as the registration of (f_2, f_1).

Additionally, it will be shown later in Sect. 4.10 that the quantity $\inf_{\gamma \in \Gamma_I} \|(q_1, \gamma) - q_2\|$ is actually proper distance, not in $\mathbb{L}^2$ but in a quotient space of $\mathbb{L}^2$. Having a distance is useful for the ensuing statistical analysis of registered functions. Therefore, the processes of registration and analysis are performed in a unified framework, rather than having two disjoint steps for each process. We summarize this discussion by the following problem statement.

Definition 4.7 (Registration Problem). For any two functions f_1 and f_2, let q_1 and q_2 denote their SRSFs, respectively. Finding the optimal registration for f_1 and f_2 is the same as solving the problem:

$$\inf_{\gamma \in \Gamma_I} E[f_1, f_2 \circ \gamma] = \inf_{\gamma \in \Gamma_I} \|q_1 - (q_2, \gamma)\| . \tag{4.22}$$

Note that the minimizer may only be approximate, depending upon the nature of f_1 and f_2, in the sense discussed in Sect. 4.7.3. While the precise optimal may not even exist, we can approximate it using the dynamic programming algorithm in most cases. This is described in the next section.

An interesting outcome of this framework is that global scaling of functions does not change their relative phases.

Lemma 4.5. *If $\gamma^* \in \arg\inf_{\gamma \in \Gamma_I} \|q_1 - (q_2, \gamma)\|$, for any $q_1, q_2 \in \mathbb{L}^2$, then the same γ^* is also in $\arg\inf_{\gamma \in \Gamma_I} \|cq_1 - (q_2, \gamma)\|$ for any $c \in \mathbb{R}_+$.*

The proof is left to the reader.

4.8.2 SRSF Alignment Using Dynamic Programming

We present an implementation of the dynamic programming algorithm (DPA) for solving the minimization stated in Eq. 4.22. While a more general description of the algorithm can be found in Appendix B, we particularize it to match our case here.

As described in the appendix, this algorithm imposes an $m \times m$, 2D grid on the square domain $[0, 1]^2$ and restricts the solution to a piecewise linear path from the origin $((0, 0))$ to the target $((1, 1))$. This path is required to change slopes only at the nodes of the grid. For each node $(i/m, j/m) \in [0, 1]^2$, one predefines a set of neighboring nodes, lying below and to the left of $(i/m, j/m)$ on the grid, and denoted by the set $\mathcal{N}_{(i,j)}$, that are allowed to reach $(i/m, j/m)$ directly using a straight line. Then, the optimal path is found using a recursive optimization, where the optimal cost $H_{i,j}$ of reaching the node $(i/m, j/m)$, starting from $(0, 0)$, is computed as follows. The cost of reach elements of $\mathcal{N}_{i,j}$ is already determined and stored. For each element of $\mathcal{N}_{i,j}$, one computes the additional cost of joining that element to (i, j) using the line segment and forms a cost of reaching (i, j). The minimum over all elements of $\mathcal{N}_{i,j}$ results in $H_{i,j}$:

$$H_{i,j} = \min_{(k/m, l/m) \in \mathcal{N}_{i,j}} \left(H_{k,l} + \int_{k/m}^{i/m} (q_1(t) - q_2(\gamma(t)) \sqrt{\dot{\gamma}(t)})^2 \, dt \right), \qquad (4.23)$$

where γ is a straight line joining the nodes $(k/m, l/m)$ and $(i/m, j/m)$ on the graph.

Algorithm 3 (SRSF Alignment Using DPA).

1. *Set $\mathcal{N}_{i,j}$, the neighbors of a lattice point $(i/m, j/m)$ allowed to reach $(i/m, j/m)$ directly using a straight line.*
2. *Set the value of H for the first row and the first column of the grid, except $(0,0)$ to ∞, i.e., $H_{0,:} = H_{:,0} = \infty$. Set $H_{0,0} = 0$.*
3. **Computation of H:**
   For each $(0,0) < (i/m, j/m) \leq (1,1)$ in the grid, compute $H_{i,j}$ using Eq. 4.23. This requires approximating the integral in Eq. 4.23 using numerical techniques. Let $((\hat{k}/m)_{i,j}, (\hat{l}/m)_{i,j})$ denote the optimal elements of $\mathcal{N}_{i,j}$ that minimize the right side in that equation. Note these are function of i and j.
4. *Computation of optimal γ^*:*
   a. *Initialize $(i/m, j/m) = (1,1)$.*
   b. *Draw a straight line from $(i/m, j/m)$ to $((\hat{k}/m)_{i,j}, (\hat{l}/m)_{i,j})$.*
   c. *Set $(i/m, j/m) = ((\hat{k}/m)_{i,j}, (\hat{l}/m)_{i,j})$. If $(i/m, j/m) = (0,0)$, then stop. Else, go to the previous step.*
   γ^* is the resulting piecewise linear curve from $(0,0)$ to $(1,1)$.

4.8.3 Examples of Functional Alignments

In this section, we demonstrate the effectiveness of this alignment method through some examples taken from real applications.

Figure 4.20 shows three examples of aligning pairs of functions. In each row, depicting a different case, we see the original functions f_1 and f_2 in the left panel and the warping $\tilde{f}_2 = f_2 \circ \gamma^*$ in the middle panel, while the optimal warping function γ^* is shown in the rightmost panel. In the top two cases, f_1 and f_2 are unimodal and γ^* results in aligning the two peaks. In the last case, where one peak is already aligned and the second one differs, the warping function γ^* only moves the second peak of f_2 to align it with the second peak of f_1. Note that the warping function takes a rather predictable form in these cases. It is a piecewise linear function with three different slopes: one for the initial region with zero values, one for the nonzero region, and one for the last region of zero values.

The next example, shown in Fig. 4.21, is taken from the famous Berkeley growth data where heights of human subjects were recorded from birth to age 19. This example uses a smoothed version of the time-derivative of the height functions, instead of the height functions themselves, as functional data. The peaks in these functions denote growth spurts and an important goal in this problem is to align the growth spurts of different subjects in order to make inferences about the number and placement of such spurts for the underlying population. Figure 4.21 shows two sets of examples in the two rows. The left panels in each row show the original functions and the middle panels show the warped versions of f_2 after their alignments with f_1. The final plot in each row shows the optimal time warping function

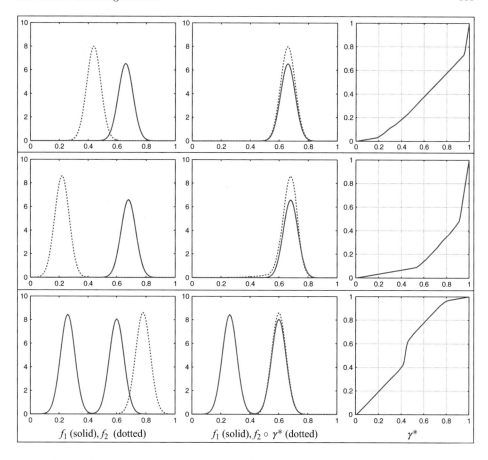

Fig. 4.20 Optimal warping of f_2 to align with f_1 using γ^*

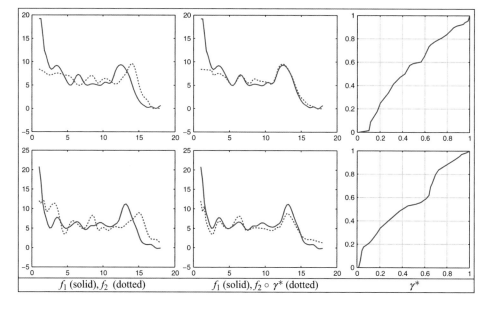

Fig. 4.21 An example of matching growth spurts for two subjects from the Berkeley growth dataset

γ^* used in that alignment. Note the nice alignment of peaks and valleys in both the cases after warpings. It becomes much easier for an analyst to estimate the location and size of the growth spurts in observed subjects and make inferences about the general population, through this alignment procedure.

Another important application of functional alignment is in mass spectrometry where one characterizes materials using their metabolite composition observed as chromatograms. The collection of data from different machines, or even same machine at different calibrations, leads to the problem of misalignment of these chromatograms. Shown in the top panel of Fig. 4.22 are two chromatograms, of the same material, that need alignment. They exhibit the same peak pattern but the locations of these peaks have been shifted in one chromatogram relative to the other. A simple shift is not sufficient to align them since some peaks shift to the left and some to the right. Shown in the middle panel is the result of alignment using the optimization listed in Eq. 4.22. It can be seen that the peaks are aligned well; this is further illustrated using a zoom-in of the interval [80, 100] in the bottom row, which shows a remarkable improvement in the peak alignment after optimization over γ.

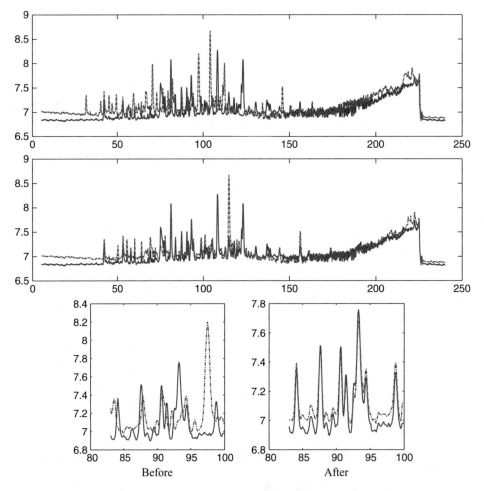

Fig. 4.22 Alignment of liquid chromatography—mass spectrometry data. The top row shows the original chromatograms while the bottom show the aligned ones after warping

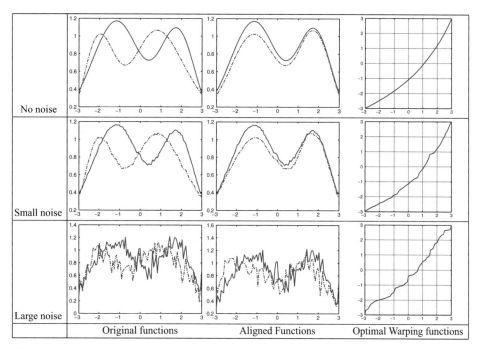

Fig. 4.23 Alignment results after adding white Gaussian noise to functions. The top row has no noise and the resulting alignment is almost perfect. The middle row has limited noise and the alignment is largely preserved. However, the bottom level is dominated by noise and the solution ends up aligning peaks and valleys, causes primarily by noise

An important issue in any data analysis is the immunity or at least some robustness to noise. Since the registration framework prescribed here is based on SRSFs, which, in turn, involve first derivatives, there is a strong possibility of noise influencing the alignment results. Figure 4.23 shows results from an experiment that increasingly adds white Gaussian noise to functions. The top shows the original two functions in the left panel, their alignment in the middle, and the optimal warping function γ^*, obtained using Eq. 4.22 in the right. Now we add a small amount of noise in the second row and observe the results. Since the level of noise is small, the alignment still looks decent although the optimal warping function shows some corruption due to noise. However, if we increase the noise level, as shown in the third row, the alignment is now governed more by noise than the original signal. The method ends up aligning peaks resulting from noise. So, how can we handle this situation?

Our goal in alignment (or registration) is to estimate the diffeomorphism γ^* that aligns the underlying signal and largely ignores the noise. Assuming that noise is high frequency and signal is low frequency, as is the case in many denoising algorithms, we can first smooth the functions to minimize noise. Then, one can use these *denoised* functions to perform alignment and apply the resulting alignment to the original functions. We illustrate this idea with the same example as in the bottom row of Fig. 4.23. We use a wavelet-based denoising and the results are shown in the left two panels of Fig. 4.24. The resulting smoothed functions are used in Eq. 4.22 to obtain the warping function shown in the third panel. This warping function, when applied to the original functions, results in the alignment

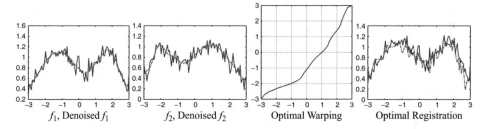

Fig. 4.24 Handling noise in given functions: denoise the given functions using a smoothing technique, obtain registration with denoised functions, and apply that registration

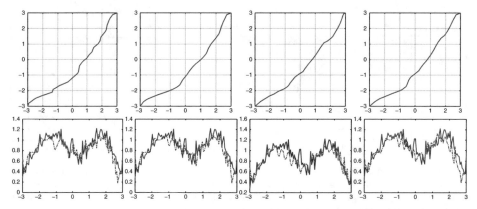

Fig. 4.25 Denoising-based alignment under different levels of denoising: from left to right, the top panels show results of γ^* obtained after increasing levels of smoothing applied to the noisy functions. The bottom row shows the corresponding alignments. The results are quite stable with respect to the amount of smoothing

shown in the last panel. If we compare it with the last row of Fig. 4.23, we can see that this result is much better in terms of alignment of the underlying signal (macro-structures).

The next issue is, of course, the amount of influence of the denoising step in the eventual registration result between the noisy functions. While the amount and type of denoising certainly affects the functions themselves, the eventual alignment remains relatively robust to these choices. Figure 4.25 illustrates this with an example. Taking the same two functions as earlier, we denoise them using wavelets as earlier, but with increasing levels of smoothness, and perform registration. The resulting optimal γ^*s are shown in the top row, with increasing smoothness from left to right. The corresponding registrations of the original functions are shown in the bottom panel. It can be seen that broad alignment of the noisy signals remains relatively unchanged despite a large change in the denoising level.

4.9 Connection to the Fisher-Rao Metric

The framework we have introduced in the previous section is intimately related to a well-known tool in Riemannian analysis of probability density functions. This is based on a Riemannian metric, namely, the **Fisher-Rao metric** that was introduced in 1945 by C. R. Rao [92] where he used the Fisher information matrix to compare different probability distributions. This metric was studied rigorously in the 1970s and 1980s by Amari [4], Efron [30], Kass [47], Cencov [24], and others. While those earlier efforts were focused on analyzing parametric families, our discussion is based on the *nonparametric* version of the Fisher-Rao Riemannian metric. There are several forms of the nonparametric version and we will start with the nonparametric version that is closest to our context. To define this metric, we will have to restrict to the set $\mathscr{F}_0 = \{f \in \mathscr{F} | \dot{f} > 0\}$.

Definition 4.8 (Fisher-Rao Metric). For any $f \in \mathscr{F}_0$ and $v_1, v_2 \in T_f(\mathscr{F})$, the Fisher-Rao Riemannian metric is defined as the inner product:

$$\langle\langle v_1, v_2 \rangle\rangle_f = \frac{1}{4} \int_0^1 \dot{v}_1(t) \dot{v}_2(t) \frac{1}{\dot{f}(t)} dt \ . \tag{4.24}$$

One of the most celebrated properties of this metric is that the action of the warping group on $\mathscr{F}_0$ is by isometries under this metric. To see this, recall the right action of Γ_I on the space of functions (Eq. 4.13), this time restricted to $\mathscr{F}_0$, given by $\mathscr{F}_0 \times \Gamma_I \to \mathscr{F}_0$, $(f, \gamma) = f \circ \gamma$. For a $\gamma \in \Gamma_I$, define the mapping $L_\gamma : \mathscr{F}_0 \to \mathscr{F}_0$ by $L_\gamma(f) = (f, \gamma) = f \circ \gamma$. Since L_γ is a linear map, its differential $L_\gamma^* : T_f(\mathscr{F}_0) \to T_{f\circ\gamma}(\mathscr{F}_0)$ is the same as L_γ. That is, for any $v \in T_f(\mathscr{F})$, $L_\gamma^*(v) = v \circ \gamma$.

Lemma 4.6. *The action of Γ_I on $\mathscr{F}_0$, endowed with the Fisher-Rao metric, is by isometries.*

Proof. We need to show that $\langle\langle v_1, v_2 \rangle\rangle_f = \langle\langle L_\gamma^*(v_1), L_\gamma^*(v_2) \rangle\rangle_{f\circ\gamma}$. Starting with the R.H.S.:

$$\langle\langle L_\gamma^*(v_1), L_\gamma^*(v_2) \rangle\rangle_{f\circ\gamma} = \langle\langle L_\gamma(v_1), L_\gamma(v_2) \rangle\rangle_{f\circ\gamma}$$

$$= \frac{1}{4} \int_0^1 (\frac{d}{dt} v_1(\gamma(t)))(\frac{d}{dt} v_2(\gamma(t))) \frac{1}{\dot{f}(\gamma(t))\dot{\gamma}(t)} dt$$

$$= \frac{1}{4} \int_0^1 \dot{v}_1(\gamma(t))\dot{v}_2(\gamma(t))\dot{\gamma}(t) \frac{1}{\dot{f}(\gamma(t))} dt$$

$$= \langle\langle v_1, v_2 \rangle\rangle_f \ . \square$$

The Fisher-Rao metric is intimately connected with the SRSF framework introduced in the previous section. In fact, the Fisher-Rao metric can be viewed as a special case of the SRSF framework when restricted to $\mathscr{F}_0$, or, in other words, SRSF representation endowed with the $\mathbb{L}^2$ metric, is a generalization of the Fisher-Rao metric to $\mathscr{F}$. To see this connection, we establish the following results.

Lemma 4.7. *Under the SRSF representation, the Fisher-Rao Riemannian metric on $\mathscr{F}_0$ becomes the standard $\mathbb{L}^2$ metric.*

Proof. As illustrated in Fig. 4.26, the mapping from f to q is as follows: $f(t) \overset{\frac{d}{dt}}{\to} \dot{f}(t) \overset{Q}{\to} q(t)$. For any $v \in T_f(\mathscr{F})$, the differential of this mapping is $v(t) \overset{\frac{d}{dt}}{\to} \dot{v}(t) \overset{Q_{*,\dot{f}(t)}}{\to} w(t)$. To evaluate the expression for w, we need the expression for Q_*. Since $x > 0$ in this context, we have $Q(x) = \sqrt{x}$ and its directional derivative

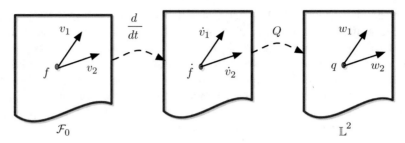

Fig. 4.26 Mapping from the function space $\mathscr{F}_0$ to $\mathbb{L}^2$ in two steps

in the direction of $y \in \mathbb{R}$ is $y/(2\sqrt{x})$. Now, to apply this result to our situation, consider two tangent vectors $v_1, v_2 \in T_f(\mathscr{F})$ and define their mappings under Q_* as $w_i(t) \equiv Q_{*,f(t)}(\dot{v}_i(t)) = \dot{v}_i(t)/(2\sqrt{\dot{f}(t)})$. Taking the $\mathbb{L}^2$ inner product between the resulting tangent vectors, we get $\langle w_1(t), w_2(t) \rangle = \int_0^1 w_1(t)w_2(t)dt = \frac{1}{4}\int_0^1 \dot{v}_1(t)\dot{v}_2(t)\frac{1}{\dot{f}(t)}dt$. The R.H.S. is compared with the expression in Definition 4.8 to complete the proof. □

We can take this relationship further and write geodesic distances under the Fisher-Rao metric explicitly. By Definition 3.5, the geodesic distance d_{FR} is given by:

$$d_{FR}(f_1, f_2) = \inf_{\{\alpha:[0,1] \mapsto \mathscr{F} | \alpha(0)=f_1, \alpha(1)=f_2\}} L[\alpha], \qquad (4.25)$$

where $L[\alpha] = \int_0^1 \sqrt{\langle\langle \dot{\alpha}(\tau), \dot{\alpha}(\tau) \rangle\rangle_{\alpha(\tau)}}d\tau$. In other words, among all differentiable paths in $\mathscr{F}$, going from f_1 to f_2, we have to find one with the shortest length, with the length measured using the Fisher-Rao metric. Given the nature of Fisher-Rao metric, this is a difficult task. Since this metric changes from point to point (note the dependence of $\langle\langle v_1, v_2 \rangle\rangle_f$ on f in Definition 4.8), it seems difficult to reach an explicit expression for the geodesic path or the geodesic distance under this metric, directly in $\mathscr{F}$. The minimization stated above is nontrivial and only some numerical algorithms are known to attempt this problem. However, using Lemma 4.7, this can be done as follows: Simply compute the $\mathbb{L}^2$ distance between the corresponding SRSFs and set d_{FR} to the value $d_{FR}(f_1, f_2) = \|q_1 - q_2\|$. Also, a geodesic path between q_1 and q_2 is a straight line: $\alpha(\tau)(t) = (1-\tau)q_1(t) + \tau q_2(t)$. For any τ, we can map the SRSF $\alpha(\tau)$ back to the function space using integration $f(\tau)(t) = \int_0^t \alpha(\tau)(s)|\alpha(\tau)(s)| \, ds$ and obtain a geodesic in $\mathscr{F}$ under the Fisher-Rao metric. This is a tremendous simplification over the minimization problem posed in Eq. 4.25.

Remark 4.1. In summary, the Fisher-Rao metric is defined only on a subset $\mathscr{F}_0 \subset \mathscr{F}$ but we can extend it to the larger space $\mathscr{F}$ using SRSF representation endowed with the $\mathbb{L}^2$ metric. We will loosely call this framework the Fisher-Rao framework, and the $\mathbb{L}^2$ metric on SRSF representation space (the full $\mathbb{L}^2$ space) as the **extended Fisher-Rao metric.**)

4.10 Phase and Amplitude Distances

Now that we have defined the notions of phase and amplitude, we seek metrics for quantifying differences between functions based on these properties. These metrics will play an important role in defining and computing summary statistics of function data, and in developing statistical models for handling such data. First, we define a metric to compare the amplitudes of two functions, ignoring their relative phase. (The reader will notice that the concept of amplitude is very similar to the notion of shape studied for 2D curves later, except there are no rotations in case of the functional data.)

4.10.1 Amplitude Space and a Metric Structure

We have defined the orbit of a function f, with SRSF $q \in \mathbb{L}^2$, under $\tilde{\Gamma}$ as the $[q] = \{(q \circ \gamma)\sqrt{\dot{\gamma}} | \gamma \in \tilde{\Gamma}_I\}$, and the amplitude space $\mathscr{A}$ as the quotient space $\mathbb{L}^2/\tilde{\Gamma}_I$. This definition assumes that $\dot{f} \neq 0$ a. e. Since the amplitude of a function is represented by its orbit, a comparison of amplitudes naturally implies a comparison of the corresponding orbits. Stated differently, the comparison should lead to the same answer irrespective of which elements from the two orbits are picked to measure the difference. We would also like to characterize geodesics, or shortest paths, between elements of $\mathscr{A}$, such that lengths of these paths match the chosen distance.

It should be noted that neither the orbits $[q]$ nor the quotient space $\mathscr{A}$ here are differentiable manifolds! So we cannot use the techniques of Riemannian geometry, such as orthogonal sections (Sect. 3.7), to analyze this situation. We can still view $\mathscr{A}$ it as a metric space, with the distance inherited from the larger space $\mathbb{L}^2$. We use Eq. 4.18 to define a proper distance on the quotient space.

Definition 4.9 (Amplitude Distance). For any two functions f_1, $f_2 \in \mathscr{F}$ and the corresponding SRSFs, $q_1, q_2 \in \mathbb{L}^2$, we define the amplitude distance d_a to be:

$$d_a([q_1], [q_2]) = \inf_{\gamma_1, \gamma_2 \in \tilde{\Gamma}} (\|(q_1, \gamma_1) - (q_2, \gamma_2)\|) , \qquad (4.26)$$

which makes $\mathscr{A} \equiv \mathbb{L}^2/\sim$ a metric space.

Lemma 4.8. *The distance d_a is a proper distance on $\mathscr{A}$.*

Proof. We need to establish the following properties of d_a.

1. **Symmetry**: The symmetry of d_a comes directly from the symmetry of $\mathbb{L}^2$ norm.
2. **Positive Definite**: For positive definiteness, we need to show that $d_a([q_1], [q_2]) = 0 \Rightarrow [q_1] = [q_2]$. Suppose that $d_a([q_1], [q_2]) = 0$. By definition, it follows immediately that for all $\epsilon > 0$, there exists a $\gamma \in \Gamma_I$ such that $\|q_1 - (q_2, \gamma)\| < \epsilon$. From this, it follows that q_1 is in the orbit of q_2. Since we are assuming that orbits are closed, it follows that $q_1 \in [q_2]$, so $[q_1] = [q_2]$.
3. **Triangle Inequality**: To establish the triangle inequality, we need to prove $d_a([q_1], [q_3]) \leq d_a([q_1], [q_2]) + d_a([q_2], [q_3])$, for any $q_1, q_2, q_3 \in \mathbb{L}^2$. Seeking contradiction, suppose that $d_a([q_1], [q_3]) > d_a([q_1], [q_2]) + d([q_2], [q_3])$. Let

$\epsilon = \frac{1}{3}(d_a([q_1],[q_3]) - d_a([q_1],[q_2]) - d_a([q_2],[q_3]))$; by our supposition, $\epsilon > 0$. From the definition of ϵ, it follows that $d_a([q_1],[q_3]) = d_a([q_1],[q_2]) + d_a([q_2],[q_3]) + 3\epsilon$. By the definition of d_a, we can choose $\gamma_1, \gamma_2 \in \tilde{\Gamma}_I$, such that $\|(q_1, \gamma_1) - q_2\| \le d_a([q_1],[q_2]) + \epsilon$ and $\|q_2 - (q_3, \gamma_2)\| \le d_a([q_2],[q_3]) + \epsilon$. Now by the triangle inequality for the $\mathbb{L}^2$-norm, we know that $\|(q_1, \gamma_1) - (q_3, \gamma_2)\| \le \|(q_1, \gamma_1) - q_2\| + \|q_2 - (q_3, \gamma_2)\| \le d_a([q_1],[q_2]) + d_a([q_2],[q_3]) + 2\epsilon$. It follows that $d_a([q_1],[q_3]) \le d_a([q_1],[q_2]) + d_a([q_2],[q_3]) + 2\epsilon$. But this contradicts that fact that $d_a([q_1],[q_3]) = d_a([q_1],[q_2]) + d_a([q_2],[q_3]) + 3\epsilon$. Hence our supposition that $d_a([q_1],[q_3]) > d_a([q_1],[q_2]) + d_a([q_2],[q_3])$ must be false. The triangle inequality follows.   □

A simple consequence of this definition is that for any γ_1, $\gamma_2 \in \Gamma_I$, and q_1, $q_2 \in \mathbb{L}^2$, $d_a([q_1],[q_2]) = d_a([(q_1,\gamma_1)],[(q_2,\gamma_2)])$. We can simplify the computation of d_a, using Eq. 4.19, to obtain:

$$d_a([q_1],[q_2]) = \inf_{\gamma_1, \gamma_2 \in \tilde{\Gamma}_I} \|(q_1,\gamma_1) - (q_2,\gamma_2)\|$$
$$= \inf_{\gamma_1, \gamma_2 \in \Gamma_I} \|(q_1,\gamma_1) - (q_2,\gamma_2)\| = \inf_{\gamma \in \Gamma_I} \|q_1 - (q_2,\gamma)\| .$$

We illustrate this idea using a simple example.

Example 4.2. Consider the two functions shown in Fig. 4.27; the nonzero parts of these functions is actually a single cycle of the sine function. These functions are:

$$f_i(t) = \begin{cases} 0, & 0 \le t < \epsilon_i \\ \sin\left(2\pi \frac{(t-\epsilon_i)}{(1-2\epsilon_i)}\right), & \epsilon_i \le t < 1 - \epsilon_i \\ 0, & 1 - \epsilon_i \ge t \le 1 \end{cases} .$$

In this example, $\epsilon_1 = 0.33$ and $\epsilon_2 = 0.01$. Let's calculate the relative phase, the amplitude distance $d_a(f_1, f_2)$, and the phase distance $d_p(f_1, f_2)$ between them. The optimal alignment comes from:

$$\gamma_1^*(t) = \begin{cases} \frac{\epsilon_2}{\epsilon_1} t, & 0 \le t < \epsilon_1 \\ \epsilon_2 + \frac{(t-\epsilon_1)}{1-2\epsilon_1}(1 - 2\epsilon_2), & \epsilon_1 \le t < 1 - \epsilon_1 \\ (1 - \epsilon_2) + \frac{\epsilon_2}{\epsilon_1}(t - (1 - \epsilon_1)), & 1 - \epsilon_1 \le t \le 1 \end{cases} .$$

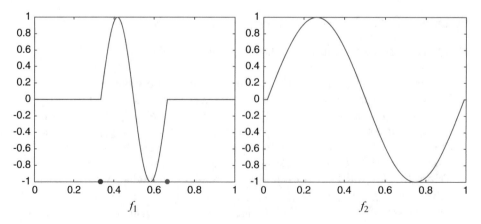

Fig. 4.27 Two functions f_1 and f_2 used in Example 4.2

and $\gamma_2^*(t) = t$. Since $\|(q_1, \gamma_1^*) - (q_2, \gamma_2^*)\| = 0$, i.e., the two functions happen to lie in the same orbit, their amplitude distance $d_a([q_1], [q_2]) = 0$.

We also note that if at least one of the functions f_1 and f_2 is piecewise linear, then there exist elements of $[q_1]$ and $[q_2]$ that achieve the minimum distance between these equivalence classes. If both are piecewise linear, than it is possible to provide an algorithm that calculates these elements [94]. (Bruveris [20] has recently proved that, if f_1 and f_2 are both C^1, then an optimal matching exists. He has also presented a pair of functions $f_1, f_2 : [0, 1] \to \mathbb{R}^2$ such that no matching exists.) For completely general, absolutely continuous functions f_1 and f_2, it is not known when the optimal matching exists. However, this is not too important for practical applications, since we generally only know a finite number of points on functions when analyzing real data.

Even though $\mathscr{A}$ is only a metric space, as opposed to a Riemannian manifold, we can still define the notion of a geodesic in this set in a way that is consistent with the definition for Riemannian manifolds.

Definition 4.10 (Geodesics in $\mathscr{A}$). A *geodesic* in $\mathscr{A}$ is a function $\alpha : [0, 1] \to \mathscr{A}$ with the property that there exists a number $R > 0$ such that, for all $s_0 \in [0, 1]$, there exists $\epsilon > 0$ with the property that for all $s \in [0, 1]$ satisfying $|s - s_0| < \epsilon$, $d_a(\alpha(s), \alpha(s_0)) = R|s - s_0|$.

Essentially, this definition says that a path in $\mathscr{A}$ is a geodesic if it gives the shortest path between any two nearby points on the path and if the speed is constant (in the definition, the speed is the number R).

If q_0 and q_1 are elements of $\mathbb{L}^2$, $\alpha : [0, 1] \to \mathbb{L}^2$ is a constant-speed path satisfying $\alpha(0) \in [q_0]$, $\alpha(1) \in [q_1]$, and the length of the path α in $\mathbb{L}^2$ is equal to $d_a([q_1], [q_2])$, then it's easy to verify that the map from $[0, 1] \to \mathscr{F}/\Gamma_I$ given by $s \mapsto [\alpha(s)]$ is a geodesic in $\mathscr{F}/\Gamma_I$. There are certain technical issues in this formulation that a reader should be aware of. Given $q_1, q_2 \in \mathbb{L}^2$, there does not necessarily exist an element $\gamma \in \Gamma_I$ such that $\|q_1 - (q_2, \gamma)\| = d_a([q_1], [q_2])$. However, it is true that for every $\epsilon > 0$, there exists a $\gamma \in \Gamma_I$ such that $\|q_1 - (q_2, \gamma)\| < d_a([q_1], [q_2]) + \epsilon$. Similarly we cannot generally expect to find actual geodesics in $\mathscr{F}/\Gamma_I$, unless the two functions are piecewise linear; however, we will approximate these geodesics by paths from $[q_1]$ to $[q_2]$ whose lengths approximate the distance $d_a([q_1], [q_2])$.

Now we look carefully at the quantity in Eq. 4.22 and investigate its role as a valid distance. Clearly, it is not a proper distance in $\mathscr{F}$, since two different functions, e.g., any f and $f \circ \gamma$ for any γ, will have zero distance between them. The correct space where this quantity is a distance is a quotient space of $\mathscr{F}$, or equivalently $\mathbb{L}^2$ using the SRSF representation, under the action of Γ_I. Later we will discuss some of the subtleties of this quotient space and its metric.

4.10.2 Phase Space and a Metric Structure

Earlier, in Sect. 4.3.3, we introduced the warping group Γ_I that was used to generate group actions on a function space $\mathscr{F}$. Now we take a careful look at the geometry of Γ_I itself and consider the problem of computing geodesics and geodesic distances between elements of Γ_I under certain chosen metrics. We remind the reader that the tangent space:

$$T_{\gamma_{id}}(\Gamma_I) = \{v : [0,1] \mapsto \mathbb{R} | v(0) = 0, v(1) = 0, \quad v \text{ is smooth}\} \ .$$

In later chapters, we will also be interested in some finite-dimensional approximations of elements of $T_{\gamma_{id}}(\Gamma_I)$. Toward that goal, we will write down some orthonormal bases of this space that can be conveniently truncated for finite-dimensional
approximations. The orthonormality, of course, will depend on the choice of metric.
There are several possibilities for the Riemannian metric:

1. $\mathbb{L}^2$ **Metric**: One obvious choice of metric to compare warping functions is the
   $\mathbb{L}^2$ metric. As in the previous section, we can ask the question: What are the
   geodesics between elements of Γ_I under this metric? The geodesic path under
   the $\mathbb{L}^2$ metric is simply a straight line: for any two points γ_1, $\gamma_2 \in \Gamma_I$ a geodesic
   path is given by: $\alpha(\tau) = (1 - \tau)\gamma_1 + \tau\gamma_2$.
   A complete basis of $T_{\gamma_{id}}(\Gamma_I)$ under the $\mathbb{L}^2$ metric is given by:

$$\left\{ \frac{1}{\sqrt{2}} \sin(\pi nt), n = 1, 2, 3 \ldots \right\} \ . \tag{4.27}$$

2. **First-Order Palais Metric**: Sometimes it is more appropriate to use a metric that involves derivatives rather than the functions themselves. Since the
   Sobolev metrics typically involve both the functions and their derivatives, we
   prefer using the Palais metrics that involve only the derivatives and function
   evaluations at fixed points. The first-order Palais metric is given by:

$$\langle v_1, v_2 \rangle_s = v_1(0)v_2(0) + \int_0^1 \dot{v}_1(t)\dot{v}_2(t) \ dt \ . \tag{4.28}$$

Later on this metric will play a prominent role in deriving a numerical approach for computing geodesics in certain Riemannian manifolds (See Eq. 6.29,
for example). An orthonormal basis of $T_{\gamma_{id}}(\Gamma_I)$ under the Palais metric is given
by:

$$\{\frac{1}{\sqrt{2}\pi n} \sin(2\pi nt), \frac{1}{\sqrt{2}\pi n}(\cos(2\pi nt) - 1)|n = 1, 2, 3 \ldots\} \ . \tag{4.29}$$

Due to the presence of n in the denominators, the contributions from higher-
order harmonics decay naturally and this provides a reason for truncating the
basis at some large value of n.

3. **Fisher-Rao Metric**: The Fisher-Rao metric introduced in Definition 4.8, for
   general functions, can be particularized to Γ_I. For any $v_1, v_2 \in T_\gamma(\Gamma_I)$, it takes
   the form:

$$\langle\langle v_1, v_2 \rangle\rangle_\gamma = \int_0^1 \dot{v}_1(t)\dot{v}_2(t)\frac{1}{\dot{\gamma}(t)}dt \ . \tag{4.30}$$

This is a special case of the Definition 4.8 when restricted to the set Γ_I. How
are the geodesics in Γ_I under the Fisher-Rao metric computed? For a $\gamma \in \Gamma_I$,
the derivative $\dot{\gamma}$ is an element of $\mathscr{P}$ discussed later in Section 4.11.2. Therefore,
we have a natural mapping $\gamma \to \dot{\gamma} \to \sqrt{\dot{\gamma}}$, termed SRSF, from Γ_I to $\mathscr{P}$ to
$\mathscr{Q}$. Once again, the Fisher-Rao metric becomes the $\mathbb{L}^2$ metric on $\mathscr{Q}$, which is
the positive orthant of $\mathbb{S}_\infty$, and the geodesics are simply arcs on great circles
(Eq. 4.9). Each point on this geodesic in $\mathscr{Q}$ can be mapped back to Γ_I to obtain
the desired geodesic. For any $\tau \in [0,1]$, the point $\alpha(\tau) \in \mathscr{Q}$ gets mapped to a
warping function using $\alpha(\tau) \mapsto \gamma(t) = \int_0^t \alpha(\tau)(s)^2 ds$. Figure 4.28 shows some

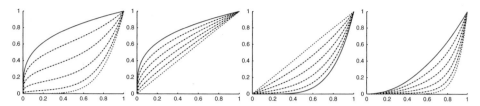

Fig. 4.28 Examples of geodesic paths between warping functions of the type $\gamma(t) = t^a$ for different values of a

examples of pairwise geodesic paths between several warping functions of the type $\gamma(t) = t^a$ for $a = 0.1, 1, 5, 10$.

An orthogonal basis of $T_{\gamma_{id}}(\Gamma_I)$, under the Fisher-Rao metric, can be easily written using the geometry of $\mathbb{S}_\infty$. Since γ_{id} maps to a constant function $\mathbf{1}$, the tangent space $T_{\mathbf{1}}(\mathbb{S}_\infty) = \{w : [0,1] \to \mathbb{R}| w \text{ is smooth }, \int_0^1 w(t)dt = 0\}$. Elements of this space can be mapped back to $T_{\gamma_{id}}(\Gamma_I)$ using the square-integral mapping mentioned above. Thus, any orthonormal basis of $T_{\mathbf{1}}(\mathbb{S}_\infty)$ under the $\mathbb{L}^2$ metric results in, through this mapping, an orthonormal basis for $T_{\gamma_{id}}(\Gamma_I)$.

In order to compare the phase components of two functions, f_1 and f_2, we look at the relative phase between them.

Definition 4.11 (Phase Distance). Let (γ_1^*, γ_2^*) be the relative phase of f_1 w.r.t f_2. Then, define the phase distance between f_1 and f_2 as $d_p(f_1, f_2) = \cos^{-1}(\langle (\dot{\gamma}_1^*)^{1/2}, (\dot{\gamma}_2^*)^{1/2}\rangle)$.

This definition is based on the Fisher-Rao distance on $\widetilde{\Gamma_I}$ measured using SRSFs of the warping functions. It was shown in Sect. 4.10.2 that SRSFs of warping functions on $[0, 1]$ are elements of a unit Hilbert sphere, so the arc length between points on the sphere provides a proper metric for their comparison. The distance d_p is exactly the arc-length distance between SRSFs of γ_1^* and γ_2^*. It is easy to check the following properties of d_p:

1. d_p is zero if and only if $\gamma_1^* = \gamma_2^*$.
2. $d_p(f_1, f_2) = d_p(f_2, f_1)$.
3. For any $a \in \mathbb{R}_+$ and $b \in \mathbb{R}$, we have $d_p(f_1, f_2) = d_p(f_1, af_2 + b)$.
4. The distance d_p satisfies the triangle inequality: $d_p(f_1, f_3) \leq d_p(f_1, f_2) + d_p(f_2, f_3)$, for any $f_1, f_2, f_3 \in \mathscr{F}$.

The proofs of these properties have been left out as exercises to the reader.

Using the approximation presented in Sect. 4.7.3, the Fisher-Rao phase distance can be approximated by $d_p(f_1, f_2) \approx \cos^{-1}(\int_0^1 \sqrt{\dot{\gamma}_{12}^*(t)}dt)$.

Example 4.3. Continuing with the two functions used in Example 4.2, we compute the relative phase γ_{12}^* between them. This warping function is given by:

$$\gamma_{12}^*(t) = \begin{cases} \frac{\epsilon_1}{\epsilon_2}t, & 0 \leq t < \epsilon_2 \\ \epsilon_1 + \frac{(t-\epsilon_2)}{1-2\epsilon_2}(1 - 2\epsilon_1), & \epsilon_2 \leq t < 1 - \epsilon_2 \\ (1 - \epsilon_1) + \frac{\epsilon_1}{\epsilon_2}(t - (1 - \epsilon_2)), & 1 - \epsilon_2 \leq t \leq 1 \end{cases}.$$

This is exactly the inverse of γ_1^* given in that example, and the Fisher-Rao phase distance is given by:

$$d_p(f_1, f_2) = \cos^{-1}(2\sqrt{\epsilon_1 \epsilon_2} + \sqrt{(1 - 2\epsilon_1)(1 - 2\epsilon_2)}) \ .$$

4.11 Different Warping Actions and PDFs

4.11.1 Listing of Different Actions

In the previous section, we have relied heavily on the action of Γ_I (or equivalently of $\widetilde{\Gamma_I}$) on $\mathbb{L}^2$ that preserves the norm of functions. What other actions are possible and what properties do they preserve? Here is a short list:

1. **Value-Preserving Warping**: The classical action of Γ_I on a space of functions, say $\mathbb{L}^2$, is by composition: $(f, \gamma) = f \circ \gamma$. The main property of this action is that no values of $f(t)$ are created or destroyed; they are simply moved from one ordinate to another. In other words, it is not possible to have $f(t_0)$, for some t_0 such that $f(t_0) \neq f(\gamma(t))$ for all $t \in [0, 1]$. This type of warping is most commonly used in matching functions as described earlier in this chapter. In the context of images and higher-dimensional signals, this warping is used to *deform* data by displacing the pixels values.

2. **Norm-Preserving Warping**: Here we have $\|f\| = \|(f, \gamma)\|$ for all $f \in \mathbb{L}^2$ and γ, and where the chosen norm is $\mathbb{L}^2$. The action of Γ_I on the space of SRVFs falls in this category. Accordingly, an example of a norm-preserving action is $(f, \gamma) = (f \circ \gamma)\sqrt{\dot{\gamma}}$. In fact, one can define an operator on the function space as $D_\gamma : \mathbb{L}^2 \to \mathbb{L}^2$, according to $D_\gamma(f) = (f \circ \gamma)\sqrt{\dot{\gamma}}$. D_γ is unitary operator and can be interpreted as a *rotation* in the Hilbert space $\mathbb{L}^2$. But it is not onto if $\dot{\gamma} = 0$ on a set of positive measure. Note that $D_\gamma(a_1 f_1 + a_2 f_2) = a_1 D_\gamma(f_1) + a_2 D_\gamma(f_2)$. This interpretation will be useful later on in shape analysis where both Euclidean rotations and warping will both play a role in registration of shapes.

3. **Area-Preserving Warping**: Another possibility is to preserve the area below the curve, $\int_0^1 f(t)dt$, rather than the function values or the $\mathbb{L}^2$ norm. Here, one is interested in defining an action (f, γ) such that $\int_0^1 (f, \gamma)(t)dt = \int_0^1 f(t)dt$. It can be verified that the action $(f, \gamma) = (f \circ \gamma)\dot{\gamma}$ satisfies this property. Since probability density functions are characterized by unit area, among other properties, we will use this action in analyzing probability density functions in the next section.

We summarize different actions of Γ_I on different Riemannian manifolds and the corresponding Fisher-Rao metric that we have seen in this chapter. This table is only for the restricted set $\mathscr{F}_0$ in order to be able to define the first two columns. The third column is extendable to the full $\mathbb{L}^2$, but with more general definition of SRSF.

	Case 1 (function)	Case 2 (density)	Case 3 (SRSF)
Function space	Absolutely continuous $\mathscr{F}_0$	Integrable $\mathbb{L}^1$	Square integrable $\mathbb{L}^2$
	f	$g = \dot{f}$	$q = \sqrt{g} = \sqrt{\dot{f}}$
Group action	$(f, \gamma) = f \circ \gamma$	$(g, \gamma) = (g \circ \gamma)\dot{\gamma}$	$(q, \gamma) = (q \circ \gamma)\sqrt{\dot{\gamma}}$
Preservation	Value preserving	Area preserving	Norm preserving
Fisher-Rao metric	$\langle v_1, v_2 \rangle_f = \int_0^1 \dot{v}_1(t)\dot{v}_2(t)\frac{1}{\dot{f}(t)}dt$	$\langle u_1, u_2 \rangle_f = \int_0^1 v_1(t)v_2(t)\frac{1}{g(t)}dt$	$\langle w_1, w_2 \rangle_f = \int_0^1 w_1(t)w_2(t)dt$
Fisher-Rao distance	Difficult to compute directly	Difficult to compute directly	$\|q_1 - q_2\|$

4.11.2 Probability Density Functions

A very interesting and important application of Fisher-Rao Riemannian metric is in a (*nonparametric*) analysis of probability densities. Consider the set of probability density functions on the interval $[0, 1]$:

$$\mathscr{P} = \{g : [0, 1] \mapsto \mathbb{R}_{\geq 0}| \int_0^1 g(t)dt = 1\} \, .$$

Geometry of $\mathscr{P}$: To prove that $\mathscr{P}$ is a manifold, consider the mapping $\phi : \mathbb{L}^1([0, 1], \mathbb{R}) \to \mathbb{R}_{\geq 0}$ given by $\phi(g) = \int_0^1 g(x)dx$. The differential of ϕ in the direction of $h \in T_g(\mathbb{L}^1([0, 1], \mathbb{R}))$ is given by $d\phi(h) = \int_0^1 h(x)dx$. Since this derivative is a surjective linear transformation, the inverse image $\phi^{-1}(1) = \mathscr{P}$ is a submanifold of $\mathbb{L}^1([0, 1], \mathbb{R})$. We point out that $\mathbb{L}^1([0, 1], \mathbb{R})$ is a Banach space and not a Hilbert space and, thus, $\mathscr{P}$ is a Banach manifold. Another interesting aspect of $\mathscr{P}$ is that it is a manifold with boundary. The constraint that $g(x) \geq 0$ for all $x \in [0, 1]$ implies that if a probability density g has at least one zero, then that g is an element of the boundary $\partial \mathscr{P}$.

We would like a metric on $\mathscr{P}$ for use in comparing probability densities using geodesic distances. In the past, there have been several quantities proposed for comparing densities but none of them are quite satisfactory in terms of their use in a statistical analysis on $\mathscr{P}$. For instance, the Kullback-Leibler divergence is frequently used to quantify differences between densities but it is not a proper distance. In fact, it is not even symmetric. Similarly, although one can use $\mathbb{L}^1$ or $\mathbb{L}^2$ for measuring differences between density functions, it is still less than satisfactory in the following sense. The action of Γ_I is not by isometries under any of these metrics. While the literature on differential geometry and Riemannian analysis of $\mathscr{P}$ is rather rich (see, e.g., [5]), there seems a big disconnect between a "differential geometric viewpoint" and traditional statistical methods. In particular, a surprisingly large fraction of techniques in statistics, especially in estimation, modeling, and inferences, have been developed without any consideration of a metric or a distance on $\mathscr{P}$. We will show that the Fisher-Rao metric defined in the previous section applies to this problem with a simple change of variable. The tangent space $T_g(\mathscr{P})$ is given by:

$$T_g(\mathscr{P}) = \{w \in \mathbb{L}^1([0, 1], \mathbb{R})| \int_0^1 w(t)dt = 0\} \, .$$

Riemannian Structure: We will use the restriction of the Fisher-Rao metric to this case. For this purpose, we restrict to the set of non-negative functions: $\mathscr{P}_0 = \{g \in \mathscr{P} | g > 0\}$.

Definition 4.12 (Nonparametric Fisher-Rao Metric for Densities). For a $g \in \mathscr{P}_0$ and vectors $v_1, v_2 \in T_g(\mathscr{P})$, the Fisher-Rao metric is defined to be:

$$((w_1, w_2))_g = \int_0^1 w_1(t) w_2(t) \frac{1}{g(t)} dt . \tag{4.31}$$

In order to compare this definition with the earlier definition involving general functions (Definition 4.8), we note that g here is equivalent to $\dot{f}$ and, hence, $w_i = \dot{v}_i$.

When restricted to a parametric family, the Fisher-Rao metric provides a quantification of "information" present in the given data and helps impose a lower bound on the expected error associated with any parameter estimation procedure. We describe this connection briefly. Let $\mathscr{P}_e$ be the subset of $\mathscr{P}_0$ containing a specific parametric family of interest:

$$\mathscr{P}_e = \{g_\theta \in \mathscr{P} | \theta \in \mathbb{R}^n\} .$$

Here $g_\theta(t)$ is a functional form that depends on $\theta \in \mathbb{R}^n$ in a smooth way, and for each $\theta \in \mathbb{R}^n$, we have $\int_0^1 g_\theta(t) dt = 1$. In estimation theory, one is interested in estimating θ using observations of t, denoted by $t_1, t_2, \ldots, t_n$, under the maximum likelihood framework according to $\hat{\theta} = \text{argmax}_{\theta \in \Theta} \prod_i g_\theta(t_i)$. One way to analyze the performance of this or any other estimator of θ is to compare the expected squared error with the **Cramer-Rao error bound** given by:

$$E[|(\hat{\theta} - \theta)(\hat{\theta} - \theta)^T|] \geq J^{-1}(\theta),$$

where $J(\theta)$ is an $n \times n$ matrix with entries given by:

$$J_{ij}(\theta) = \int_0^1 \left(\frac{\partial}{\partial \theta_i} \log(g_\theta(t)) \right) \left(\frac{\partial}{\partial \theta_j} \log(g_\theta(t)) \right) g_\theta(t) dt .$$

This matrix $J(\theta)$ is called the **Fisher information matrix**. Using the fact that $\frac{1}{g_\theta(t)} \frac{\partial g_\theta(t)}{\partial \theta_i} = \frac{\partial \log(g_\theta(t))}{\partial \theta_i}$, $J(\theta)$ can also be written as:

$$J_{ij}(\theta) = \int_0^1 \left(\frac{\partial g_\theta(t)}{\partial \theta_i} \right) \left(\frac{\partial g_\theta(t)}{\partial \theta_j} \right) \frac{1}{g_\theta(t)} dt . \tag{4.32}$$

Note that the conditions assumed here work for the exponential families often studied in an information geometry.

What is the restriction of the nonparametric Fisher-Rao metric to the parametric family $\mathscr{P}_e$? The tangent vectors in this case are given by the partial derivatives: $w_i = \frac{\partial}{\partial \theta_i} g_\theta$. In other words, these partial derivatives form a basis of the tangent space of $\mathscr{P}_0$ at g_θ. Since any other tangent vector can be expressed as a linear combination to these basis elements, it suffices to define the Riemannian metric on these elements.

Definition 4.13 (Parametric Fisher-Rao Metric for Densities). Expressing the Fisher-Rao metric given in Eq. 4.31 in the parametric terms, we get:

$$\left\langle \frac{\partial}{\partial \theta_i} g_\theta, \frac{\partial}{\partial \theta_j} g_\theta \right\rangle_{g_\theta} = \int_0^1 (\frac{\partial}{\partial \theta_i} g_\theta(t))(\frac{\partial}{\partial \theta_j} g_\theta(t)) \frac{1}{g_\theta(t)} \, dt \ . \tag{4.33}$$

Comparing this with Eq. 4.32, we can see that this is nothing but the i, j^{th} element of the Fisher information matrix. Due to this connection, the underlying Riemannian metric carries the name Fisher-Rao!

Remark 4.2. The computation of geodesic paths and geodesic distances under the parametric Fisher-Rao metric, i.e., when restricted to parametric families of probability density functions, is quite difficult. This is in contrast to the nonparametric case where the expressions for geodesic paths and geodesic distances are readily available.

Another interesting aspect of this metric is its relationship with the commonly used Kullback-Leibler (K-L) divergence. For any two densities $g_1, g_2 \in \mathscr{P}_0$, the K-L divergence is defined to be:

$$KL(g_1 \| g_2) = \int_0^1 g_1(t) \log(\frac{g_1(t)}{g_2(t)}) dt \ .$$

It can be shown that, infinitesimally, KL divergence is equivalent to the Fisher-Rao metric.

Proposition 4.1. *Let $g \in \mathscr{P}_0$ and $v \in T_g(\mathscr{P}_0)$ and $((\cdot, \cdot))_g$ denote the nonparametric Fisher-Rao metric on $\mathscr{P}_0$ (Eq. 4.31). Then, we have:*

$$\lim_{\epsilon \to 0} \left(\frac{KL(g \| g + \epsilon v)}{((\epsilon v, \epsilon v))_g} \right) = \frac{1}{2} \ .$$

Proof. The Kullback-Leibler divergence between g and a perturbation $g + \epsilon v$, is given by:

$$\begin{aligned}
KL(g \| g + \epsilon v) &= \int_0^1 g(t) \log(\frac{g(t)}{g(t) + \epsilon v(t)}) dt \\
&= -\int_0^1 g(t) \log(\frac{g(t) + \epsilon v(t)}{g(t)}) dt \\
&= -\int_0^1 g(t) \left(\frac{\epsilon v(t)}{g(t)} - \frac{\epsilon^2}{2} \frac{v(t)^2}{g(t)^2} + \frac{\epsilon^3}{3} \frac{v(t)^3}{g(t)^3} + O(\epsilon^4) \right) dt \\
&= \epsilon^2 \int_0^1 \left(\frac{v(t)^2}{2g(t)} - \epsilon \frac{v(t)^3}{3g^2(t)} + O(\epsilon^2) \right) dt \ .
\end{aligned}$$

Here we have used the Taylor's expansion of $\log(1+x) = (x - \frac{x^2}{2} + \frac{x^3}{3} - \dots)$, and the fact that $\int v(t) dt = 0$. On the other hand, the Fisher-Rao norm of the tangent vector ϵv at g is $((\epsilon v, \epsilon v))_g = \epsilon^2 \int_0^1 v(t)^2 \frac{1}{g(t)} \, dt$. Therefore, the ratio:

$$\frac{KL(g \| g + \epsilon v)}{((\epsilon v, \epsilon v))_g} = \frac{1}{2} - \epsilon \left(\frac{\int_0^1 (v(t)^3 / 2g^2(t)) dt}{\int_0^1 (v(t)^2 / g(t)) dt} \right) + \dots$$

In the limit, the ratio converges to $\frac{1}{2}$. $\square$

This infinitesimal relationship further underscores the importance of Fisher-Rao metric for use in analysis of probability density functions.

Similar to the functional analysis studied in the last section, a convenient way to study the Riemannian structure of $\mathscr{P}$ is through the square-root representation. Let $q(t) = \sqrt{g(t)}$ denote the point-by-point, positive square root of a probability density function $g(t)$. Since g integrates to one, we have $\int_0^1 q(t)^2 dt = \int_0^1 g(t) dt = 1$. That is, q is an element of the positive orthant $\mathbb{S}_\infty^+$ of the infinite-dimensional sphere $\mathbb{S}_\infty$ introduced in Example A.11 and studied earlier in Sect. 4.3.2. This mapping identifies $\mathscr{P}$ with the positive orthant of $\mathbb{S}_\infty$, including the boundaries. The function q is also referred to as the **half density** of g since its square is a full probability density. For any half density q, we can easily obtain the probability density function using $g(t) = q(t)^2$. Let $\mathscr{Q} \subset \mathbb{S}_\infty^+$ be the set of all square-root forms or half-densities of probability density functions on $[0, 1]$. If a function q is zero for some subset of the domain, then it lies in the boundary of $\mathscr{Q}$.

Lemma 4.9. *The Fisher-Rao metric for probability densities transforms to the* $\mathbb{L}^2$ *metric under the square-root mapping, up to a constant.*

Proof. This is analogous to Lemma 4.7 for general functions. Since $q(t) = \sqrt{g(t)}$, a tangent vector w in $T_q(\mathscr{Q})$ is related to the corresponding vector v in $T_g(\mathscr{P})$ by $v(t) = 2\sqrt{g(t)}w(t)$. In the new coordinates, the Fisher-Rao metric becomes:

$$((v_1, v_2))_g = \int_0^1 v_1(t) v_2(t) \frac{1}{g(t)} dx = 4 \int_0^1 \sqrt{g(t)} w_1(t) \sqrt{g(t)} w_2(t) \frac{1}{g(t)} dt$$

$$\propto \int_0^1 w_1(t) w_2(t) dt = \langle w_1, w_2 \rangle_{\mathscr{Q}} , \tag{4.34}$$

the $\mathbb{L}^2$ metric in $\mathscr{Q}$. $\square$

Once again this is an important result since, for the $\mathbb{L}^2$ structure on $\mathscr{Q}$, we already have expressions for computing geodesics, exponential maps, etc., for probability density functions, and these can be used for a direct Riemannian analysis of probability density functions. (Furthermore, in case a probability density function has zero values, this does not pose any problem in this analysis.) For example, if we have two probability density functions g_1 and g_2 on the domain $[0, 1]$ and we want to compute the Riemannian distance, i.e., geodesic length distance, between them, we can do so simply using:

$$d(g_1, g_2) = \cos^{-1} \left(\int_0^1 \sqrt{g_1(t)} \sqrt{g_2(t)} dt \right) . \tag{4.35}$$

Also, the geodesic path between them can also be computed easily using Eq. 4.9. Figure 4.29 shows two examples of computing geodesic paths between elements of $\mathscr{P}$ under the Fisher-Rao metric. First, we compute their half-densities using element-wise square root, use Eq. 4.9 to compute the great circle $\alpha(\tau)$ connecting the half-densities in $\mathscr{Q}$, and finally compute the squares of the points along the great circle to get the geodesic in $\mathscr{P}$. For each path, we show nine equally spaced points $\{\alpha(\tau), \tau = 0, 1/9, \ldots, 1\}$ in vertical fashion from bottom to top.

Note that the use of the $\mathbb{L}^2$ metric as the Riemannian metric does not imply that the geodesic distances between points in $\mathscr{Q}$ are given by $\mathbb{L}^2$ norms of their differences. This is because $\mathscr{Q}$ is not all of $\mathbb{L}^2$ but only a subset of it. The geodesic

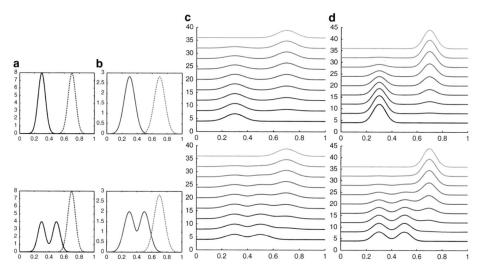

Fig. 4.29 Two examples of geodesics under Fisher-Rao metric. In each row: (**a**) density functions $g_1, g_2 \in \mathscr{P}$, (**b**) the corresponding half-densities $q_1, q_2 \in \mathscr{Q}$, (**c**) the geodesic path between q_1 and q_2 in $\mathscr{Q}$, and (**d**) the corresponding path between g_1 and g_2 in $\mathscr{P}$

distance between any two points is given by the length of the shortest arc connecting them, while the latter is simply the chord length. Incidentally, the chord-length distance between any two elements of $\mathscr{Q}$ is known as the **Hellinger distance**:

$$d_h(g_1, g_2) = \int_0^1 \| \sqrt{g_1(x)} - \sqrt{g_2(x)} \|^2 dx \ .$$

The square-root mapping from $\mathscr{P}$ to $\mathscr{Q}$ was helpful in two ways: (i) the invariant Fisher-Rao metric on $\mathscr{P}$ becomes the $\mathbb{L}^2$ metric on $\mathscr{Q}$ and (ii) $\mathscr{Q}$ is a positive orthant of a unit sphere with known differential geometry. Together these two items enable us to write down geodesics between probability densities in a rather simple fashion.

Area-Preserving Warping : Consider the warping group Γ_I introduced in the last section. This group acts on $\mathscr{P}$ in the following way (note the difference of this action from its action on $\mathscr{F}$ used in the previous section):

$$\mathscr{P} \times \Gamma_I \to \mathscr{P}, \quad (g, \gamma) = (g \circ \gamma)\dot{\gamma} \ . \tag{4.36}$$

To show that this is a group action, we only point out that $\int_0^1 g(\gamma(t))\dot{\gamma}(t)dt = \int_0^1 g(t)dt = 1$. The remaining steps are left as an exercise. It is easy to establish that the action of Γ_I on $\mathscr{P}$ under the Fisher-Rao metric is by isometries.

Lemma 4.10. *The Fisher-Rao metric, given in Eq. 4.31, is invariant under area-preserving warping, i.e.:*

$$((\tilde{v}_1, \tilde{v}_2))_{\tilde{g}} = \langle v_1, v_2 \rangle_g \ ,$$

where $\tilde{v}_i = (v_i \circ \gamma)\dot{\gamma}$ and $\tilde{g} = (g \circ \gamma)\dot{\gamma}$.

The proof is similar to that of Lemma 4.6 and left as an exercise to the reader.

4.12 Exercises

4.12.1 Theoretical Exercises

1. Derive the expression for the $R \in \mathbb{R}^{K \times K}$ matrix in penalized least-squares estimation of f (Eq. 4.3) when the penalty term is given by $\int_0^1 \frac{d^4 f(t)}{dt^4} \, dt$.
2. By deriving an expression for $(\frac{d}{d\tau} \alpha(\tau))|_{\tau=0}$, for α given in Eq. 4.9, verify the formula for inverse exponential map on S_∞ as given in Eq. 4.12.
3. Prove that Γ_I, the set of all boundary-preserving diffeomorphisms from $[0,1]$ to itself, forms a group under composition. Additionally, if we define $\gamma_a(t) = te^a/(1 - t + te^a)$, then the set $\{\gamma_a | a \in R\}$ forms a subgroup of Γ_I.
4. Let $\mathscr{F} = \{f : [0,1] \to \mathbb{R}\}$ and Γ_I be the group of boundary-preserving diffeomorphisms of $[0,1]$, as above. Show that the mapping $(f, \gamma) \mapsto f \circ \gamma$ defines an action of Γ_I on $\mathscr{F}$. (The notion of a group action is defined in Sect. 3.5.)
5. Find two functions f_1 and f_2 in $\mathbb{L}^2$, such that $\inf_{\gamma \in \Gamma_I} \|f_1 - f_2 \circ \gamma\|$ is different from $\inf_{\gamma \in \Gamma_I} \|f_2 - f_1 \circ \gamma\|$. (One such pair is presented in Example 4.1.)
6. Prove that the property of invariance to identical warping implies the property of inverse symmetry. That is, assume that for a certain cost function E we have $E[f_1, f_2] = E[f_1 \circ \gamma, f_2 \circ \gamma]$ for a pair f_1, f_2, and all $\gamma \in \Gamma_I$. Then, let $\gamma^* = \text{argmin}_{\gamma \in \Gamma_I} E[f_1, f_2 \circ \gamma]$ and show that $\gamma^{*-1} \in \text{argmin}_{\gamma \in \Gamma_I} E[f_1 \circ \gamma, f_2]$.
7. Let $\mathscr{F}$ be the set of absolutely continuous functions on $[0,1]$. Show that $\mathscr{F}$ is a vector space.
8. Define a parametric family of warping functions given by $\gamma_a(t) = te^a/(1 - t + te^a)$, for $a \in R$. First, show that $\gamma_a \in \Gamma_I$, the set of all boundary-preserving group of diffeomorphisms from $[0,1]$ to itself, for all $a > 0$. Show that the limit $\lim_{a \to 0} \gamma_a(t)$ is not in Γ_I. This helps establish the fact that Γ_I is not a closed set under the usual metric.

9. a. Let f be a real-valued function on $[0,1]$ that is absolutely continuous and has $\dot{f} > 0$. Show that the resulting SRSF $q(t) = \text{sign}(\dot{f}(t))\sqrt{|\dot{f}(t)|}$ is square integrable.
   b. Now prove the same result without assuming that $\dot{f} > 0$.

10. Calculate and display the SRSFs of the functions shown in Fig. 4.30.
11. For any two smooth and positive functions $q_1, q_2 \in \mathbb{L}^2$, show that there exists a $\gamma^* \in \Gamma_I$ such that $(q_1, \gamma^*) = cq_2$ for some constant c. Prove that this γ^* lies in the set $\text{argmin}_{\gamma \in \Gamma_I} \|q_1 - (q_2, \gamma)\|$.
12. Let $q_1, q_2 \in \mathbb{L}^2$ be any two functions. Then, for any constant $c > 0$, show that if $\text{argmin}\|q_1 - (q_2, \gamma)\|$ exists, then $\text{argmin}_{\gamma \in \Gamma_I} \|cq_1 - (q_2, \gamma)\| = \text{argmin}_{\gamma \in \Gamma_I} \|q_1 - (q_2, \gamma)\|$.

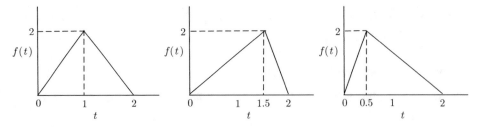

Fig. 4.30 Compute the SRSFs of these functions

13. For any two SRSFs q_1 and q_2, let $\gamma^* \in \Gamma_I$ be a solution of the problem:

$$\gamma^* = \operatorname*{arginf}_{\gamma \in \Gamma_I} \|q_1 - (q_2, \gamma)\| .$$

Then, prove that, for any $\gamma_0 \in \Gamma_I$, we have

$$(\gamma^* \circ \gamma_0) \in \operatorname*{arginf}_{\gamma \in \Gamma_I} \|(q_1, \gamma_0) - (q_2, \gamma)\| .$$

14. For any two $q_1, q_2 \in \mathbb{L}^2$, show that:

$$\inf_{\gamma_1, \gamma_2 \in \tilde{\Gamma}_I} \|(q_1, \gamma_1) - (q_2, \gamma_2)\| = \inf_{\gamma \in \tilde{\Gamma}} \|q_1 - (q_2, \gamma)\| .$$

15. If f_1 and f_2 are smooth functions $[0, 1] \to \mathbb{R}$ with both $\dot{f}_1 > 0$ and $\dot{f}_2 > 0$ on $[0, 1,$ and their relative phase is (γ_1^*, γ_2^*), then show that $f_1 \circ \gamma_1^*$ is a scalar multiple of $f_2 \circ \gamma_2^*$ plus a constant. (In other words, their derivatives are a scalar multiple of each other.)

16. Show that the mapping $\mathbb{L}^2 \times \Gamma_I \to \mathbb{L}^2$ given by $(q, \gamma) = (q \circ \gamma)\sqrt{\dot{\gamma}}$ defines an action of Γ_I on $\mathbb{L}^2$.

17. Define a relation $\sim$ on $\mathbb{L}^2$ as follows. Set $q_1 \sim q_2$ if and only if q_1 is an element of the $\mathbb{L}^2$ closure of $[q_2]$. Show that this is an equivalence relation.

18. Show that the following relation is an equivalence relation: for any two functions f and g in $\mathscr{F}$, with $\mathscr{F}$ being the set of absolutely continuous functions, define them to be related if the closures of their orbits $[f] = f\Gamma_I$ and $[g] = g\Gamma_I$ are identical.

19. Show that if two functions belong to two different equivalence classes in the previous problem, then, under d_a, the distance defined on the quotient space $\mathscr{F}/\Gamma_I$, their distance is strictly greater than zero.

20. Show that the set $\widetilde{\Gamma_I}$ as defined in Definition 4.3 is a monoid. That is, elements of this set satisfy all other properties of a group except that some elements may not have an inverse.

21. For any two functions $f_1, f_2 \in \mathscr{F}$, if there exists a $\gamma \in \Gamma_I$ such that $f_1 = f_2 \circ \gamma$, then show that $[f_1] = [f_2]$.

22. For the relative phase defined in Eq. 4.18, verify the following properties:

    a. The relative phase of f_1 w.r.t f_2 is exactly the inverse of the relative phase of f_2 w.r.t f_1, i.e., the relative phase of f_2 w.r.t f_1 is given by (γ_2^*, γ_1^*).
    b. Two functions are said to have zero phase between them if their relative phase is $(\gamma_{id}, \gamma_{id})$. Show that this is the case when $f_1 = f_2$.
    c. The relative phase of f_1 w.r.t $f_2 \circ \gamma$, for any $\gamma \in \Gamma_I$, is $(\gamma_1^*, \gamma^{-1} \circ \gamma_2^*)$. Similarly, the relative phase of $f_1 \circ \gamma$ w.r.t f_2 is $(\gamma^{-1} \circ \gamma_1^*, \gamma_2^*)$. Combining these results, the relative phase of $f_1 \circ \gamma_a$ w.r.t $f_2 \circ \gamma_b$ is $(\gamma_a^{-1} \circ \gamma_1^*, \gamma_b^{-1} \circ \gamma_2^*)$.
    d. As a corollary to the last property, the relative phase of $f_1 \circ \gamma$ w.r.t $f_2 \circ \gamma$, for any $\gamma \in \Gamma_I$, is $(\gamma^{-1} \circ \gamma_1^*, \gamma^{-1} \circ \gamma_2^*)$.
    e. The relative phase between any two functions is unchanged by vertical translations and scalar multiplications. That is, the relative phase of $(c_1 f_1 + e_1)$ relative to $(c_2 f_2 + e_2)$ is also (γ_1^*, γ_2^*), where $c_1, c_2 \in \mathbb{R}_+$ and $e_1, e_2 \in \mathbb{R}$.

23. For the amplitude distance d_a defined in Eq. 4.26, and denoting $d_a([q_1], [q_2])$ by $d_a(f_1, f_2)$, prove the following properties:

    a. $d_a(f_1, f_2) = d_a(f_2, f_1)$ and $d_a(f_1, f_3) \leq d_a(f_1, f_2) + d_a(f_2, f_3)$ for all f_1, f_2, $f_3 \in \mathscr{F}$.
    b. For any two f_1, $f_2 \in [f]$ for any $f \in \mathscr{F}$, $d_a(f_1, f_2) = 0$.
    c. For any γ_1, $\gamma_2 \in \Gamma_I$ and f_1, $f_2 \in \mathscr{F}$, $d_a(f_1, f_2) = d_a(f_1 \circ \gamma_1, f_2 \circ \gamma_2)$.
    d. The distance d_a is a proper distance on the set of amplitudes $\mathscr{A}$.

24. For the phase distance d_p as defined in this chapter, prove that:

    a. d_p is zero if and only if $\gamma_1^* = \gamma_2^*$.
    b. $d_p(f_1, f_2) = d_p(f_2, f_1)$.
    c. For any $a \in \mathbb{R}_+$ and $b \in \mathbb{R}$, we have $d_p(f_1, f_2) = d_p(f_1, af_2 + b)$.
    d. d_p satisfies the triangle inequality, i.e., $d_p(f_1, f_3) \leq d_p(f_1, f_2) + d_p(f2, f_3)$ for any $f_1, f_2, f_3 \in \mathscr{D}$.

25. *Derive a proof for Theorem 4.1.

26. Let $\mathscr{P}$ be the set of probability densities on $[0, 1]$. Show that for any $g \in \mathscr{P}$, the tangent space $T_g(\mathscr{P})$ is given by:

$$T_g(\mathscr{P}) = \{w \in \mathbb{L}^1([0, 1], \mathbb{R}) | \int_0^1 w(t)dt = 0\} \ .$$

27. a. Show that the mapping $g \mapsto (g \circ \gamma)\dot{\gamma}$, for a $\gamma \in \Gamma_I$, is area preserving. In other words, the integral of the function on the interval $[0, 1]$ remains the same.
    b. Show that the mapping $\mathscr{P} \times \Gamma_I \to \mathscr{P}$ given by $(g, \gamma) = (g \circ \gamma)\dot{\gamma}$ defines an action of Γ_I on $\mathscr{P}$.

28. Prove Lemma 4.10: Show that under the Fisher-Rao Riemannian metric, the action of Γ_I on $\mathscr{P}$, as given in Eq. 4.36, is by isometries.

29. Show that the set

$$\left\{\frac{1}{\sqrt{2}\pi n}\sin(2\pi nt), \frac{1}{\sqrt{2}\pi n}(\cos(2\pi nt) - 1)|n = 1, 2, 3 \dots\right\}$$

forms an orthonormal basis for $T_{\gamma_{id}}(\Gamma_I)$ under the first-order Palais metric.

30. (This example can be found, e.g., in http://arxiv.org/pdf/1210.2354.pdf) Consider the family of univariate normal densities, parameterized by two parameters: the mean $\mu \in \mathbb{R}$ and the variance $\sigma^2 \in \mathbb{R}_+$. Each element f of this family $\mathscr{N}(\mu, \sigma^2)$ can be mapped to a point (μ, σ) in the upper-half-plane H:

$$f(x; \mu, \sigma) = \frac{1}{\sqrt{2\pi\sigma^2}}e^{-\frac{1}{2}(x-\mu)^2/2\sigma^2} \ .$$

    a. Show that the Fisher information matrix is given by:

$$g(\mu, \sigma) \equiv \begin{bmatrix} E[\frac{\partial^2 f(x;\mu,\sigma)}{\partial \mu^2}] & E[\frac{\partial^2 f(x;\mu,\sigma)}{\partial \mu \partial \sigma}] \\ E[\frac{\partial^2 f(x;\mu,\sigma)}{\partial \mu \partial \sigma}] & E[\frac{\partial^2 f(x;\mu,\sigma)}{\partial \sigma^2}] \end{bmatrix} = \begin{bmatrix} \frac{1}{\sigma^2} & 0 \\ 0 & \frac{2}{\sigma^2} \end{bmatrix} \ ,$$

    where $E[g(x, \mu, \sigma)] = \int_x g(x, \mu, \sigma)f(x; \mu, \sigma)dx$.

b. Using Eq. 3.1, show that the 2×2 matrix representing the hyperbolic metric is given by (using the coordinates as (μ, σ) instead of (p_1, p_2) used earlier) :

$$g_H = \begin{bmatrix} \frac{1}{\sigma^2} & 0 \\ 0 & \frac{1}{\sigma^2} \end{bmatrix} .$$

c. Therefore, show that:

$$d_{FR}(f(x; \mu_1, \sigma_1), f(x; \mu_2, \sigma_2)) = \sqrt{2} d_H (f(x; \mu_1/\sqrt{2}, \sigma_1), f(x; \mu_2/\sqrt{2}, \sigma_2)) .$$

d. Using Example 3.4, derive an expression for the geodesic path and geodesic distance d_{FR} between any two univariate normal densities.

31. The Wasserstein distance between any two non-negative probability density functions $g_1, g_2 \in \mathscr{P}_0$ is given by: $d_w(g_1, g_2) = \|f_1^{-1} - f_2^{-1}\|$, where f_i is the cumulative distribution function of g_i, $i = 1, 2$. Check if this metric is preserved under the action of Γ on $\mathscr{P}_0$. That is, check if $d_w(g_1, g_2) = d_w((g_1 \circ \gamma)\dot{\gamma}, (g_2 \circ \gamma)\dot{\gamma})$ for all $\gamma \in \Gamma$.

4.12.2 Computational Exercises

1. Given a function $f : [0, 1] \to \mathbb{R}$:

   a. Write a computer program to sample a given function f at some given points $\{t_i | i = 1, 2, \ldots, n\}$ in $[0, 1]$. Call these sampled values $\{y_i\}$.
   b. Using a set of sampled points $\{(t_i, y_i) | i = 1, 2, \ldots, n\}$, estimate the function f on the set $\{\frac{i}{5n} | i = 0, 1, \ldots, 5n\}$ using interpolation.

   Test your program using $f(t) = 2sin(2\pi t) + cos(4\pi t)$ and $\{t_i = i/10 | i = 0, 1, \ldots, 10\}$.

2. Write a program to compute and display a discrete geodesic path $\{\alpha(\tau) | \tau \in \{0, 1/10, 2/10, \ldots 1\}\}$ between any two non-antipodal points on a unit Hilbert sphere $\mathbb{S}_\infty$ using Eq. 4.9.

3. Write a program to compute the exponential and inverse exponential maps on $\mathbb{S}_\infty$. For this computation, any function f on $[0, 1]$ can be represented by a uniform discrete sampling $\{f(k/T) | k = 0, 1, ; T\}$.

4. Write a program to compute fPCA for a given function dataset. In other words, given $\{f_i(t) | i = 1, 2, \ldots, n, \ t = 0, \frac{1}{T}, \frac{2}{T}, \ldots, 1\}$, compute and display the mean function $\hat{\mu}_f$, the singular values σ_i, and the first three principal directions $\hat{b}_1$, $\hat{b}_2$, and $\hat{b}_3$. (Use the earlier program for interpolation of functions, in between the sampled points, to display them as functions.)

5. Verify that the optimal warping functions given in Example 4.1 are as stated in the example.

6. Write a program to compute the SRSF q of a given function f that is given to you in the form of a set of sampled points. You will need some form of approximation for the derivative $\dot{f}$ from the discrete samples of f. Also, write a program to compute an inverse of this map from q to f (assuming $f(0) = 0$). Here you will need an approach to perform numerical integration.

7. The SRSF of a warped function $f \circ \gamma$ is given by $(q \circ \gamma)\sqrt{\dot{\gamma}}$, where q is the SRSF of f. This provides two ways to compute this SRSF: (1) Compute SRSF of the original function and compute $(q \circ \gamma)\sqrt{\dot{\gamma}}$, and (2) compute $f \circ \gamma$ and then

compute the SRSF of this resulting function. Write a program for implementing each of this approach. Assuming that the original function is given only on a uniform grid, you will need to use the earlier program for interpolation of functions. Comment on the numerical accuracies of these two approaches, relative to each other.

8. Implement the dynamic programming algorithm to solve the discrete version of the registration problem (Definition 4.7):

$$\underset{\gamma}{\text{argmin}} \left(\sum_{t=0}^{T} |q_1(t/T) - q_2(\gamma(t/T))\sqrt{\dot{\gamma}(t/T)}|^2 \right) .$$

9. Combine elements from the previous three problems to write a program for pairwise alignment under Definition 4.7 for any two functions defined discretely on a uniform grid on $[0, 1]$. Display the original functions, the warped function, and the optimal warping function.

10. Write a program for computing a geodesic path between any two functions, f_1 and f_2, computed under the $\mathbb{L}^2$ metric on their SRSF representations q_1 and q_2. Display the path in the original function space $\mathcal{F}$. Compare this path with the geodesic computed under the $\mathbb{L}^2$ metric directly on the function space $\mathcal{F}$.

11. Write a program to compute the geodesic and the Fisher-Rao distance between a pair of probability density function. Evaluate this distance for two Gaussian densities with same variance σ^2 and means μ_1 and $\mu_2 = \mu_1 + \epsilon$, respectively. Plot the distance as ϵ goes to zero.

12. Write a program to compute discrete geodesic paths and geodesic distances between elements of Γ_I under the Fisher-Rao metric.

13. As mentioned above, the Wasserstein distance between any two non-negative probability density functions $g_1, g_2 \in \mathcal{P}_0$ is given by $d_w(g_1, g_2) = \|f_1^{-1} - f_2^{-1}\|$, where f_i is the cumulative distribution function of g_i, $i = 1, 2$. Write a problem to compute the sample mean of a given set of densities $g_1, g_2, \ldots, g_n$, under the Wasserstein metric. Use the following steps:

   a. Compute the cumulative distribution functions f_i for each g_i, $i = 1, 2, \ldots, n$.
   b. Compute the inverses of these f_is and find their cross-sectional average $\bar{f}$.
   c. Take the inverse of $\bar{f}$ and find the desired mean density using $\frac{d}{dt}(\bar{f}^{-1}(t))$.

4.13 Bibliographic Notes

Fisher-Rao Riemannian metric was introduced in 1945 by C. R. Rao [92], where he used the Fisher information matrix to compare different probability distributions. This metric was studied rigorously in the 1970s and 1980s by Amari [4], Efron [30], Kass [47, 119], Cencov [24], Amari [5], and others [6]. While those earlier efforts were focused on analyzing parametric families, we use the *nonparametric* version of the Fisher-Rao Riemannian metric in this chapter. (This nonparametric form has found an important use in shape analysis of curves [106].) An important attribute of this metric is that it is preserved under warping [24]. It is difficult to compute the distance d_{FR} directly under this metric but Bhattacharya [16] introduced a square-root representation that greatly simplifies this calculation. More mathematical treatments on the nature of diffeomorphism group can be found in [29, 3, 3]. [108]

were the first paper to develop elastic registration methods for functional data, although similar ideas already existed in shape analysis literature for curves.

FPCA has been described well in several places including [91, 38]. A large number of papers have dealt with the problem of registration of functional data but mostly using $\mathbb{L}^2$ norm on the functions directly [71, 113, 67, 90, 91, 55].

Chapter 5
Shapes of Planar Curves

In this chapter we introduce a framework for analyzing shapes of curves that lie in a two-dimensional plane. Using the mathematical tools introduced in the previous Chaps. 3 and 4 and Appendices A and A.2, we will develop approaches for comparing, deforming, and statistical modeling of shapes of curves. The main topics discussed here are the following: (i) parametric representations of curves using velocity-based functions, (ii) imposition of certain constraints on these functions to reach desired invariances, (iii) utilization of differential geometries of the resulting constrained spaces to impose Riemannian structures and to compute geodesics, and (iv) removal of re-parameterization variability from the representations, using the quotient method, to define shape spaces, and the construction of geodesics in these shape spaces. In particular, we will illustrate these ideas using two representations of curves—angle functions and square-root velocity functions—with appropriate metrics to obtain inelastic deformations (which allow only bending) and elastic deformations (which allow stretching as well as bending), respectively, for shape analysis.

5.1 Goals and Challenges

Our main goals in this chapter are:

- To provide several mathematical representations of planar curves with the purpose of studying their shapes.
- To provide some Riemannian metrics for jointly registering, deforming, measuring, and comparing shapes of curves.
- To provide efficient algorithms for performing above tasks.

What are the main challenges in reaching these goals? First, we are interested in continuous curves that are represented by functions. Consequently, we will use tools for functional analysis that have been summarized in the previous chapter. Second, in contrast to linear spaces of functions, the shape spaces are characterized by nonlinear constraints that shall be utilized and preserved in our analysis. For instance, in order to standardize the length, it is intuitively clear that one should rescale the curves to make them a certain fixed length. This rescaling, however, makes the set of allowable curves a nonlinear manifold and one has to perform calculus on that manifold. Third, shape is a geometric property that is invariant

© Springer-Verlag New York 2016
A. Srivastava, E.P. Klassen, *Functional and Shape Data Analysis*,
Springer Series in Statistics, DOI 10.1007/978-1-4939-4020-2_5

to certain additional transformations. In our formulation, these transformations will be translation, rotation, and scaling. Since we have chosen to work with the parametric forms of curves, we will also have to deal with the variability generated by different re-parameterizations of curves, as a re-parameterization of a curve does not change its shape. So, the challenge before us is to formulate a shape space (with a metric) such that the curves that are within these transformations of each other are deemed identical in terms of their shapes and, thus, have zero shape metric between them. It is possible that several representations and metrics achieve this goal. Among them, we would like to emphasize the ones that provide physical interpretations. Summarizing the main challenges, we need to be able to handle (i) infinite dimensionality and (ii) nonlinearity and (iii) obtain the desired invariances in shape representations.

5.2 Parametric Representations of Curves

Let β be a continuous curve in $\mathbb{R}^2$. It is most naturally represented in a parameterized form:

$$\beta : [a, b] \to \mathbb{R}^2, \quad \beta(t) = (x(t), y(t)) ,$$

where x, y are scalar-valued functions called the *coordinate functions* of β and t is the parameter. As t varies from a to b, the point $\beta(t)$ traces a path from $\beta(a)$ to $\beta(b)$ in $\mathbb{R}^2$. This is called a *parameterization* of β and it dictates the rate (or speed if t is time) at which one traces the curve.

This is one parameterization of this curve, but many more are possible. For example, define a function:

$$\gamma_\alpha : [a, b] \to [a, b] \text{ as } \gamma_\alpha(t) = t + \alpha(t - a)(t - b) , \quad \frac{-1}{b - a} < \alpha < \frac{1}{b - a} . \quad (5.1)$$

Note that $\gamma_\alpha(a) = a$ and $\gamma_\alpha(b) = b$ and $\gamma_\alpha(t)$ changes, monotonically from a to b in between. The middle row of Fig. 5.1 shows some examples of γ_α for different values of α and for $a = 0$ and $b = 2\pi$. Such functions are interesting because we can use them to define new specifications of curves of the type $\tilde{\beta}(t) = \beta(\gamma_\alpha(t))$. What does the curve $\tilde{\beta}$ look like? In fact, it looks exactly the same as β, since it passes through the same points as β, and in the same order (see Fig. 5.1 top). However, the difference lies in the parameter values associated with these points. Thus, re-parameterization is a shape-preserving transformation. Figure 5.1 (bottom row) shows some examples of re-parameterizations. Except for the case when $\alpha = 0$, the curves will have different rates of traversal along the curve. For $\alpha > 0$ the rate of traversal is slow at the start and fast in the end, while for $\alpha < 0$ it is the opposite. The function γ_α is an example of a *re-parameterization* function, and $\tilde{\beta}(t)$ is a re-parameterization of the curve β.

Definition 5.1. For a continuous curve $\beta : [a, b] \to \mathbb{R}^n$, a function γ is called a **re-parameterization** function if it is an orientation-preserving diffeomorphism from $[a, b] \to [a, b]$. Note that "orientation-preserving" means $\dot{\gamma}(t) > 0$ for all $t \in [a, b]$.

Please refer to Sect. 4.10.2 for an introduction of diffeomorphisms. As an example, the γ_α given in Eq. 5.1 is actually a diffeomorphism for all allowable α's. Another example of re-parameterization function is $t \mapsto a + (b - a) \left(\frac{t - a}{b - a} \right)^n$ for some positive

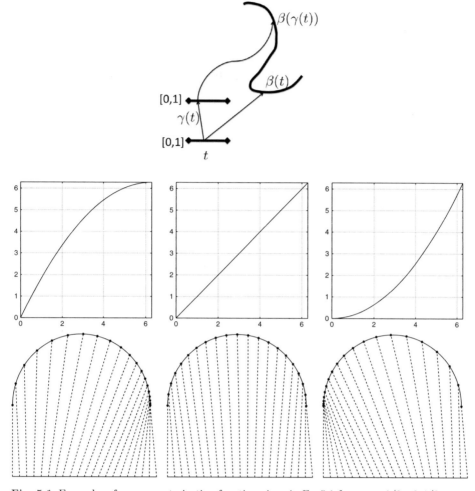

Fig. 5.1 Examples of re-parameterization function given in Eq. 5.1 for $\alpha = -1/2\pi, 0, 1/2\pi$

integer n. (Technically these are not diffeomorphisms because their derivatives at $t = a$ are zero, however they serve well enough as illustrations). A cartoon illustration of the re-parameterization of a curve is shown in the top panel of Fig. 5.1.

For a differentiable, parameterized curve $\beta : [a, b] \rightarrow \mathbb{R}^2$, $|\dot{\beta}(t)| = \sqrt{\left\langle \dot{\beta}(t), \dot{\beta}(t) \right\rangle}$ is its instantaneous speed and $\frac{\dot{\beta}(t)}{|\dot{\beta}(t)|}$ is the instantaneous direction. The total length of β can be computed as (using Eq. 3.2):

$$L[\beta] = \int_a^b \sqrt{\left\langle \dot{\beta}(t), \dot{\beta}(t) \right\rangle} dt \ ,$$

where the inner product is the Euclidean inner product in $\mathbb{R}^2$. Let $\tilde{\beta}(t) = \beta(\gamma(t))$ be a re-parameterized version of β. Then, we have:

$$L[\tilde{\beta}] = \int_a^b \sqrt{\left\langle \dot{\tilde{\beta}}(t), \dot{\tilde{\beta}}(t) \right\rangle} dt = \int_a^b \sqrt{\left\langle \dot{\beta}(\gamma(t))\dot{\gamma}(t), \dot{\beta}(\gamma(t))\dot{\gamma}(t) \right\rangle} dt$$

$$= \int_a^b \sqrt{\left\langle \dot{\beta}(\gamma(t)), \dot{\beta}(\gamma(t)) \right\rangle} \dot{\gamma}(t) \ dt = \int_a^b \sqrt{\left\langle \dot{\beta}(s), \dot{\beta}(s) \right\rangle} ds = L[\beta] \ ,$$

where we have used the change of variable $s = \gamma(t)$ in the last equality. So, we conclude that a re-parameterization of a curve does not change its length. Furthermore, as noted previously, re-parameterization does not change its appearance or shape either. A particular class of parameterizations is that of *constant-speed* parameterizations. Here the distance along the curve between any two points $\beta(t_1)$ and $\beta(t_2)$ is $c(t_2 - t_1)$, for all $t_1, t_2 \in [a, b]$, where $c > 0$ is a constant. Mathematically, this implies that

$$\int_a^s \sqrt{\left\langle \dot{\beta}(t), \dot{\beta}(t) \right\rangle} dt = c(s - a) , \ \forall s.$$

Such a situation is possible if and only if we have $\left\langle \dot{\beta}(t), \dot{\beta}(t) \right\rangle = c^2$ for all $t \in [a, b]$. That is, the instantaneous speed of the parameterization is constant and, therefore, the name constant-speed parameterization. In this case, the length of the curve is simply the speed multiplied by the length of the interval, $L[\beta] = c(b-a)$. A special element of this class is when $c = 1$. It is called the *arc-length* parameterization; in this case, one traverses the curve at the unit speed and $L[\beta] = (b - a)$.

Example 5.1. Let β be a circle of unit radius, $\beta(t) = (\cos(t), \sin(t)) \in \mathbb{R}^2$ for $t \in [0, 2\pi]$. Since

$$\left\langle \dot{\beta}(t), \dot{\beta}(t) \right\rangle = \langle (-\sin(t), \cos(t)), (-\sin(t), \cos(t)) \rangle = 1,$$

β is an arc-length parameterized curve with the total length given by 2π.

5.3 General Framework

In this section, we develop a general framework that will allow us to represent, measure, and deform shapes of planar curves.

5.3.1 Mathematical Representations of Curves

When analyzing the shape of a curve β, we will be interested in the effects of rotations and translations on β and we will do so using the notion of group action (see Sect. 3.5 for a definition of group action). The translation group $\mathbb{R}^2$ acts on the set of parameterized curves according to

$$(x, \beta) \mapsto \beta + x \quad (\text{for } x \in \mathbb{R}^2)$$

and the rotation group $SO(2)$ acts according to

$$(O, \beta) = O\beta \quad (\text{for } O \in SO(2)).$$

Several mathematical representations of β are possible:

1. **Coordinate Functions**: For each curve β, we can consider β_x and β_y coordinates as functions of the parameter t along the curve. This pair (β_x, β_y) contains

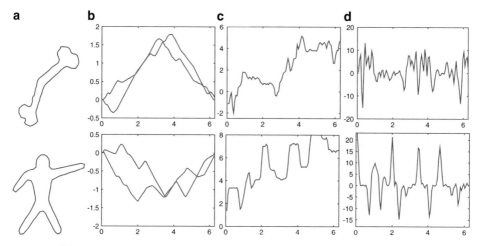

Fig. 5.2 (a) A simple closed curve β, (b) the two coordinate functions x and y plotted against the arc length, (c) the angle function θ, and (d) the curvature function κ

all the information about β and can be used for analyzing its shape. For the curve shown in Fig. 5.2a, the plot in (b) shows the corresponding coordinate functions. It is important to note that these coordinate functions *change* when we rotate and translate the original curve. That is, the coordinate functions of $\beta + x$ or $O\beta$ are different from those of β, as long as $x \neq 0$ and $O \neq I$. Similarly, a re-parameterization of β also changes its coordinate functions. One can handle the translations and the rotations by "standardizing" curves, using some alignment techniques, but the variability due to re-parameterization is more difficult to remove. Therefore, from the perspective of shape invariance, this is not considered a good representation.

2. **Angle Function**: Assuming that β is differentiable, $\dot{\beta}$ is continuous, and β is parameterized by the arc length parameter s[1]; we have $|\dot{\beta}(s)| = 1$ for all $s \in [a, b]$, where $|\cdot|$ is the two-norm. That is, the speed of traversal along β has a constant value of one. Therefore, the only information contained in $\dot{\beta}(s)$ is the direction of $\dot{\beta}(s)$. This is better understood using complex analysis and, for that purpose, let us identify $\mathbb{R}^2$ with the complex space $\mathbb{C}$ in the usual way: any point $\beta(s)$ on the curve is a point $\beta_x(s) + i\,\beta_y(s)$ in $\mathbb{C}$. In this notation, the velocity vector $\dot{\beta}(s)$ can be written as a unit complex number $e^{i\theta(s)}$, where $\theta(s)$ is the angle formed by this vector with the x axis or the real axis. Note that for each s, $\theta(s)$ is only well defined up to the addition of an integer multiple of 2π. Since we are assuming that $\dot{\beta}(s)$ is continuous, we can choose $\theta(s)$ to be continuous. Once we have chosen such a continuous angle function $\theta(s)$, all other angle functions for β will be of the form $\theta(s) + 2\pi n$, where n is an integer. The function θ, with the parameter s as the variable, contains all the information about the shape of β and can be used as a mathematical representation of that shape.

In terms of the desired invariances, θ is invariant to translations of β but it does change with rotations. That is, the angle function of $\beta + x$ is same as that of β, for all $x \in \mathbb{R}^2$, while that of $O\beta$ differs from the angle function of β as long

---

[1] We use s for the arc-length parameter and t for a general parameter of a curve.

as $O \neq I$. If we rotate the curve β counterclockwise around the origin by an angle θ_0, then the angle function of the rotated curve is $\theta(s) + \theta_0$. Figure 5.2c shows the angle function of the curve in (a), plotted versus the parameter s. In summary, the angle function of β is invariant to translation but not to rotation.

3. **Curvature Function**: When we take the derivative of the angle function, with respect to the arc-length parameter, we obtain the curvature function of β. Denoting this function by κ, we have that $\kappa(s) = \dot{\theta}(s)$ for all $s \in [a, b]$. Since a rotation of β adds a constant to its angle function, the curvature function κ remains unchanged. Also, it is already invariant to translation. In other words, the curvature functions of $\beta + x$ and $O\beta$, for any $x \in \mathbb{R}^2$ and $O \in SO(2)$, are the same as that of β. Although the curvature function has the desired invariance to rotation and translation, its use in practical situations is quite susceptible to observation noise. If the original curve β is observed under noisy conditions, then κ, which involves the second derivative of β, shows a noticeable amplification of this noise. Shown in Fig. 5.2d is the curvature function of the curve in panel (a).

In case the curves of interest are parameterized using arc length, these representations—the angle function and the curvature function—are individually sufficient to analyze the shapes of the original curves. However, in many situations we get shape data in the form of curves that are arbitrarily parameterized. Also, as we shall see later, it is advantageous to include the variability associated with different parameterizations of curves in the process of comparing their shapes. This is because the points across curves may match better if we allow nonlinear re-parameterizations, (i.e. variable speed parameterizations). In this situation, the representation space needs to be augmented to include this additional information. In fact, the inclusion of parameterization variability becomes a tool for improving registration across curves, similar to how warping was used to register functions in the previous chapter. So, now let $\beta : [a, b] \to \mathbb{R}^2$ be a curve with arbitrary parameterization. Often, we will assume that the curve β is scaled to be of certain length l:

$$L[\beta] = \int_a^b |\dot{\beta}(t)| dt = l \ . \tag{5.2}$$

Due to arbitrary parameterizations of curves, the speed $|\dot{\beta}(t)|$ is not constant anymore, and we need an additional function to keep track of its variation along the curve. Now the function $|\dot{\beta}(t)|$ denotes the instantaneous speed of the parameterization at the point $\beta(t)$. Inclusion of $|\dot{\beta}(t)|$ augments the representation space and gives rise to the following possibilities:

1. **Coordinate Functions**: As earlier, one can use the two coordinate functions to represent the curve, forming the double (β_x, β_y), where $\beta_x, \beta_y : [a, b] \to \mathbb{R}$ are two real-valued functions. Note that these coordinate functions also specify the speed of traversal along the curve. So, no additional function is required to represent the speed function.

2. **Angle and Log-Speed Function**: For this representation, we first impose a condition that the parameterization is non-singular, or $|\dot{\beta}(t)| \neq 0$ for all t. Now we can write the velocity function $\dot{\beta}(t)$ in the complex notation as $|\dot{\beta}(t)| e^{i\theta(t)}$, where the first term is the speed function and the second term is the direction

function . Since the speed is constrained to be non-negative, it poses a challenge during shape analysis. To get around this constraint, one can use its logarithm $\phi(t) = \log(|\dot{\beta}(t)|)$ and work with the real-valued function ϕ. Since β is a curve of length l, we have that $\int_a^b e^{\phi(t)} dt = \int_a^b |\dot{\beta}(t)| dt = l$. We can now represent the curve β with the pair (ϕ, θ) as they completely describe both the shape and the parameterization of β. Here, $\phi, \theta : [a, b] \to \mathbb{R}$.

3. **Curvature and Log-Speed Function**: Repeat the previous idea with the curvature function, instead of the angle function. This representation is based on the pair (ϕ, κ), each of them being a real-valued function on $[a, b]$ with appropriate constraints.

4. **Square-Root Velocity Function** (SRVF): There is an interesting possibility of merging the two variabilities—shape and parameterization—together to form a single (vector-valued) function, while keeping all the relevant information. This is done by forming a function, called the *square-root velocity function* of β. This is a natural extension of the square-root slope functions seen in the previous chapter. Let $F : \mathbb{R}^2 \to \mathbb{R}^2$ be a mapping given by

$$F(v) = \begin{cases} v/\sqrt{|v|} & \text{when } |v| \neq 0 \\ 0 & \text{when } |v| = 0 \end{cases} . \tag{5.3}$$

Then, define the square-root velocity function (SRVF) to be:

$$q(t) \equiv F(\dot{\beta}(t)) = \frac{\dot{\beta}(t)}{\sqrt{|\dot{\beta}(t)|}} . \tag{5.4}$$

It is thus named because the vector $q(t)$ has the same direction as the velocity vector $\dot{\beta}(t)$ but its magnitude is given by the square root of the instantaneous speed:

$$|q(t)| = \frac{|\dot{\beta}(t)|}{\sqrt{|\dot{\beta}(t)|}} = \sqrt{|\dot{\beta}(t)|} , \text{ and } \frac{q(t)}{|q(t)|} = \frac{\dot{\beta}(t)}{|\dot{\beta}(t)|} .$$

We make a few remarks about this definition. Firstly, since F is a continuous map, even where $|\dot{\beta}(t)| = 0$, the SRVF is well defined. In other words, SRVF makes sense for all absolutely continuous functions (defined later). This representation exists even for curves with singular parameterizations. In fact, this definition of the SRVF is valid for curves in $\mathbb{R}^n$, not just $\mathbb{R}^2$. (We will use it to study the shapes of curves in higher dimensions later, in Chap. 10.) Finally, while this SRVF can be seen as an extension of the SRSF mentioned earlier in Sect. 4.6, from $\mathbb{R}$ to $\mathbb{R}^n$, it is not the only possibility. In fact, $q_w(t) = \sqrt{\rho(t)} e^{iw\theta(t)}$, for any positive real number w, is potentially useful in shape analysis, although some values of w are better than others. The case of $w = 0.5$ has been studied in more detail by Younes and others (see citations in bibliography). For $w = 0.5$, $q_{0.5}(t)$ is simply a complex square root of when $\dot{\beta}(t)$ is written as a complex number $\dot{\beta}_x(t) + i\dot{\beta}_y(t)$. To gain some insight in this representation, we identify $\dot{\beta}(t)$ with $\rho(t) e^{i\theta(t)}$ and obtain $q_{0.5}(t) = \sqrt{\rho(t)} e^{i\theta(t)/2}$. There are certain problems that arise in the exponential term. Firstly, since $\theta(t)$ can be replaced by $\theta(t) \pm 2n\pi$ for any integer n, this division by 2 can lead to quite different answers. For instance, note that while ϵ and $2\pi + \epsilon$ are the same angles, their halves $\epsilon/2$ and $\pi + \epsilon/2$ are not the same. So, one has to be careful about this operation. This problem can lead to a real ambiguity near the points where $\theta(t) = 0$. Secondly, this definition of $q_{0.5}$ relies completely on complex analysis and, therefore, is difficult to extend to curves in higher dimensions.

In this setup, the SRVF (Eq. 5.4) results from the value $w = 1$, i.e., $q(t) = \sqrt{\rho(t)}e^{i\theta(t)}$, and it does not suffer from the problems mentioned above. Since the exponential term remains unchanged, there is no phase ambiguity in making the transformation. Furthermore, the corresponding definition of q applies to curves in higher dimensions also. Any other choice of w restricts one to planar curves only.

If the curve β is of length l, the SRVF q satisfies $\int_a^b |q(t)|^2 dt = \int_a^b |\dot{\beta}(t)| dt = l$. One can reconstruct the curve β from q, up to a translation, using the formula $\beta(t) = \int_a^t q(\tau)|q(\tau)| d\tau$. In contrast to the (ϕ, θ) representation, the angle and speed functions have been mixed together in this square-root velocity representation $q(t) = \sqrt{\rho(t)}e^{i\theta(t)}$. Note that we are defining $\rho(t) = |\dot{\beta}(t)|$. The number of functions remains the same; $q(t) = [q_x(t) \ \ q_y(t)]$ are still two scalar-valued functions, but the mixing leads to two important simplifications later. Firstly, if we restrict the length of the curve to be l, then the function q lies on a sphere in the space of all (square-integrable) $\mathbb{R}^2$-valued functions on $[a, b]$ (this is due to the fact that $\int_a^b |q(t)|^2 dt = l$). Since the differential geometry of a sphere is completely understood, we can use it explicitly to simplify the ensuing shape analysis. Later sections will have more on that topic. Secondly, for our choice of metric on this representation of curves, the action of the re-parameterization group on curves will be by isometries! This will be important in defining a proper distance in shape analysis of curves.

5.3.2 Shape-Preserving Transformations

We now have a few mathematical representations of parameterized curves in $\mathbb{R}^2$. For the purpose of shape analysis, we need to make our representation invariant to all the shape-preserving transformations—translation, rotation, scaling, and re-parameterization. By scaling the curves to be of a fixed length, the scale variability has been removed. All curves are of the same length and, therefore, are compared fairly. Similarly, by using the velocity function $\dot{\beta}$, and functions derived from it, to represent the shape of β, we have also removed the translation variability. In other words, if β is translated in any fashion, its representative that depends only on $\dot{\beta}$ will not change. Thus, all representations mentioned previously, except for (β_x, β_y), are already invariant to translations and uniform scalings of β. However, the rotation transformation has not been addressed yet. A rigid rotation of β will change its representation and that is undesirable. This will result in many elements of the representation set, each with differing rotation, having the same shape but different mathematical representations, with possibly nonzero distances between them. In order to remove this redundancy and achieve a uniqueness of representation (of shape), we want to reach a rotational invariance. This is done differently for different representations. In case of angle function representation, we use the notion of a section (Sect. 5.4), but for the SRVF representation, we use the notion of orbits (Sect. 5.5).

So far we have discussed the "standard" shape-preserving transformations—translation, rotation, and scaling. However, due to the parametric nature of our representations of curves, we have introduced an extra variability that will need to be addressed. If we re-parameterize β, within the constraints imposed on it, we get a new parametric form whose shape remains unchanged. An example was presented in Sect. 5.2 using Eq. 5.1. In the later part on shape analysis, we will account for this variability using the actions of re-parameterization groups on spaces of curves.

5.4 Pre-shape Spaces

In this section we start restricting to the sets that are relevant for our shape analysis. In particular, we will focus on two of the representations—angle function θ for arc-length parameterized curves and SRVF q for arbitrarily parameterized curves—and will investigate their use in analyzing shapes of curves. For this discussion, we will focus on the sets of fixed-length curves and study the differential geometries of the resulting sets. These spaces are termed *pre-shape spaces* since we have not accounted for all the variability generated by shape-preserving transformations. For instance, in case of square-root velocity representations, there are points in the corresponding set that are simply re-parameterizations of each other and, hence, have the same shape. The defining characteristic of pre-shape spaces is that the shape representations have been identified but not unified. Later on, when we reach shape spaces, we will see a unification of all points in pre-shape spaces that represent the same shape. Thus, shape spaces will be quotient spaces of the pre-shape spaces.

Although many more pre-shape spaces can be constructed, we will deal with the following two:

1. **Case 1**: As the simplest example of shape analysis of curves, we will consider fixed-length curves under the arc-length parameterization, each represented by its angle function. By rescaling the curves to a fixed length, we get rid of the scale variability. Similarly, by restricting to the arc-length parameterization, we get rid of different parameterizations associated with curves. Loosely speaking, these restrictions amount to forming a section of the space of curves for the scaling and re-parameterization group actions. Let $[0, 2\pi]$ be the domain of parameterization, and we will restrict to those angle functions that are absolutely continuous on this domain. We will use the $\mathbb{L}^2$ metric on this set to compare and analyze angle functions. By the virtue of arc-length parameterization over $[0, 2\pi]$, the curves represented by these angle functions are of length 2π. The reader may notice that $\mathbb{L}^2([0, 2\pi], \mathbb{R})$ is actually a topological completion of our space of curves.
   Orthogonal Section Under the Rotation Group: The rotation of a curve is represented by the translation of its angle function. This translation group $\mathbb{R}$ acts on $\mathbb{L}^2([0, 2\pi], \mathbb{R})$ according to:

$$(\theta_0, \theta)(s) = (\theta(s) + \theta_0) .$$

We can remove this translation variability by forming the quotient space $\mathbb{L}^2([0, 2\pi], \mathbb{R})/\mathbb{R}$ and studying points in that space. However, let us consider the set:

$$\mathscr{C}_1 = \{\theta \in \mathbb{L}^2([0, 2\pi], \mathbb{R}) |\ \frac{1}{2\pi} \int_0^{2\pi} \theta(s)ds = \pi\ \}\quad \subset \mathbb{L}^2([0, 2\pi], \mathbb{R}). \qquad (5.5)$$

The constraint on the right side fixes the average orientation to be π. It can be shown that $\mathcal{C}_1$ is an orthogonal section (see Definition 3.19) of the Hilbert manifold $\mathbb{L}^2([0, 2\pi], \mathbb{R})$ under the action of $\mathbb{R}$ by translation. Furthermore, since $\mathbb{R}$ acts on $\mathbb{L}^2$ by isometries, we can form an isometric map between $\mathcal{C}_1$ and $\mathbb{L}^2/\mathbb{R}$ (under the inherited metric). This allows us to study elements of the quotient space by studying the corresponding elements of $\mathcal{C}_1$ under the $\mathbb{L}^2$ metric.

The set $\mathcal{C}_1$ is not a vector space but an affine space, since it does not contain the zero function. (An affine space is a linear subspace of a vector space that has been translated so that it does not pass through the origin.) This space has the property that if it contains two points, then it contains the entire straight line between the two points. Hence, the shortest geodesic between any two points is simply the straight line connecting them.

2. **Case 2**: Secondly, we will study the pre-shape space of fixed-length, absolutely continuous curves under arbitrary parameterizations, using the square-root velocity representations. In this case we will assume the parameter interval to be $[0, 1]$, with the resulting space being given by:

$$\mathcal{C}_2 = \{q \in \mathbb{L}^2([0,1], \mathbb{R}^2) | \int_0^1 |q(t)|^2 dt = 1\} . \qquad (5.6)$$

Here $\mathbb{L}^2([0,1], \mathbb{R}^2) = \{q : [0,1] \to \mathbb{R}^2 | \int_0^1 |q(t)|^2 dt < \infty\}$. Recall that $\|q\|^2$ also equals the length of the curve, so the constraint on the right side in Eq. 5.6 restricts the curves to be of unit length.

Orthogonal Section Under the Scaling Group: Similar to the case of the scaling group action in Example 3.14, we can show that $\mathcal{C}_2$ is an orthogonal section of the scaling group on $\mathbb{L}^2([0,1], \mathbb{R}^2)$. Furthermore, under the scaled-$\mathbb{L}^2$ metric as in Example 3.16, we can show that the action of the scaling group $\mathbb{R}^\times$ is by isometries and construct an isometric map between the orthogonal section $\mathcal{C}_2$ and the quotient space $\mathbb{L}^2([0,1], \mathbb{R}^2)/\mathbb{R}^\times$. Therefore, from now on we will use the orthogonal section $\mathcal{C}_2$ for analyzing elements of the quotient space. Since the scaled-$\mathbb{L}^2$ metric is the standard $\mathbb{L}^2$ metric on $\mathcal{C}_2$, we will continue with the standard metric on that space. As mentioned in Example 3.14, the action of the rotation group on the space of landmarks does not admit an orthogonal section. The same holds here and we will have to use more general techniques for removing the rotation group.

$\mathcal{C}_2$ is the set of all square-root velocity functions with $\mathbb{L}^2$ norm one and, hence, is a sphere in $\mathbb{L}^2([0,1], \mathbb{R}^2)$. Using the Example A.11, $\mathcal{C}_2$ is a Hilbert submanifold of $\mathbb{L}^2([0,1], \mathbb{R}^2)$ with codimension 1. Since a lot is known about the geometry of a sphere, including geodesics, exponential map, etc, the shape analysis of curves under this representation (without additional constraints) is relatively simple.

At this point, it is worthwhile to point out exactly which classes of curves have representatives in $\mathcal{C}_2$. This representation makes use of the first derivative, so the curves represented must of course satisfy some sort of differentiability requirement. It turns out, however, that the class of curves having representatives in $\mathcal{C}_2$ includes curves that are not differentiable in the usual sense but indeed satisfy the weaker condition of *absolute continuity*. For our purposes the following definition will suffice.

Definition 5.2. A function f is absolutely continuous if and only if there exists an integrable function $g : [0,1] \to \mathbb{R}^2$ such that

$$f(t) - f(0) = \int_0^t g(\tau)d\tau, \quad \text{for all } t \in [0,1] .$$

This is sometimes referred to as the *fundamental theorem of Lebesgue integral calculus.* Thus function g is unique up to sets of Lebesgue measure zero. Further, in this situation, f is differentiable almost everywhere, and $\dot{f}(t) = g(t)$ almost everywhere. Note that although the result is stated for functions from an interval $[0,1]$ to $\mathbb{R}^2$, it immediately generalizes to functions $[0,1] \mapsto \mathbb{R}^n$ since integration is a componentwise operation. Now let $\mathscr{F}_2$ denote the set of all functions $\beta : [0,1] \to \mathbb{R}^2$ satisfying: (1) β is absolutely continuous, (2) $\beta(0) = 0$, and (3) $L[\beta] = \int_0^1 \|\dot{\beta}(t)\| dt = 1$. Let $\mathscr{Q} : \mathscr{F}_2 \to \mathscr{C}_2$ denote the map that sends each curve to its SRVF. It is straightforward to show that $\mathscr{Q}$ is a bijection (left as an exercise for the reader), with inverse given by

$$\mathscr{Q}^{-1}(q)(t) = \int_0^t q(u)|q(u)|du, \quad \text{for all } t \in [0,1] .$$

Thus, $\mathscr{F}_2$ is precisely the set of curves with representatives in $\mathscr{C}_2$.

We have now identified these two pre-shape spaces, denoted by $\mathscr{C}_1 \subset \mathbb{L}^2([0,2\pi],\mathbb{R})$ and $\mathscr{C}_2 \subset \mathbb{L}^2([0,1],\mathbb{R}^2)$, as the spaces of interest for further analysis.

5.4.1 Riemannian Structure

An important step in our shape analysis is to define a Riemannian structure on a shape space and to compute geodesic paths between shapes with respect to the chosen metric. Since shape spaces are quotient spaces of pre-shape spaces, with the Riemannian metric inherited from the larger set, the geodesics in the shape spaces are closely related to those in the pre-shape spaces. In this section, we present our choices of Riemannian structures on the two pre-shape spaces $\mathscr{C}_1$ and $\mathscr{C}_2$. In each case we will establish the tangent spaces and will define inner products on these spaces to impose Riemannian structures.

Note that both the pre-shape spaces are level sets of functionals on Hilbert spaces. In the first case, the Hilbert space is $\mathbb{L}^2([0,2\pi],\mathbb{R})$, while in the second case it is $\mathbb{L}^2([0,1],\mathbb{R}^2)$. This fact can be used to characterize the tangent spaces of these pre-shape spaces as follows. (Please refer to Sect. A.1.1 for definitions and examples of tangents and normals to manifolds.) We start with the general notion. Let $\Phi : \mathbb{H} \to \mathbb{R}^n$ be a differentiable mapping, with $\mathbb{H}$ being one of the earlier Hilbert spaces, and let $x \in \mathbb{R}^n$ be a regular value of Φ (i.e., $d\Phi_p : \mathbb{H} \to \mathbb{R}^n$ is onto for all $p \in \Phi^{-1}(x)$). By Theorem A.1, we know that $M = \Phi^{-1}(x)$ is a submanifold of $\mathbb{L}^2$, and the tangent space $T_p(M)$ is just the kernel of $d\Phi_p$ for all $p \in \Phi^{-1}(x)$. As a consequence, $d\Phi_p$ decomposes $\mathbb{H}$ into a direct sum: $\mathbb{H} = T_p(M) \oplus N_p(M)$, where $N_p(M)$ is the set of vectors normal to M at p in $\mathbb{H}$. Since $d\Phi_p$ is surjective, the normal space $N_p(M)$ has the dimension n and can be specified using a finite basis. This is in contrast to $T_p(M)$, which is an infinite-dimensional space. Therefore, in this context, it is much easier to specify the normal space as compared to the tangent space. We now use these ideas to derive the tangent spaces of $\mathscr{C}_1$ and $\mathscr{C}_2$. These spaces are level sets of $\mathbb{R}^n$-valued functionals on Hilbert spaces, and we will follow the practice of specifying their normal spaces in the larger Hilbert spaces.

1. **Case 1**: First, we verify that $\mathscr{C}_1$ is indeed a manifold. To do this, define a map $\Phi_1 : \mathbb{L}^2([0, 2\pi], \mathbb{R}) \to \mathbb{R}$ by

$$\Phi_1(\theta) = \frac{1}{2\pi} \int_0^{2\pi} \theta(s) \, ds \ . \tag{5.7}$$

Then, $\mathscr{C}_1 = \Phi_1^{-1}(\pi) \subset \mathbb{L}^2([0, 2\pi], \mathbb{R})$. Because Φ_1 is a bounded linear transformation to a finite-dimensional vector space, it is automatically smooth, and, furthermore, it is its own derivative, i.e.,

$$d\Phi_{1,\theta}(f) = \Phi_1(f) = \frac{1}{2\pi} \int_0^{2\pi} f(s) \, ds = \frac{1}{2\pi} \langle f, 1 \rangle \tag{5.8}$$

Clearly, $d\Phi_{1,\theta}$ is surjective at every function $\theta \in \mathbb{L}^2$ (for example, $d\Phi_{1,\theta}(1) = 1$), which shows that $\mathscr{C}_1$ is a submanifold of $\mathbb{L}^2$ of codimension one. Furthermore, Eq. 5.8 shows that the kernel of $d\Phi_{1,\theta}$ is simply $\{f \in \mathbb{L}^2([0, 2\pi], \mathbb{R}) : f \perp 1\}$. It follows immediately that the normal space of $\mathscr{C}_1$ at θ is given by $N_\theta \mathscr{C}_1 = \mathrm{span}\{1\}$, the one-dimensional vector space spanned by a constant function 1. Hence the tangent space is given by:

$$
\begin{aligned}
T_\theta(\mathscr{C}_1) = \ & \text{null space of } d\Phi_{1,\theta} \\
= \ & \{f \in \mathbb{L}^2([0, 2\pi], \mathbb{R}) | f \perp N_\theta(\mathscr{C}_1)\}, \\
& \text{where } \ N_\theta(\mathscr{C}_1) = \text{all constant functions} \ , \\
= \ & \{f \in \mathbb{L}^2([0, 2\pi], \mathbb{R}) | \int_0^{2\pi} f(s) ds = 0\} \ .
\end{aligned}
$$

Since the tangent space $T_\theta(\mathscr{C}_1)$ is a subspace of $\mathbb{L}^2$, we can restrict the $\mathbb{L}^2$-inner product to define a Riemannian structure on $\mathscr{C}_1$. For any two elements $g_1, g_2 \in T_\theta(\mathscr{C}_1)$, the Riemannian metric $\langle g_1, g_2 \rangle$ is given by $\langle g_1, g_2 \rangle = \int_0^{2\pi} g_1(s) g_2(s) ds$. There is a nice physical interpretation associated with the use of this Riemannian metric. We claim that this metric, when measuring distortions in shapes of curves, quantifies the amount of *bending* needed to obtain one curve from the other. Since $\theta(s)$ at the point $\beta(s)$ on the curve is the angle made by the vector $\dot{\beta}(s)$ with the X axis, $g_1(s)$ and $g_2(s)$ denote infinitesimal changes in that angle. A change in the angle of $\dot{\beta}(s)$ results in bending of the curve at that point and, hence, an $\mathbb{L}^2$ metric for measuring changes in angle functions is called the **bending** metric. This terminology does not have the same formal definition as in physics but carries the same connotation. Refer to Fig. 5.3 for an example where the angle function θ of the ellipse is perturbed by a small function $f \in T_\theta(\mathscr{C}_1)$, resulting in the bending of the ellipse. The squared $\mathbb{L}^2$ norm of f measures the amount of bending in the ellipse.

Using the Riemannian structure, it becomes possible to define lengths of paths on $\mathscr{C}_1$ (See Sect. 3.2 for the definition). Let $\alpha : [0, 1] \mapsto \mathscr{C}_1$ be a parameterized path on $\mathscr{C}_1$ that is differentiable everywhere on $[0, 1]$. Then $\frac{d\alpha}{d\tau}$, the velocity vector at τ, is an element of the tangent space $T_{\alpha(\tau)}(\mathscr{C}_1)$ and its length is

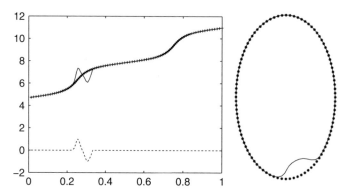

Fig. 5.3 Left panel shows the perturbation of an angle function θ (*marked line*) by f (*broken line*) into $\theta + f$, and the right panel shows the corresponding bending of the curve. The quantity $\|f\|^2$ measures the bending energy

defined to be $\sqrt{\left\langle \frac{d\alpha}{d\tau}, \frac{d\alpha}{d\tau} \right\rangle}$ with the inner product being $\mathbb{L}^2$. The length of the path α is then given by:

$$L[\alpha] = \int_0^1 \sqrt{\left\langle \frac{d\alpha}{d\tau}, \frac{d\alpha}{d\tau} \right\rangle} d\tau . \qquad (5.9)$$

This is the integral of the speed along α and, hence, is the length of the whole path α. For any two points θ_1, $\theta_2 \in \mathscr{C}_1$, one can define the distance between them as the infimum of the lengths of all smooth paths on $\mathscr{C}_1$ that start at θ_1 and end at θ_2:

$$d_{\mathscr{C}_1}(\theta_1, \theta_2) = \inf_{\{\alpha:[0,1] \mapsto \mathscr{C}_1 | \alpha(0)=\theta_1, \alpha(1)=\theta_2\}} L[\alpha] . \qquad (5.10)$$

A path $\hat{\alpha}$ that achieves the above minimum, if it exists, is a geodesic between θ_1 and θ_2 in $\mathscr{C}_1$. We know that $\mathscr{C}_1$ is an affine space. As mentioned earlier, given any two points in this space, the entire straight line joining them is still in the space. Hence, the geodesics in $\mathscr{C}_1$ are simply straight lines.

2. **Case 2**: For the SRVF representation, we select a Riemannian structure as follows. Given a curve represented by $q \in \mathbb{L}^2([0,1], \mathbb{R}^2)$, and the tangent vectors $u, v \in T_q(\mathbb{L}^2([0,1], \mathbb{R}^2))$, respectively, the standard $\mathbb{L}^2$ inner product between u, v is defined as

$$\langle u, v \rangle = \int_0^1 \langle u(t), v(t) \rangle \, dt . \qquad (5.11)$$

To begin analyzing $\mathscr{C}_2$, define a mapping $\Phi_2 : \mathbb{L}^2([0,1], \mathbb{R}^2) \to \mathbb{R}$ by: $\Phi_2(q) = \int_0^1 |q(t)|^2 dt$. Then $\mathscr{C}_2$ can also be written as $\Phi_2^{-1}(1)$. To check that $\mathscr{C}_2$ is a manifold, we need to calculate the linear transformation $d(\Phi_2)_q : \mathbb{L}^2([0,1], \mathbb{R}^2) \to \mathbb{R}$. Hence,

$$d(\Phi_2)_q(w) = \frac{d}{d\epsilon}\Big|_{\epsilon=0} \Phi_2(q + \epsilon w) = 2 \int_0^{2\pi} \langle w(t), q(t) \rangle \, dt = 2 \langle q, w \rangle .$$

Clearly, $d(\Phi_2)_q$ is surjective for every $q \neq 0$, and, hence, it is surjective for every $q \in \Phi_2^{-1}(1)$. It follows by Theorem A.2 that $\mathscr{C}_2$ is a submanifold of $\mathbb{L}^2([0,1], \mathbb{R}^2)$ of codimension one. (In fact, it is simply the sphere of radius one centered at the origin in this Hilbert space.) An explicit calculation of the tangent space $T_q(\mathscr{C}_2)$ is given by

$$T_q(\mathscr{C}_2) = \ker(d(\Phi_2)_q) = \{w \in \mathbb{L}^2([0,1], \mathbb{R}^2) : \langle q, w \rangle = 0\}.$$

From this, it follows that the normal space of $\mathscr{C}_2$ at q is given by

$$N_q(\mathscr{C}_2) = \mathrm{span}(q), \tag{5.12}$$

i.e., the one-dimensional linear space spanned by q. Of course, this is another way of saying that the tangent space of a sphere at a point q is the orthogonal complement of the radial vector from the center to q! The metric defined in Eq. 5.11 has a nice physical interpretation in being an elastic metric, as explained later in Sect. 5.6.

Similar to the previous case, we can now discuss the geodesics on $\mathscr{C}_2$. Let $\alpha : [0,1] \mapsto \mathscr{C}_2$ be a parameterized path on $\mathscr{C}_2$ that is differentiable everywhere on $[0,1]$. The length of this path is given by $L[\alpha] = \int_0^1 \sqrt{\langle \frac{d\alpha}{d\tau}, \frac{d\alpha}{d\tau} \rangle} \, d\tau$, with the inner product in the integral defined by Eq. 5.11. We have already seen that $\mathscr{C}_2$ is a unit sphere in $\mathbb{L}^2([0,1], \mathbb{R}^2)$. Now we see that it is a Riemannian manifold with the metric inherited from the larger Hilbert space. With this Riemannian structure, the geodesics between points on a sphere are known to be the arcs of great circles containing those points, where a great circle is defined to be the intersection of the sphere with any 2-dimensional plane through the origin.

5.4.2 Geodesics in Pre-shape Spaces

One of the most important tools in shape analysis is the construction of geodesic paths. For any two given elements of a space, pre-shape or shape, we want to be able to construct a geodesic path connecting them in that space. We start the discussion with pre-shape spaces. For the pre-shape spaces studied in this chapter, it is possible to write down the geodesics analytically, as discussed next.

1. **Case 1**: Since $\mathscr{C}_1$ is an affine space, the geodesic between any two elements θ_1 and θ_2 of $\mathscr{C}_1$ is given by a straight line:

$$\alpha(\tau) = (1 - \tau)\theta_1 + \tau\theta_2, \quad \tau \in [0,1] . \tag{5.13}$$

This geodesic has a constant-speed parameterization, i.e., $\|\dot{\alpha}(\tau)\| = $ constant, and it starts from θ_1 and ends at θ_2. The geodesic distance between θ_1 and θ_2 is given by $d_{\mathscr{C}_1}(\theta_1, \theta_2) = \|\theta_1 - \theta_2\|$.

Some examples of geodesic paths between elements of $\mathscr{C}_1$ are presented next. These geodesics are actually computed using angle functions as elements in $\mathscr{C}_1$ for all τ, but are displayed using coordinate functions and are drawn at convenient displacements.

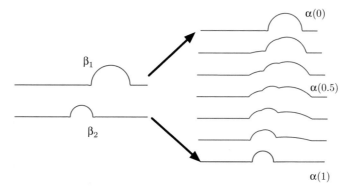

Fig. 5.4 Example of a geodesic path in $\mathscr{C}_1$: The first two panels show curves β_1 to β_2 while the third shows the geodesic path from β_1 to β_2 in $\mathscr{C}_1$

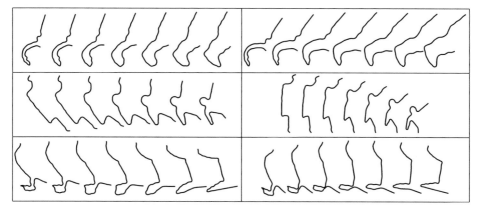

Fig. 5.5 Comparisons of geodesics in pre-shape spaces—$\mathscr{C}_1$ (*left*) and $\mathscr{C}_2$ (*right*)—for the same pairs of curves

Figure 5.4 shows an example of geodesic between two curves shown on the left, at the points $\alpha(\tau)$ for $\tau = 0, \ 1/6, \ 2/6, \ldots, 1$. Both the original curves have a bump each but at different locations. A look at the geodesics between them underscores the nature of this representation. In going from one curve to another, the deformation effectively removes one bump and creates another one. This has the appearance of bending a steel wire from one shape into another, without allowing for any stretching or compression of the wire. This is exactly what the bending framework suggests.

The left column of Fig. 5.5 shows a set of geodesic paths between arbitrary open curves computed using Eq. 5.13. In each row of this column the first and the last curves are given and intermediate curves denote equally sampled points on the geodesic connecting those two curves in $\mathscr{C}_1$. Once again the geodesics have the appearance of bending curves from one into another without allowing for stretching.

A third set of examples is presented in Fig. 5.6. What is special here is that the original two curves β_1 and β_2 are closed curves. However, these curves are represented as elements of $\mathscr{C}_1$, ignoring the fact that they are closed. It is interesting to note that the intermediate points on $\alpha(\tau)$ are not necessarily closed curves! This is visible in the two examples shown in this figure. The issue

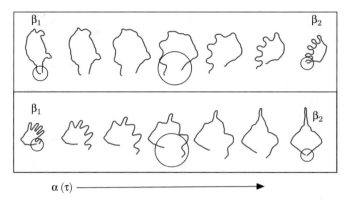

$\alpha(\tau)$ ──────────────────────────▶

Fig. 5.6 Examples of geodesics in $\mathscr{C}_1$ between closed curves. The original curves are closed, but the intermediate curves are far from closed. The openings are highlighted using circles in some curves

of closure is an important one and is addressed in the next chapter. Another, even more difficult, issue is that of curves crossing themselves. The question is: even if the original curves do not cross themselves, will the geodesic pass through curves that are self-intersecting? Experiments indicate that in case of highly twisted shapes, the curves along the geodesics can cross themselves. This is an issue that is not handled in this textbook.

The exponential map and its inverse on $\mathscr{C}_1$ are straightforward. For θ_1, $\theta_2 \in \mathscr{C}_1$, and an $f \in T_{\theta_1}(\mathscr{C}_1)$, we have:

$$\exp_{\theta_1}(f) = \theta_1 + f$$
$$\exp_{\theta_1}^{-1}(\theta_2) = \theta_2 - \theta_1 \ .$$

2. **Case 2**: The space of all SRVFs of curves in $\mathbb{R}^2$ is $\mathscr{C}_2$, a hypersphere in $\mathbb{L}^2([0,1], \mathbb{R}^2)$ with the inherited Riemannian metric. The minimal geodesics in $\mathscr{C}_2$ are given by the shorter arcs on great circles. Given any two parameterized curves β_1 and β_2 in $\mathbb{R}^2$, we can compute geodesic paths between their representations q_1 and q_2 using Eq. 3.5 (see Example 3.4 for details):

$$\alpha(\tau) = \frac{1}{\sin(\vartheta)} \left[\sin(\vartheta(1-\tau))q_1 + \sin(\tau\vartheta)q_2 \right] \tag{5.14}$$

where $\vartheta = \cos^{-1}(\langle q_1, q_2 \rangle)$. This geodesic starts at q_1, at $\tau = 0$, and reaches q_2 at $\tau = 1$ while traveling at a constant speed. The geodesic distance between any two points in $\mathscr{C}_2$ is given by:

$$d_{\mathscr{C}_2}(q_1, q_2) = \vartheta = \cos^{-1}(\langle q_1, q_2 \rangle) \ . \tag{5.15}$$

A geodesic on $\mathscr{C}_2$ can also be characterized in terms of a tangent direction $w \in T_q(\mathscr{C}_2)$:

$$\alpha_t(w) = \cos(t\|w\|)q + \sin(t\|w\|)\frac{w}{\|w\|}. \tag{5.16}$$

This equation gives the constant-speed parameterization of the geodesic passing through q with velocity vector w at $t = 0$. According to Example 3.5, the exponential map, $\exp : T_q(\mathscr{C}_2) \mapsto \mathscr{C}_2$ is given by:

$$\exp_q(w) = \cos(\|w\|)q + \sin(\|w\|)\frac{w}{\|w\|} \qquad (5.17)$$

The right column of Fig. 5.5 shows examples of geodesic paths between given in curves in $\mathscr{C}_2$. Recall that the left column shows the geodesic paths between the same curves in the pre-shape space $\mathscr{C}_1$. (The curves have been rescaled for display purposes. Recall that elements of $\mathscr{C}_1$ have length 2π while elements of the set $\mathscr{C}_2$ are of length one.) Thus, we can compare the nature of geodesics for the same curves under the two metrics. Although the differences between the two sets of geodesics are not drastic, they are still perceptible in some cases. It can be seen that for geodesics in $\mathscr{C}_1$, the corners and bends in the intermediate shapes are smoother than those for shapes along geodesics in $\mathscr{C}_2$. For example, in the topmost example, the corner remains sharper on the right side, as it deforms from one shape to other.

For any $q_2 \in \mathscr{C}_2$, the inverse of the exponential map at $q_1 \in \mathscr{C}_2$, denoted by $\exp_{q_1}^{-1} : \mathscr{C}_2 \to T_{q_1}(\mathscr{C}_2)$, is computed as follows: $\exp_{q_1}^{-1}(q_2) = v$, where:

$$v = \frac{\vartheta}{\sin(\vartheta)}(q_2 - \cos(\vartheta)q_1), \quad \text{where} \quad \vartheta = \cos^{-1}(\langle q_1, q_2 \rangle) . \qquad (5.18)$$

5.5 Shape Spaces

Of the different shape-preserving transformations—translation, rotation, scale, and re-parameterization—we have taken care of translation and scale in our representations. Additionally, the angle-function representation removes the rotation variability by fixing the average angle associated with a curve. The next step is to take care of variability introduced by different re-parameterizations of curves; the re-parameterization of curves was introduced in Sect. 5.2. We will also address the effect of different rotations in the SRVF representation. A mathematically elegant and convenient way to account for these variabilities is to define a re-parameterization group (and a rotation group, where needed) and to use the actions of these groups on pre-shape spaces established earlier. The orbits generated by such group actions will contain curves with the same shapes and, thus, will define equivalence classes of shapes. The resulting quotient spaces, preshape spaces modulo the re-parameterization group, will form the shape spaces for our analysis. Next we elaborate on this basic framework.

5.5.1 Removing Parameterization

We consider the two representations in order:

1. **Case 1**: This case is easy since the curves have been assumed to be arc-length parameterized, and, hence, no re-parameterization of a curve is possible. Additionally, we have fixed the rotation of a curve by imposing a constraint on its average angle function. With no variability to remove, the shape space is

identical to the pre-shape space, $\mathscr{S}_1 = \mathscr{C}_1$. All the previous discussions about the geometry, the Riemannian metric, and the construction of geodesics on the pre-shape space $\mathscr{C}_1$ apply verbatim to the shape space $\mathscr{S}_1$. Consequently, we have $d_{\mathscr{S}_1}(\theta_1, \theta_2) = d_{\mathscr{C}_1}(\theta_1, \theta_2)$.

2. **Case 2**: The task of unifying re-parameterization and defining equivalence classes is more complicated when arbitrary parameterizations are included in the representation. In this case where the curves are represented using their SRVFs, the variability introduced by having different parameterizations needs to be addressed.

Let Γ_I denote the set of all orientation-preserving diffeomorphisms $[0,1] \to [0,1]$, as introduced in Sect. 4.3.3. ("orientation-preserving" means the diffeomorphism preserves direction, i.e., 0 maps to 0 and 1 maps to 1.) As described in Sect. 5.2, the re-parameterization of a curve $\beta : [0,1] \to \mathbb{R}^2$ by a $\gamma \in \Gamma_I$ is given by $\beta \circ \gamma$. In terms of the SRVFs, what is the effect of re-parameterization? Similar to the case of SRSFs in Chap. 4, it is given by the following group action. Define: $\mathscr{C}_2 \times \Gamma_I \to \mathscr{C}_2$ by

$$(q, \gamma) \mapsto (q \circ \gamma)\sqrt{\dot{\gamma}} \,. \tag{5.19}$$

The reason this is the correct action of Γ_I is as follows. Suppose we are given two curves β_1 and β_2, where each $\beta_i : [0,1] \to \mathbb{R}^2$. Assume they are related by an element $\gamma \in \Gamma_I$, so $\beta_2 = \beta_1 \circ \gamma$. If q_1 and q_2 are the corresponding representative functions, what is the relation between q_1 and q_2? First, recall that $q_1(t) = \dot{\beta}_1(t)/\sqrt{|\dot{\beta}_1(t)|}$. Likewise, $q_2(t) = \dot{\beta}_2(t)/\sqrt{|\dot{\beta}_2(t)|}$. Using the chain rule, we then compute that

$$q_2(t) = \frac{\dot{\beta}_1(\gamma(t))\dot{\gamma}(t)}{\sqrt{|\dot{\beta}_1(\gamma(t))\dot{\gamma}(t)|}} = \frac{\dot{\beta}_1(\gamma(t))\dot{\gamma}(t)}{\sqrt{|\dot{\beta}_1(\gamma(t))|}\sqrt{\dot{\gamma}(t)}} = \frac{\dot{\beta}_1(\gamma(t))}{\sqrt{|\dot{\beta}_1(\gamma(t))|}}\sqrt{\dot{\gamma}(t)}$$
$$= q_1(\gamma(t))\sqrt{\dot{\gamma}(t)}.$$

Thus, we have justified the action of Γ_I defined above.

Similarly, the rotations of a curve can be represented by the action of $SO(2)$: $SO(2) \times \mathscr{C}_2 \to \mathscr{C}_2$:

$$(O, q) \mapsto \{t \mapsto Oq(t)\} \,. \tag{5.20}$$

How do the two group actions interact with each other?

Lemma 5.1. *The actions of $SO(2)$ and Γ_I on $\mathscr{C}_2$ commute.*

Proof. If we apply a rotation O and a re-parameterization γ to a curve β, the SRVF of the resulting curve $\tilde{\beta}(t) = O\beta(\gamma(t))$ becomes:

$$\tilde{q}(t) = \frac{\dot{\tilde{\beta}}(t)}{\sqrt{|\dot{\tilde{\beta}}(t)|}} = \frac{O\dot{\beta}(\gamma(t))\dot{\gamma}(t)}{\sqrt{|O\dot{\beta}(\gamma(t))\dot{\gamma}(t)|}} = O\frac{\beta(\gamma(t))}{\sqrt{|\dot{\beta}(\gamma(t))|}}\sqrt{\dot{\gamma}(t)} = O(q \circ \gamma)(t)\sqrt{\dot{\gamma}(t)} \,.$$

The last term is the same if we apply γ and O in any order. Hence, the two group actions commute. $\square$

This provides the equivalence class, or orbit, associated with a curve q under the actions of Γ_I and $SO(2)$:

$$[q] = \{Oq(\gamma(t))\sqrt{\dot{\gamma}(t)}|(\gamma, O) \in \Gamma_I \times SO(2)\} \ . \tag{5.21}$$

The shape space, using the SRVF representation, is defined by:

$$\mathscr{S}_2 = \mathscr{C}_2/(\Gamma_I \times SO(2)) \ . \tag{5.22}$$

As discussed previously in Sect. 4.10, there are some theoretical difficulties with this quotient construction. We would like to use the basic fact that if a compact Lie group G acts freely (i.e., no elements of M are fixed by $g \in G$ unless g is the identity) on a Riemannian manifold M by isometries, and the orbits are closed, then the quotient M/G is a manifold and inherits a Riemannian metric from M. The trouble is that while we have the product group $\Gamma_I \times SO(2)$ acting by isometries, the orbits are not closed. This because the space of diffeomorphisms is not compact and is not even closed with respect to either the $\mathbb{L}^2$ or the Palais metric, since a sequence of diffeomorphisms might approach a map that is not a diffeomorphism under either of these two metrics. Please refer to Sect. 4.10 for a discussion of this issue in the case of real-valued functions.

Similar to Sect. 4.10, we resolve this difficulty by using the closures of Γ_I-orbits, rather than by Γ_I-orbits themselves. (This step is related to the fact that, if $q \neq 0$ almost everywhere, then the closure of a Γ_I-orbit of q is actually the orbit of q under the monoid $\widetilde{\Gamma_I}$ of absolutely continuous, weakly-increasing functions.) Thus, if there is a sequence q_i in the orbit $[q_1]$, and this sequence converges to a function $\tilde{q}$ in $\mathscr{C}_2$ (with respect to the $\mathbb{L}^2$-metric), then we identify q_1 with $\tilde{q}$ in this quotient construction. As evidence that this method of resolving the problem has merit, it is not difficult to prove that in this situation, if we let β_1 and $\tilde{\beta}$ be the curves corresponding to q_1 and $\tilde{q}$, both β_1 and $\tilde{\beta}$ contain exactly the same points. (This is assuming that we normalize both of these paths to start at the same point.) We will henceforth refer to the quotient space as $\mathscr{C}_2/(\Gamma_I \times SO(2))$, and suppress the fact that we have to take the closure of the orbits. Although this quotient space will not be a Riemannian manifold, it can still be treated as a metric space with the distance inherited from $\mathscr{C}_2$.

Lemma 5.2. *The action of the product group $\Gamma_I \times SO(2)$ on $\mathscr{C}_2$ is by isometries with respect to the chosen metric.*

Proof. For a $q \in \mathscr{C}_2$, let $u, v, \in T_q(\mathscr{C}_2)$. Since $\langle Ou(t), Ov(t)\rangle = \langle u(t), v(t)\rangle$, for all $O \in SO(2)$ and $t \in [0, 1]$, the proof for $SO(2)$ follows. Now, fix an arbitrary element $\gamma \in \Gamma_I$ and define a map $\phi : \mathscr{C}_2 \to \mathscr{C}_2$ by $\phi(q) = (q, \gamma)$. A glance at the formula for (q, γ) confirms that ϕ is a linear transformation. Hence, its derivative $d\phi$ has the same formula as ϕ. In other words, the mapping $d\phi : T_q(\mathscr{C}_2) \to T_{(\gamma,q)}(\mathscr{C}_2)$ is given by: $u \mapsto \tilde{u} \equiv (u \circ \gamma)\sqrt{\dot{\gamma}}$. The Riemannian metric after the transformation is:

$$\begin{aligned}
\langle \tilde{u}, \tilde{v}\rangle &= \int_0^1 \langle \tilde{u}(t), \tilde{v}(t)\rangle \, dt \\
&= \int_0^1 \left\langle u(\gamma(t))\sqrt{\dot{\gamma}(t)}, v(\gamma(t))\sqrt{\dot{\gamma}(t)}\right\rangle dt = \int_0^1 \langle u(r), v(r)\rangle \, dr, \quad r = \gamma(t) \ .
\end{aligned}$$

Putting these two results together, the joint action of $\Gamma_I \times SO(2)$ on $\mathscr{C}_2$ is by isometries with respect to the chosen metric. $\square$

The lemma just proved is perhaps the most important reason for using the SRVF representation of curves.

5.6 Motivation for SRVF Representation

The main representation put forward for analyzing shapes of curves in this textbook is the SRVF. It is actually the main representation proposed for an *elastic* analysis of shapes of curves. So a natural question is: What is the fundamental need for this peculiar representation? Another question that is closely related is: What is an elastic analysis of shapes? In this section we try to answer these two and some other questions about choices of representations and metrics. As we shall see, there are three main reasons for selecting the SRVF representation in shape analysis of curves. These are:

1. The re-parameterization group, under the elastic metric, acts by isometries on the representation space of SRVFs.
2. If one uses an elastic metric for shape analysis of curves, then, under the SRVF representation, the elastic metric reduces to the simple $\mathbb{L}^2$ metric.
3. The space of SRVFs of fixed-length curves is a unit sphere under the $\mathbb{L}^2$ metric.

All these three points are fundamental and interrelated. We explain these items next, starting with a definition of the elastic metric.

5.6.1 What Is an Elastic Metric?

In elastic analysis of shapes, we consider deformations of curves resulting not just from bending but from stretching as well. The amount of deformation is quantified using a Riemannian metric called the *elastic metric*. Before we present that metric, we discuss the motivation behind it in general terms. Consider the two curves shown in Fig. 5.8a; each of these has two bumps but these bumps have slightly different shapes and placements on the two curves. If we have to deform the top curve into the bottom curve, what is a good way of doing it? It is clear that the deformation could be efficient if it matches the corresponding bumps on the two curves—first bump on the first curve to the first bump on the second curve, and the same for the second bumps. This way we can simply modify the shapes and locations of the bumps from one to another without creating or destroying these bumps. This matching requires compressing the long horizontal piece between the two bumps on the top curve into a shorter piece present on the bottom curve. Also, since the two curves have the same length, some part of the first curve has to be stretched to compensate for this compression. This motivation is similar to that presented in Sect. 4.4 where peaks and valleys are to be matched across real-valued functions. Figure 5.8b shows this desired matching of points across the two curves—the bumps register with the bumps and the horizontal pieces register with the horizontal pieces. Not only does this matching seem intuitively efficient but the actual path of deformation, from one curve to the other, also looks more natural, as shown in Fig. 5.9a. Notice that this deformation preserves

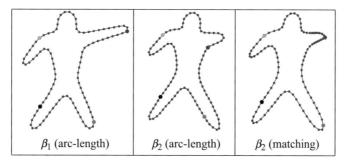

Fig. 5.7 Registration of points across two curves using the arc length and a convenient nonuniform sampling. Nonuniform sampling allows a better matching of features between β_1 and β_2

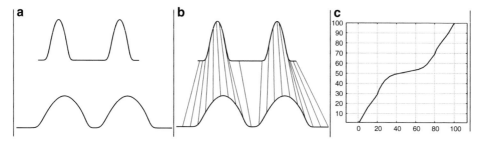

Fig. 5.8 Illustration of elastic metric. In order to compare shapes of the two curves shown in (**a**), some combination of bending and stretching seems efficient. The elastic metric measures the amounts of these deformations in a way that is invariant to the parameterizations of curves. An optimal matching of points along the curves, under a specific elastic metric, and the resulting optimal matching function, are shown in (**b**) and (**c**), respectively

the two bumps in all the curves along the deformation path while simultaneously changing their shapes and positions. This is termed an *elastic deformation*. On the other hand, if one uses a non-elastic metric, one based completely on bending one curve into the other, the resulting deformation will look unnatural, as shown in Fig. 5.9b. The non-elastic path shows additional bends introduced in the curves since the algorithm is trying to bend the first curve into another and the bumps are located at different points along the curves. This non-elastic deformation is computed using Eq. 5.13, i.e., linear interpolation of angle functions of the two curves (under the arc-length parameterization).

As another motivation, consider the two human silhouettes shown in the left two panels of Fig. 5.7. Each one is parameterized by arc length, a fact highlighted by uniformly placing points along those curves. If we start from the top of the head and walk down the left side on those two curves, the corresponding points fall on the same features until we reach the right side of the bodies. By the time we reach the right hand, the tip of the hand in the left panel corresponds to the armpit of the second panel. This is another example to highlight the fact that arc length, or any other pre-determined parameterization, does not provide natural matching of features across curves. The right panel shows a nonlinear parameterization of the second silhouette where the points are perfectly matched with the left panel.

This example shows that a central step in the process of elastic deformation is *elastic matching*. We explain this important point further. Consider again the two curves shown in Fig. 5.8a. Let us fix the parameterization of the top curve to be arc length. That is, we are going to traverse that curve with speed equal to one. The question is: At what rate should we move along the second curve so that the matching points, the points reached on the two curves at the same times, are as close as possible under some geometric criterion? For this to happen, the rate of traversal on the second curve should be such that we reach the peaks and valleys on both the curves at the same time. A function that accomplishes this is shown in Fig. 5.8c. This is an increasing function whose derivative at any point gives the instantaneous speed of traversal on the second curve. This curve has a higher slope on the two corners but is very flat in the middle, denoting a slowing down in the middle when traversing the second curve. This is, of course, justified because, in order to match the horizontal pieces in the middle across curves, we have to travel slowly on the second curve. This rate function incorporates the desired matching between curves and any deformation that respects this matching will necessarily be an elastic deformation.

Now we make these ideas more precise using a "polar" representation of curves. Let $\beta : [0,1] \to \mathbb{R}^2$ be a curve in $\mathbb{R}^2$ and assume that for all $t \in [0,1]$, $\dot{\beta}(t) \neq 0$. Recall the log-speed, angle representation introduced earlier in Sect. 5.3.1. Here $\phi : [0,1] \to \mathbb{R}$ is a function defined as $\phi(t) = \log(|\dot{\beta}(t)|)$, and $\theta : [0,1] \to \mathbb{R}$ is defined such that $\dot{\beta}(t)/|\dot{\beta}(t)| = e^{i\theta(t)}$, $i = \sqrt{-1}$. (The complex plane is being identified with $\mathbb{R}^2$ in the standard way for this comparison.) Clearly, ϕ and θ completely specify $\dot{\beta}$, since for all t, $\dot{\beta}(t) = e^{\phi(t)}e^{i\theta(t)}$. Thus, we have defined a map from the space of planar curves to $\Phi \times \Theta$, where $\Phi = \{\phi : [0,1] \to \mathbb{R}\}$ and $\Theta = \{\theta : [0,1] \to \mathbb{R}\}$. This map is surjective; it is not injective, but two curves are mapped to the same pair (ϕ, θ) if and only if they are translates of each other, i.e., if they differ by an additive constant. Intuitively, ϕ tells us the (log of the) speed of traversal of the curve, while θ tells us the direction of the curve at each time t. We should note here that the curve β does not completely specify the function θ; if we add any integer multiple of 2π to θ, it will not change the original curve. However this ambiguity will not in any way interfere with the Riemannian metric that we are using on θ and ϕ.

In order to quantify the magnitudes of perturbations of β (and enable ourselves to do geometry on the space of these curves), we wish to impose a Riemannian metric on the space of curves that is invariant under translation, and we will do this by putting a metric on $\Phi \times \Theta$. First, we note that the tangent space of $\Phi \times \Theta$ at any point (ϕ, θ) is given by

$$T_{(\phi,\theta)}(\Phi \times \Theta) = \{(u,v) : u \in \Phi \text{ and } v \in \Phi\} = \Phi \times \Theta .$$

Suppose (u_1, v_1) and (u_2, v_2) are both elements of $T_{(\phi,\theta)}(\Phi \times \Theta)$. Let a and b be positive real numbers.

Definition 5.3 (Elastic Riemannian Metric for Planar Curves). For every point $(\phi, \theta) \in (\Phi \times \Theta)$, define an inner product on the tangent space $T_{(\phi,\theta)}(\Phi \times \Theta)$ as

$$\langle (u_1, v_1), (u_2, v_2) \rangle_{(\phi,\theta)} = a^2 \int_0^1 u_1(t)u_2(t)e^{\phi(t)}\, dt + b^2 \int_0^1 v_1(t)v_2(t)e^{\phi(t)}\, dt. \quad (5.23)$$

This defines a Riemannian metric on the manifold $\Phi \times \Theta$.

This metric has the interpretation that the first integral measures the amount of "stretching," since u_1 and u_2 are variations of the log-speed ϕ of the curve,

Fig. 5.9 Deformations between curves using (**a**) elastic and (**b**) non-elastic matchings

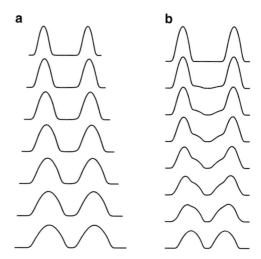

while the second integral measures the amount of "bending," since v_1 and v_2 are variations of the angle function θ of the curve. The constants a^2 and b^2 are weights that we choose depending on how much we want to penalize these two types of deformations. Therefore, this inner product has been called the *elastic metric*. Notice, this is not just one metric but rather a family of metrics parameterized by a and b.

Perhaps the most important property of this Riemannian metric is that the groups $SO(2)$ and Γ_I both act by isometries. To elaborate on this, recall that $O \in SO(2)$ can be identified with an angle $\theta_0 \in \mathbb{S}^1$, and, in our current notation, it acts on a curve β by $(\theta_0, \beta)(t) = e^{i\theta_0}\beta(t)$, and $\gamma \in \Gamma_I$ acts on β by $(\beta, \gamma)(t) = \beta(\gamma(t))$. Using our identification of the set of curves with the space $\Phi \times \Theta$ results in the following actions of these groups. $\theta_0 \in SO(2)$ and $\gamma \in \Gamma_I$ act on (ϕ, θ) by

$$(\theta_0, (\phi, \theta)) = (\phi, \theta_0 + \theta) \tag{5.24}$$

$$(\gamma, (\phi, \theta)) = (\phi \circ \gamma + \log \circ \dot{\gamma}, \theta \circ \gamma) \ . \tag{5.25}$$

We now need to understand the differentials of these group actions on the tangent spaces of $\Phi \times \Theta$. $SO(2)$ is easy, since each $\theta_0 \in \mathbb{S}^1$ acts by an additive translation, so its differential is the identity. The action of $\gamma \in \Gamma_I$ given in Eq. 5.24 is not linear, but affine linear, because of the additive term $(\log \circ \dot{\gamma})$. Hence, its action on the tangent space is the same, but without this additive term: $(\gamma, (u, v)) = (u \circ \gamma, \theta \circ \gamma)$, where $(u, v) \in T_{(\phi, \theta)}(\Phi \times \Theta)$, and $(u \circ \gamma, \theta \circ \gamma) \in T_{(\gamma, (\phi, \theta))}(\Phi \times \Theta)$. Combining these actions of $SO(2)(\equiv \mathbb{S}^1)$ and Γ_I with the above inner product on $\Phi \times \Theta$, it is an easy verification that these actions are by isometries, i.e.,

$$\langle (\theta_0, (u_1, v_1)), (\theta_0, (u_2, v_2)) \rangle_{(\theta_0, (\phi, \theta))} = \langle (u_1, v_1), (u_2, v_2) \rangle_{(\phi, \theta)}$$

$$\langle (\gamma, (u_1, v_1)), (\gamma, (u_2, v_2)) \rangle_{(\gamma, (\phi, \theta))} = \langle (u_1, v_1), (u_2, v_2) \rangle_{(\phi, \theta)}. \tag{5.26}$$

This result is related to Lemma 5.2.

Since we have identified the space of planar curves with $\Phi \times \Theta$, we may identify the space of planar shapes with the quotient space $(\Phi \times \Theta)/(\Gamma_I \times SO(2))$. Furthermore, since these group actions are by isometries with respect to all the metrics we have introduced above (no matter what values we assign to a and b), we get a corresponding two-parameter family of metrics on the quotient space $(\Phi \times \Theta)/(\Gamma_I \times SO(2))$. Note that in distinguishing between the structures (for example, geodesics) associated with these metrics, only the ratio of b to a is important, since if you multiply both by the same real number you just rescale the metric, which results in the same geodesics and all distances multiplied by the same constant. Given two shapes, we can find the geodesics between the corresponding elements of $(\Phi \times \Theta)$ with respect to any of these metrics (i.e., with respect to any choice of a and b). The choice of relative weights will determine whether these deformations are biased toward allowing more stretching or more bending.

5.6.2 Significance of the Square-Root Representation

Now that we have established a family of elastic Riemannian metrics, dictated by the ratio of weights a and b, and have motivated their use in shape comparisons, we ask the next logical question: Is there a particular choice of weights that will be especially natural and that will result in the geodesics being easier to compute? We now show that the SRVF representation is a natural answer to this question. Recall that the SRVF is the function $q : [0,1] \to \mathbb{R}^2$ given by $q(t) = \frac{\dot{\beta}(t)}{\sqrt{|\dot{\beta}(t)|}}$. Relating this to the (ϕ, θ) representation of the curve gives $q(t) = e^{\frac{1}{2}\phi(t)} \begin{bmatrix} \cos(\theta(t)) \\ \sin(\theta(t)) \end{bmatrix}$. A couple of simple differentiations show that if $(u, v) \in T_{(\phi,\theta)}(\Phi \times \Theta)$, then the corresponding tangent vector f to $\mathbb{L}^2([0,1], \mathbb{R}^2)$ at q is given by

$$ f = \frac{1}{2} e^{\frac{1}{2}\phi} u \begin{bmatrix} \cos(\theta) \\ \sin(\theta) \end{bmatrix} + e^{\frac{1}{2}\phi} \begin{bmatrix} -\sin(\theta) \\ \cos(\theta) \end{bmatrix} v \ . $$

Theorem 5.1. *The $\mathbb{L}^2$ metric on the space of SRVFs corresponds to the elastic metric on $\Phi \times \Theta$ with $a = \frac{1}{2}$ and $b = 1$.*

Proof. Let (u_1, v_1) and (u_2, v_2) denote two elements of $T_{(\phi,\theta)}(\Phi \times \Theta)$, and let f_1 and f_2 denote the corresponding tangent vectors to $\mathbb{L}^2([0,1], \mathbb{R}^2)$ at q. Computing the $\mathbb{L}^2$ inner product of f_1 and f_2 yields: $\langle f_1, f_2 \rangle =$

$$ \int_0^1 \left\langle \left(\frac{1}{2} e^{\frac{1}{2}\phi} u_1 \begin{bmatrix} \cos(\theta) \\ \sin(\theta) \end{bmatrix} + e^{\frac{1}{2}\phi} \begin{bmatrix} -\sin(\theta) \\ \cos(\theta) \end{bmatrix} v_1 \right), \left(\frac{1}{2} e^{\frac{1}{2}\phi} u_2 \begin{bmatrix} \cos(\theta) \\ \sin(\theta) \end{bmatrix} + e^{\frac{1}{2}\phi} \begin{bmatrix} -\sin(\theta) \\ \cos(\theta) \end{bmatrix} v_2 \right) \right\rangle dt $$

$$ = \int_0^1 \frac{1}{4} e^{\phi} u_1 u_2 + e^{\phi} v_1 v_2 \, dt \ . \tag{5.27} $$

This is the same as the right side of Eq. 5.23 with $a = \frac{1}{2}$ and $b = 1$. This implies that the $\mathbb{L}^2$ metric on the space of SRVF representations (i.e., q-functions) corresponds precisely to the elastic metric on $\Phi \times \Theta$, with those values of a and b. $\square$

There are several advantages of using this representation of curves for elastic shape analysis. Firstly, expressed in terms of the q-functions, the $\mathbb{L}^2$-metric is the "same" at every point of $\mathbb{L}^2([0,1], \mathbb{R}^2)$ (it is simply $\int_0^1 \langle f_1, f_2 \rangle \, dt$ and does not depend on the q-function at which these tangent vectors are defined), and we will thus have available more efficient ways of computing geodesics in our pre-shape and shape spaces. Secondly, under this representation and the elastic metric, the pre-shape space $\mathscr{C}_2$ is simply a unit hypersphere. This further simplifies the geodesic calculations to the extent that analytical formulas are available.

Connections with Fisher-Rao Metric Interestingly, the elastic metric in Eq. 5.23 is intimately connected to the extension of the nonparametric Fisher-Rao metric given in Chap. 4 for scalar-valued functions. In case of curves in $\mathbb{R}^1$, the second term disappears and this metric is identical to the metric discussed in the previous chapter. Conversely, the elastic metric in Eq. 5.23 can be viewed as an extension of the framework in Chap. 4 to curves in $\mathbb{R}^2$ (and higher dimensions).

We have chosen the SRVF $q(t) = e^{\frac{1}{2}\phi(t)} \begin{bmatrix} \cos(\theta(t)) \\ \sin(\theta(t)) \end{bmatrix}$ to represent a curve β for elastic shape analysis. This representation provides certain advantages when using the elastic metric with $a = \frac{1}{2}$ and $b = 1$. Are there other representations that provide similar advantages? It turns out that there is a whole family of representations, namely, $q_b = e^{\frac{1}{2}\phi} \begin{bmatrix} \cos(b\theta(t)) \\ \sin(b\theta(t)) \end{bmatrix}$, under which (i) the elastic metric becomes the $\mathbb{L}^2$-metric, (ii) the space of q_b functions is a subset of $\mathbb{L}^2$, and consequently (iii) the geodesics between shapes of curves are easy to establish. The reparameterization group continues to act on the corresponding spaces by isometries (this can be established as earlier). It is remarkable that this desirable structure is present for any real value of b. This implies that the geodesic computations can be simplified for any ratio of the two terms in the elastic metric. Now, the specific choice of $b = 1/2$ is even more special; it results in an additional structure that is very useful for studying closed curves. It can be shown that the space of $q_{1/2}$ functions representing unit-length, planar, closed curves is a Stiefel manifold. We will return to this issue when we discuss the shape analysis of closed curves.

5.7 Geodesic Paths in Shape Spaces

So far we have studied two pre-shape spaces of parameterized curves, have imposed Riemannian structures on them, and have computed geodesic paths under those metrics. Now we consider the problem of computing shortest paths in the corresponding shape spaces, under the inherited metrics. We will use the framework discussed in Sect. 3.6 that uses the inherited distance (Definition 3.16) for constructing geodesics in a quotient space. It should be noted that the shape space may not be a manifold, as discussed earlier in Sect. 4.10, but we can still impose a distance on it and talk about shortest paths, i.e., paths that achieve those distances. We will call these paths *geodesics* even though they do not originate from a Riemannian structure on the quotient space. We remind the reader of the general setup. According to Definition 3.17, if a group G acts on a Riemannian manifold M by isometries and the orbits under G are closed, then the inherited distance on M/G is given by

$$d_{M/G}([p_1], [p_2]) = \inf_{r_1 \in [p_1], r_2 \in [p_2]} d_M(r_1, r_2) = \inf_{r_2 \in [p_2]} d_M(p_1, r_2) \ .$$

If there exists a shortest path between $[p_1]$ and $[p_2]$ in the quotient space M/G, then this path may be obtained by forming geodesics (in M) between all possible crosspairs in the sets $[p_1]$ and $[p_2]$, and selecting the shortest. If no shortest path

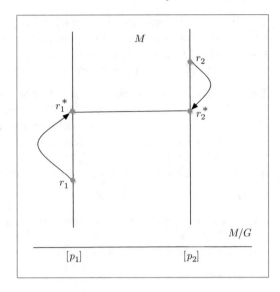

Fig. 5.10 A cartoon illustration of computing geodesics in the quotient space M/G. We keep one point $r_1 \in [p_1]$ fixed and search over the second orbit $[p_2]$ to find the shortest geodesic. This geodesic is perpendicular to all the orbits it intersects

exists in M/G, then this procedure will only yield a path in M/G whose length is within a small ϵ of $d_{M/G}([p_1], [p_2])$. Since the algorithms we are aiming at are numerical, we will be content with the approximate nature of this solution. Because the group G acts by isometries, we may fix p_1 in the first orbit, and search over all $r_2 \in [p_2]$ to find the shortest path (again, within ϵ). The geodesic between p_1 and r_2^*, the optimal point on the second orbit, is perpendicular to all the G-orbits that it meets (if these orbits are submanifolds). This last property can be instrumental in our search for the shortest geodesic. A cartoon illustration of this idea is presented in Fig. 5.10. The roles of M and M/G are played by the pre-shape and shape spaces, respectively. Summarizing this idea, the tool for computing geodesics in shape spaces between two orbits is based on repeatedly applying the geodesic computation in the pre-shape space in order to reach the shortest geodesic between all pairs formed by elements of the two orbits.

There are two cases to investigate—one for angle function representations and one for the SRVF representations. In the former case, the pre-shape space $\mathscr{C}_1$ is identical to the shape space $\mathscr{S}_1$ and need not be repeated. This leaves only the SRVF representation under the elastic metric.

As described earlier, the orbits of $SO(2) \times \Gamma_I$ in pre-shape space $\mathscr{C}_2$ are given by

$$[q] = \text{closure}\{O\sqrt{\dot{\gamma}}(q \circ \gamma)|(\gamma, O) \in \Gamma_I \times SO(2))\} \ .$$

The resulting shape space is a collection of these orbits:

$$\mathscr{S}_2 = \{[q] : q \in \mathscr{C}_2\} \ .$$

The shape space $\mathscr{S}_2$ is a metric space with the distance inherited from the pre-shape space $\mathscr{C}_2$. This distance between any two orbits $[q_1]$ and $[q_2]$ is given by

$$
\begin{aligned}
d_{\mathscr{S}_2}([q_0], [q_1]) &= \inf_{\tilde{q}_0 \in [q_0], \tilde{q}_1 \in [q_1]} d_{\mathscr{C}_2}(\tilde{q}_0, \tilde{q}_1) \\
&= \inf_{\gamma_1 \in \tilde{\Gamma}_I, (\gamma_2, O) \in \tilde{\Gamma}_I \times SO(2)} d_{\mathscr{C}_2}((q_0, \gamma_1), O(q_1 \circ \gamma_2)\sqrt{\dot{\gamma}}) \\
&= \inf_{(\gamma, O) \in \Gamma_I \times SO(2)} d_{\mathscr{C}_2}(q_0, O(q_1 \circ \gamma)\sqrt{\dot{\gamma}}) \ , \quad (5.28)
\end{aligned}
$$

where $d_{\mathscr{C}_2}$ is the distance function in the pre-shape space, which, in turn, is given by Eq. 5.15. (The last equality comes from the same approximation arguments as those used earlier in Sect. 4.7.3.) Thus, computing d_s requires solving an optimization problem over the space $\Gamma_I \times SO(2)$. Once the optimal pair (γ^*, O^*) is known, the actual geodesic between $[q_1]$ and $[q_2]$ in $\mathscr{S}_2$ is given by $[\alpha_t]$, where α_t is the geodesic in $\mathscr{C}_2$ between q_1 and $\sqrt{\dot{\gamma}^*}O^*(q_2 \circ \gamma^*)$ (using Eq. 5.14):

$$\alpha(\tau) = \frac{1}{\sin(\theta)} \left[\sin(\theta(1 - \tau))q_1 + \sin(\tau\theta)O^*(q_2 \circ \gamma^*) \right] \tag{5.29}$$

where $\theta = \cos^{-1}(\langle q_1, O^*(q_2 \circ \gamma^*) \rangle)$.

At this point, we pause to ask a fundamental question: for a given pair of orbits, $[q_1]$ and $[q_2]$, does there necessarily exist an optimal element $(\gamma^*, O^*) \in \Gamma_I \times SO(2)$ such that the distance $d_{\mathscr{C}_2}(q_1, \sqrt{\dot{\gamma}^*}O^*(q_2 \circ \gamma^*))$ realizes the infimum in the above equation? Since equivalence classes $[\cdot]$ are defined using closures of the orbits under $\Gamma_I \times SO(2)$, the optimal points may actually be on the boundaries of these classes. However, the numerical procedures we will use for this optimization will provide points in $\Gamma_I \times SO(2)$ that are arbitrarily close to the optimal points. For example, in certain cases, the optimal matching may be achieved by collapsing a whole section of the first curve to a single point of the second curve. Clearly the corresponding re-parameterization would not be a diffeomorphism. In cases like this, our methods will produce a correspondence that matches a large section of the first curve to a very small section of the second curve, rather than a single point.

So, the remaining problem is to solve the joint optimization on $\Gamma_I \times SO(2)$. For each of the components—rotation and re-parameterization—there exist certain optimization techniques that can be applied here. Our strategy is to develop algorithms for optimizing over each of these groups individually and then to iterate between the individual solutions. A closer look at that distance function reveals the following:

$$\underset{\gamma \in \Gamma_I, O \in SO(2)}{\arg\min} \cos^{-1}\left\langle q_1, \sqrt{\dot{\gamma}}O(q_2 \circ \gamma) \right\rangle = \underset{\gamma \in \Gamma_I, O \in SO(2)}{\arg\min} \|q_1 - \sqrt{\dot{\gamma}}O(q_2 \circ \gamma)\|^2,$$
$$\tag{5.30}$$

where the last norm is simply the $\mathbb{L}^2$ norm on the space $\mathbb{L}^2([0, 1], \mathbb{R}^2)$. This equality says that minimizing the arc length on a sphere is the same as minimizing the square of the chord length. If one is minimized then so is the other. Therefore, for the purpose of finding the minimizer, we can use the $\mathbb{L}^2$ norm, and that opens up the possibility of a computationally efficient solution. Now we have the problem:

$$(\gamma^*, O^*) = \underset{\gamma \in \Gamma_I, O \in SO(2)}{\arg\min} \|q_1 - \sqrt{\dot{\gamma}}O(q_2 \circ \gamma)\|^2$$
$$= \underset{\gamma \in \Gamma_I, O \in SO(2)}{\arg\max} \left\langle q_1, \sqrt{\dot{\gamma}}O(q_2 \circ \gamma) \right\rangle. \tag{5.31}$$

This uses the fact that $\|O(q_2 \circ \gamma)\sqrt{\dot{\gamma}}\| = \|q_2\|$.

First, we consider the optimization over $SO(2)$. For a fixed $\gamma \in \Gamma_I$, the optimization problem in Eq. 5.31 over $SO(2)$ is solved as shown in Example 3.13. Let $\tilde{q}_2 = \sqrt{\dot{\gamma}}(q_2 \circ \gamma)$ and define:

$$O^* = \underset{O \in SO(2)}{\arg\max} \langle q_1(t), O\tilde{q}_2(t) \rangle \, dt$$

$$= \begin{cases} UV^T & \text{if } \det(A) > 0 \\ U \begin{bmatrix} 1 & 0 \\ 0 & -1 \end{bmatrix} V^T & \text{otherwise} \end{cases} \qquad (5.32)$$

Here $U\Sigma V^T = \text{svd}(A)$, for $A = \int_0^1 q_1(t)\tilde{q}_2^T(t)dt$. The more difficult problem of optimizing over the re-parameterization group Γ_I is described in the next section.

5.7.1 Optimal Re-Parameterization for Curve Matching

For the current rotation $O \in SO(2)$, let $\tilde{q}_2 = Oq_2$ and define a cost function: $H : \Gamma_I \to \mathbb{R}_{\geq 0}$ by

$$H(\gamma) = \int_0^1 \|q_1(t) - \sqrt{\dot{\gamma}(t)}\tilde{q}_2(\gamma(t))\|^2 dt . \qquad (5.33)$$

Our goal is to find a minimum of H in Γ_I. There are at least two ways of solving this problem. One is to use the dynamic programming algorithm (DPA) given in Appendix B and another is to use a gradient approach. The use of DPA seems better in this problem, both for getting an optimal solution and from a computational perspective. Therefore, we will focus on that approach first.

Since H is additive over the path $(t, \gamma(t))$, the DPA can be used to solve the minimization problem in Eq. 5.33. A standard version of the DPA has already been described in Appendix B and need not be repeated here. Instead, we will just provide a short discussion on its implementation. This algorithm is based on forming an $N \times N$ square-grid in $[0,1]^2$ and approximating re-parameterization (i.e., γ) functions by piecewise linear paths through the grid points. Furthermore, we restrict to those paths that (1) start at $(0,0)$, (2) end at $(1,1)$, and (3) have positive slope p/q, where p and q are positive integers, and an upper bound has been chosen for p and q. The DPA provides an iterative mechanism for searching over all such paths and results in an approximation to the optimal γ. We will denote that estimated piecewise linear graph by γ^*:

Figure 5.11 shows an example of minimizing the cost function H in Eq. 5.33 using the DPA. In this figure, panels (a) and (b) show the two curves β_1 and β_2 that are represented by their respective SRVFs q_1 and q_2 in Eq. 5.33. Shown in panel (d) is the optimal re-parameterization γ^* of the second curve that minimizes the function H. Another way to show this function is to mark the corresponding points on the two curves $\beta_1(t)$ and $\beta_2(\gamma^*(t))$; this is shown in panel (c).

5.7.2 Geodesic Illustrations

The computations of geodesic paths in shape spaces are illustrated here using simple examples. We start with Fig. 5.12 that shows a geodesic path between two curves we have seen earlier in Fig. 5.4. This time the geodesic is computed between their representations in $\mathscr{S}_2$. In the top row we plot the two curves. In the bottom row, first we display the geodesic path between these two curves in the non-elastic

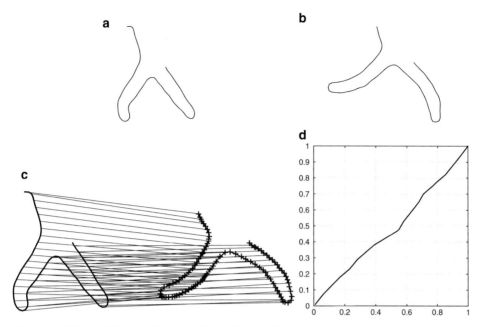

Fig. 5.11 (**a**) and (**b**) show the curves β_1 and β_2, (**c**) shows the optimal correspondence between points on two curves, and (**d**) shows the optimal γ^*

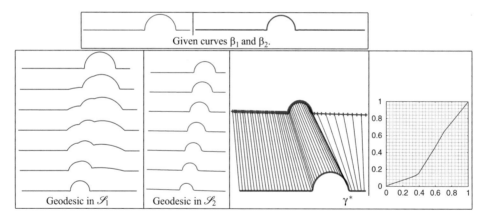

Fig. 5.12 Examples of geodesic paths in $\mathscr{S}_1$ and $\mathscr{S}_2$

space $\mathscr{S}_1$. Next, we see the corresponding example in the shape space $\mathscr{S}_2$, and, finally, we see the optimal registration between the two curves in two different ways. The first of the two panels uses thin lines to connect $\beta_1(t)$ with $\beta_2(\gamma^*(t))$ for some t while the right panel shows a plot of γ^* itself. These results suggest that the matching of points across the curves is nonlinear; some intervals have been stretched and others have been compressed in order to better match the points. The result, as seen in the geodesic path, is that one curve is elastically deformed into the other curve. This effect is further highlighted when we compare the geodesic between the same two curves in $\mathscr{S}_1$. This comparison clearly shows the bending-only nature of the deformation for geodesics in $\mathscr{S}_1$ and the elastic nature of the deformations in $\mathscr{S}_2$.

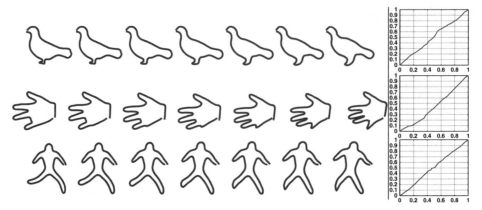

Fig. 5.13 Examples of geodesic paths between shapes. The left part shows the geodesic while the right part shows the optimal γ^* function used in finding that geodesic

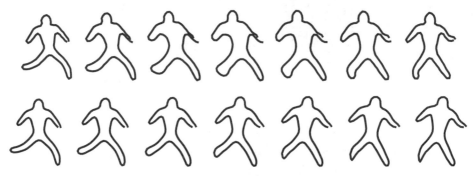

Fig. 5.14 Comparisons of geodesics between the same two curves in spaces $\mathscr{S}_1$ (*top*) and $\mathscr{S}_2$ (*bottom*), respectively

Figures 5.13 and 5.14 show some additional examples of these geodesics. Figure 5.13 shows geodesic paths between pairs of similar shapes in $\mathscr{S}_2$. Although the original shapes are closed curves, we treat them as elements of $\mathscr{S}_2$ (without the closure condition, which is handled in the next chapter). As a point of comparison, Fig. 5.14 shows geodesic paths between the same pair of shapes in both $\mathscr{S}_1$ and $\mathscr{S}_2$.

Figure 5.15 shows an example of a geodesic path between two face profiles. These profiles belong to the same person but with two different facial expressions: one with the mouth closed and other with the mouth open. The two individual profiles are simply the contours of the faces looking sideways and are shown in the leftmost panel. This panel also shows the optimal matching between the two faces obtained using the DPA. It is interesting to note that parts in one profile match the corresponding parts in the second, even though the second profile has the mouth open. The nose matches the nose, the lips match the lips, and so on. A geodesic path between them in $\mathscr{S}_2$ is shown in the middle panel and the optimal re-parameterization $\gamma*$ is shown in the right panel. This γ^* shows that we need to stretch the first face in only one place—around the lips—with a jump in the corresponding location of γ^* and the rest of this function is quite linear.

To demonstrate a simple application of these geodesic distances in a practical application, we consider the problem of analyzing signatures. While the signatures

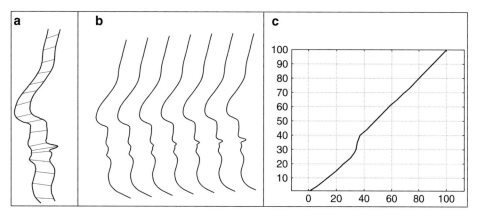

Fig. 5.15 Elastic geodesics between two facial profiles in $\mathscr{S}_2$: left panel shows two facial profiles, middle panel shows the geodesic path between them on $\mathscr{S}_2$, and right panel shows the optimal γ

Fig. 5.16 *Top two rows*: elastic geodesics between two handwritten letters in $\mathscr{S}_2$. Each row shows the geodesic path between the letters, the corresponding points, and the optimal γ. *Bottom row*: elastic matching and geodesic between two words

are quite complicated, consisting of several letters and symbols, we take some individual letters of a person's signature and study their shape variations. Using the methods described in this section, we compute geodesics between each pair in $\mathscr{S}_2$ and note the pairwise geodesic distances. Figure 5.16 shows examples of elastic geodesic paths between handwritten letters.

5.8 Gradient-Based Optimization Over Re-Parameterization Group

Even though the DPA provides a reasonable solution for optimization of H over Γ_I, such an algorithm will not be applicable in all circumstances. As an example, in the next chapter we will study shape analysis of closed curves and the additional constraints present in that problem results on a cost function that is not additive over the graph $(t, \gamma(t))$. Thus the DP algorithm does not apply directly. In situations such as these, it will be good to have a more general methodology that is more universally applicable. Since we are trying to solve a minimization problem, a gradient-based approach seems broadly applicable.

To develop a framework for a gradient approach, we develop some useful terminology first. We remark at the outset that this development overlooks certain mathematical technicalities in order to reach efficient numerical implementations of the alignment process. The issue here is that the shape space does not have a manifold structure even though the pre-shape space is a manifold. Due to this situation, we view the shape space only as a set with a distance inherited from the pre-shape space. However, in the context of deriving numerical procedures, such as the gradient-based approach for alignment, we will utilize ideas such as the tangent spaces of orbits and the tangent spaces of the quotient space to help out conceptually.

Let $[q]_{\Gamma_I}$ denote the orbit of $q \in \mathscr{C}_2$ under the action of Γ_I (as apposed to $[q]$, which denotes the closure of this orbit). We want to develop our understanding of the tangent space $T_q([q]_{\Gamma_I})$. The tangent space $T_q([q]_{\Gamma_I})$ can be identified to the space $T_{\gamma_{id}}(\Gamma_I)$, where $T_{\gamma_{id}}(\Gamma_I)$ is the set of all smooth functions that are zero at the boundaries:

$$T_{\gamma_{id}}(\Gamma_I) = \{v : [0,1] \to \mathbb{R} | v(0) = 0, v(1) = 0, \quad v \text{ is a smooth function}\} .$$

To formalize this identification, let ϕ be a mapping from Γ_I to $[q]_{\Gamma_I}$ defined as: $\phi(\gamma) = \sqrt{\dot{\gamma}}(q \circ \gamma)$. Note that ϕ is simply defined using the action of Γ_I on $\mathscr{C}_2$. We are interested in the differential of ϕ at the identity, denoted by $d\phi_{\gamma_{id}}$. We know that $d\phi_{\gamma_{id}}$ is a linear mapping from $T_{\gamma_{id}}(\Gamma_I)$ to $T_q([q]_{\Gamma_I})$, but what form does it take? This is answered by the following lemma.

Lemma 5.3. *For any point $\gamma \in \Gamma_I$, the differential of $\phi : \Gamma_I \to [q]_{\Gamma_I}$ is the linear transformation $d\phi_\gamma : T_\gamma(\Gamma_I) \to T_{\sqrt{\dot{\gamma}}(q \circ \gamma)}([q]_{\Gamma_I})$ defined by the equation*

$$(d\phi_\gamma(v))(s) = \sqrt{\dot{\gamma}(s)}\dot{q}(\gamma(s))v(s) + \frac{1}{2\sqrt{\dot{\gamma}(s)}}\dot{v}(s)q(\gamma(s)) . \tag{5.34}$$

Proof. Let $\Psi : (-\epsilon, \epsilon) \times [0, 2\pi] \to [0, 2\pi]$ be a mapping such that $\Psi(t, \cdot)$ is a differentiable path in Γ_I passing through γ at $t = 0$. Let the velocity of this path at $t = 0$ be given by $v \in T_\gamma(\Gamma_I)$. In other words, this path satisfies the following initial conditions:

$$\Psi(0, s) = \gamma(s), \quad \Psi_t(0, s) = v(s), \quad \Psi_s(0, s) = \dot{\gamma}(s), \quad \Psi_{ts}(0, s) = \dot{v}(s) .$$

Since $\Psi(t, \cdot)$ is a path in Γ_I, $\phi(\Psi(t, \cdot))$ is the corresponding path in $[q]_{\Gamma_I}$, and since the initial velocity of $\Psi(t, \cdot)$ is v, the initial velocity of the corresponding path in $[q]_{\Gamma_I}$ is $d\phi_\gamma(v)$. Using the fact that

$$\phi(\Psi(t, s)) = \sqrt{\Psi_s(t, s)}q(\Psi(t, s)) ,$$

we get:

$$\frac{d}{dt}|_{t=0}(\phi(\Psi(t,s)) = \left(\sqrt{\Psi_s(t,s)}\dot{q}(\Psi(t,s))\Psi_t(t,s) + \frac{1}{2\sqrt{\Psi_s(t,s)}}\Psi_{t,s}q(\Psi(t,s))\right)|_{t=0}$$

$$= \sqrt{\dot{\gamma}(s)}\dot{q}(\gamma(s))v(s) + \frac{1}{2\sqrt{\dot{\gamma}(s)}}\dot{v}(s)q(\gamma(s)) ,$$

which is the desired formula for $d\phi_\gamma(v)$. □

For $\gamma = \gamma_{id}$, we have

$$(d\phi_{\gamma_{id}}(v))(s) = \dot{q}(s)v(s) + \frac{1}{2}q(s)\dot{v}(s), \quad s \in [0,1] . \tag{5.35}$$

This result can be used to specify the tangent space $T_{[q]}(\mathscr{S}_2)$. An orthonormal basis for $T_{\gamma_{id}}(\Gamma_I)$ under the Palais metric is given by $\{(\frac{1}{n\pi\sqrt{2}}\sin(2\pi nt),$ $\frac{1}{n\pi\sqrt{2}}(\cos(2\pi nt) - 1))|n = 1, 2, \dots\}$. Then, using Lemma 5.3, the set $\{d\phi_{\gamma_{id}}(v_1),$ $d\phi_{\gamma_{id}}(v_2), \dots\}$ spans the vector space $T_q([q]_{\Gamma_I})$. The elements of this set span $T_q([q]_{\Gamma_I})$, but they do not necessarily form an orthonormal basis, since the mapping $d\phi_{\gamma_{id}}$ is not an isometry. In case one needs an orthonormal basis, we can use the Gram-Schmidt algorithm to make these elements orthogonal. Also, note that the space tangent to the $SO(2)$ orbit of q is a one-dimensional space spanned by the function

$$u(t) = \begin{bmatrix} 0 & -1 \\ 1 & 0 \end{bmatrix} q(t) .$$

Consequently, we can define the tangent space:

$$T_{[q]}(\mathscr{S}_2) = \{w \in \mathbb{L}^2([0,1], \mathbb{R}^2)| \langle w, q\rangle = 0, \langle w, d\phi_{\gamma_{id}}(v_i)\rangle = 0, \forall i, \text{ and } \langle w, u\rangle = 0\} . \tag{5.36}$$

With these results in hand, we return to the problem of minimizing H using a gradient method, with the following setup. In each iteration, we seek an incremental γ for minimizing H according to its negative gradient. We will denote the incremental group element by γ_k, for the k-th iteration, and $\gamma^{(k)}$ to denote the cumulative re-parameterization, i.e., $\gamma^{(k)} = \gamma_1 \circ \gamma_2 \circ \cdots \circ \gamma_k$. In this notation, at the $(k+1)$-st iteration, we seek the increment γ_{k+1} that minimizes $H(\gamma^{(k+1)})$.

Let $\tilde{q}_2$ denote the current element of the orbit $[q_2]$, i.e., $\tilde{q}_2 = (q_2 \circ \gamma^{(k)})\sqrt{\dot{\gamma}^{(k)}}$. For any $v \in T_{\gamma_{id}}(\Gamma_I)$, we need to find the directional derivative of H at γ_{id} in the direction of v. (This result can also be proven using the more general result given in Lemma 5.3.)

Theorem 5.2. *The directional derivative of H in the direction of $v \in T_{\gamma_{id}}(\Gamma_I)$ is given by*

$$\nabla_v H = -2\int_0^1 \left\langle q_1(t) - \tilde{q}_2(t), (\dot{\tilde{q}}_2(t)v(t) + \frac{1}{2}\tilde{q}_2(t)\dot{v}(t))\right\rangle dt . \tag{5.37}$$

Proof. Let $\alpha : (-\epsilon, \epsilon) \to \Gamma_I$ be a path such that $\alpha(0) = \gamma_{id}$ and $\dot{\alpha}(0) = v$. We can rewrite the cost function as

$$H(\alpha(\tau)) = \int_0^1 \langle q_1(t) - (\alpha(\tau), \tilde{q}_2)(t), q_1(t) - (\alpha(\tau), \tilde{q}_2)(t)\rangle dt .$$

Taking the derivative with respect to τ and setting $\tau = 0$, we obtain

$$-2 \int_0^1 \left\langle q_1(t) - \tilde{q}_2(t), \frac{d}{d\tau}|_{\tau=0}((\alpha(\tau), \tilde{q}_2)(t)) \right\rangle dt \ .$$

Substituting for the last term from Eq. 5.35, we obtain the desired result. □

Let $\{v_i, i = 1, 2, \dots\}$ form an orthonormal basis of the vector space $T_{\gamma_{id}}(\Gamma_I)$. As an example, we can take the set:

$$\{\frac{1}{\sqrt{2}\pi n} \sin(2\pi nt)|n = 1, 2, 3, \dots\} \bigcup \{\frac{1}{\sqrt{2}\pi n}(\cos(2\pi nt) - 1)|n = 1, 2, 3 \dots\} \ .$$

$$(5.38)$$

Then, the full gradient of the cost function H is approximated by

$$\nabla H = \sum_{i=1}^N (\nabla_{v_i} H) v_i \ , \ , \tag{5.39}$$

for a large N and can be used to find the incremental γ_{k+1} using a small step size $\delta > 0$.

Algorithm 4. *1. Set $k = 0$, $\gamma_k = \gamma^{(k)} = \gamma_{id}$.*
2. Compute $\tilde{q}_2 = (\gamma^{(k)}, q_2)$.
3. Compute the gradient of H with respect to γ according to Eq. 5.39.
4. Set $\gamma_{k+1} = \gamma_{id} + \delta \nabla H$ and compute $\gamma^{(k+1)} = \gamma^{(k)} \circ \gamma_{k+1}$.
5. If $\|\nabla H\|$ is small, then stop. Else, set $k = k + 1$ and return to Step 2.

One issue in this implementation is the choice of step size δ. If δ is very small, then of course the algorithm converges very slowly. If, however, it is too large, then the incremental function γ_{k+1} may not be a proper re-parameterization function. For instance, since $\dot{\gamma}_{k+1} = 1 + \delta \frac{d}{dt}(\nabla H)$ and the last term can have negative values, it is possible to have $\dot{\gamma}_{k+1}(t) < 0$ for some t. One can avoid this problem by having a small enough δ but the optimal choice of δ is not known in advance.

To overcome this issue, one possibility is to use a SRSF representation of γ. Define $h_k = \sqrt{\dot{\gamma}_k}$ and use h_k to represent γ_k in the optimization process. Note that the space of all h-functions is a subset of a sphere of unit radius; denote it by $\mathcal{H}$. We initialize with $\gamma_{k+1}(t) = t$, with the corresponding representation being $\mathbf{1}$, the constant function with value one. Now take the gradients of H, with respect to h_{k+1}, and update these individually. The directional derivative of H in a direction $c \in T_1(\mathcal{H})$ is given by

$$\nabla_h H(c) = \int_0^1 \left\langle q_1(t) - q_2(t), \left(2\dot{\tilde{q}}_2(t)\tilde{c}(t) + \tilde{q}_2(t)c(t)\right) \right\rangle dt, \quad \tilde{c}(t) = \int_0^t c(s)ds \ .$$

Form an approximate basis for the tangent space $T_1(\mathcal{H}) = \{f : [0, 1] \to \mathbb{R} | \langle f, \mathbf{1} \rangle = 0\}$ using

$$\{(\frac{1}{\sqrt{2\pi}} \sin(2\pi nt), \frac{1}{\sqrt{2\pi}} \cos(2\pi nt))|n = 1, 2, \dots\} \ ,$$

and approximate the gradient using $\sum_{i=1}^m \nabla_h H(c_i)c_i$, where the c_is are the basis elements. Then, update the h function according to

$$\mathbf{1} \mapsto h_{k+1} \equiv \cos(\|c\|)\mathbf{1} + \sin(\|c\|)\frac{c}{\|c\|} \ .$$

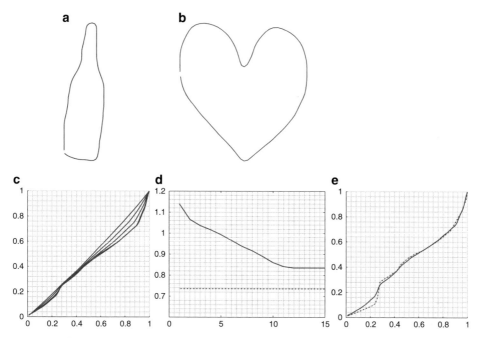

Fig. 5.17 (**a**), (**b**) The original two shapes represented by q_1 and q_2, respectively. (**c**) Gradient-based update of $\gamma^{(k)}$ versus k. (**d**) evolution of H in matching $\tilde{q}_2$ with q_1, (**e**) Estimated optimal γ using the gradient method (*solid line*) and the DPA (*broken line*)

This h_{k+1} in turn gives $\gamma_{k+1}(t) = \int_0^t h_{k+1}(s)^2 ds$ and thus $\gamma^{(k+1)}$.

Figures 5.17 and 5.18 show examples of the optimization of H using the function h. The top row shows the curves that are to be matched. They are represented by their SRVFs q_1 and q_2 in Eq. 5.33. The first panel in the bottom row shows the evolution of $\gamma^{(k)}$ as a function of the iteration index k. The initial condition for the iteration is chosen here to be γ_{id}. The middle panel shows the evolution of H as a function of k (solid line) and the right panel shows the optimal γ obtained using the gradient approach (drawn in solid line). To compare this procedure with the DPA, we show the optimal H value and the optimal re-parameterization function obtained using that algorithm using broken lines in the last two panels.

Looking at the results we can make the following remarks:

1. It is possible that the gradient method is caught in local solutions, and does not reach the globally optimal re-parameterization functions. The latter can be obtained using the DPA. This is a major limitation of gradient approaches in general.
2. We think that this problem can be mitigated by using two ideas. One is to use an adaptable step size in the gradient update. A fixed update seems too rigid to allow a convergence of the algorithm to the global solution. Second is improvise so that the gradient algorithm is allowed to move freely on the boundaries of the search domain. It seems that the algorithm reaches the boundary and becomes stuck there rather than continuing to update along the boundary.

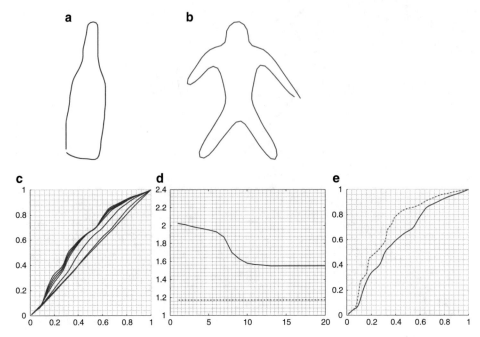

Fig. 5.18 Same as Fig. 5.17

3. In terms of the computational cost, the gradient method is an order or magnitude faster than the DPA in the current setup. So, the actual choice of a method is a choice essentially between speed and accuracy.

5.9 Summary

Here we provide a tabulated summary of the important elements of shape analysis of planar open curves (Table 5.1).

5.10 Exercises

5.10.1 Theoretical Exercises

1. Prove that for a continuously differentiable, arc-length parameterized curve $\beta : [0, 2\pi] \to \mathbb{R}^2$, there exists a unique angle function $\theta : [0, 2\pi] \to \mathbb{R}$ that is continuous and $\theta(0) \in [0, 2\pi)$.
2. Consider the angle function representation of arc-length parameterized curves and the set $\mathbb{L}^2([0, 2\pi], \mathbb{R})$ of all angle functions.

   a. Establish that $\mathbb{R}$ acts on $\mathbb{L}^2([0, 2\pi], \mathbb{R})$ according to the map $(\theta_0, \theta)(s) = \theta(s) + \theta_0$.

Table 5.1 Summary of shape analysis of planar curves

Property	Bending-only planar curves	Elastic planar curves
Representation	Angle function θ (under arc-length param.) $\dot{\beta}(s) = e^{i\theta(s)}$ with $|\dot{\beta}(s)| = 1$	SRVF q (arbitrary param.) $q(t) = \frac{\dot{\beta}(t)}{\sqrt{|\dot{\beta}(t)|}}$
Metric/larger space	$\mathbb{L}^2$ metric/$\mathbb{L}^2([0, 2\pi], \mathbb{R})$	$\mathbb{L}^2$ metric/$\mathbb{L}^2([0, 1], \mathbb{R}^2)$
Pre-shape space		
Pre-shape space	$\mathscr{C}_1 = \{\theta \in \mathbb{L}^2([0, 2\pi], \mathbb{R}) | \frac{1}{2\pi} \int_0^{2\pi} \theta(t)dt = \pi\}$ angle functions of curves of length 2π with a fixed orientation π	$\mathscr{C}_2 = \{q \in \mathbb{L}^2([0, 1], \mathbb{R}^2) | \int_0^1 \|q(t)\|^2 dt = 1\}$ SRVFs of curves of unit length
Normal space	$N_\theta(\mathscr{C}_1) = \text{span}\{\mathbf{1}\}$	$N_q(\mathscr{C}_2) = \text{span}\{q\}$
Tangent space	$T_\theta(\mathscr{C}_1) = \{v \in \mathbb{L}^2([0, 2\pi], \mathbb{R}) | \langle v, \mathbf{1} \rangle = 0\}$	$T_q(\mathscr{C}_2) = \{w \in \mathbb{L}^2([0, 1], \mathbb{R}^2) | \langle w, q \rangle = 0\}$
Geodesic distance	$d_{\mathscr{C}_1} = \|\theta_1 - \theta_2\|$	$d_{\mathscr{C}_2} = \vartheta = \cos^{-1}(\langle q_1, q_2 \rangle)$
Geodesics	straight line $\alpha(\tau) = \tau\theta_2 + (1 - \tau)\theta_1$	arc on a great circle $\alpha(\tau) = \frac{1}{\sin(\vartheta)}(\sin(\vartheta(1 - \tau))q_1 + \sin(\tau\vartheta)q_2)$
Shape Space		
Shape-preserving transformations	None	Rotation $SO(2)$, action: $(O, q) = Oq$ Re-parameterization Γ_I, action: $(\gamma, q) = \sqrt{\dot{\gamma}}(q \circ \gamma)$
Shape space	$\mathscr{S}_1 = \mathscr{C}_1$	$\mathscr{S}_2 = \mathscr{C}_2/(SO(2) \times \Gamma_I)$
Tangent space	$T_\theta(\mathscr{S}_1) = T_\theta(\mathscr{C}_1)$	$T_q(\mathscr{S}_2) = \{w \in T_q(\mathscr{C}_2) | w \perp T_q([q])\}$ $T_q([q]) = \{E(\sqrt{\dot{\gamma}}(\dot{q} \circ \gamma)v + \frac{1}{2\sqrt{\dot{\gamma}}}(q \circ \gamma)\dot{v})| v \in T_\gamma(\Gamma_I), E \in T_O(SO(2))\}$
Optimization	None	Iterate over the two alignment steps until convergence: 1. Find optimal rotation O^* in $SO(2)$ 2. Find optimal re-parameterization γ^* in Γ_I using DP Set $q_2^* = \sqrt{\dot{\gamma}^*}O^*(q_2 \circ \gamma^*)$
Geodesic distance	$d_{\mathscr{S}_1} = \|\theta_1 - \theta_2\|$	$d_{\mathscr{S}_2} = \vartheta = \cos^{-1}\langle q_1, q_2^* \rangle)$
Geodesics	Same as $\mathscr{C}_1$	$\alpha(\tau) = \frac{1}{\sin(\vartheta)}(\sin(\vartheta(1 - \tau))q_1 + \sin(\tau\vartheta)q_2^*)$

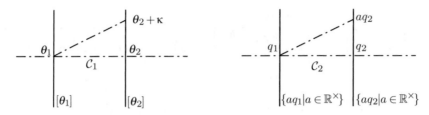

Fig. 5.19 $\mathscr{C}_1$ and $\mathscr{C}_2$ are orthogonal sections of respective group actions on relevant Hilbert spaces

b. Show that $\mathscr{C}_1$ is an orthogonal section (see Definition 3.19) of $\mathbb{L}^2([0, 2\pi], \mathbb{R})$ under this action of $\mathbb{R}$. See the left panel of Fig. 5.19.

3. Let θ_1, θ_2 be any two elements of $\mathscr{C}_1$. Show that the minimizer in $\mathrm{argmin}_{\kappa \in [0, 2\pi]} \|\theta_1 - (\theta_2 + \kappa)\|$ is zero. (This is actually a consequence of the previous result. If θ_1 is an element of an orthogonal section, and the action is by isometries, then the nearest element on θ_2 orbit is also an element of the same section.)

4. Find the angle function of the following half circles centered at the origin and having the following angles (with respect to the positive x axis: (1) 0^{deg} to 180^{deg}, (2) 90^{deg} to 270^{deg}, and (3) 180^{deg} to 360^{deg}. Note that we need the radius of the circle to be two so that these half circles have length 2π. Find the intersections of their orbits with the section $\mathscr{C}_1$.

5. Show that the mapping $\mathscr{Q} : \mathscr{F}_2 \to \mathscr{C}_2$ is a bijection. Recall that $\mathscr{F}_2$ is the set of absolutely continuous, unit-length curves with starting value zero.

6. Consider the SRVF representation of parameterized curves.

   a. Prove that the multiplication group $\mathbb{R}^\times$ acts on $\mathbb{L}^2([0, 1], \mathbb{R}^2)$ by the mapping $(a, q)(t) = aq(t)$.
   b. Then, verify that this action is not by isometries under the $\mathbb{L}^2$ metric.

7. For a point $q \in \mathbb{L}^2$ such that $\|q\| \neq 0$, define the *scaled-$\mathbb{L}^2$ metric* between two functions $v, w \in \mathbb{L}^2$ at a point q as

$$\langle\langle v, w \rangle\rangle_q = \frac{\langle v, w \rangle}{\langle q, q \rangle} .$$

Prove the following statements:

   a. The scaling group $\mathbb{R}^\times$ acts on $\mathbb{L}^2$ by isometries under the scaled-$\mathbb{L}^2$ metric.
   b. $\mathscr{C}_2$ is an orthogonal section of $\mathbb{L}^2$ under the scaling group, under the scaled-$\mathbb{L}^2$ metric. See the right side of Fig. 5.19.
   (We comment that as a consequence of $\mathscr{C}_2$ being an orthogonal section of $\mathbb{L}^2$ under the scaling group, we have that for any $q_1, q_2 \in \mathscr{C}_2$, $\mathrm{argmin}_{c \in \mathbb{R}_+} d_{sc}(q_1, cq_2) = 1$, where d_{sc} is the geodesic distance under the scaled-$\mathbb{L}^2$ metric.)

8. Show that the elastic metric, given in Definition 5.3, reduces to the Fisher-Rao metric, given in Definition 4.8, in the case of curves in $\mathbb{R}^1$.

9. Shown that the expression given in Eq. 5.23 provides a Riemannian metric on the space of curves under the joint log-speed and angle function representation.

10. a. Let (ϕ, θ) represent the log-speed and the angle functions of a parameterized curve β. For any re-parameterization function $\gamma \in \Gamma_I$, show that the corresponding functions of $\beta \circ \gamma$ are $(\phi \circ \gamma + \log \circ \dot{\gamma}, \theta \circ \gamma)$.

   b. Show that the re-parameterization group Γ_I acts on the space of curves according to the previous mapping.

   c. Prove that this group action is by isometries under the metric given in Eq. 5.23.

11. Using the (ϕ, θ) representation of a curve, define a new representation according to

$$h(t) = e^{\frac{1}{2}\phi(t)} \begin{bmatrix} \cos(\frac{\theta(t)}{k}) \\ \sin(\frac{\theta(t)}{k}) \end{bmatrix} ,$$

for any $k > 0$. Let (u, v) be an element of the tangent space $T_{(\phi, \theta)}(\Phi \times \Theta)$.

   a. Show that the tangent vector f to $\mathbb{L}^2([0, 1], \mathbb{R}^2)$ at h, corresponding to (u, v), is given by

$$f = \frac{1}{2}e^{\frac{1}{2}\phi}u \begin{bmatrix} \cos(\theta/k) \\ \sin(\theta/k) \end{bmatrix} + \frac{1}{k}e^{\frac{1}{2}\phi} \begin{bmatrix} -\sin(\theta/k) \\ \cos(\theta/k) \end{bmatrix} v .$$

   b. Next, compute the $\mathbb{L}^2$ inner product between such tangent vectors—$\langle f_1, f_2 \rangle$. Show that the result is equivalent to the elastic metric on $\Phi \times \Theta$ with $a = 1/4$ and $b = 1/k^2$.

   c. Plot the coordinate functions of h (with $k = 2$) for a unit circle. Compare that with the coordinate functions of the SRVF of the same curve.

5.10.2 Computational Exercises

1. Write a program to calculate the length of a piecewise linear curve $\beta : [a, b] \to \mathbb{R}^n$ under a given Riemannian metric on $\mathbb{R}^n$.

2. Compute the length of the curve $t \mapsto \beta(t) = (\sin(2\pi t), \cos(2\pi t))$ on the interval $[0, 1]$ using the Euclidean metric. Is this a constant-speed parameterization?

3. Given a discrete curve $\{\beta(t) \in \mathbb{R}^2 | t \in [0, \frac{1}{T}, \ldots, 1]\}$ and a discrete representation of a warping function $\{\gamma(t) | t \in [0, \frac{1}{T}, \ldots, 1]\}$, write a program to estimate the warped curve $\{\tilde{\beta}(t) \equiv \beta(\gamma(t)) | t \in [0, \frac{1}{T}, \ldots, 1]\}$. Since you will need a technique to interpolate between given values of β, the program should be able to choose one of the two options—piecewise linear and spline—for that purpose. Draw the following plots:

   a. Plot the two curves β and $\tilde{\beta}$ over each other to highlight differences in their shapes.

   b. Plot the coordinate functions before—$\beta_1(t)$ and $\beta_2(t)$ versus t—and after $\tilde{\beta}_1(t)$ and $\tilde{\beta}_2(t)$ versus t—warping.

   c. Plot the triple before $(t, \beta_1(t), \beta_2(t))$ in $\mathbb{R}^3$ and after $(t, \tilde{\beta}_1(t), \tilde{\beta}_2(t))$ warping.

4. Write a program to resample a given curve (in a discrete form) by the arc-length parameterization.

5. Write a program that (1) rescales a given curve β to length 2π, (2) resamples it by the arc-length parameterization (Problem 4), and (3) computes its angle function θ. Furthermore, add a constant to θ such that it becomes an element

of $\mathscr{C}_2$. Additionally, write a program that takes an angle function θ and a starting point $\beta(0)$ and reconstructs the curve β over the interval $[0, 2\pi)$.

6. Write a program to compute a discrete geodesic path between any two given curves in $\mathscr{S}_1$. Note that the set $\mathscr{C}_1$ is the same as $\mathscr{S}_1$, so the geodesics are the same in the two cases.

7. Write a program that (1) rescales a curve β to unit length and (2) computes its SRVF to form an element of $\mathscr{C}_2$. Additionally, write a program to reconstruct the curve from its SRVF using the initial condition $\beta(0) = 0$. Illustrate your results on a unit circle.

8. Given any two curves, β_1 and β_2, use the program written for Problem 7 to find their SRVFs q_1, q_2. Show that they satisfy the property $\|q_1 - q_2\| = \|(q_1, \gamma) - (q_2, \gamma)\|$ up to numerical precision, for any $\gamma \in \Gamma_I$.

9. Write a program to compute a discrete geodesic path between any two given curves in $\mathscr{C}_2$ and display these results by mapping SRVFs to the corresponding curve space.

10. Write a program to rotationally align one curve with another using Eq. 5.32 and display the results.

11. Implement the dynamic programming algorithm to minimize the cost function given in Eq. 5.33 and display the results.

12. Write a program that computes a discrete geodesic path between any two given curves in $\mathscr{S}_2$:

    a. Computes the scaled SRVFs of the given curves using Problem 7.
    b. Iteratively solve for the optimal re-parameterization (Problem 11) and optimal rotation (Problem 10) of the second curves to obtain the optimal alignment.
    c. Compute geodesic path between the SRVFs of the aligned curves using Problem 9.

13. Write a program to compute the shooting vector associated with a uniform-speed geodesic from one curve to another under the SRVF representation.

5.11 Bibliographic Notes

The earliest work in *elastic* shape analysis of planar curves, using complex square-roots of coordinates, appears in [125]. (We should mention that there are several other papers on alignment of curves, albeit without Riemannian considerations, such as [100].) A summary of different Sobolev metrics for shape analysis of planar curves can be found in [77]. Klassen et al. [54] was the first paper to focus on spaces formed by closed curves. Later on, several papers studied the use of elastic metrics for planar curves using complex representations [124, 126, 112].

The general elastic metric for planar curves was studied in [79]. Srivastava et al. [42, 43, 106] introduced the SRVF representation curves in general Euclidean spaces. The existence results for optimal re-parameterizations under SRVF representation has been discussed in [20], and some generalization of SRVF for variable relative weight between stretching/bending term is developed in [11]. The use of higher-order Sobolev metrics for elastic shape analysis is motivated in [10]. White [121] studied the extensions of complex square-root approaches to analysis of curves in more than two dimensions. The use of Riemannian optimization technique for speeding up the registration of curves, in lieu of the dynamic programming algorithm, has been studied in [39]. A transform similar to SRVF, but with arbitrary elasticity, has been introduced in [12]. While it has the strength of encoding arbitrary elasticity in the metric, it does not transform to a fully Euclidean space as is the case for SRVFs.

Some applications of planar, elastic shape analysis can be found in [70] (protein structure analysis), [61] (plant leaves), [2] (video coding), [22] (SONAR image analysis), and [1] (human activity recognition).

Chapter 6
Shapes of Planar Closed Curves

6.1 Goals and Challenges

In this chapter we are interested in analyzing shapes of planar curves that are closed. A closed curve is a curve that starts and ends at the same point. If differentiability is needed (for example, in the case of the angle function representation), then we will require the derivative at the initial point of the curve to agree with the derivative at the ending point. However, if only absolute continuity is required (for example, in the case of the square-root velocity representation), then it will be enough to assume the curve starts and ends at the same point. Consequently, these curves do not have any boundary and every point has neighbors on both sides. The boundaries of objects in images typically form simple closed curves in the image plane (simple curves are those curves that do not cross themselves) and shape analysis of imaged objects motivates the study of simple, closed curves. We will treat closure as an additional constraint on the sets of curves studied in the previous chapter and the question arises: how does this constraint alter the differential geometries of the previously discussed pre-shape and shape spaces? Further, how can we modify the current ideas for computing geodesics in these spaces, under the non-elastic and elastic representations of curves, in order to accommodate the closure constraint? Our main goals in this chapter are:

- To modify the previous two mathematical representations of planar curves, based on the angle functions and the square-root velocity functions (SRVFs), for restricting the analysis to planar, closed curves.
- To utilize the same Riemannian metrics for registering, deforming, measuring, and comparing shapes of such curves.
- Finally, to provide algorithms for performing tasks such as geodesic computations in the resulting pre-shape and shape spaces.

The main challenge in shape analysis of closed curves comes from the added closure restriction. In the previous chapter, planar curves have been represented by functions that are elements of a vector space (when using angle functions) or a unit sphere (when using SRVFs). Since these spaces have simple geometries, we could obtain explicit expressions for geodesics, at least in the pre-shape spaces. The closure constraint introduces a strong nonlinearity in the representation. As a quick example, let θ_1 and θ_2 be the angle functions of two arc-length parameterized *closed*

© Springer-Verlag New York 2016
A. Srivastava, E.P. Klassen, *Functional and Shape Data Analysis*,
Springer Series in Statistics, DOI 10.1007/978-1-4939-4020-2_6

curves. It will be made clear later on that their linear combination $c_1\theta_1 + c_2\theta_2$, for arbitrary c_1, $c_2 \in \mathbb{R}$, will not represent a closed curve. Two pictorial examples of this idea were presented in the previous chapter (see Fig. 5.6). The manifolds resulting from the closure constraint become relatively more complicated. Consequently, explicit expressions for geodesics are no longer available and we have to resort to numerical approaches for constructing geodesics.

6.2 Representations of Closed Curves

Let $\beta : [a, b] \to \mathbb{R}^2$ be a parameterized curve that is differentiable with respect to the parameter. The exact restrictions on the curves will depend on their representations and are clarified later in those sections. For studying shapes of closed curves, we impose an additional condition that $\beta(a) = \beta(b)$. In the case of the angle function representation, we also require that $\dot{\beta}(a) = \dot{\beta}(b)$ so that the derivative is continuous at a. In view of this condition, it is natural to have the domain of parameterization be the unit circle $\mathbb{S}^1$ for closed curves. If we think of the point $(1, 0) \in \mathbb{R}^2$ as the "starting point" or "origin" of $\mathbb{S}^1$, then $\mathbb{S}^1$ can be identified with the interval $[a, b]$ using the function $t \mapsto (\cos(2\pi \frac{(t-a)}{(b-a)}), \sin(2\pi \frac{(t-a)}{(b-a)})) \in \mathbb{R}^2$, under which $t = a$ and $t = b$ are both identified with the point $(1, 0)$. If we assume that the curve is continuously parameterized by $\mathbb{S}^1$, then our curve is automatically closed!

The main difference in parameterizations of open and closed curves is that the choice of starting point, or origin, for open curves is limited to just two points. In the following, assume that the interval of parameterization is $[a, b] = [0, 1]$. One end of the curve is designated $\beta(0)$ and the other $\beta(1)$. Once an end is chosen as a starting point, all re-parameterizations will have to respect that choice (recall that $\gamma(0) = 0$ and $\gamma(1) = 1$, for a $\gamma \in \Gamma_I$ in the previous chapter). In contrast, the number of choices for the placement of origin on a closed curve is infinite. A re-parameterization of a closed curve can not only change the rate of traversal along the curve but also the placement of the origin. Thus, a re-parameterization function for a closed curve is a certain diffeomorphism of $\mathbb{S}^1$ to itself. We will use Γ_S to denote the set of such γs, with the subscript S denoting that the underlying domain is $\mathbb{S}^1$ and not an interval.

Definition 6.1 (Re-parameterization of Closed Curve). If $\beta : \mathbb{S}^1 \to \mathbb{R}^n$ is a closed curve, and $\gamma : \mathbb{S}^1 \to \mathbb{S}^1$ is an orientation-preserving diffeomorphism ($\gamma \in \Gamma_S$), then we say that the composition $\beta \circ \gamma$ is a *non-singular re-parameterization* of β.

The visualization of such a γ is a little tricky since both the domain and the range are a unit circle and the graph of γ is actually a closed path on a torus. To visualize this function, we use the fact that there is a natural bijection between Γ_S and $\Gamma_I \times \mathbb{S}^1$. Consequently, we arbitrarily select points on the domain and range as origins, and then γ can be plotted as a function from $[0, 1]$ to itself.

Example 6.1. Let $\beta(t) = (\cos(t), \sin(t))$ be the arc-length parameterized unit circle, and define $\gamma : \mathbb{S}^1 \to \mathbb{S}^1$ by

$$\gamma(t) = t + \alpha \sin(t), \quad -1 < \alpha < 1 . \tag{6.1}$$

Then, $\tilde{\beta} = \beta \circ \gamma$ is a re-parameterized circle where the rate of traversal is no longer constant.

In Chap. 5, we discussed several mathematical representations of parameterized curves in $\mathbb{R}^2$ and selected two of them—the angle functions and the SRVFs—for further consideration. Also, we presented formulas and algorithms for computing geodesics between curves in shape spaces, under the chosen Riemannian metrics. We continue to use these representations and metrics, but this time focusing on closed curves:

1. **Angle Function Representation**: Suppose we have a curve $\beta : [0, 2\pi] \to \mathbb{R}^2$ that is para- meterized by arc length (i.e., $|\dot{\beta}(s)| = 1$ for all $s \in [0, 2\pi]$) and, in addition, assume that $\dot{\beta}$ is continuous. Then the only information contained in $\dot{\beta}(s)$ is its direction. The velocity vector $\dot{\beta}(s)$ can be identified with a unit complex number $e^{i\theta(s)}$, where $\theta(s)$ is the angle formed by this vector with the positive x-axis (i.e., the real axis). Note that the angle function is not affected by a rigid translation of the curve β. Also note that for each s, $\theta(s)$ is only well defined up to the addition of an integer multiple of 2π. Since we are assuming $\dot{\beta}(s)$ is continuous, we will choose $\theta(s)$ to be continuous.

   For representing the shape of a closed curve, one has to clarify one more characteristic, namely, its rotation index. As we traverse β, the velocity vector $\dot{\beta}$ goes through an integer number of full rotations. We call the number of these rotations the *rotation index* of the curve. It is well known that for a simple closed curve (recall that a *simple* closed curve is one that never meets itself), the rotation index is always plus or minus one; it is plus one for counterclockwise parameterizations (see, e.g., [23], p. 396). Due to our interest in the *simple* closed curves, we will restrict to the set of all closed curves with rotation index plus one. While this set is larger than the set of simple closed curves, it contains the set of simple closed curves as an open subset. Also, we will see that this larger set is a *complete* Riemannian manifold, a property that will be important later when we start computing geodesic paths between points on this set.

   Although we have fixed the direction and the speed of traversal along the curve, the starting point $\beta(0)$, referred to as the *origin*, is still free to vary along the curve. Since β is a closed curve, there is no natural choice of origin on β, an issue that we will deal with later. If we place the origin on a certain point of β, then θ becomes a function from $[0, 2\pi]$ to $\mathbb{R}$. Often it is convenient to extend θ to the entire real line, both in the domain and the range, using the definition: $\theta(2n\pi + s) = 2n\pi + \theta(s)$, for all $n \in \mathbb{N}$. (For this extension to be continuous, it is required that the rotation index is 1, which we have already assumed to be true.) Note that for a closed curve having rotation index 1, we have $\theta(2\pi) - \theta(0) = 2\pi$. In summary, the angle function of β is invariant to translation but it varies with rotation and the choice of origin on β.

   If the curve β is closed, its angle function θ also satisfies the additional conditions:

$$\int_0^{2\pi} \cos(\theta(s))ds = 0 \quad \text{and} \quad \int_0^{2\pi} \sin(\theta(s))ds = 0 \ . \tag{6.2}$$

These conditions are nonlinear in the sense that if θ_1 and θ_2 are two angle functions that satisfy these conditions, then their linear combination $c_1\theta_1 + c_2\theta_2$, for arbitrary $c_1, c_2 \in \mathbb{R}$, will, in general, not satisfy them.

2. **Square-Root Velocity Representation (SRVF)**: The second representation described in Sect. 5.3.1 is based on the square-root velocity function q of an absolutely-continuous curve $\beta : \mathbb{S}^1 \to \mathbb{R}^2$ given by $q(t) = \frac{\dot{\beta}(t)}{\sqrt{|\dot{\beta}(t)|}}$. Observe that we do not need the condition $\dot{\beta}(t) \neq 0$, and this representation exists even for curves with singular parameterizations. Note that when using the SRVF representation on closed curves, it is our practice to consider the domain of the functions β and q as the unit interval $[0, 1]$. Thus we are essentially identifying the unit interval $[0, 1]$ with the unit circle S^1, in a way that matches both endpoints of $[0, 1]$ with the single point $(1, 0)$ of S^1. In what follows, when the domain of an integral is indicated to be S^1, it should actually be *computed* as an integral over $[0, 1]$. If the curve β is of unit length, then q satisfies $\int_{\mathbb{S}^1} |q(t)|^2 dt = \int_{\mathbb{S}^1} |\dot{\beta}(t)| dt = 1$. Also, one can reconstruct the curve β from q, up to a translation, using the formula $\beta(t) = \int_0^t q(u)|q(u)| du$. If we choose to represent a curve by its SRVF, then the closure condition becomes

$$\int_{\mathbb{S}^1} \dot{\beta}(t) dt = \int_{\mathbb{S}^1} q(t)|q(t)| dt = 0 \ . \tag{6.3}$$

Once again, this closure condition is a nonlinear condition. If q_1 and q_2 are two SRVFs that satisfy Eq. 6.3, then their linear combination $c_1 q_1 + c_2 q_2$, for arbitrary $c_1, c_2 \in \mathbb{R}$, will, in general, not satisfy that condition.

6.2.1 Pre-shape Spaces

Next, we look at how the closure condition changes the pre-shape and the shape spaces studied in Chap. 5.

1. **Case 1**: Consider closed curves of length 2π with continuous first derivatives and rotation index 1, under arc-length parameterization, each represented by its angle function. The pre-shape space of such angle functions is given by

$$\mathscr{C}_1^c = \{\text{continuous } \theta \in \mathbb{L}^2([0, 2\pi], \mathbb{R})| \ \frac{1}{2\pi} \int_0^{2\pi} \theta(s) ds = \pi, \ \theta(2\pi) = \theta(0) + 2\pi,$$

$$\int_0^{2\pi} \sin(\theta(s)) ds = 0, \ \int_0^{2\pi} \cos(\theta(s)) ds = 0 \ \} \ \subset \mathscr{C}_1 \ \subset \mathbb{L}^2([0, 2\pi], \mathbb{R}). \tag{6.4}$$

Compared to Eq. 5.5, this set has three additional constraints; the two coming from Eq. 6.2 ensure the closure and the one $(\theta(2\pi) = \theta(0) + 2\pi)$ coming from the condition that the curve has rotation index one. Note that the condition of average rotation removes the rotation variability from the curves.

2. **Case 2**: This case involves unit-length, closed curves under arbitrary parameterizations, using SRVF representation:

$$\mathscr{C}_2^c = \{q \in \mathbb{L}^2(\mathbb{S}^1, \mathbb{R}^2)| \int_{\mathbb{S}^1} |q(t)|^2 dt = 1, \ \int_{\mathbb{S}^1} q(t)|q(t)| dt = 0\} \subset \mathbb{L}^2([0, 1], \mathbb{R}^2). \tag{6.5}$$

Compared to Eq. 5.6, the last term has been added to ensure closure of curves. In this representation we do not explicitly impose the condition of average rotation and will deal with it later when forming shape spaces.

We have identified these two pre-shape spaces of closed curves as $\mathscr{C}_1^c \subset \mathscr{C}_1 \subset \mathbb{L}^2([0, 2\pi], \mathbb{R})$ and $\mathscr{C}_2^c \subset \mathscr{C}_2 \subset \mathbb{L}^2(\mathbb{S}^1, \mathbb{R}^2)$ as spaces of interest for further analysis.

6.2.2 Riemannian Structures

By analogy with Sect. 5.4.1, we describe tangent spaces to $\mathscr{C}_1^c$ and $\mathscr{C}_2^c$ and endow them with Riemannian metrics.

1. **Case 1**: Let us start with the pre-shape space $\mathscr{C}_1^c$. For any two elements $u, v \in \mathbb{L}^2$, we have the usual inner product:

$$\langle u, v \rangle = \int_0^{2\pi} u(s)v(s)ds \ . \tag{6.6}$$

In order to define a Riemannian structure on $\mathscr{C}_1^c$, we will restrict this inner product to the tangent spaces of $\mathscr{C}_1^c$.

Theorem 6.1. *The pre-shape space $\mathscr{C}_1^c$ is a manifold.*

Proof. Let's begin by defining

$$\mathscr{D} = \{\text{continuous } \theta \in \mathbb{L}^2([0, 2\pi], \mathbb{R}) : \theta(2\pi) = \theta(0) + 2\pi\} \ . \tag{6.7}$$

$\mathscr{D}$ is an affine space, i.e., it is a linear subspace that has been translated so that it no longer passes through the origin. Note that $\mathscr{D}$ contains $\mathscr{C}_1^c$ as a subset. The tangent space of $\mathscr{D}$ is the vector space:

$$\mathscr{B} = \{\text{continuous } \theta \in \mathbb{L}^2([0, 2\pi], \mathbb{R}) : \theta(2\pi) = \theta(0)\}.$$

This space can also be characterized as the set of elements of $\mathbb{L}^2([0, 2\pi], \mathbb{R})$ that can be expressed as the difference between two elements of $\mathscr{D}$. An element of $\mathscr{D}$ needs to satisfy three additional conditions in order to be in $\mathscr{C}_1^c$. Hence we define the following map $\Phi_1 = (\Phi_1^1, \Phi_1^2, \Phi_1^3): \mathscr{D} \to \mathbb{R}^3$ by

$$\Phi_1^1(\theta) = \frac{1}{2\pi} \int_0^{2\pi} \theta(s) \, ds,$$

$$\Phi_1^2(\theta) = \int_0^{2\pi} \cos(\theta(s)) \, ds,$$

$$\Phi_1^3(\theta) = \int_0^{2\pi} \sin(\theta(s)) \, ds \ . \tag{6.8}$$

In terms of Φ_1, $\mathscr{C}_1^c$ can be rewritten as the pullback set $\Phi_1^{-1}(\pi, 0, 0)$. Using the $\mathbb{L}^2$ inner product, the directional derivative $d\Phi_1$, at a point $\theta \in \mathscr{D}$ and in the direction of an $f \in \mathscr{B}$, is given by

$$d\Phi_1^1(f) = \frac{1}{2\pi} \int_0^{2\pi} f(s) \, ds = \frac{1}{2\pi} \langle f, 1 \rangle,$$

$$d\Phi_1^2(f) = -\int_0^{2\pi} \sin(\theta(s))f(s)ds = -\langle f, \sin \circ \theta \rangle,$$

$$d\Phi_1^3(f) = \int_0^{2\pi} \cos(\theta(s))f(s)ds = \langle f, \cos \circ \theta \rangle \ . \tag{6.9}$$

We would like to conclude that $\mathscr{C}_1^c$ is a codimension-3 smooth submanifold of $\mathscr{D}$ using Theorem A.2. To do this, we need to verify that $d\Phi_1 : \mathscr{B} \to \mathbb{R}^3$ is surjective at every $\theta \in \Phi_1^{-1}(\pi, 0, 0)$ (posed as an exercise). Once we have that $d\Phi_1$ is surjective, we conclude that $\mathscr{C}_1^c$ is a codimension-3 smooth submanifold of $\mathscr{D}$.

$\square$

It follows that the tangent space $T_\theta(\mathscr{C}_1^c)$ is given by

$$T_\theta(\mathscr{C}_1^c) = \text{null space of } d\Phi_1$$
$$= \{f \in \mathscr{B} | f \perp N_\theta(\mathscr{C}_1^c)\}, \text{ where } N_\theta(\mathscr{C}_1^c) = \text{span}\{1, \cos \circ \theta, \sin \circ \theta\} .$$

Since the tangent space $T_\theta(\mathscr{C}_1^c)$ is a subspace of $\mathbb{L}^2$, we can restrict the $\mathbb{L}^2$ inner product to define a Riemannian structure on $\mathscr{C}_1^c$. For any two elements $u, v \in T_\theta(\mathscr{C}_1^c)$, the Riemannian metric is given in Eq. 6.6. The resulting geodesic distance on $\mathscr{C}_1^c$ will be denoted by $d_{\mathscr{C}_1^c}$.

2. **Case 2**: For the SRVF representation, we select a Riemannian structure as follows. Given a curve represented by $q \in \mathbb{L}^2(\mathbb{S}^1, \mathbb{R}^2)$, and the tangent vectors $u, v \in T_q(\mathbb{L}^2(\mathbb{S}^1, \mathbb{R}^2))$, the inner product between u, v is defined as

$$\langle u, v \rangle = \int_{\mathbb{S}^1} \langle u(t), v(t) \rangle \, dt . \tag{6.10}$$

Define a mapping $\Phi_2 : \mathbb{L}^2(\mathbb{S}^1, \mathbb{R}^2) \to \mathbb{R}^3$ by

$$\Phi_2^1(q) = \int_{\mathbb{S}^1} |q(t)|^2 dt,$$

$$\Phi_2^2(q) = \int_{\mathbb{S}^1} q^1(t)|q(t)|dt,$$

$$\Phi_2^3(q) = \int_{\mathbb{S}^1} q^2(t)|q(t)|dt , \tag{6.11}$$

where $q(t) = (q^1(t), q^2(t))$. Then $\mathscr{C}_2^c$ is defined as $\Phi_2^{-1}(1, 0, 0)$. The directional derivative of the map Φ_2 at a point q in the direction of $w \in \mathbb{L}^2(\mathbb{S}^1, \mathbb{R}^2)$ is given by

$$d\Phi_2^1(w) = \int_{\mathbb{S}^1} \langle w(t), q(t) \rangle \, dt$$

$$d\Phi_2^2(w) = \int_{\mathbb{S}^1} \left\langle w(t), \frac{q^1(t)}{|q(t)|} q(t) + |q(t)| \mathbf{e}^1 \right\rangle dt$$

$$d\Phi_2^3(w) = \int_{\mathbb{S}^1} \left\langle w(t), \frac{q^2(t)}{|q(t)|} q(t) + |q(t)| \mathbf{e}^2 \right\rangle dt ,$$

where $\mathbf{e}^1 = [1 \ 0]^T$ and $\mathbf{e}^2 = [0 \ 1]^T$. We note that at the points where $|q(t)| = 0$, we assign $\frac{q^1(t)}{|q(t)|} q(t)$, $\frac{q^2(t)}{|q(t)|} q(t)$ equal to zero. Therefore, having $|q(t)| = 0$ in the denominator is not a problem in these integrands.

Theorem 6.2. *The pre-shape space $\mathscr{C}_2^c$ is a manifold.*

Proof. Since $\mathscr{C}_2^c = \Phi_2^{-1}(1, 0, 0)$, we can show that it is a submanifold of $\mathbb{L}^2(\mathbb{S}^1, \mathbb{R}^2)$ if $d\Phi_2$ is surjective, using Theorem A.2. To establish that, we consider the three functions appearing in the above integrals:

$$q(t), \left(\frac{q^1(t)}{|q(t)|} q(t) + |q(t)| \mathbf{e}^1 \right), \text{ and } \left(\frac{q^2(t)}{|q(t)|} q(t) + |q(t)| \mathbf{e}^2 \right).$$

To show that $d\Phi_2$ is surjective, we need to show these three functions are linearly independent. The proof is as follows: if they are dependent then there exist real constants a, b, and c such that

$$aq(t) + b(A(t)q(t) + B(t)\mathbf{e}^1) + c(C(t)q(t) + B(t)\mathbf{e}^2) = 0 ,$$

where A, B, and C are the scalar functions given by $A(t) = \frac{q^1(t)}{|q(t)|}$, $B(t) = |q(t)|$, and $C(t) = \frac{q^2(t)}{|q(t)|}$. This simplifies to $-B(t)(b\mathbf{e}^1 + c\mathbf{e}^2) = (a + bA(t) + cC(t))q(t)$. Clearly this implies that for all t, $q(t)$ is in the same direction as the constant vector $b\mathbf{e}^1 + c\mathbf{e}^2$. This proves that for any q-function that doesn't lie completely in a single one-dimensional subspace, the function Φ_2 is surjective, so the pre-shape space is a manifold except at those points. These exceptional functions correspond to curves that lie entirely in a straight line in $\mathbb{R}^2$. We have proven that $\mathscr{C}_2^c$ is a manifold at every function q whose values do not lie entirely in a one-dimensional subspace of $\mathbb{R}^2$.

So now consider a function $q \in \mathscr{C}_2^c$ whose values *do* lie entirely in a one-dimensional subspace. Without loss of generality, we may assume that $q(t) = (r(t), 0)$, so $q^1(t) = r(t)$ and $q^2(t) = 0$. Then, the three functions we are considering become $(r(t), 0)$, $(2|r(t)|, 0)$, and $(0, |r(t)|)$. Since $\int_{\mathbb{S}^1} q(t)|q(t)|dt = 0$ and $\int_{\mathbb{S}^1} |q(t)|^2 dt = 1$, it follows that $r(t)$ must take both positive and negative values. From this it follows easily that the first two functions are linearly independent, and even more easily that the third is independent of the first two. Thus, $\mathscr{C}_2^c$ is a codimension-three submanifold of $\mathbb{L}^2(\mathbb{S}^1, \mathbb{R}^2)$. $\square$

In order to define the space of tangent vectors to $\mathscr{C}_2^c$, we derive the normal space of $\mathscr{C}_2^c$ in $\mathbb{L}^2(\mathbb{S}^1, \mathbb{R}^2)$ at q first. For any $q \in \mathscr{C}_2^c$, the space of normals to $\mathscr{C}_2^c$ in $\mathbb{L}^2(\mathbb{S}^1, \mathbb{R}^2)$ is

$$N_q(\mathscr{C}_2^c) = \text{span} \left\{ q(t), \left(\frac{q^1(t)}{|q(t)|} q(t) + |q(t)| \mathbf{e}^1 \right), \left(\frac{q^2(t)}{|q(t)|} q(t) + |q(t)| \mathbf{e}^2 \right) \right\}.$$
$$(6.12)$$

and the space of tangent vectors is

$$T_q(\mathscr{C}_2^c) = \left\{ w : \mathbb{S}^1 \to \mathbb{R}^2 | w \perp N_q(\mathscr{C}_2^c) \right\}. \tag{6.13}$$

We reiterate that while it may appear that the terms $\frac{q^1(t)}{|q(t)|}$, $\frac{q^2(t)}{|q(t)|}$ are not well defined in situations where $q(t)$ vanishes, it is not the case. Note that $q^1(t), q^2(t) \leq |q(t)|$, so we can set $\frac{q^1(t)}{|q(t)|}q(t)$, $\frac{q^2(t)}{|q(t)|}q(t)$ to be zero for the values of t at which $|q(t)| = 0$. As another remark we note that since the three normal vectors vary in a continuous way with the function q, one can show that $\mathscr{C}_2^c$ is a C^1-manifold [64].

We will use $d_{\mathscr{C}_2^c}$ to denote the resulting geodesic distance on $\mathscr{C}_2^c$, associated with the $\mathbb{L}^2$ Riemannian metric.

6.2.3 Removing Parameterization

In Sect. 5.5 we introduced the idea of using group actions to study re-parameterization variability of general curves. Now we apply this idea to closed curves.

1. **Case 1**: We start with the case of arc-length parameterized curves, represented by their angle functions and viewed as elements of the pre-shape space $\mathscr{C}_1^c$. Since the speed of traversing curves is fixed to be one, the only remaining way to re-parameterize a closed curve is to change the placement of the origin on it. In other words, a "translation" of origin along $\mathbb{S}^1$ is the same as a rotation of the points in $\mathbb{R}^2$. Since the net effect of these re-parameterizations is simply translations of points along $\mathbb{S}^1$ by a constant distance, these re-parameterizations are also called *translation diffeomorphisms* . Incidentally, they are also called *distance-preserving diffeomorphisms* since they preserve distances between points on $\mathbb{S}^1$. A translation of the origin by u is given by $s \mapsto (s-u)_{\mathrm{mod}\ 2\pi}$. Here, $u \in \mathbb{S}^1$ stands for the angle associated with a point on $\mathbb{S}^1$, expressed in radians. So, the space of all possible translations along $\mathbb{S}^1$ is $\mathbb{S}^1$ itself and it becomes the set of all translation diffeomorphisms of $\mathbb{S}^1$.

   The process of changing the origin on a closed curve can be expressed mathematically as an action of $\mathbb{S}^1$ on $\mathscr{C}_1^c$. Define the mapping $\mathbb{S}^1 \times \mathscr{C}_1^c \to \mathscr{C}_1^c$ as follows. For any $u \in \mathbb{S}^1$ and $\theta \in \mathscr{C}_1^c$, let

$$(u, \theta)(s) = \theta((s-u)_{\mathrm{mod}\ 2\pi}) + u, \qquad s \in [0, 2\pi] \tag{6.14}$$

   Addition of u in this equation ensures that the average value of (u, θ) is π and this shifted function remains an element of $\mathscr{C}_1^c$. To verify this, we first extend θ to the function $\theta_e : \mathbb{R} \to \mathbb{R}$ using $\theta_e(2n\pi + s) = 2n\pi + \theta(s)$ for all $s \in [0, 2\pi]$ and $n \in \mathbb{Z}$. The extended function θ_e agrees with θ on $[0, 2\pi]$ but has extended θ by repeatedly translating its graph along the 45° line. The corresponding translation action of $\mathbb{R}$ on θ_e is simply $(u, \theta_e)(s) = \theta_e(s-u) + u$, which when restricted back to $[0, 2\pi]$ gives the right side of Eq. 6.14. Pictorially, one can think of this action as translation of the graph of θ_e in a direction parallel to the graph of $y = x$. To understand the reason for this translation, note that first we are translating horizontally by a distance of u (this has the effect of changing the location of the origin of our curve), and then we are translating vertically by u (this has the effect of rotating the curve to where its "average angle" is once again π).

   A pictorial illustration of this group action is given in Fig. 6.1. The columns (a) and (b) show closed curves, and the corresponding angle functions θ for certain placements of the origin while the columns (c) and (d) show the same curve but different placements of the origin and the corresponding angle functions (u, θ), respectively. In terms of shapes, the shape of the curve represented by the function (u, θ) is same as that of the curve represented by θ. To unify all such representations, we define an equivalence relation based on this group action. We define two functions $\theta_1, \theta_2 \in \mathscr{C}_1$ to be equivalent if and only if there exists a $u \in \mathbb{S}^1$ such that $(u, \theta_1) = \theta_2$. The orbit associated with a representation $\theta \in \mathscr{C}_1^c$, given by

$$[\theta] = \{(u, \theta) | u \in \mathbb{S}^1\} \ ,$$

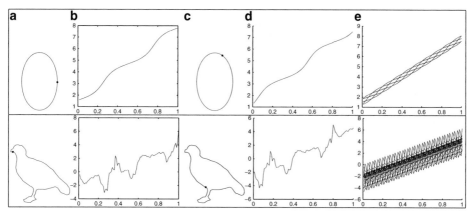

Fig. 6.1 Example of action of $\mathbb{S}^1$ on $\mathscr{C}_1^c$: (**a**) a curve β, (**b**) its angle function θ, (**c**) curve with shifted origin $\beta(s-u)$, and (**d**) resulting angle function (u, θ). (**e**) shows some examples of elements in $[\theta]$

is the equivalence class of shapes containing θ. Figure 6.1e shows several elements (u, θ) of the orbit $[\theta]$, for the θ shown in Fig. 6.1b. This equivalence relation partitions $\mathscr{C}_1^c$ into disjoint equivalence classes and the set of all such classes is denoted by $\mathscr{S}_1^c$

$$\mathscr{S}_1^c = \{[\theta] | \theta \in \mathscr{C}_1^c\} \ . \tag{6.15}$$

$\mathscr{S}_1^c$ is the quotient space of $\mathscr{C}_1^c$ modulo the group $\mathbb{S}^1$, i.e., $\mathscr{S}_1^c = \mathscr{C}_1^c/\mathbb{S}^1$ and is the *shape space* of closed, arc-length parameterized curves in $\mathbb{R}^2$. Every element of $\mathscr{S}_1^c$ has a different shape and, conversely, two curves with the same shape map to the same element of $\mathscr{S}_1^c$ as long as they are arc-length parameterized. The cartoon illustration in Fig. 6.3a shows this idea where each vertical line denotes an orbit and the quotient space is a collection of such orbits.

Next, we look at the problem of computing distances between points in $\mathscr{S}_1^c$. We have already established a Riemannian metric on $\mathscr{C}_1^c$ in Eq. 6.6 and can inherit it to $\mathscr{S}_1^c$. Since $\mathscr{S}_1^c$ is the quotient space of $\mathscr{C}_1^c$, under the action of $\mathbb{S}^1$, we need to verify the condition (stated in Definition 3.16) that the action of $\mathbb{S}^1$ is by isometries.

Theorem 6.3. *The action of $\mathbb{S}^1$ on $\mathscr{C}_1^c$ given by Eq. 6.14 is an action by isometries.*

We will leave the proof as an exercise for the reader.

So, the action of $\mathbb{S}^1$ on $\mathscr{C}_1^c$ is by isometries and, by Definition 3.16, the Riemannian metric on $\mathscr{C}_1^c$ descends to $\mathscr{S}_1^c$. Consequently, the distance between any two points on $\mathscr{S}_1^c$, say $[\theta_1], [\theta_2]$, according to Definition 3.17, is:

$$d_{\mathscr{S}_1^c}([\theta_1], [\theta_2]) = \min_{\tau \in \mathbb{S}^1} d_{\mathscr{C}_1^c}(\theta_1, (\tau, \theta_2)) \ .$$

This *min* exists because $\mathbb{S}^1$ is a compact set. This is depicted pictorially on the right side of Fig. 6.3.

2. **Case 2**: The task of unifying re-parameterization and defining equivalence classes for open curves represented by SRVFs was studied in Sect. 5.5. The case for closed curves is similar, except this time the domain of parameterization

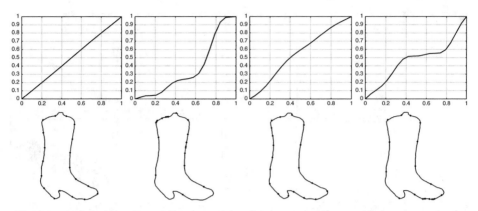

Fig. 6.2 *Top row*: examples of (origin-preserving) increasing diffeomorphisms on a circle, displayed by splitting $\mathbb{S}^1$ into $[0,1]$ at an arbitrary point. The first function is identity. *Bottom row*: the corresponding parameterization of β demonstrated using some grid points on $[0,2\pi]$

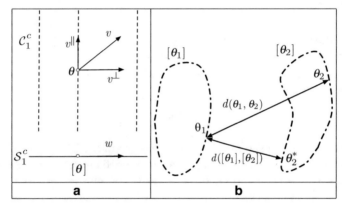

Fig. 6.3 (a) The *vertical lines* denote orbits in the pre-shape spaces $\mathcal{C}_1^c$ ($\mathcal{C}_2^c$). The set of these orbits forms the shape space $\mathcal{S}_1^c$ ($\mathcal{S}_2^c$). (b) The distances between orbits are computed by fixing one of the shapes, say θ_1 (q_1), and optimizing over the other orbit, say $[\theta_2]$ ($[q_2]$)

is $\mathbb{S}^1$, instead of $[0,1]$. Now the re-parameterization group, Γ_S, is the group of all orientation-preserving diffeomorphisms from $\mathbb{S}^1$ to itself. For a curve $\beta : \mathbb{S}^1 \to \mathbb{R}^2$ and $\gamma \in \Gamma_S$, a re-parameterization function, $\beta \circ \gamma$, is the re-parameterized curve. Some examples of re-parameterization functions are shown in the top row of Fig. 6.2; for the purpose of visualization we select an origin on the curve and then the re-parameterization function can be thought of as a diffeomorphism from some closed unit interval in $\mathbb{R}$, say $[0,1]$, to itself. The bottom row shows the corresponding re-parameterized curves with ticks on them. In the first figure, where the ticks are equally spaced, the parameterization is by arc length.

For elements of $\mathcal{C}_2^c$, the variability associated with re-parameterizations of curves is represented using the following group action. The group action $\mathcal{C}_2^c \times \Gamma_S \to \mathcal{C}_2^c$ is the same as in Eq. 5.19: $(q,\gamma) \mapsto (q \circ \gamma)\sqrt{\dot{\gamma}}$. The function $\dot{\gamma}$ here should be thought of as a function from $\mathbb{S}^1$ to the positive real numbers and is defined by first thinking of γ as a diffeomorphism from $[0,1]$ to itself and then differentiating this function.

Similarly, the rotations of a curve can be represented by the action $SO(2) \times \mathscr{C}_2^c \to \mathscr{C}_2^c$ defined by

$$(O, q) \mapsto \{Oq(t) | t \in \mathbb{S}^1\} \ . \tag{6.16}$$

How do the two group actions interact with each other?

Lemma 6.1. *The actions of $SO(2)$ and Γ_S on $\mathscr{C}_2^c$ commute.*

Proof. Same as Lemma 5.1 in Chap. 5.

Since these actions commute, they combine to give an action of the product group $\Gamma_S \times SO(2)$ on $\mathscr{C}_2^c$. We can then obtain the equivalence class, or orbit, associated with a curve q under the actions of Γ_S and $SO(2)$:

$$[q] = \text{closure}\{(Oq, \gamma) | (\gamma, O) \in \Gamma_S \times SO(2)\} \ , \tag{6.17}$$

and the shape space using the SRVF representation is defined by

$$\mathscr{S}_2^c = \{[q] | q \in \mathscr{C}_2^c\} \ .$$

Once again, the cartoon illustration in Fig. 6.3a shows this idea where each vertical line denotes an orbit and the quotient space is a collection of such orbits. The reason for including the closure of these orbits, rather than the orbits themselves, is the same as in Sect. 5.5 and is not repeated here.

The next consideration is the metric structure on $\mathscr{S}_2^c$. To see that $\mathscr{S}_2^c$ inherits a metric structure from $\mathscr{C}_2^c$, we establish the following important result.

Lemma 6.2. *The action of the product group $\Gamma_S \times SO(2)$ on $\mathscr{C}_2^c$ is by isometries with respect to the metric defined in Eq. 6.10.*

Proof. Same as Lemma 5.2 in Chap. 5.

Since the action of $\Gamma_S \times SO(2)$ is by isometries, the distance resulting from the Riemannian metric in Eq. 6.10 descends to the quotient space $\mathscr{S}_2^c$. Therefore, the distance between any two orbits, say $[q_1]$ and $[q_2]$, (using Definition 3.17) is

$$\begin{aligned}
d_{\mathscr{S}_2^c}([q_1], [q_2]) &= \inf_{(\gamma_1, O_1), (\gamma_2, O_2) \in \Gamma_S \times SO(2)} d_{\mathscr{C}_2^c}(O_1(q_1, \gamma_1), O_2(q_2, \gamma_2)) \\
&= \inf_{(\gamma, O) \in \Gamma_S \times SO(2)} d_{\mathscr{C}_2^c}(q_1, ((O, q_2), \gamma)) \ .
\end{aligned}$$

We have narrowed our representation of shapes to the elements of these two shape spaces - $\mathscr{S}_1^c$ and $\mathscr{S}_2^c$. What are the properties of these spaces? First, an important property is that two distinct elements of a shape space denote two different shapes. We have taken care of unifying all possible equivalences—all curves obtained by rotating, translating, scaling, and re-parameterizing the same curve β are in the same equivalence class as β, and they all map to the same point in the shape space. The first representation is quite restricted where the curves are assumed to be arc-length parameterized, while the second representation allows for more general parameterizations of curves. Secondly, the differential geometries of shape spaces closely relate to those of the corresponding pre-shape spaces. As a result, the construction of geodesics in shape spaces will be similar to construction of geodesics in the corresponding pre-shape spaces.

6.3 Projection on a Manifold

We are going to develop numerical approaches for finding geodesics on pre-shape and shape spaces of closed planar curves. Since the pre-shape spaces are nonlinear manifolds embedded inside larger Hilbert spaces, it is reasonable to obtain points, through observations or processing, that are not in the desired manifolds but in the larger spaces. So, we would like have a general numerical tool for projecting points of the larger space into the manifolds of interest. We point out that for some manifolds, there are analytical expressions to perform this projection. For example, in case of a unit sphere $\mathbb{S}^2$, the projection of a point x in $\mathbb{R}^3$ to the nearest point in $\mathbb{S}^2$ is simply

$$x \to \frac{x}{|x|} .$$

However, for the pre-shape spaces defined in this chapter we will have to resort to a numerical technique. Since this tool will be quite useful, we first try to understand the basic idea behind it with a specific example. Consider the first pre-shape space, $\mathscr{C}_1^c$; its elements satisfy the following constraints:

$$\frac{1}{2\pi} \int_0^{2\pi} \theta(s)ds = \pi, \quad \int_0^{2\pi} \sin(\theta(s))ds = 0, \quad \int_0^{2\pi} \cos(\theta(s))ds = 0 ,$$

and the constraint $\theta(2\pi) = \theta(0) + 2\pi$. If we have an angle function that is in the larger space $\mathbb{L}^2([0, 2\pi], \mathbb{R})$, but does not satisfy some of these conditions, it will not be in $\mathscr{C}_1^c$. If the first condition is not satisfied, then one can simply rotate the underlying curve to get the correct orientation. However, if the closure condition is not satisfied, there is no obvious fix to close an open curve. More precisely, the question that we are asking is: for the given $\theta \in \mathbb{L}^2([0, 2\pi], \mathbb{R})$ and $\theta \notin \mathscr{C}_1^c$, how to find the closest point in $\mathscr{C}_1^c$?

Here, we derive a projection technique for a general manifold $M \subset \mathscr{H}$, where $\mathscr{H}$ is a Hilbert space and M has been defined as a level set of $\Phi : \mathscr{H} \to \mathbb{R}^n$, much like our two pre-shape spaces, $\mathscr{C}_1^c$ and $\mathscr{C}_2^c$.

Problem Statement: Let Φ be a smooth mapping such that $d\Phi$ is surjective at each point on its domain. Let the level set of Φ at $\mathbf{a}$, i.e., $M = \Phi^{-1}(\mathbf{a})$, be a manifold of interest, for some $\mathbf{a} \in \mathbb{R}^n$. For any given $p \in \mathscr{H}$ (but not in M), we want a tool to project it into M. In other words, we want to change p as efficiently as possible in such a way that $\Phi(p)$ reaches $\mathbf{a}$.

Solution: Our approach is to move along a path perpendicular to the level sets of Φ such that the images under Φ of points in this path move toward $\mathbf{a}$ in $\mathbb{R}^n$ along a straight line. For the starting point p, let $\Phi(p) = \mathbf{b}$ ($\mathbf{b} \neq \mathbf{a}$), which means that p is in the level set $\Phi^{-1}(\mathbf{b})$. To update p, there are two orthogonal spaces: the tangent space $T_p(\Phi^{-1}(\mathbf{b}))$ and the normal space $N_p(\Phi^{-1}(\mathbf{b}))$. Since moving p along a tangent direction will not change its Φ-value, the most efficient way to change the value of $\Phi(p)$ is to move in the normal direction. As Φ maps $\mathscr{H}$ to $\mathbb{R}^n$, the derivative $d\Phi_p$ at p also maps $\mathscr{H}$ to $\mathbb{R}^n$. It follows that the restriction of $d\Phi_p$ induces an isomorphism $N_p(\Phi^{-1}(\mathbf{b})) \to \mathbb{R}^n$. Since $N_p(\Phi^{-1}(\mathbf{b}))$ is an n-dimensional vector space, we can express this isomorphism ($d\Phi_p$ restricted to $N_p(\Phi^{-1}(\mathbf{b}))$) as an $n \times n$ matrix J with respect to an orthonormal basis $\{w_i, i = 1, 2, \ldots, n\}$ of $N_p(\Phi^{-1}(\mathbf{b}))$.

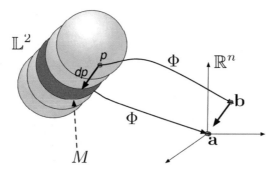

Fig. 6.4 A cartoon illustration of projecting an arbitrary point $p \in \mathbb{L}^2$ into the level set $M = \Phi^{-1}(\mathbf{a})$ using the Φ map and its Jacobian J

We want to find a displacement vector $p_\delta \in N_p(\Phi^{-1}(\mathbf{b}))$ at p such that $\Phi(p + p_\delta) \approx \mathbf{a}$, up to the first order. If we define the residual vector $\mathbf{r}(p) = \mathbf{a} - \Phi(p)$, the displacement vector is given by $p_\delta = \delta \sum_{i=1}^n \beta_i w_i$, where $\beta = J^{-1} r(p)$ and $\delta > 0$ is an empirically chosen step size. Update the point p using $p = p + p_\delta$, and iterate until the norm $\|r(p)\|$ converges to zero. These steps are summarized below:

Algorithm 5. *1. Compute the residual vector* $\mathbf{r}(p) = \mathbf{a} - \Phi(p)$.
2. Compute the $n \times n$ Jacobian matrix J of $d\Phi_p$ with respect to a basis $\{w_1, w_2, \ldots, w_n\}$ of $N_p(\Phi^{-1}(\boldsymbol{b}))$, where $\Phi(p) = \mathbf{b}$.
3. Compute $\beta = J^{-1}\mathbf{r}(p)$ and set $p_\delta = \delta \sum_{i=1}^n \beta_i w_i$, for a small $\delta > 0$.
4. Update p to $p + p_\delta$.

We will use this tool repeatedly to map arbitrary points into pre-shape spaces. Examples of this algorithm for projecting curves into pre-shape spaces $\mathscr{C}_1^c$ and $\mathscr{C}_2^c$ are presented later (Figs. 6.8 and 6.14).

6.4 Geodesic Computation

The next task is to compute geodesics between elements of pre-shape and shape spaces of closed curves. It was possible to determine geodesics in the previous chapter using analytical expressions, at least on the pre-shape spaces, but the current cases are more complicated. For these cases we shall develop two computational approaches for finding geodesics that are broadly applicable. This discussion applies to any Riemannian manifold, although our interest remains in the pre-shape and shape spaces, $\mathscr{C}_i^c$s and $\mathscr{S}_i^c$s.

There are two main ideas in numerical construction of geodesic paths on manifolds. In the first approach, called the *shooting method*, one tries to "shoot" a geodesic from the first point, iteratively adjusting the shooting direction until the resulting geodesic passes through the second point. A cartoon illustration of this idea is shown in the left panel of Fig. 6.5. The second method, called the *path-straightening method*, one initializes with an arbitrary path between the given two points on the manifold and then iteratively "straightens" it until a geodesic is reached. The right panel of Fig. 6.5 shows a cartoon illustration of this approach. In the next two sections, we describe these mathematical constructions with a focus on our pre-shape spaces $\mathscr{C}_1^c$ and $\mathscr{C}_2^c$:

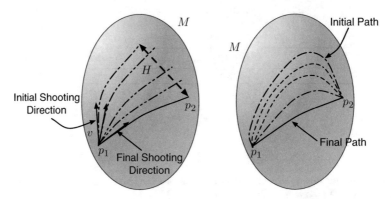

Fig. 6.5 Numerical approaches for construction of geodesics: shooting method (*left*) and path-straightening method (*right*)

6.5 Geodesic Computation: Shooting Method

Assume that M is a submanifold of a Hilbert space $\mathcal{H}$, with the Riemannian structure that is inherited from $\mathcal{H}$. Our goal is to compute geodesics between any two points p_1 and $p_2 \in M$, with respect to that metric. The basic approach is as follows:

1. Select one of the points, say p_1, as the starting point and the other, p_2, as the target point.
2. Construct a geodesic starting from p_1, in an arbitrary direction $v \in T_{p_1}(M)$; denote it by $\alpha_\tau(p_1; v)$ where τ is the time parameter for the geodesic flow.
3. If this geodesic reaches p_2 in unit time, i.e., $\alpha_1(p_1; v) = p_2$, then we are done. If not, measure the amount of miss, the discrepancy between $\alpha_1(p_1; v)$ and p_2, using a simple measure, e.g., the distance in $\mathcal{H}$. Call this discrepancy function F.
4. Iteratively, update the shooting direction v to reduce this discrepancy F to zero. We will use the gradient of F to update v in the tangent space $T_{p_1}(M)$, but there are other ideas that are equally effective.

Implementation of these algorithms requires two important pieces. First, given a point $p_1 \in M$ and a direction $v \in T_{p_1}(M)$, we need to construct the geodesic path $\alpha_\tau(p_1; v)$, and we will do so in a general setting using numerical approximations. Second, given two points p_1 and p_2, we need to find a direction $v \in T_{p_1}(M)$ such that $\alpha_1(p_1; v) = p_2$. This ordering of the problem is important because the second piece will use the first piece. Actually, the solution of the first piece can be used to evaluate the exponential map and that of the second piece to evaluate the inverse exponential map. Please refer to Sect. 3.2 for definitions and examples of the exponential map and its inverse for some simple manifolds.

1. **Evaluation of the Exponential Map**: Given a point $p \in M$ and a tangent vector $v \in T_P(M)$, the goal is to find the value of the exponential map $\exp_p(v) \in M$. This point $\exp_p(v)$ is the point reached by the constant-speed geodesic starting from p and with the initial velocity v at time $\tau = 1$. Additionally, we want the geodesic path that goes from p to $\exp_p(v)$. Our construction is based

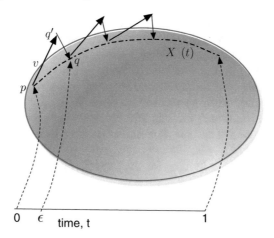

Fig. 6.6 A numerical approximation of the exponential map

on numerically solving the differential equation $\frac{d}{dt}\alpha(\tau) = V(\alpha(\tau))$, where V is a vector field on M obtained by parallel transport of v, along geodesic paths, to all of M. The parallel transport of a vector along a geodesic was discussed in Sect. 3.4 (Definition 3.8) but the transport along more general paths will be defined later.

A numerical solution to this problem is illustrated in Fig. 6.6. The idea is to start at $\alpha(0) = p$ and follow the vector $V(p) = v$ for a small time in the larger space $\mathscr{H}$. At $\tau = \epsilon$, we reach $\alpha(0) + \epsilon v$, but this new point $\tilde{p}_\epsilon \equiv \alpha(0) + \epsilon v$ may not be in M, since M is a nonlinear manifold and our increment is linear. Assuming that we have a mechanism (an analytical expression or a numerical procedure of the type presented in Algorithm 5) to project $\tilde{p}_\epsilon$ to a nearby point in M, using the structure of M, we can obtain that point, say p_ϵ. This way we get an approximation of the geodesic's location at time $t = \epsilon$. We repeat this process for the next time interval $[\epsilon, 2\epsilon]$, this time starting from p_ϵ. For the tangent direction at p_ϵ we use $V(p_\epsilon)$, obtained by parallel transport of v from p to p_ϵ. For this part we can assume that ϵ was small enough so that the path from p to p_ϵ is approximately a geodesic. There are two steps that are important in this implementation: (i) projection of a point $\tilde{p}_\epsilon \in \mathscr{H}$ to the nearest point in M and (ii) parallel transport of a vector $v \in T_p(M)$ to a point $p_\epsilon \in M$ that is not far from p. To extend the geodesic to a desired time τ, we repeat these steps, taking time increments of size ϵ until we reach τ. In this setup, the geodesic is actually computed only at the times $\alpha_{k\epsilon}(p_1; v)$, for $k = 1, 2, \ldots$. Since we interested in the value of at time $\tau = 1$, we choose ϵ to be $\frac{1}{T}$ and take T steps of the geodesic to compute $\alpha_1(p_1; v)$. Using straightforward calculus, one can show that the value of $\alpha_{k\epsilon}(p_1; v)$ converges to the desired continuous geodesic $\alpha_\tau(p_1; v)$ as $T \to \infty$. Also, the resulting path approximates a constant-speed geodesic.

2. **Evaluation of Inverse Exponential Map**: Now that we have a tool for computationally approximating a geodesic path from a point p with the initial velocity $v \in T_p(M)$, we can tackle the problem of finding a geodesic path between any two given points $p_1, p_2 \in M$. We can rephrase this as follows: find the initial velocity $v \in T_{p_1}(M)$ such that the constant-speed geodesic from p_1 in the direction v passes through p_2 at time $\tau = 1$. If we choose v arbitrarily,

this condition will not be satisfied unless we get very lucky. So the alternative is to start with an initial condition for v and iteratively update it in such a way that the end point of the geodesic gets closer and closer to p_2 and ultimately reaches it. By end point, we mean the point reached by the geodesic at time $\tau = 1$. This update can be performed in many ways and we will introduce a simple gradient method to perform these updates.

One can treat the search for this initial velocity as an optimization problem on the space $T_{p_1}(M)$. The cost function we wish to minimize is given by the functional: $F : T_{p_1}(M) \to \mathbb{R}$, $F[v] = \|\alpha_1(p_1; v) - p_2)\|^2$, under the norm associated with $\mathcal{H}$, and we are looking for that $v \in T_{p_1}(M)$ for which (i) $F[v]$ is zero and (ii) $\sqrt{\langle v, v \rangle}$ (under the chosen Riemannian metric) is minimum among all such tangents.

Remark. In case M and $T_{p_1}(M)$ are infinite dimensional, this optimization is not straightforward, and we need to use a finite-dimensional approximation of $T_{p_1}(M)$ to find the optimal direction. Take the case when $\mathcal{H} = \mathbb{L}^2([0, 2\pi], \mathbb{R})$ and $M \subset \mathbb{L}^2([0, 2\pi], \mathbb{R})$; now $v \in \mathbb{L}^2$ is a real-valued function. There are several ways to approximate elements of $T_{p_1}(M)$, one of them being the Fourier approximation described in Sect. A.3.1. The function v can be approximated using a finite number of terms in the Fourier expansion. In other words, replace $v \in T_{p_1}(M)$ by $\pi(\sum_{n=0}^{m}(a_n \cos(nt) + b_n \sin(nt)))$ for a large positive integer m. The mapping π here denotes a projection from the space $\mathbb{L}^2$ to the tangent space $T_{p_1}(M)$. (This mapping π becomes the third requirement for implementing the shooting method, in addition to the abovementioned needs for projection into M and parallel transport on M.) With this approximation, the cost function modifies to $\tilde{F} : \mathbb{R}^{2m+1} \to \mathbb{R}_+$:

$$\tilde{F}(a, b) = \|\alpha_1(p_1; \pi(\sum_{n=0}^{m} a_n \cos(ns) + b_n \sin(ns))) - p_2\|^2 . \tag{6.18}$$

Now one can use a gradient-based approach for minimizing $\tilde{F}$ as a function of the Fourier coefficients a_n, b_n. The gradient, in general, will be computed numerically for the shape manifolds. The tangent vector can then be updated in the direction of negative gradient to update the shooting direction.

Another idea for iteratively refining the shooting direction, this time without the use of the gradient of F, is to compute the *miss vector* $w \equiv (p_2 - \alpha_1(p_1; v)) \in \mathcal{H}$, for the current α, project it into the tangent space of $\alpha_1(p_1; v)$, and parallel transport it to p_1. Call this transport $\tilde{w}$ and update v according to $v + \delta\tilde{w}$ for a small δ.

We illustrate the shooting method on two spaces: (1) $\mathbb{S}^2$, as a simple example and (2) $\mathscr{C}_1^c$, as the main tool for constructing geodesics between non-elastic planar closed curves.

6.5.1 Example 1: Geodesics on $\mathbb{S}^2$

As a simple example of the shooting method, we consider the problem of constructing geodesics between points on $\mathbb{S}^2$ under the standard Euclidean metric. Even though this problem has a well-known analytical solution, we solve it using the shooting method for the purpose of illustration. As stated previously, we need the following three items to implement the shooting method:

1. **Projection on** $\mathbb{S}^2$: Any nonzero point $x \in \mathbb{R}^3$ can be mapped to the nearest point in $\mathbb{S}^2$ using the projection:

$$x \mapsto \frac{x}{|x|}. \tag{6.19}$$

2. **Parallel Transport of Tangents**: For points x and y in $\mathbb{S}^2$, $x \neq -y$, the parallel transport of a vector $v \in T_x(\mathbb{S}^2)$ along the shortest geodesic (i.e., great circle) from x to y, is

$$v \longmapsto \left(v - \frac{2 \langle v, y \rangle (x + y)}{|x + y|^2}\right) \in T_y(\mathbb{S}^2) . \tag{6.20}$$

   This result has been derived earlier in Eq. 3.9.
3. **Projection on Tangent Space**: A vector $w \in \mathbb{R}^3$ can be projected on to the tangent space $T_x(\mathbb{S}^2)$ using

$$w \longmapsto (w - \langle w, x \rangle x) . \tag{6.21}$$

The shooting method solves for the inverse exponential map on a manifold and, in turn, requires a tool for evaluating the exponential map. An algorithm for numerically evaluating the exponential map $\exp_x(v)$ on $\mathbb{S}^2$ follows. Once again, we already have an analytical form for this map but present a numerical method only for illustration.

Algorithm 6 (Exponential Map on $\mathbb{S}^2$).
Let $x \in \mathbb{S}^2$ and $v \in T_x(\mathbb{S}^2)$. Fix an $\epsilon = 1/T$ small. Set $i = 0$, $\alpha_0 = x$ and $w(i\epsilon) = v$.

1. *Compute $\alpha_{i\epsilon} + \epsilon w(i\epsilon)$ in $\mathbb{R}^3$ and project it on $\mathbb{S}^2$ using Eq. 6.19. Call this point $\alpha_{(i+1)\epsilon}$.*
2. *Transport $w(i\epsilon)$ at $\alpha_{i\epsilon}$ to the new point $\alpha_{(i+1)\epsilon}$ using the Eq. 6.20.*
3. *If $i = T$, stop and return $\alpha_{T\epsilon}$. Else, set $i = i + 1$ and go to Step 1.*

Now the algorithm for finding the initial velocity for going from x_1 to x_2 in $\mathbb{S}^2$, or for evaluating the inverse of the exponential map $\exp_{x_1}^{-1}(x_2)$, is as follows.

Algorithm 7 (Shooting Method on $\mathbb{S}^2$).
Let x_1, $x_2 \in \mathbb{S}^2$. Let $\mathbf{e}_{x_1}^1$, $\mathbf{e}_{x_1}^2$ be an orthonormal basis of $T_{x_1}(\mathbb{S}^2)$. Initialize a direction $v \in T_{x_1}(\mathbb{S}^2)$, perhaps using

$$v = w - \langle w, x_1 \rangle x_1, \quad w = (x_2 - x_1) .$$

Select small numbers ϵ, $\delta > 0$.

1. **Current Cost**: *Compute the exponential map $\exp_{x_1}(v)$ using Algorithm 6. Compute the cost function of reaching this point: $F(v) = |x_2 - \exp_{x_1}(v)|^2$.*
2. **Gradient of Cost**: *For $v_j' = v + \delta \mathbf{e}_{x_1}^j$, $j = 1, 2$, Compute the exponential map $\exp_{x_1}(v_j')$ using Algorithm 6 and the cost function $F(v_j') = \|x_2 - \exp_{x_1}(v_j')\|^2$.*

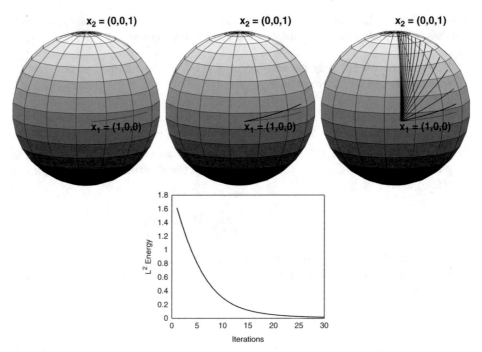

Fig. 6.7 Shooting method on $\mathbb{S}^2$. *Top*: iterative improvements in the shooting direction. *Bottom*: the evolution of F during minimization

Approximate the partial derivatives $\frac{\partial F}{\partial v(j)} \approx \frac{1}{\delta}(F(v_j') - F(v))$ for $j = 1, 2$. The gradient of F at v is then given by $g = \frac{\partial F}{\partial v(1)}\mathbf{e}_{x_1}^1 + \frac{\partial F}{\partial v(2)}\mathbf{e}_{x_1}^2$.

3. **Update Direction***: Replace v by $v - \epsilon g$.*

4. If the norm of the gradient g is small, then stop. Else, return to Step 1.

Shown in Fig. 6.7 is an example of computing a geodesic between $x_1 = (1, 0, 0)$ and $x_2 = (0, 0, 1)$ using the shooting method. The first panel shows an initial shooting direction v going horizontally and the geodesic shot in that direction. The next two panels show iterative updates on v so that the geodesic hits x_2 at time $\tau = 1$. Eventually, the shooting direction becomes vertical and reaches x_2. The plot on the bottom shows the decrease in the value of F as the shooting direction is iteratively improved.

This experiment also highlights a major limitation of the shooting method. On a sphere, there are two geodesics connecting any two given points, both along the same great circle containing the two points, but one forming the shorter arc and the other forming the longer arc. If the initial shooting direction v is closer to the longer arc, this gradient-based search for geodesic will result in the longer geodesic. Thus, this approach has the possibility of reaching a local solution to the optimization problem. It will find a geodesic but there is no guarantee that it will be the *minimal geodesic*. In some cases, for example, in the sphere, one can choose an initial shooting direction carefully so as to be closer to the desired solution. Of course, for more general manifolds this may not be as straightforward.

6.5.2 Example 2: Geodesics in Non-elastic Pre-shape Space

The second example that we consider is $\mathscr{C}_1^c$, the pre-shape space of closed, planar curves represented by their angle functions and endowed with the non-elastic metric. Given any two points θ_1, $\theta_2 \in \mathscr{C}_1^c$, we want to construct a geodesic path between them in $\mathscr{C}_1^c$ under the chosen Riemannian metric. As in the case of $\mathbb{S}^2$, we need the following three items to implement the shooting method.

1. **Projection on $\mathscr{C}_1^c$**: Recall that $\mathscr{C}_1^c$ is a codimension-3 submanifold of the ambient space

$$\mathscr{D} = \{\text{continuous } \theta \in \mathbb{L}^2([0, 2\pi], \mathbb{R}) : \theta(2\pi) = \theta(0) + 2\pi\},$$

   and that the tangent space of $\mathscr{D}$ is given by

$$\mathscr{B} = \{\text{continuous } \theta \in \mathbb{L}^2([0, 2\pi], \mathbb{R}) : \theta(2\pi) = \theta(0)\}.$$

   Unlike $\mathbb{S}^2$, we do not have an analytical expression for projecting points from $\mathscr{D}$ into the manifold $\mathscr{C}_1^c$. So we will use the technique described in Algorithm 5. Recall that $\mathscr{C}_1^c$ is the level set of $\Phi_1 : \mathscr{D} \to \mathbb{R}^3$ for the value $(\pi, 0, 0)$. Thus, to project a given $\theta \in \mathscr{D}$ to $\mathscr{C}_1^c$, we need to alter θ as efficiently as possible to arrange that $\Phi_1(\theta)$ becomes $(\pi, 0, 0)$. The first condition, namely, $\Phi_1^1(\theta) = \frac{1}{2\pi} \int_0^{2\pi} \theta(s)ds = \pi$, is easy to impose. Simply change θ by $\theta - \frac{1}{2\pi} \int_0^{2\pi} \theta(s)ds + \pi$. The remaining two conditions require an iterated procedure. We take the current point θ and define a displacement θ_δ as follows. Recall that the normal space $N_\theta(\mathscr{C}_1^c)$ is spanned by the functions $\mathbf{1}$, $\sin(\theta(s))$, and $\cos(\theta(s))$, ignoring the constant function $\mathbf{1}$, since the corresponding constraint is already taken care of, we focus on the other two functions. The 2×2 Jacobian matrix J_1 of $[d\Phi_1^2 \ d\Phi_1^3]$ is given by

$$J_1 = \begin{bmatrix} -\langle \sin \circ \theta, \sin \circ \theta \rangle & -\langle \sin \circ \theta, \cos \circ \theta \rangle \\ \langle \cos \circ \theta, \sin \circ \theta \rangle & \langle \cos \circ \theta, \cos \circ \theta \rangle \end{bmatrix} . \qquad (6.22)$$

   Define the residual vector as $\mathbf{r}(\theta) = [0 \ \ 0] - [\Phi_1^2(\theta) \ \ \Phi_1^3(\theta)] \in \mathbb{R}^2$. Then, the desired increment is given by

$$\theta_\delta = \delta \sum_{i=1}^2 \beta_i w_i, \qquad (6.23)$$

   where $\beta = J_1(\theta)^{-1} \mathbf{r}(\theta)$ and $(w_1(s), w_2(s)) = (\sin(\theta(s)), \cos(\theta(s)))$. This leads to the following procedure.

   Algorithm 8. *Project a given $\theta \in \mathscr{D}$ into $\mathscr{C}_1^c$*
   a. Set $\theta = \theta - \frac{1}{2\pi} \int_0^{2\pi} \theta(s)ds + \pi$.
   b. Compute θ_δ according to Eq. 6.23.
   c. Update the angle function $\theta \to \theta + \theta_\delta$. If $|\theta_\delta| > \epsilon$, go to Step 1.

   Shown in Fig. 6.8 are some examples of projecting curves, represented by their angle functions in $\mathscr{D}$, to the nearest points in $\mathscr{C}_1^c$. The original points $\theta \in \mathscr{D}$ are shown in terms of their corresponding curves, drawn in solid lines. The projected angle functions are displayed by their curves in marked lines.

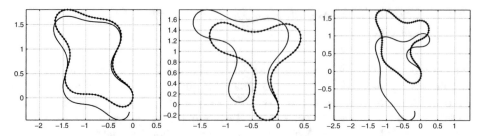

Fig. 6.8 Examples of open curves in represented by points in $\mathbb{L}^2([0, 2\pi], \mathbb{R})$ (*solid lines*) and their projections in $\mathscr{C}_1^c$ (*marked lines*) using Algorithm 8

2. **Parallel Translation of Vectors**: The second requirement is a tool for transporting a vector $v \in T_\theta(\mathscr{C}_1^c)$ to a nearby point $\theta' \in \mathscr{C}_1^c$. Since θ and θ' are close, this transport can be approximated, up to the first order, by simply projecting v into $T_{\theta'}(\mathscr{C}_1^c)$ and re-scaling it to be of the correct length.

   If $w_i, i = 1, 2, 3$ are orthonormal basis elements of the normal space $N_{\theta'}(\mathscr{C}_1^c) = \text{span}\{\mathbf{1}, \sin(\theta(s)), \cos(\theta(s))\}$ in $\mathbb{L}^2([0, 2\pi], \mathbb{R})$, then this approximation is given by

$$v \rightarrow |v| \left(\frac{v - \sum_{i=1}^3 \langle v, w_i \rangle w_i}{|v - \sum_{i=1}^3 \langle v, w_i \rangle w_i|} \right). \tag{6.24}$$

3. **Projection into Tangent Spaces**: The last requirement is a formula for projecting elements of $\mathscr{B}$ into tangent spaces. For any function $v \in \mathscr{B}$, it can be projected into the tangent space $T_\theta(\mathscr{C}_1^c)$ using the equation:

$$v \mapsto v - \sum_{i=1}^3 \langle v, w_i \rangle w_i , \tag{6.25}$$

   where w_i's are the orthonormal basis functions mentioned in the previous item.

Using these three items, we can write down an algorithm for finding geodesics in $\mathscr{C}_1^c$ using the shooting method. We start with the forward problem of computing the geodesic when the starting point and the starting direction are given.

Algorithm 9 (Exponential Map $\exp_\theta(v)$).
Let $\theta \in \mathbb{L}^2([0, 2\pi], \mathbb{R})$ and $v \in T_\theta(\mathscr{C}_1^c)$. Fix an $\epsilon = 1/T$ small. Set $\tau = 0$, $\alpha_{i\epsilon} = \theta$ and $w_{\tau\epsilon} = v$.

1. *Compute $\alpha_{\tau\epsilon} + \epsilon w_{\tau\epsilon}$ and project it on $\mathscr{C}_1^c$ using Algorithm 8. Call this point $\alpha_{(\tau+1)\epsilon}$.*
2. *Transport $w_{\tau\epsilon}$ at $\alpha_{\tau\epsilon}$ to the new point $\alpha_{(\tau+1)\epsilon}$ using the Eq. 6.24.*
3. *If $\tau = T$, stop and return $\alpha_{T\epsilon}$. Else, set $\tau = \tau + 1$ and go to Step 1.*

With this procedure for approximating the exponential map on $\mathscr{C}_1^c$, we can try to solve the optimization problem for finding the desired initial velocity. Recall the earlier remark that when the underlying manifold is infinite dimensional, one has to restrict the search for optimal direction to a finite-dimensional subspace of the tangent space at the starting point. We can particularize the resulting cost function, given in Eq. 6.18, for $\mathscr{C}_1^c$ resulting in

$$\tilde{F}\left(\mathbf{a}, \mathbf{b}\right) = \| \exp_{\theta_1} \left(\pi \left(\sum_{n=1}^{m} a_n \cos(ns) + b_n \sin(ns) \right) \right) - \theta_2 \|^2 . \qquad (6.26)$$

Here, π refers to the projection of an arbitrary element of $\mathscr{B}$ into the tangent space $T_{\theta_1}(\mathscr{C}_1^c)$, given in Eq. 6.25. We point out that the two Fourier terms for $n = 0$ have not been included in the summation because (i) the cosine term for $n = 0$ is a constant that will be readjusted in the projection π and (ii) the sine term is zero anyway.

For the gradient-based update of the shooting direction, we will need to approximate the gradients of $\tilde{F}$ with respect to the coefficient vectors a and b. One simple approximation is given by finite differences: for $n = 1, 2, \ldots m$,

$$\frac{\partial \tilde{F}}{\partial a_n} \approx \frac{1}{2\epsilon} \left(\tilde{F}(\mathbf{a} + \epsilon\mathbf{e}_n, \mathbf{b})) - \tilde{F}(\mathbf{a} - \epsilon\mathbf{e}_n, \mathbf{b})) \right)$$

$$\frac{\partial \tilde{F}}{\partial b_n} \approx \frac{1}{2\epsilon} \left(\tilde{F}(\mathbf{a}, \mathbf{b} + \epsilon\mathbf{e}_n)) - \tilde{F}(\mathbf{a}, \mathbf{b} - \epsilon\mathbf{e}_n)) \right) . \qquad (6.27)$$

Here $\{\mathbf{e}_n\}$ is the canonical basis vector of $\mathbb{R}^m$: $\mathbf{e}_n$ has zeros everywhere except in the nth location where it is one.

Using these ideas, we can now sketch an algorithm for finding the geodesics in $\mathscr{C}_1^c$ using the shooting method.

Algorithm 10 (Shooting Method on $\mathscr{C}_1^c$).
Given points $\theta_1, \theta_2 \in \mathscr{C}_1^c$, initialize the coefficients $\mathbf{a} = \{a_1, \ldots, a_m\}$, $\mathbf{b} = \{b_1, \ldots, b_m\}$. Fix a $\delta > 0$ small.

1. **Current Cost**: *Find the current shooting direction v by projecting $\sum_{n=1}^{m} a_n \cos(ns) + b_n \sin(ns)$ into $T_{\theta_1}(\mathscr{C}_1)$ using Eq. 6.25. Compute the exponential map $\exp_{\theta_1}(v)$ using Algorithm 9 and evaluate the cost function $F(v) = \|\theta_2 - \exp_{\theta_1}(v)\|^2$.*
2. **Compute Gradient**: *Approximate the partial derivatives $\frac{\partial \tilde{F}}{\partial a_n}$ and $\frac{\partial \tilde{F}}{\partial b_n}$, $n = 1, 2, \ldots, m$, using Eq. 6.27.*
3. **Update**: *Update $\mathbf{a}$, $\mathbf{b}$ using the negative gradient of $\tilde{F}$:*

$$\mathbf{a} \to \mathbf{a} - \delta\nabla_a \tilde{F}, \quad \mathbf{b} \to \mathbf{b} - \delta\nabla_b \tilde{F} .$$

4. **Stopping Criterion**: *If the norms of the two gradient vectors are both small, then stop. Otherwise, return to Step 1.*

In Figs. 6.9, 6.10, we present some examples of geodesic paths in $\mathscr{C}_1^c$ obtained using Algorithm 10. Each row in these figures shows a geodesic path: the first and the last curves correspond to the given angle functions θ_1 and θ_2, respectively, and the intermediate curves denote equally spaced points along the geodesic $\alpha_\tau(\theta_1; v)$ in $\mathscr{C}_1^c$, for $\tau = 0, 1/T, 2/T, \ldots, 1$, that passes through θ_2 at $\tau = 1$. Figure 6.9 shows some geodesic paths between some polygons and curvilinear objects. For each of the curves shown in these examples, the corresponding angle function results

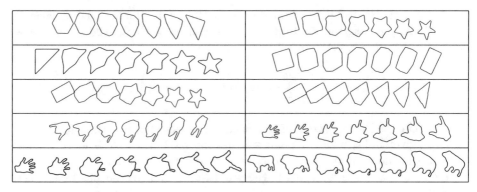

Fig. 6.9 Examples of geodesics paths between some curves in $\mathscr{C}_1^c$

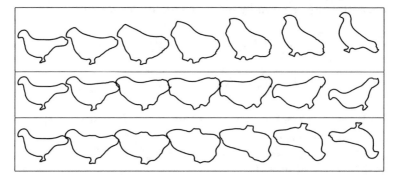

Fig. 6.10 Geodesics between sample two curves in $\mathscr{C}_1^c$, with the difference that the origin has been placed differently in the target curves

from a certain placement of origin on that curve. For the initial and the target curves, these placements are assumed given while for the intermediate curves, the algorithm dictates the placements of origins. We draw the attention of the reader to an earlier statement about the chosen Riemannian metric on $\mathscr{C}_1^c$ and its interpretation as allowing only bending in shapes. These geodesics represent the paths of minimum bending energy in going from θ_1 to θ_2, and this effect can be seen in the examples presented here.

Figure 6.10 highlights a limitation of geodesics in $\mathscr{C}_1^c$ in terms of shape analysis. In this example, all the target curves have the same shape as the initial curve except that the origin has been placed differently. These targets are different elements of an orbit of $\mathbb{S}^1$ in $\mathscr{C}_1^c$, and, as a consequence, the geodesics are visibly different and have nonzero path lengths. We emphasize that the re-parameterization group for this case, $\mathbb{S}^1$, is still to be removed, and thus the optimal placement of origin on the target curve (for optimally aligning the two curves) has not been reached. This part will be accomplished when we consider geodesics in the quotient space $\mathscr{S}_1^c$.

6.6 Geodesic Computation: Path-Straightening Method

The second numerical approach for finding geodesic paths between points on a Riemannian manifold is based on a procedure called *path straightening*. The main idea here is the following: initialize a path between the two given points on the manifold and iteratively straighten it, using the gradient of an appropriate energy function, until it cannot be straightened any further. The resulting path will be a geodesic path. A cartoon illustration of this approach is shown in the right panel of Fig. 6.5. Relative to the shooting method, there are several advantages to this method:

1. The solution is, by construction, guaranteed to start and end at the desired points in the manifold. In contrast, the shooting method can result in geodesic paths that do not quite reach the target point because the miss function did not reach zero.
2. In case the gradient of the chosen energy can be written analytically, which is the case in this chapter, the computational cost of implementing it is relatively low.
3. There is no need to approximate the tangent spaces (in case of infinite-dimensional manifolds) with finite-dimensional subspaces. This was done when using the shooting method on a pre-shape space so that the optimization problem could be restricted to a finite-dimensional space.

At first we explain the theoretical setup for the path-straightening method and then present numerical procedures for implementing it.

6.6.1 Theoretical Background

We describe this procedure on a general Riemannian manifold M with the following restriction. We will assume that the manifold M is a submanifold of a Hilbert space, with the Riemannian structure inherited from that larger space. We will denote this larger ambient space by $\mathscr{H}$. This condition certainly holds for our pre-shape spaces. (In those cases, the ambient vector space is $\mathscr{H} = \mathbb{L}^2$.) Now we pose the formal problem of finding geodesics on M. Say we are given two points p_1 and p_2 in M that we want to join using a geodesic path in M. Let $\mathscr{A}$ be the set of all differentiable paths in M, whose first derivatives are $\mathbb{L}^2$ functions, parameterized by $\tau \in [0, 1]$:

$$\mathscr{A} = \{\alpha : [0, 1] \to M | \ \alpha \text{ is differentiable and } \dot{\alpha} \in \mathbb{L}^2([0, 1], M)\} \ ,$$

and $\mathscr{A}_0$ be the subset of $\mathscr{A}$ consisting of those paths that start at p_0 and end at p_1:

$$\mathscr{A}_0 = \{\alpha \in \mathscr{A} | \alpha(0) = p_1 \text{ and } \alpha(1) = p_2\} \ .$$

The desired geodesic is an element of $\mathscr{A}_0$. For elements of $\mathscr{A}$, define an energy function $E : \mathscr{A} \to \mathbb{R}_+$ by

$$E[\alpha] = \frac{1}{2} \int_0^1 \langle \dot{\alpha}(\tau), \dot{\alpha}(\tau) \rangle \; d\tau \; . \tag{6.28}$$

We have some remarks about this definition of E:

- Note that for each τ, $\dot{\alpha}(\tau)$ is an element of $T_{\alpha(\tau)}(M)$, and, by our assumption on $\mathscr{A}$, $\dot{\alpha} : [0,1] \to TM$ is an $\mathbb{L}^2$ function. The inner product appearing inside the integral sign comes, of course, from the Riemannian metric on M.
- E is not the length of the path α, although it is closely related. If we use the square-root of the integrand (and remove the $1/2$ factor from the front), we obtain the length of the path α (see Eq. 5.9). Thus, E is $1/2$ the integral of the square of the instantaneous speed along the curve (the integral is taken with respect to the parameter τ).
- Later we will show that the critical points of E on the space $\mathscr{A}_0$ are precisely the constant-speed geodesic paths on M between p_1 and p_2. Therefore, one way to find a geodesic is to use the gradient of E to reach its critical points. This is the method we will describe.

We will be using the gradient of E to find its critical points on $\mathscr{A}_0$. So we start with the differential structure of $\mathscr{A}_0$. The tangent spaces of $\mathscr{A}$ and $\mathscr{A}_0$ are

$$T_\alpha(\mathscr{A}) = \{ w : [0,1] \to TM | \; \frac{Dw}{d\tau} \in \mathbb{L}^2 \text{ and } \forall \tau \in [0,1], w(\tau) \in T_{\alpha(\tau)}(M) \} \; ,$$

where $T_{\alpha(\tau)}(M)$ is the tangent space of M at the point $\alpha(\tau) \in M$, and

$$T_\alpha(\mathscr{A}_0) = \{ w \in T_\alpha(\mathscr{A}) | w(0) = w(1) = 0 \} \; .$$

Note that the tangent space element w is, by definition, a vector field along the path α tangent to M at each point of α. Next, we introduce some tools from covariant calculus, the calculus dealing with differentiation and integration of tangent vector fields along paths on manifolds. More specifically, we will define covariant derivatives and integrals of vector fields. A vector field w along α implies a time-indexed collection of tangent vectors along α:

$$\{ w(\tau) \in T_{\alpha(\tau)}(M), \; \tau \in [0,1] \} \; .$$

Refer to the Fig. 6.11 for an example each of the elements of $T_\alpha(\mathscr{A})$ and $T_\alpha(\mathscr{A}_0)$ for a path on a unit sphere.

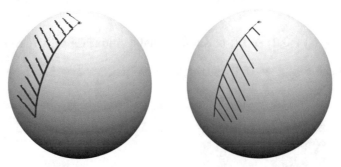

Fig. 6.11 Examples of path α and tangent vector field w on it. The left pictures show an example of $w \in T_\alpha(\mathscr{A})$ while the right picture shows a $w \in T_\alpha(\mathscr{A}_0)$

Definition 6.2 (Covariant Derivative). For a given path $\alpha \in \mathscr{A}$ and a vector $w \in T_\alpha(\mathscr{A})$ (a vector field along α), define the covariant derivative of w along α, denoted $\frac{Dw}{d\tau}$, to be the vector field obtained by projecting $\frac{dw}{d\tau}(\tau)$ onto the tangent space $T_{\alpha(\tau)}(M)$, for all τ. Note that given $w \in T_\alpha(\mathscr{A})$, $\frac{Dw}{d\tau}$ is an $\mathbb{L}^2$ vector field along α.

Since $M \subset \mathscr{H}$ is not a linear submanifold, the vector $\frac{dw}{d\tau}$, for any τ, is an element of $\mathscr{H}$, but not necessarily of $T_{\alpha(\tau)}(M)$. The projection ensures that the resulting set of vectors form a vector field along α that is tangent to M at each point. (Since this definition assumes that M is a submanifold of a Hilbert space, with the inherited metric, this is not the most general definition for the covariant derivative.) The opposite of a covariant derivative is the covariant integral.

Definition 6.3 (Covariant Integral). A vector field $u \in T_\alpha(\mathscr{A})$ is called a *covariant integral* of w along α if the covariant derivative of u is w, i.e., $\frac{Du}{d\tau} = w$.

Since we are interested in the gradient of E on $\mathscr{A}$, we need to impose a Riemannian structure on $\mathscr{A}$ also. The most obvious way to do this would be to define the inner product of w_1 and w_2 (where w_1 and w_2 are vector fields along α) by $\int_0^1 \langle w_1(\tau), w_2(\tau) \rangle \, d\tau$. However, this obvious inner product would have some fatal disadvantages for us. The main one is that $T_\alpha(\mathscr{A}_0)$ would not be a closed subspace of $T_\alpha(\mathscr{A})$. (That's because one can easily construct a sequence of elements of $T_\alpha(\mathscr{A}_0)$ that converges to an element of $T_\alpha(\mathscr{A}) - T_\alpha(\mathscr{A}_0)$ with respect to this metric.) To remedy this problem, we instead make $\mathscr{A}$ a Riemannian manifold using the first-order Palais metric on $\mathscr{A}$ [87]: for $w_1, w_2 \in T_\alpha(\mathscr{A})$, define

$$\langle\langle w_1, w_2 \rangle\rangle = \langle w_1(0), w_2(0) \rangle + \int_0^1 \left\langle \frac{Dw_1}{d\tau}(\tau), \frac{Dw_2}{d\tau}(\tau) \right\rangle d\tau \, , \qquad (6.29)$$

where $Dw/d\tau$ denotes the covariant derivative of w along α. With respect to the Palais metric, $T_\alpha(\mathscr{A}_0)$ is a closed linear subspace of $T_\alpha(\mathscr{A})$, and $\mathscr{A}_0$ is a closed subspace of $\mathscr{A}$.

The next step is to calculate the gradient of E, with respect to α, as an element of $T_\alpha(\mathscr{A}_0)$. To do this, we first find the gradient of E in the larger space $T_\alpha(\mathscr{A})$ (with respect to the Palais metric) and then orthogonally project it into $T_\alpha(\mathscr{A}_0)$ (where, again, the projection is with respect to the Palais metric).

Theorem 6.4. *The gradient vector of E in $T_\alpha(\mathscr{A})$ is given by a vector field u along α satisfying $Du/d\tau = d\alpha/d\tau$ and $u(0) = 0$. In other words, u is the covariant integral of the vector field $\dot{\alpha}(\tau)$, with zero initial value at $\tau = 0$.*

Proof. Define a *variation* of α to be a function $h : [0,1] \times (-\epsilon, \epsilon) \to \mathscr{H}$ (with $\mathbb{L}^2$ first derivative) such that $h(\tau, 0) = \alpha(\tau)$ for all $\tau \in [0,1]$. The variational vector field corresponding to h is given by $v(\tau) = h_s(t, 0)$ where s denotes the second argument in h (subscripts imply partial derivative here). Thinking of h as a path of curves in $\mathscr{A}$, we define $E(s)$ as the energy of the curve obtained by restricting h to $[0,1] \times \{s\}$. That is,

$$E(s) = \frac{1}{2} \int_0^1 \langle h_\tau(\tau, s), h_\tau(\tau, s) \rangle \, d\tau \, .$$

We now compute:

$$\dot{E}(0) = \int_0^1 \left\langle \frac{Dh_\tau}{ds}(\tau, 0), h_\tau(\tau, 0) \right\rangle d\tau = \int_0^1 \left\langle \frac{Dh_s}{d\tau}(\tau, 0), h_\tau(\tau, 0) \right\rangle d\tau$$

$$= \int_0^1 \left\langle \frac{Dv}{d\tau}(\tau), \frac{d\alpha}{d\tau}(\tau) \right\rangle d\tau \ ,$$

since $h_\tau(\tau, 0)$ is simply $\frac{d\alpha}{d\tau}(\tau)$. Now, the gradient of E should be a vector field u along α such that $\dot{E}(0) = \langle\langle v, u \rangle\rangle$. That is,

$$\dot{E}(0) = \langle v(0), u(0) \rangle + \int_0^1 \left\langle \frac{Dv}{d\tau}, \frac{Du}{d\tau} \right\rangle d\tau \ .$$

Clearly, if u satisfies the conditions that $u(0) = 0$ and $\frac{Du}{d\tau} = \frac{d\alpha}{d\tau}$, these two expressions for $\dot{E}(0)$ are equal to each other, proving the theorem. $\square$

Given a velocity vector field, $d\alpha/d\tau$, its covariant integral u is either obtained analytically or using numerical approximations, depending on the nature of M. If we apply the negative of the vector field u to update the path α, we expect the value of E to decrease. Furthermore, because of the condition $u(0) = 0$, the initial point of the curve α will not change and hence will remain at p_1. However, there is no reason to believe $u(1) = 0$, so the value of the terminal point $\alpha(1)$ will be perturbed and will not remain at p_2.

This is not satisfactory because we want to reduce the value of E on the space $\mathscr{A}_0$ (i.e., keeping the endpoints of α fixed), not on $\mathscr{A}$. To do this, we need to calculate the gradient of E thought of as a function on $\mathscr{A}_0$. To obtain this gradient, we simply take the gradient of E on $\mathscr{A}$, which is a vector in $T_\alpha \mathscr{A}$, and project it (using the Palais metric) into $T_\alpha \mathscr{A}_0$. When we use this projected gradient vector to update α, we will decrease E as rapidly as possible while keeping the endpoints of α fixed.

To accomplish this projection, we need the following definitions.

Definition 6.4 (Covariantly Constant). A vector field w along the path α is called *covariantly constant* if $Dw/d\tau$ is zero at all points along α.

Definition 6.5 (Geodesic). A path α on M is called a *geodesic* if its velocity vector field $d\alpha/d\tau$ is covariantly constant. That is, α is a geodesic if $\frac{D}{d\tau}(\frac{d\alpha}{d\tau}) = 0$ for all τ.

Recall that we have seen a definition of the geodesic in Sect. 3.2, but the definition presented here is more general.

Definition 6.6 (Covariantly Linear). A vector field w along the path α is called *covariantly linear* if $Dw/d\tau$ is a covariantly constant vector field.

Lemma 6.3. *The orthogonal complement of $T_\alpha(\mathscr{A}_0)$ in $T_\alpha(\mathscr{A})$ (with respect to the Palais metric) is the space of all covariantly linear vector fields w along α.*

Proof. Suppose that $v \in T_\alpha(\mathscr{A}_0)$ (i.e., $v(0) = v(1) = 0$), and $w \in T_\alpha(\mathscr{A})$ is covariantly linear. Then, using (covariant) integration by parts and the definition of the Palais metric given in Eq. 6.29:

$$\langle\langle v, w \rangle\rangle = \int_0^1 \left\langle \frac{Dv(\tau)}{d\tau}, \frac{Dw(\tau)}{d\tau} \right\rangle d\tau$$

$$= \left\langle v(\tau), \frac{Dw(\tau)}{d\tau} \right\rangle\Big|_0^1 - \int_0^1 \left\langle v(\tau), \frac{D}{d\tau}\left(\frac{Dw(\tau)}{d\tau}\right) \right\rangle d\tau = 0 \ .$$

The last term is zero since $\frac{D}{d\tau}\left(\frac{Dw}{d\tau}\right) = 0$ for all τ (i.e., w is covariantly linear). Hence, $T_\alpha(\mathscr{A}_0)$ is orthogonal to the space of covariantly linear vector fields along α in $T_\alpha(\mathscr{A})$. This proves that the space of covariantly linear vector fields is contained in the orthogonal complement of $T_\alpha(\mathscr{A}_0)$. To prove that these two spaces are equal, observe first that given any choice of tangent vectors at $\alpha(0)$ and $\alpha(1)$, there is a unique covariantly linear vector-field interpolating them. It follows that every vector field along α can be uniquely expressed as the sum of a covariantly linear vector field and a vector field in $T_\alpha(\mathscr{A}_0)$. The lemma then follows.

□

Definition 6.7 (Parallel Translation). A vector field $\tilde{w}$ is called the *forward parallel transport* of a tangent vector $w \in T_{\alpha(0)}(M)$, along α, if $\tilde{w}(0) = w$ and $\frac{D\tilde{w}(\tau)}{d\tau} = 0$ for all $\tau \in [0, 1]$.

Similarly, $\tilde{w}$ is called the *backward parallel translation* of a tangent vector $w \in T_{\alpha(1)}(M)$, along α, if $\tilde{w}(1) = w$ and $\frac{D\tilde{w}(\tau)}{d\tau} = 0$ for all $\tau \in [0, 1]$.

This definition is similar to the one in Chap. 3 (Definition 3.8), but in the current notation. Please refer to the cases studied in Example 3.7. It must be noted that forward or backward parallel transports along a path α lead, by definition, to vector fields along α that are covariantly constant.

Recall that we have computed the gradient u of the energy function $E : \mathscr{A} \to \mathbb{R}$ at any path α in M. According to Lemma 6.3, to project the gradient u into $T_\alpha(\mathscr{A}_0)$, we simply need to subtract off a covariantly linear vector field that agrees with u at $\tau = 0$ and $\tau = 1$. Clearly, the correct covariantly linear field is simply $\tau\tilde{u}(\tau)$, where $\tilde{u}(\tau)$ is the covariantly constant field obtained by parallel translating $u(1)$ backward along α. Hence, we have proved the following theorem.

Theorem 6.5. *Let $\alpha : [0, 1] \to M$ be a path such that $\alpha(0) = p_1$ and $\alpha(1) = p_2$, i.e., $\alpha \in \mathscr{A}_0$. Then, with respect to the Palais metric (Eq. 6.29):*

1. *The gradient of the energy function E on $\mathscr{A}$ at α is the vector field u along α satisfying $u(0) = 0$ and $\frac{Du}{d\tau} = \frac{d\alpha}{d\tau}$.*
2. *The gradient of the energy function E restricted to $\mathscr{A}_0$ is $w(\tau) = u(\tau) - \tau\tilde{u}(\tau)$, where u is the vector field defined in the previous item, and $\tilde{u}$ is the vector field obtained by parallel translating $u(1)$ backward along α.*

Example 6.2. As a simple example, consider the gradient of E for a path in the plane $M = \mathbb{R}^2$. We want to straighten this path into a geodesic while keeping the end points fixed at the current $\alpha(0)$ and $\alpha(1)$ in $\mathbb{R}^2$. Of course, a geodesic in $\mathbb{R}^2$ is the straight line joining these two points. As shown in Fig. 6.12, the particularization of Theorem 6.5 leads to a simple, intuitive result. In $\mathbb{R}^2$, the covariant derivative (integral) is replaced by the ordinary derivative (integral). Therefore, $u(\tau) = \int_0^\tau \frac{d\alpha}{d\tau}(s)ds = \alpha(\tau) - \alpha(0)$, and since $u(1) = \alpha(1) - \alpha(0)$, and backward parallel transport results in the same vector at all points, $\tau\tilde{u}(\tau)$ is simply

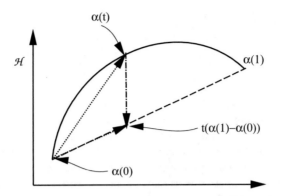

Fig. 6.12 Illustration of path-straightening update on a curve in $\mathbb{R}^2$

$\tau(\alpha(1) - \alpha(0))$. For any point $\alpha(\tau)$ on the curve, the gradient vector is given by $w(\tau) = (\alpha(\tau) - \alpha(0)) - \tau(\alpha(1) - \alpha(0))$. As shown in Fig. 6.12, it is the vector that takes the point $\alpha(\tau)$ to the corresponding point on the straight line (the geodesic) joining $\alpha(0)$ and $\alpha(1)$, and that is exactly the vector field to straighten α into a straight line in one step! Since $\mathbb{R}^2$ is a vector (flat) space, one can straighten a path in one step of gradient update. For nonlinear manifolds, iterating these updates will eventually converge to a geodesic.

Now we are ready to show that a critical point of the energy E is indeed a geodesic as we had suggested earlier.

Theorem 6.6. *For a given pair p_1, $p_2 \in M$, a path α from p_1 to p_2 is a critical point of E on $\mathscr{A}_0$ if and only if α is a geodesic from p_1 and p_2.*

Proof. Given a path α in M from p_1 to p_2, define the vector field u along α by the conditions that $Du/d\tau = \dot{\alpha}(\tau)$ and $u(0) = 0$. Recall that u is the gradient of $E : \mathscr{A} \to \mathbb{R}$ at α. We then have the following logical equivalences:

$$\alpha \text{ is a geodesic} \;\Leftrightarrow\; \frac{d\alpha}{d\tau} \text{ is covariantly constant}$$

$$\Leftrightarrow Du/d\tau \text{ is covariantly constant}$$

$$\Leftrightarrow u \text{ is covariantly linear}$$

$$\Leftrightarrow \text{ the projection of } u \text{ to } T_\alpha(\mathscr{A}_0) \text{ is } 0$$

$$\text{(because the covariantly linear vector fields are orthogonal to } T_\alpha(\mathscr{A}_0))$$

$$\Leftrightarrow \text{ the gradient of } E : \mathscr{A}_0 \to \mathbb{R} \text{ at } \alpha \text{ is } 0$$

$$\Leftrightarrow \alpha \text{ is a critical point of } E : \mathscr{A}_0 \to \mathbb{R}.$$

This completes the proof. □

6.6.2 Numerical Implementation

To implement a path straightening approach on a computer, one has to work with a discrete version of the path α. We will assume a uniform partition $\{0, \frac{1}{k}, \frac{2}{k}, \dots, 1\}$ of $[0, 1]$ on which the calculations will be performed. Step by step, we present numerical procedures for computing different quantities of interest.

The first requirement is to initialize a path in M between the two given points p_1 and p_2. The method we present here is to draw a geodesic between them in the ambient Hilbert space $\mathcal{H}$, simply taking the straight line between them and then to project this path onto M using the previous projection algorithm. (In some cases, other methods are also useful.) As we outline this process, we will mark (using †) the basic procedures that are needed repeatedly. So, when we particularize this process on a specific manifold, we will need to derive these basic procedures first.

Algorithm 11 (Initialize a path between p_1 and p_2 in M).
For all $\tau = 0, 1, \ldots, k$,

1. *Compute $(\tau/k)p_1 + (1 - (\tau/k))p_2$ in $\mathcal{H}$.*
2. *Project this value to a nearby point in M to obtain $\alpha(\tau/k)$. († Needs a mechanism to project points from $\mathcal{H}$ into M.)*

The next item is to compute the velocity vector $\frac{d\alpha}{dt}$ at points sampled along the curve. Our idea is to use a finite-difference approximation in $\mathcal{H}$ and project into the appropriate tangent space to estimate the velocity vector.

Algorithm 12 (Compute $\frac{d\alpha}{dt}$ along α).
For all $\tau = 0, 1, \ldots, k$,

1. *Compute: $c(\tau/k) = k(\alpha(\tau/k) - \alpha((\tau - 1)/k))$. This difference is computed in $\mathcal{H}$.*
2. *Project $c(\tau/k)$ into $T_{\alpha(\tau/k)}(M)$ to get an approximation for $\frac{d\alpha}{dt}(\tau/k)$. († Needs a mechanism to project vectors from $\mathcal{H}$ into $T_p(M)$.)*

Next, we want to approximate the covariant integral of $\frac{d\alpha}{dt}$ along α, using partial sums. While moving along α from $\alpha(0)$ to $\alpha(1)$, we want to add the current sum, say $u((\tau-1)/k)$, to the velocity $\frac{d\alpha}{dt}(\tau/k)$. However, these two quantities are elements of two different tangent spaces and cannot be added directly. Therefore, we parallel transport $u((\tau - 1)/k)$ to the point $\alpha(\tau/k)$ first and then add it to $\frac{d\alpha}{dt}(\tau/k)$ to estimate $u(\tau/k)$.

Algorithm 13 (Compute covariant integral of $\frac{d\alpha}{dt}$ along α).
Set $u(0) = 0 \in T_{\alpha(0)}(M)$. For all $\tau = 1, 2, \ldots, k$,

1. *Parallel transport $u((\tau - 1)/k)$ to the point $\alpha(\tau/k)$ to result in $u^{\|}((\tau - 1)/k)$. († Needs a mechanism to parallel transport tangent vectors along geodesics to nearby points in M.)*
2. *Set $u(\tau/k) = \frac{1}{k}\frac{d\alpha}{dt}(\tau/k) + u^{\|}((\tau - 1)/k)$.*

This covariant integration results in a vector field u along α that is the gradient of $E : \mathscr{A} \to \mathbb{R}$ at α. To obtain its gradient in $\mathscr{A}_0$, we need to subtract a covariantly linear component as follows. First, we compute an estimate for the backward parallel transport of $u(1)$:

Algorithm 14 (Backward parallel transport of $u(1)$).
Set $\tilde{u}(1) = u(1)$ and $l = \|u(1)\|$. For all $\tau = k - 1, k - 2, \ldots, 0$,

1. *Project $\tilde{u}((\tau + 1)/k)$ into $T_{\alpha(\tau/k)}(M)$ to obtain $c(\tau/k)$.*

2. Set $\tilde{u}(\tau/k) = lc(\tau/k)/\|c(\tau/k)\|$.

Now we can compute the desired gradient:

Algorithm 15 (Gradient vector field of E in $\mathscr{A}_0$).
For all $\tau = 1, 2, \ldots, k$, compute:

$$w(\tau/k) = u(\tau/k) - (\tau/k)\tilde{u}(\tau/k) .$$

By construction, this vector field, w, is zero at $\tau = 0$ and $\tau = k$. As a final step, we need to update the path α in direction opposite to the gradient of E.

Algorithm 16 (Path update).
Select a small $\epsilon > 0$ as the update step size. For all $\tau = 0, 1, \ldots, k$, perform

1. *Compute the gradient update $\alpha'(\tau/k) = \alpha(\tau/k) - \epsilon w(\tau/k)$. This update is performed in the ambient space $\mathscr{H}$.*
2. *Project $\alpha'(\tau/k)$ to M to obtain the updated $\alpha(\tau/k)$*

This completes a numerical recipe for computing geodesics using the path-straightening method. What are the main ingredients needed for applying this recipe for any given manifold M, given as a submanifold of an ambient Hilbert space $\mathscr{H}$? Similar to the case of the shooting method, one needs the following tools (marked by † in the algorithms above):

1. **Projection on M**: For any point $p \in \mathscr{H}$, we need an analytical or computational tool for projecting p to the nearest point in M. This item is needed in Algorithms 11 and 16.
2. **Parallel Transport**: Given a tangent vector $v \in T_p(M)$, we need a tool to transport it along a geodesic to a nearby point $q \in M$. This item is needed in Algorithm 13.
3. **Projection on Tangent Space**: Given an arbitrary vector $w \in \mathscr{H}$, we require a procedure for projecting w into $T_p(M)$. This item is needed in Algorithms 12 and 14.

If we can accomplish these three tasks on a manifold M, we can apply the path-straightening method to M, as summarized in the following algorithm.

Algorithm 17 (Path Straightening on M).
Initialize a path α between p_1 and p_2.

1. *Compute the velocity vector $\frac{d\alpha}{d\tau}$ using Algorithm 12.*
2. *Compute the covariant integral u of $\frac{d\alpha}{d\tau}$ using Algorithm 13.*
3. *Compute the backward parallel transport of $u(1)$ along α using Algorithm 14.*
4. *Compute the gradient w of E on $\mathscr{A}_0$ using Algorithm 15.*
5. *Update the path α in the direction of $w \in T_\alpha(\mathscr{A}_0)$ using Algorithm 16. If $\|w\|$ is small, then stop. Else, return to Step 1.*

6.6.3 Example 1: Geodesics on $\mathbb{S}^2$

We first illustrate the path-straightening approach for finding geodesics on the unit sphere in $\mathbb{R}^3$. In this case, the three required items are well known; these were given in Eqs. 6.19–6.21 earlier and are repeated here for convenience. Note that we are assuming that $\mathbb{S}^2$ is a subset of $\mathbb{R}^3$ with the Euclidean metric.

1. **Projection on $\mathbb{S}^2$**: Any nonzero point $x \in \mathbb{R}^3$ can be mapped to the nearest point in $\mathbb{S}^2$ using the projection (Eq. 6.19):

$$x \mapsto \frac{x}{|x|}. \tag{6.30}$$

2. **Parallel Transport of Tangents**: A vector $v \in T_x(\mathbb{S}^2)$ can be transported along a great circle to a point $y \neq -x \in \mathbb{S}^2$ using the formula (Eq. 3.10):

$$v \mapsto \left(v - \frac{2\langle v, y \rangle}{|x+y|^2}(x+y) \right). \tag{6.31}$$

This is based on Householder reflection of the vector v so that v undergoes the same rotation as one that takes x to y. If we need to parallel transport a tangent vector along a path that is not a great circle, we subdivide the path and use the above formula for each subdivision, as described in the second part of Example 3.7.

3. **Projection on Tangent Space**: Any vector $w \in \mathbb{R}^3$ can be projected on the tangent space $T_x(\mathbb{S}^2)$ using

$$w \mapsto w - \langle w, x \rangle x. \tag{6.32}$$

With these tools the path-straightening algorithm for computing a geodesic between any two points $x_1, x_2 \in \mathbb{S}^2$ can be carried out using Algorithm 17.

Figure 6.13 shows an example of obtaining a geodesic between two points using path straightening. The leftmost panel shows two points x_1 and x_2 in $\mathbb{S}^2$, the second panel shows the initial path α and the gradient vector field w obtained using Algorithm 15. Finally, the last panel shows several iterations of path straightening until α converges to the geodesic between x_1 and x_2. The convergence of α is measured using the norm of the gradient vector field w. Since the geodesics on $\mathbb{S}^2$ are great circles, one can verify that the resulting α is an arc on the great circle going through both p_1 and p_2.

6.6.4 Example 2: Geodesics in Elastic Pre-shape Space

Next we focus our attention on shape analysis and consider the problem of finding geodesics in $\mathscr{C}_2^c$, the pre-shape space of curves represented by their SRVFs (defined in Eq. 6.5). Elements of $\mathscr{C}_2^c$ are those elements of $\mathbb{L}^2([0,1], \mathbb{R}^2)$ that satisfy the conditions of unit length and closure. To provide some insight in the structure of $\mathscr{C}_2^c$, recall the set

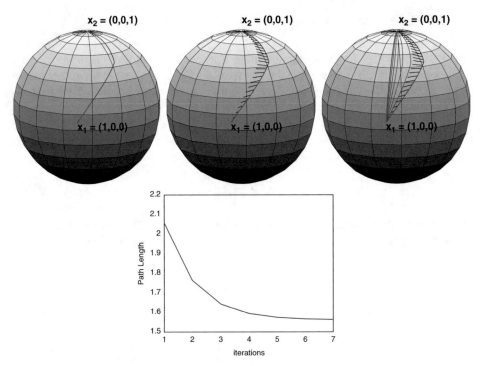

Fig. 6.13 An example of path-straightening method for computing geodesics between two points on $\mathbb{S}^2$. The right panel shows the decrease in the path length

$$\mathscr{C}_2 = \left\{ q \in \mathbb{L}^2([0,1], \mathbb{R}^2) | \int_0^1 |q(t)|^2 dt = 1 \right\},$$

defined in the previous chapter. The hierarchy of spaces is $\mathscr{C}_2^c \subset \mathscr{C}_2 \subset \mathbb{L}^2([0,1], \mathbb{R}^2)$. (Recall that functions in $\mathscr{C}_2$ correspond to curves of length 1 but are not necessarily closed.) Instead of treating $\mathscr{C}_2^c$ as a submanifold of $\mathbb{L}^2([0,1], \mathbb{R}^2)$, as was suggested in the general approach presented in Sect. 6.6.1, we will consider it as a submanifold of $\mathscr{C}_2$. Since $\mathscr{C}_2$ is a unit hypersphere inside $\mathbb{L}^2([0,1], \mathbb{R}^2)$, many aspects of its geometry—tangent spaces, projections, exponential maps, and geodesics—are already known (see Example A.11). This knowledge helps us in many ways, including a reduction in the computational cost of path straightening in $\mathscr{C}_2^c$. Although we will work with $\mathscr{C}_2$ as the larger space containing $\mathscr{C}_2^c$, the Riemannian structure of $\mathscr{C}_2^c$ remains same. We can rewrite $\mathscr{C}_2^c$ as

$$\mathscr{C}_2^c = \left\{ q \in \mathscr{C}_2 | \int_0^1 q(t)|q(t)| dt = 0 \right\} \tag{6.33}$$

i.e., $\mathscr{C}_2^c$ represents those curves in $\mathscr{C}_2$ that satisfy the closure condition. Algorithms 11–16 will have to be modified to take into account the geometry of $\mathscr{C}_2$, rather than $\mathbb{L}^2([0,1], \mathbb{R}^2)$, as the larger space. We remind the reader that if we remove the closure constraint, then the curves can be represented as elements of $\mathscr{C}_2$, a unit hypersphere, and the geodesics between them are straightforward, as discussed in the previous chapter. This implies that shape analysis of open curves is much simpler than that of closed curves.

A Riemannian structure on $\mathscr{C}_2^c$ is imposed in Sect. 5.4.1. Recall that in order to define $\mathscr{C}_2^c$ as a submanifold of $\mathbb{L}^2([0,1],\mathbb{R}^2)$, we introduced a function $\Phi_2 : \mathbb{L}^2([0,1],\mathbb{R}^2) \to \mathbb{R}^3$, where the three components of Φ_2 were denoted by $(\Phi_2^1, \Phi_2^2, \Phi_2^3)$. (The function Φ_2^1 gave the length of the curve, while the other two functions measured how far the curve was from being closed.) With respect to this function, we used the fact that $\mathscr{C}_2^c = \Phi_2^{-1}(1,0,0)$. Note that our set $\mathscr{C}_2$ can be described precisely by $\mathscr{C}_2 = (\Phi_2^1)^{-1}(1)$. Hence, to define $\mathscr{C}_2^c$ as a submanifold of $\mathscr{C}_2$, we only need to consider the remaining functions Φ_2^2 and Φ_2^3. We denote this simplified function by $\Psi : \mathscr{C}_2 \to \mathbb{R}^2$, where

$$\Psi(q) = (\Phi_2^2(q), \Phi_2^3(q)) = \left(\int_0^1 q^1(t)|q(t)|dt, \ \int_0^1 q^2(t)|q(t)|dt \right) .$$

We can redefine $\mathscr{C}_2^c$ as a subset of $\mathscr{C}_2$ by $\mathscr{C}_2^c = \Psi^{-1}(0,0)$. The tangent spaces have the hierarchy:

$$T_q(\mathscr{C}_2^c) \quad \subset \quad T_q(\mathscr{C}_2) \quad \subset \quad T_q(\mathbb{L}^2([0,1],\mathbb{R}^2)) \equiv \mathbb{L}^2([0,1],\mathbb{R}^2) .$$

Also, we can decompose $T_q(\mathscr{C}_2)$ according to $T_q(\mathscr{C}_2) = T_q(\mathscr{C}_2^c) \oplus N_q^b(\mathscr{C}_2^c)$, where $N_q^b(\mathscr{C}_2^c)$ is the set of normals at q to $\mathscr{C}_2^c$ inside $T_q(\mathscr{C}_2)$. We use the additional superscript b to distinguish it from $N_q(\mathscr{C}_2^c)$, which was used to denote the set of normals to $\mathscr{C}_2^c$ at q inside $\mathbb{L}^2([0,1],\mathbb{R}^2)$. Since we want to restrict our operations to $\mathscr{C}_2$, we seek the normal space $N_q^b(\mathscr{C}_2^c)$ inside $T_q(\mathscr{C}_2)$. From Sect. 5.4.1, the directional derivative of Ψ is given by for any $w \in T_q(\mathscr{B})$,

$$d\Psi^1(w) = \int_0^1 \left\langle w(t), \frac{q^1(t)}{|q(t)|}q(t) + |q(t)|\mathbf{e}^1 \right\rangle dt$$

$$d\Psi^2(w) = \int_0^1 \left\langle w(t), \frac{q^2(t)}{|q(t)|}q(t) + |q(t)|\mathbf{e}^2 \right\rangle dt .$$

To obtain the normal space $N_q^b(\mathscr{C}_2^c)$, we need to take the basis functions

$$\left\{ \frac{q^1(t)}{|q(t)|}q(t) + |q(t)|\mathbf{e}^1, \ \frac{q^2(t)}{|q(t)|}q(t) + |q(t)|\mathbf{e}^2 \right\} ,$$

and project them into $T_q\mathscr{C}_2$; this means we need subtract off the component of each of them in the direction of q.

Thus, define:

$$b_i(t) \equiv \frac{q^i(t)}{|q(t)|}q(t) + |q(t)|\mathbf{e}^i - q(t)\int_0^1 \left\langle q(u), \frac{q^i(u)}{|q(u)|}q(u) + |q(u)|\mathbf{e}^i \right\rangle du$$

$$= \frac{q^i(t)}{|q(t)|}q(t) + |q(t)|\mathbf{e}^i - 2q(t)\int_0^1 q^i(u)|q(u)|du. \tag{6.34}$$

To implement the path-straightening procedure for finding geodesics in $\mathscr{C}_2^c$, considered as a submanifold of $\mathscr{C}_2$, we need the following three ingredients:

1. **Projection on** $\mathscr{C}_2^c$: We particularize the general approach presented in Sect. 6.3. The idea is to define a residual vector $\mathbf{r}(q) = \Psi(q) \in \mathbb{R}^2$ and to evolve q in the direction perpendicular to the level set of Ψ so as to move its Ψ

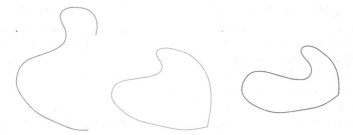

Fig. 6.14 Projection of an open curve (*left panel*), represented by its SRVF, in $\mathscr{C}_2^c$ using Algorithm 18 (*middle panel*). The display uses the coordinate functions while the computations are performed using SRVFs. The right panel shows the corresponding result using Algorithm 8

image toward the origin in $\mathbb{R}^2$. Algorithm 18 describes the procedure to project an open curve $q \in \mathscr{C}_2$ into $\mathscr{C}_2^c$. The Jacobian for this projection is a 2×2 matrix whose elements are given by $J_{ij} = \langle b_i(t), b_j(t) \rangle$.

The algorithm for projection is as follows:

Algorithm 18. *Projection of $q \in \mathscr{C}_2$ to $\mathscr{C}_2^c$. Let $\epsilon > 0$.*

a. *Compute $\mathbf{r}(q) = \Psi(q)$. If $|r(q)| < \epsilon$, stop, otherwise continue.*
b. *Calculate the Jacobian matrix $J(q)$ given above.*
c. *Solve the equation $J(q)\beta = -\mathbf{r}(q)$ for β.*
d. *Define $dq = \sum_{i=1}^{2} \beta_i b_i$, where b_1, b_2 are given in Eq. 6.34.*
e. *Update using $q \mapsto \cos(\|dq\|)q + \sin(\|dq\|)\frac{dq}{\|dq\|}$.*
f. *If converged, stop. Else, go to Step a.*

Figure 6.14 shows an example of projecting open curves $q \in \mathscr{C}_2$ onto $\mathscr{C}_2^c$ using Algorithm 18.

2. **Projection on Tangent Space**: For any given function $w \in \mathbb{L}^2([0,1], \mathbb{R}^2)$, we want a procedure for projecting it into the tangent space $T_q(\mathscr{C}_2^c)$, for any $q \in \mathscr{C}_2^c$. We will do this in two steps: first project w into the larger space $T_q(\mathscr{C}_2)$ and then project it into $T_q(\mathscr{C}_2^c)$. Although we can go straight to the second step, involving the first step improves the accuracy and stability of this numerical procedure.

   a. Start by projecting w into $T_q(\mathscr{C}_2)$ by

$$ w \mapsto w - \langle w, q \rangle q \, . \tag{6.35} $$

   If the starting tangent is already in $T_q(\mathscr{C}_2)$, then this step can be skipped.
   b. Compute an orthonormal basis of the normal space $N_q^b(\mathscr{C}_2^c)$ inside $T_q(\mathscr{C}_2)$ by performing Gram-Schmidt on b_1 and b_2; those were defined earlier in Eq. 6.34. Call the resulting elements b_1^o and b_2^o.
   c. Then, the projection of w into $T_q(\mathscr{C}_2^c)$ is given by

$$ \tilde{w} \equiv w - \sum_{i=1}^{2} \langle \tilde{w}, b_i^o \rangle b_i^o. \tag{6.36} $$

3. **Parallel Transport**: The next item we need is to transport a tangent vector $w \in T_{q_1}(\mathscr{C}_2^c)$ to the tangent space at a nearby point $q_2 \in \mathscr{C}_2^c$. This we accomplish in two steps. First, we recall that the points q_1, q_2 are also elements of the

hypersphere $\mathscr{C}_2$ since $\mathscr{C}_2^c$ is a subset of $\mathscr{C}_2$. Therefore, we can use the structure of $\mathscr{C}_2$ to transport the tangent w to the tangent space $T_{q_2}(\mathscr{C}_2)$. Secondly, we will project it into the desired space $T_{q_2}(\mathscr{C}_2^c)$.

a. For the first step, we use the following equation:

$$w \mapsto \tilde{w} \equiv \left(w - \frac{2 \langle w, q_2 \rangle}{\|q_1 + q_2\|^2}(q_1 + q_2) \right) . \tag{6.37}$$

This is the same equation we used to translate tangent vectors along geodesics in $\mathbb{S}^2$ (e.g., in Eq. 6.31), but now applied to the hypersphere $\mathscr{C}_2$.

b. Compute the norm of $\tilde{w}$, call it l. Form the projection of $\tilde{w}$ into $T_{q_2}(\mathscr{C}_2^c)$ using

$$\bar{w} = \tilde{w} - \sum_{i=1}^{2} \langle \tilde{w}, b_i^o \rangle b_i^o ,$$

where b_1^o, b_2^o form an orthonormal basis of $N_{q_2}^b(\mathscr{C}_2)$ inside $T_q(\mathscr{C}_2)$ as earlier. Rescale the resulting projection to obtain the answer: $\bar{w} \mapsto \bar{w}l\|/\|\bar{w}\|$.

Having accomplished these three tasks on $\mathscr{C}_2^c$, we can apply our path-straightening algorithm to $\mathscr{C}_2^c$. The resulting algorithm is very similar to the earlier cases but repeated here for convenience:

Algorithm 19 (Path Straightening on $\mathscr{C}_2^c$).

1. *Initialize a path α between q_1 and q_2. One way is to form a geodesic between them in $\mathscr{C}_2$ and then project each point on the path to $\mathscr{C}_2^c$. The geodesic on a hypersphere is given by for $\tau = 0, 1/k, \ldots, 1$*

$$\alpha(\tau) = \frac{1}{\sin(\theta)} \left[\sin(\theta - \tau\theta)q_1 + \sin(\tau\theta)q_2 \right] \tag{6.38}$$

where $\theta = \cos^{-1}(\langle q_1, q_2 \rangle)$. Note that $\alpha(0) = q_1$ and $\alpha(1) = q_2$. The projection into $\mathscr{C}_2^c$ is accomplished using Algorithm 18.
2. *Compute the velocity vector $\frac{d\alpha}{dt}$ using Algorithm 12.*
3. *Compute the covariant integral u of $\frac{d\alpha}{dt}$ using the Algorithm 13 and using the initial condition $u(0) = 0$.*
4. *Compute the backward parallel transport of $u(1)$ along α using the Algorithm 14.*
5. *Compute the gradient w of E on $\mathscr{A}_0$ using the Algorithm 15.*
6. *Update the path α in the direction of $w \in T_\alpha(\mathscr{A}_0)$ using Algorithm 16. If $\|w\|$ is small, then stop. Else, return to Step 2.*

We start with some examples of the path-straightening process. Shown in the top panel of Fig. 6.15 is a path that forms the initial condition for Algorithm 19. It is essentially a path from one curve (bird) to another (human) and then back to the first curve (bird). The evolution of this path under the path-straightening algorithm is shown in the remaining rows with the evolution of E shown in bottom right. The final path after eight iterations is shown in bottom left row. Since the first and the last curves of this path are the same, the results are as expected, a constant path with constant value given by that curve. The path energy E is zero for this constant path. Figure 6.16 shows two more examples of this path

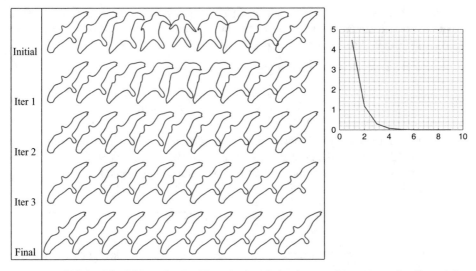

Fig. 6.15 Illustration of Algorithm 19. The initial path (*top*), several iterations of path straightening (*middle panels*), the final path (*bottom left*), and the evolution of the path energy E (*bottom right*)

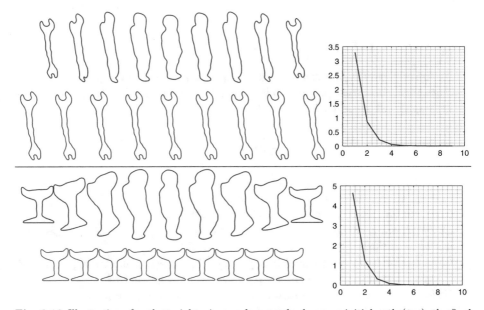

Fig. 6.16 Illustration of path straightening: each example shows an initial path (*top*), the final path (*bottom left*), and the evolution of the path energy E (*bottom right*)

straightening into constant paths to demonstrate the success of Algorithm 19 is finding geodesics in $\mathscr{C}_2^c$.

Next we show some general examples of geodesics in $\mathscr{C}_2^c$ obtained using Algorithm 19 in Figs. 6.17, 6.18. The first set of results involves some polygonal shapes. Each row in Fig. 6.17 shows geodesic paths between two polygons. The first row shows the geodesic between the hexagon and a right triangle, the second between a triangle and a star, and so on. Consider the third row of this figure that shows the

Fig. 6.17 Examples of geodesic paths in the pre-shape space $\mathscr{C}_2^c$ for some polygonal shapes

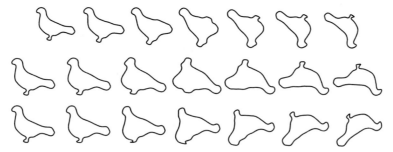

Fig. 6.18 Geodesics in $\mathscr{C}_2^c$ when the target curves have the same shape as the initial curve but differ in orientations and parameterizations

geodesic between a triangle and a star rectangle. A look at the intermediate shapes suggests that the two opposite corners are kept while the remaining two corners disappear. Instead, two corners are formed in new places so as to match the lengths of sides in the rectangle. This is an illustration of some bending in transforming one shape into another. Since this is a geodesic in the pre-shape space (and not the shape space), the resulting paths have not been optimized over $SO(2)$ and Γ_S. Later when we study a geodesic between the same pair in the shape space $\mathscr{S}_2^c$, we will return to this figure to compare results.

The underlying issue is that these are geodesics in $\mathscr{C}_2^c$ where the rotations and the parameterizations of the curves are used as given. Therefore, the features in one shape do not always match with the features in the other. To demonstrate this point further, Fig. 6.18 shows the geodesics between the same two shapes but with different placements of the starting point (origin) on the second curve is different for each row. As a result, the registration of points across the curves and the actual geodesic path is different for each case.

For future use, we are interested in computing the exponential map on $\mathscr{C}_2^c$. Given a point $q \in \mathscr{C}_2^c$ and a tangent vector $v \in T_q(\mathscr{C}_2^c)$, we want to construct a constant-speed geodesic path $\alpha : [0,1] \to \mathscr{C}_2^c$ such that $\alpha(0) = q$ and $\dot{\alpha}(0) = v$. Then, the point reached at the unit time $\alpha(1)$ is actually the desired exponential $\exp_q(v)$. The vector v is also called the *shooting vector* of the geodesic α.

Algorithm 20 (Shooting Geodesics on $\mathscr{C}_2^c$).
Initialize $\alpha(0) = q$, $w(0) = v$, set $\tau = 0$.

1. *Set $q_{temp} = \alpha(\tau/k) + \frac{1}{k}w(\tau/k)$. Project q_{temp} onto $\mathscr{C}_2^c$ and call it $\alpha((\tau+1)/k)$.*
2. *Parallel translate $w(\tau/k)$ from $\alpha(\tau/k)$ to $\alpha((\tau+1)/k)$.*
3. *If $\tau = k - 1$, stop. Else, set $\tau = \tau + 1$ and return to Step 1.*

This algorithm provides a numerical way of approximating the exponential map in $\mathscr{C}_2^c$.

6.7 Geodesics in Shape Spaces

So far we have looked at the problem of computing geodesics between given curves as elements of the pre-shape spaces $\mathscr{C}_1^c$ and $\mathscr{C}_2^c$. As demonstrated through several examples, this is not sufficient because the registration and the rotations of curves are not optimized in order to achieve the most efficient deformation from one to the other. One simply takes the given orientations and the parameterizations, and computes geodesics between the resulting curves. This is not appropriate for shape analysis on two counts. One, due to a lack of optimal alignment between the curves, the resulting geodesic paths look unnatural. Second, and more important, the resulting analysis of shapes, based on geodesics in pre-shape spaces, is not invariant to these operations: rigid transformations and re-parameterizations. Our solution to both of these problems is to go to the shape spaces.

We have seen how to compute geodesics in the pre-shape spaces. The question is: how can we extend this construction to the shape spaces? Remembering that the shape spaces are quotient spaces of the respective pre-shape spaces, we can adapt the previous tools to get the desired results. Recall that by a geodesic in a quotient space we mean the shortest geodesic between the two corresponding orbits in the pre-shape space. This was discussed previously in Sect. 5.7. Since we have used a shooting method to compute geodesics in $\mathscr{C}_1^c$ and a path-straightening method to compute geodesics in $\mathscr{C}_2^c$, the required adaptations are different in these two cases and will be discussed separately.

6.7.1 Geodesics in Non-elastic Shape Space

As described in Sect. 6.2.3, the shape space $\mathscr{S}_1^c$ is a quotient space of the pre-shape space $\mathscr{C}_1^c$ under the action of $\mathbb{S}^1$. Therefore, the problem of finding geodesics in $\mathscr{S}_1$ reduces to the problem of finding those geodesics in $\mathscr{C}_1^c$ that are orthogonal to the $\mathbb{S}^1$ orbits. The fact that $\mathbb{S}^1$ acts by isometries also implies that if a geodesic in a pre-shape space is orthogonal to one $\mathbb{S}^1$ orbit, then it is orthogonal to all $\mathbb{S}^1$ orbits that it meets and, hence, projects to a geodesic in the corresponding shape space. Let us assume arbitrary points $\theta_1, \theta_2 \in \mathscr{C}_1^c$; our goal is to find a geodesic between the orbits $[\theta_1]$ and $[\theta_2]$ in $\mathscr{S}_1^c = \mathscr{C}_1^c/\mathbb{S}^1$. The geodesic distance in $\mathscr{S}_1^c$ is given by]

$$d_{\mathscr{S}_1^c}([\theta_1], [\theta_2]) = \inf_{\tau \in \mathbb{S}^1} d_{\mathscr{C}_1^c}(\theta_1, (\tau, \theta_2)) \ .$$

Using Algorithm 10, we can find geodesics between any two points of $\mathscr{C}_1^c$, as demonstrated in Sect. 6.5.2. For finding the geodesic in $\mathscr{S}_1^c$, we will perform the following additional steps:

1. Select any point on the orbit $[\theta_1]$ in $\mathscr{C}_1^c$ and fix it. Say we select θ_1 itself.
2. We need to restrict the shooting direction at $T_{\theta_1}(\mathscr{C}_1^c)$ to ensure that the geodesic is perpendicular to $[\theta_1]$. That is, the shooting direction is perpendicular to the tangent space $T_{\theta_1}([\theta_1])$. Recall that the action of $\mathbb{S}^1$ on $\mathscr{C}_1^c$ is given by for a $\tau \in \mathbb{S}^1$

$$(\tau, \theta_1) = \theta_1((s - \tau)_{2\pi}) + \tau \ .$$

The derivative of that group action with respect to τ, evaluated at $\tau = 0$ is $1 - \dot{\theta}_1$. Therefore, the one-dimensional space $T_{\theta_1}([\theta_1])$ is spanned by $1 - \dot{\theta}_1$. Consequently, the desired shooting direction v should be perpendicular to the function $1 - \dot{\theta}_1$. (Here we restrict to those elements of $\mathscr{C}_1^c$ that have continuous first derivative.)

   In terms of the modifications to Algorithm 10, the only change from this item comes in the projection on tangent space (Eq. 6.25) that is changed as follows. The new projection of a vector $v \in \mathbb{L}^2([0, 2\pi], \mathbb{R})$ to the desired shooting space is given by

$$v \mapsto v - \sum_{i=1}^{4} \langle v, w_i \rangle \, w_i \ , \tag{6.39}$$

where w_i's form an orthonormal basis of the space spanned by $\{1, \cos \circ \theta_1, \sin \circ \theta_1, \dot{\theta}_1\}$. The last element in this set is the only change from earlier.
3. Then, we need to search over all elements of $[\theta_2]$ to find the one nearest to θ_1 in terms of the geodesic length. How will this be accomplished? Recall that in Algorithm 10, we search for the shooting direction $v \in T_{\theta_1}(\mathscr{C}_1^c)$ that minimizes the miss function $F[v] = \|\alpha_1(\theta_1; v) - \theta_2)\|^2$, where α_τ is a geodesic shot from θ_1 with the initial velocity v. We will change this definition of F to be

$$F[v] = \inf_{\tau \in \mathbb{S}^1} \|\alpha_1(\theta; v) - (\tau, \theta_1)\|^2 \ . \tag{6.40}$$

Furthermore, the initial velocity v is now restricted to be perpendicular to the orbit $[\theta_1]$, i.e., $v \perp \dot{\theta}_1$. This equation redefines the miss function to be the $\mathbb{L}^2$ distance squared from $\alpha_1(\theta; v)$ to the nearest point in the orbit $[\theta_2]$. In case the direction v is represented by its Fourier approximation $\pi(\sum_{i=0}^m (a_i \cos(is) + b_i \sin(is)))$, where π is the projection from $\mathbb{L}^2$ to the tangent space $T_{[\theta_1]}(\mathscr{S}_1^c)$, we get the miss function:

$$\tilde{F}(a, b) = \inf_{\tau \in \mathbb{S}^1} \|\alpha_1(\theta; \sum_{i=0}^m (a_i \cos(is) + b_i \sin(is)) - (\tau, \theta_1)\|^2 \ . \tag{6.41}$$

We can modify Algorithm 10 to take these three steps into account and reach the following algorithms for finding geodesics between $[\theta_1]$ and $[\theta_2]$ in $\mathscr{S}_1^c$.

Algorithm 21 (Shooting Method on $\mathscr{S}_1^c$).
Given points $\theta_1, \theta_2 \in \mathscr{C}_1^c$, initialize the coefficients $\mathbf{a} = \{a_1, \ldots, a_m\}$ and $\mathbf{b} = \{b_1, \ldots, b_m\}$ and find the tangent direction $v \in T_{\theta_1}(\mathscr{S}_1)$ by projecting $\sum_{i=1}^m a_i \cos(is) + b_i \sin(is)$ as described in (Eq. 6.39). Fix a $\delta > 0$ small.

1. **Current Cost**: *Compute the exponential map* $\exp_{\theta_1}(v)$ *using Algorithm 9. Compute the cost function* $F(v) = \inf_{u \in \mathbb{S}^1} \|(u, \theta_2) - \exp_{\theta_1}(v)\|^2$ *as given in Eq. 6.40. This infimum is estimated by evaluating the function on the right for* u *values on a dense grid on* $\mathbb{S}^1$ *and taking the smallest value.*
2. **Gradient of Cost Function**: *Use difference quotients to approximate the gradient* $\frac{\partial \tilde{F}}{\partial a_i}$ *and* $\frac{\partial \tilde{F}}{\partial b_i}$.
3. *Update* **a** *and* **b** *using the negative gradients of* $\tilde{F}$, *form* $v(s) = \sum_{i=1}^{m} a_i \cos(is) + b_i \sin(is)$ *and project it into* $T_{\theta_1}(\mathscr{S}_1^c)$ *using Eq. 6.39.*
4. *If the added norms of the vectors* $\nabla_a \tilde{F}$ *and* $\nabla_b \tilde{F}$ *are small, then stop. Else, return to Step 1.*

We present some results on the shape space $\mathscr{S}_1^c$ generated using this algorithm. To highlight the advantage of removing the re-parameterization group $\mathbb{S}^1$, we want to compare these results with those obtained for the larger space $\mathscr{C}_1^c$. As a simple comparative example, consider the two end shapes shown in the first row of Fig. 6.19. Both the curves are perturbations of a heart shape with the difference that the perturbations (little dents) are on different sides. The top row shows a geodesic path between these two curves in $\mathscr{C}_1^c$, for some arbitrarily chosen origins on these curves. Since the origins are fixed, the correspondences between points on the two curves are not very natural and the resulting deformation along the geodesic path in $\mathscr{C}_1^c$ looks awkward. However, we should point out that for the given placements of origins this is the path of least bending energy spent in going from one shape to another. Next, consider the geodesic shown in the lower row of this figure. In this case, the geodesic is computed in the quotient space $\mathscr{S}_1^c$, which means that the geodesic is perpendicular to the orbits of $\mathbb{S}^1$. In other words, the algorithm automatically searches over all possible placements of origin on the second curve (through the infimum definition of $\tilde{F}$ in Eq. 6.41) to find the best one. As a result, the correspondence of points across the two shapes looks more appropriate and the geodesic path shows a natural deformation of the first shape into the second one.

Figure 6.20, 6.21 show some additional examples of bending-only geodesics between shapes that we have already seen in Sect. 6.5.2. The first case is that of some polygonal shapes. Figure 6.20 shows geodesic paths between polygons that are drawn as end shapes in each row. The intermediate shapes represent equally spaced points along the geodesic ψ_t between those end shapes in $\mathscr{S}_1^c$. For comparison with geodesics in $\mathscr{C}_1^c$, look at the corresponding geodesics in Fig. 6.9. In particular, compare the second rows of both the figures where a square is being

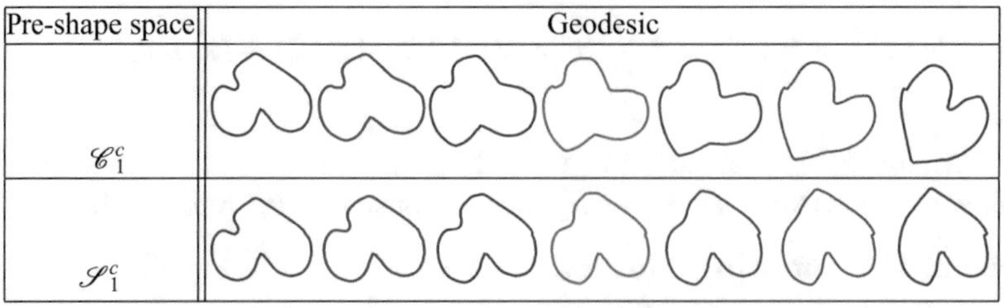

Fig. 6.19 Geodesics between the same two curves under the non-elastic representation. The top geodesic is computed in $\mathscr{C}_1^c$ with the given orientation and placement of origins, while the bottom geodesic is computed in $\mathscr{S}_1^c$ where one finds the optimal alignment of shapes

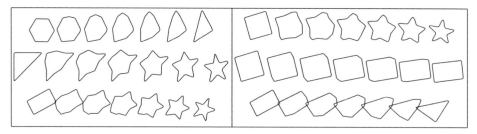

Fig. 6.20 Examples of geodesic paths between polygons in $\mathscr{S}_1^c$. Compare with the corresponding paths in $\mathscr{C}_1^c$ shown in Fig. 6.9

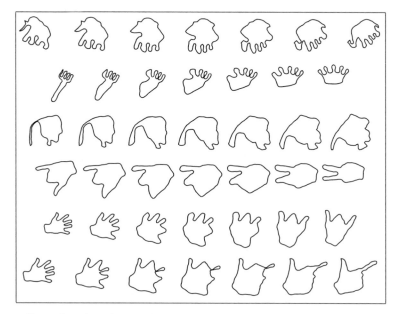

Fig. 6.21 Examples of geodesic paths between some curves in $\mathscr{S}_1^c$. Compare with the corresponding paths in $\mathscr{C}_1^c$ shown in Fig. 6.9

deformed into a rectangle. Just by comparing these two paths visually, one can convince oneself that bending-wise, the result for $\mathscr{S}_1^c$ seems more efficient than the one for $\mathscr{C}_1^c$. In the latter case, the given parameterization is such that all four corners of the square disappear and new corners appear to make the rectangle, while in the former case two of the corners remain unchanged and the remaining two are modified to go from the square to the rectangle. Similar observations can be made for all the cases shown in these two figures:

Figure 6.21 shows more examples of geodesics in $\mathscr{S}_1^c$, this time using more natural shapes. In each case, we can see the effect of the algorithm finding the optimal placement of origin (on the second curve) so as to best match the two curves. The net effect seems to be that the algorithm picks up the most dominant feature (a protrusion or a corner) in both the shapes and matches them first and it matches the remaining points accordingly. With this matching, it bends the first shape into the second in the most efficient way. For example, in the first row, some legs of the two elephant shapes are matched to each other and the remaining points match accordingly. So one leg of the first shape has to deform into the trunk of the second shape.

6.7.2 Geodesics in Elastic Shape Space

Next we consider the task of finding geodesics between shapes in the space $\mathscr{S}_2^c$. As described in Sect. 6.2.3, this set of elastic shapes is realized as a quotient space $\mathscr{C}_2/(SO(2) \times \Gamma_S)$. So, the equivalence class of an element $q \in \mathscr{C}_2^c$ is given by Eq. 5.21 (repeated here for convenience):

$$[q] = \text{closure}\{O(q \circ \gamma)\sqrt{\dot{\gamma}} | \gamma \in \tilde{\Gamma}_S, O \in SO(2)\} \ .$$

For any two planar closed curves, β_1 and β_2, let their SRVFs be given by q_1 and q_2, respectively. Then, our goal is to find a geodesic path between the orbits $[q_1]$ and $[q_2]$ in $\mathscr{S}_2^c$. In principle, a geodesic in $\mathscr{S}_2^c = \mathscr{C}_2^c/(\Gamma_S \times SO(2))$ is obtained by forming geodesics between all possible pairs in the set $([q_1] \times [q_2]) \subset \mathscr{C}_2^c$ and then selecting the shortest. Thus, the geodesic distance in $\mathscr{S}_2^c$ is given by

$$
\begin{aligned}
d_{\mathscr{S}_2^c}([q_1],[q_2]) &= \inf_{r_1 \in [q_1], r_2 \in [q_2]} d_{\mathscr{C}_2^c}(r_1, r_2) \\
&= \inf_{O \in SO(2), \gamma \in \Gamma_S} d_{\mathscr{C}_2^c}(q_1, \sqrt{\dot{\gamma}} O(q \circ \gamma)) \ .
\end{aligned}
\tag{6.42}
$$

The last inequality comes from the fact that the action of $SO(2) \times \Gamma_S$ on $\mathscr{C}_2^c$ is by isometries and Γ_S is dense in $\tilde{\Gamma}_S$. So, one needs to minimize only over one orbit and not two. We remind the reader that we have already developed a path-straightening approach for finding geodesics between any two points in $\mathscr{C}_2^c$ (Algorithm 19) and we will modify it to reach an algorithm for finding geodesics in $\mathscr{S}_2^c$.

Before we proceed further, we remind the reader this development overlooks certain mathematical technicalities in order to reach efficient numerical implementations of the alignment process. Similar to the previous chapter, the issue here is that the shape space does not have a manifold structure. Due to this situation, we view the shape space only as a set with a distance inherited from the pre-shape space. However, in the context of deriving numerical procedures, such as the gradient-based approach for alignment, we will utilize ideas such as the tangent spaces of orbits and the tangent spaces of the quotient space to help out conceptually.

Equation 6.42 outlines a problem of optimization over the joint space $\Gamma_S \times SO(2)$ and we will use an iterative numerical approach to find the solution. Each iteration of this search takes the following form: take an arbitrary pairing (q_1, r) from the set $(\{q_1\} \times [q_2])$ and form a geodesic α between them in $\mathscr{C}_2^c$. If this geodesic happens to be orthogonal to orbit $[q_2]$, as measured by computing the inner product between tangent to the geodesic at r and basis elements of $T_{\gamma_{id}}(\Gamma_S)$ (when placed at r), then we can stop, as we already have a solution to the optimization problem. In other words, if the norm of the projection of the tangent to α at r into the set $T_{\gamma_{id}}(\Gamma_S)$ (when placed at r) is not zero, then we update r in such a way that the norm of this projection is reduced. Naturally, the direction in which to update r arises from the projection itself. Apply this idea repeatedly until the projection becomes zero. A schematic illustration of this idea is shown in Fig. 6.22. A word of caution here for the reader—this particular discussion on iterative optimization seeks a local solution to the optimization problem. That is, depending on the initial point r, we can end up at a point that is a local minimizer of the cost function. To search over a larger space, some other ideas are needed.

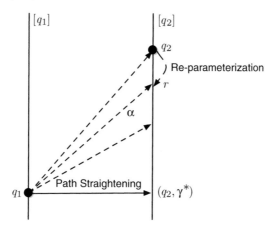

Fig. 6.22 Gradient-based update of elements in $[q_2]$, while keeping q_1 fixed, to find the shortest geodesic between the orbits of $[q_1]$ and $[q_2]$

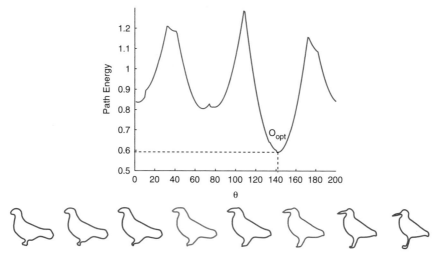

Fig. 6.23 *Top*: The cost function $d_{\mathscr{C}_2^c}(q_1, Oq_2)$ vs. the angle of rotation in O. *Bottom*: Geodesic Path corresponding to the optimal alignment

Next we describe the projection-based iterative update of the element $r \in [q_2]$. Since the two groups $SO(2)$ and Γ_S commute, we can present the updates under each action separately.

1. **Optimal Rotation**: The orbit of r generated by the action of $SO(2)$ on r is given by $\{Or | O \in SO(2)\}$. In Fig. 6.23, we show the plot of path energy, defined in Eq. 6.28, for the geodesic path between q_1 and elements of this orbit, for the two bird shapes shown at the ends of the bottom row. In this example, r corresponds to arc-length parameterization of the second curve. This example supports what we already know, that the energy of a path between q_1 and Or depends on O and there are several local minima in this optimization problem. Depending on the computational resources, we can put a coarse grid on $SO(2)$, evaluate the cost function over the grid points, and select the best point among those candidates. We can use this as an initial condition for the gradient

iteration. To set up the projection-based or the gradient update, we need to establish the tangent space to the rotation orbit. For $r \in [q_2]$, this is a one-dimensional space spanned by the function Er, where $E = \frac{1}{\sqrt{2}} \begin{bmatrix} 0 & -1 \\ 1 & 0 \end{bmatrix}$ is the basis element of the tangent space $T_I(SO(2))$ (see Sect. A.1.2 for a discussion on geometry of $SO(2)$).

We are going to optimize over the rotation O iteratively and let r be the current estimate of the closest point to q_1 in $[q_2]$. We already have a geodesic path between q_1 and r in $\mathscr{C}_2^c$, denoted by α, such that w is the initial velocity vector to the geodesic α at r. That is, $w \equiv \exp_r^{-1}(q_1)$. Taking the projection of w in the rotation orbit of q_2, we obtain $x = \langle w, Er \rangle$. The iterative update of r on rotation orbit of q_2 is given by $r \mapsto e^{-x\epsilon E}r$, with $\epsilon > 0$ being a small step size.

In summary, an iteration of the optimization over $SO(2)$ is as follows: Let α be the geodesic path in $\mathscr{C}_2^c$ from r to q_1 and let w be its initial velocity vector.

a. Compute the gradient at $r \in [q_2]$ using $x = \langle w, Er \rangle$, where $E = \frac{1}{\sqrt{2}} \begin{bmatrix} 0 & -1 \\ 1 & 0 \end{bmatrix}$.

b. Update r using $\tilde{q}_2 \mapsto e^{-x\epsilon E}r$.

c. Stop if $|x|$ is small, otherwise return to Step a.

In some cases, where the speed of optimization is an important issue, we may choose a fast, but approximate, optimization over $SO(2)$ as follows. Here we change the cost function for optimization over $SO(2)$, from the one in Eq. 6.42 to the $\mathbb{L}^2$ distance between q_1 and Oq_2. In other words, instead of using the geodesic distance in $\mathscr{C}_2^c$, we use the geodesic distance in the larger Hilbert space $\mathbb{L}^2([0, 2\pi], \mathbb{R}^2)$, resulting in the optimization problem:

$$\hat{O} = \underset{O \in SO(2)}{\arg\min} \|q_1 - O \cdot r\|^2 , \tag{6.43}$$

where the norm is $\mathbb{L}^2$. For matching some kinds of shapes, this approximation may be good enough. The solution to this problem is available analytically:

$$\hat{O} = \begin{cases} UV^T, & \text{if } \det(A) > 0 \\ U \begin{bmatrix} 1 & 0 \\ 0 & -1 \end{bmatrix} V^T, & \text{otherwise} \end{cases} .$$

Here $A = U\Sigma V^T$ is the SVD of the 2×2 matrix $A = \int_0^1 q_1(t)r(t)^T dt$. Recall that in our notation $q(t)$ is a 2×1 vector for any t. Again, the disadvantage of this method is that the distance function we are minimizing will not be precisely the one based on geodesics in $\mathscr{C}_2^c$. However, the advantages of avoiding the gradient descent procedure and thereby avoiding getting stuck in local minima are considerable.

2. **Optimal Re-Parameterization**: The main problem in finding geodesics in $\mathscr{S}_2^c$ is the optimization over the re-parameterization group Γ_S. Before we present our gradient-based approach, we briefly discuss the nature of elements of Γ_S.

a. We first point out a couple of important subgroups of Γ_S. First, we can think of $\mathbb{S}^1$ itself as a subgroup. To see this, consider $\mathbb{S}^1$ as consisting of the unit complex numbers. Then, given any $w \in \mathbb{S}^1$, we obtain the diffeomorphism $z \mapsto wz$. Thus, $\mathbb{S}^1$ itself (as a Lie group) is isomorphic to the subgroup of Γ_S consisting of rotations. As a re-parameterization, such a diffeomorphism corresponds to choosing a different point of a closed curve as a "starting point". Another important subgroup of Γ_S is the group of "Moebius transformations". Given any two complex numbers a and b satisfying $|a| > |b|$, we define a function:

$$z \mapsto \frac{az + b}{\bar{b}z + \bar{a}} . \tag{6.44}$$

This function is a diffeomorphism $\mathbb{S}^1 \to \mathbb{S}^1$ and the set of all these functions is actually a subgroup of Γ_S. While it may appear that this subgroup has a real dimension of four, since it depends on the choice of two complex numbers, its real dimension is three, since multiplying both a and b by the same real scalar does not affect the action of the diffeomorphism.

b. As discussed in Chap. 4, Sect. 4.10.2, Γ_S is a differentiable manifold with the tangent space at $\gamma_{id} \in \Gamma_S$ given by

$$T_{\gamma_{id}}(\Gamma_S) = \mathbb{L}^2(\mathbb{S}^1, \mathbb{R}) = \{\text{smooth } g : \mathbb{S}^1 \to \mathbb{R} | \int_{\mathbb{S}^1} g(s)^2 ds < \infty\} .$$

We will need a basis of this set to perform gradient-based minimizations on it. One possible basis is to use the Fourier components:

$$\{\cos(nt), \sin(nt), \ n = 0, 1, 2, \ldots\} .$$

As we know, these functions form an orthonormal basis of $T_{\gamma_{id}}(\Gamma_S)$ under the $\mathbb{L}^2$ metric. If we take the first component, the constant function, and map its span to Γ_S using the exponential map, then the resulting set, $\exp_{id}(\text{span}(\mathbf{1}))$, is simply the set of all rotations $\mathbb{S}^1$, described in the previous paragraph. Similarly, if we take the first three components and form their span, $\text{span}\{\mathbf{1}, \cos(s), \sin(s)\}$, then the exponential map of this set gives the well-known 3-dimensional subgroup of Γ_S known as the "Moebius transformations", also described in the previous paragraph. Another basis of $T_{\gamma_{id}}(\Gamma_S)$ that can be useful is

$$\left\{ 1, \frac{\sin(nt)}{n\pi}, \frac{\cos(nt) - 1}{n\pi}, n = 1, 2, 3, \ldots \right\} . \tag{6.45}$$

These functions form an orthonormal basis of the set $T_{\gamma_{id}}(\Gamma_S)$ under the Palais metric (Eq. 6.29). Often there are some numerical advantages of using this basis in the gradient process used to optimize over Γ_S, which is described next.

Basic Idea: The idea underlying this gradient-based approach, for optimizing the cost function in Eq. 6.42 over Γ_S, is as follows. We want to search over the Γ_S-orbit of q_2 and find the point that minimizes the distance from q_1 in $\mathscr{C}_2^c$. Since this orbit has a one to one correspondence with Γ_S, we can transfer the problem to Γ_S. (This is actually how the problem is written in Eq. 6.42.) We are going to solve the problem iteratively and let $r \in [q_2]$ be the current estimate, an element of $[q_2]$. Define a mapping $\Upsilon : \Gamma_S \to [q_2]$ by $\gamma \mapsto (r, \gamma) = (r \circ \gamma)\sqrt{\dot{\gamma}}$ that establishes a correspondence between Γ_S and the Γ_S-orbit of $[q_2]$. Note that this mapping identifies r with the identity map γ_{id}. With the help of the differential of Υ, we will write the derivative of the cost function with respect to γ. Under the steepest descent scheme, this gradient provides a direction to update γ_{id}; call this update $d\gamma$. Applying this incremental $d\gamma$ to γ provides γ for

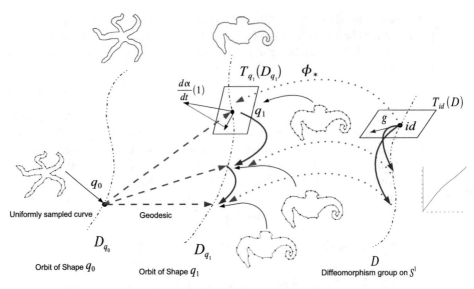

Fig. 6.24 Schematic view of the computation of geodesic in $\mathscr{S}_2^c$

the next iteration and the process is repeated until the norm of this increment is small enough to stop. Shown in Fig. 6.24 is a cartoon diagram illustrating this process.

The important piece here is the gradient of the cost function in Eq. 6.42 and that is derived next.

Theorem 6.7. *Let $v \in T_{\gamma_{id}}(\Gamma_S)$ be a tangent vector. The directional derivative of the cost function $d_{\mathscr{C}_2^c}(q_1, \Upsilon(\gamma))^2$ in the direction of v is given by $\langle w, d\Upsilon_{\gamma_{id}}(v)\rangle$ where $w = exp_{\Upsilon(\gamma)}^{-1}(q_1)$ and*

$$d\Upsilon_{\gamma_{id}}(v) = \dot{r}(t)v(t) + \frac{1}{2}r(t)\dot{v}(t) \ . \tag{6.46}$$

Proof. We prove this theorem using the chain rule that includes two partial derivatives. The first partial derivative is obtained using a result established later in the book (Theorem 7.1). This theorem states that on a Riemannian manifold M, the gradient of the square of the geodesic distance between $p_1, p_2 \in M$, with respect to p, is given by the inverse exponential map of p_2 onto the tangent space at p_1. Since the cost function here is exactly the geodesic distance squared in $\mathscr{C}_2^c$, this theorem applies directly. Hence, the gradient of $d_{\mathscr{C}_2^c}(q_1, \gamma)^2$ with respect to r is given by the vector $w = \exp_r^{-1}(q_1)$; note that w is an element of $T_r(\mathscr{C}_2^c)$.

The second partial derivative in the chain rule comes from Lemma 5.3. It states that for any point $\gamma \in \Gamma_S$, the differential of $\Upsilon : \Gamma_S \to [q_2]$ is the linear transformation $d\Upsilon_\gamma : T_\gamma(\Gamma_S) \to T_{\Upsilon(\gamma)}([q_2])$ defined by the equation:

$$(d\Upsilon_\gamma(v))(s) = \sqrt{\dot{\gamma}(s)}\dot{r}(\gamma(s))v(s) + \frac{1}{2\sqrt{\dot{\gamma}(s)}}\dot{v}(s)r(\gamma(s)) \ . \tag{6.47}$$

Applying this equation at $\gamma = \gamma_{id}$ and combining the two partial derivatives proves the theorem.

We summarize steps for an iterative optimization over Γ_S: let r be the current element of $[q_2]$ and let $\gamma \in \Gamma_S$ such that $r = (q_2, \gamma)$. Also, let α be the geodesic between r and q_1 in $\mathscr{C}_2^c$, obtained using Algorithm 19. The initial velocity of this geodesic is the desired tangent vector $w = \exp_r^{-1}(q_1)$. Let $\{b_i\}, i = 1, \ldots, d$ be an orthonormal basis of the set $T_{\gamma_{id}}(\Gamma_S)$. We suggest using the basis given in Eq. 6.45.

a. Compute the gradient of the cost function according to

$$d\gamma = \sum_{i=1}^{n} \langle w, d\Upsilon_{\gamma_{id}}(b_i) \rangle \, b_i \; . \tag{6.48}$$

where $d\Upsilon_{\gamma_{id}}$ is defined in Eq. 6.46.
b. If $\|d\gamma\| < \epsilon$, then stop. Else, continue.
c. Re-parameterize r using $d\gamma$:

$$r \mapsto \sqrt{\dot{d\gamma}}(r \circ d\gamma) \; .$$

In practice, there is some computational advantage to keeping track of the cumulative γ and applying it to the original curve β_2. That is, update $\gamma \mapsto \gamma + d\gamma$ and find $\tilde{\beta}_2 = \beta_2 \circ \gamma$ and compute r from $\tilde{\beta}_2$.

Combining the iterative optimization over both Γ_S and $SO(2)$, we obtain the following algorithm for finding geodesics in the elastic shape space $\mathscr{S}_2^c$.

Algorithm 22 (Geodesics on Shape Space of Elastic Closed Curves). *Let q_1 and q_2 be two elements of $\mathscr{C}_2^c$, and let β_2 be the parameterized path corresponding to q_2, i.e., $\beta_2(t) = \int_0^t |q_2(s)| q_2(s) ds$. Set $r = q_2$, $\gamma = \gamma_{id}$, and $O = I_2$. Choose the step sizes δ_1 and δ_2 for the gradient updates of the rotation and the re-parameterization, respectively:*

1. *Compute a geodesic path from γ to q_1 using Algorithm 19. Let w be the initial velocity of this geodesic path. That is, $w = \exp_r^{-1}(q_1)$.*
2. *Compute the two gradients as follows:*
   a. *Compute the gradient of the cost function with respect to the rotation O using $x = \langle w, E\gamma \rangle$, where $E = \begin{bmatrix} 0 & -1 \\ 1 & 0 \end{bmatrix}$. Set $dO = \exp(\delta_1 x E)$.*
   b. *Compute the gradient of the cost function with respect to the re-parameterization according to*

$$d\gamma = \delta_2 \sum_{i=1}^{n} \langle w, d\Upsilon_{\gamma_{id}}(b_i) \rangle \, b_i \; . \tag{6.49}$$

   where $d\Upsilon_{\gamma_{id}}$ is defined in Eq. 6.46 and $\{b_i\}, i = 1, \ldots, d$ are as given in Eq. 6.45.
3. *Update $\gamma \mapsto \gamma \circ (\gamma_{id} + d\gamma)$ and $O \mapsto O.dO$. Caution: In this step it is important that δ_2 was chosen small enough that $\gamma_{id} + d\gamma$ is a diffeomorphism (i.e., has positive derivative at each t).*
4. *Update β_2 according to $\tilde{\beta}_2 = O(\beta_2 \circ \gamma)$. Compute $r(t) = SRVF(\tilde{\beta}_2)$.*
5. *If $|x|$ and $\|d\gamma\|$ are small, then stop. Otherwise, return to Step 1.*

We present some results to demonstrate this algorithm. For any two shapes, the optimal orientations and re-parameterizations of the second shape are determined

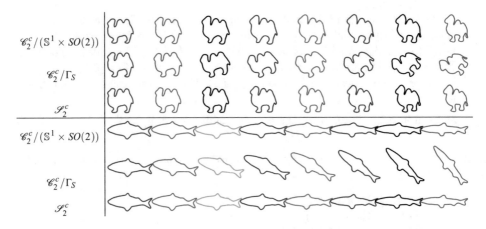

$\mathscr{C}_2^c/(\mathbb{S}^1 \times SO(2))$

$\mathscr{C}_2^c/\Gamma_S$

$\mathscr{S}_2^c$

$\mathscr{C}_2^c/(\mathbb{S}^1 \times SO(2))$

$\mathscr{C}_2^c/\Gamma_S$

$\mathscr{S}_2^c$

Fig. 6.25 Geodesics in elastic pre-shape and shape spaces

using Algorithm 22. In the process we also obtain the geodesic paths using path straightening in the shape space $\mathscr{S}_2^c$. Figure 6.25 displays two sets of examples. In both the cases, the top row shows the geodesic path when the optimal rotation and only the subgroup $\mathbb{S}^1 \subset \Gamma_S$ have been optimized for the second shape. That is, the path shown is a geodesic in the space $\mathscr{C}_2^c/(SO(2) \times \mathbb{S}^1)$. The second row shows the resulting geodesic when only the re-parameterization group Γ_S (but not the rotation group $SO(2)$) has been removed. Finally, the third row shows the case when both the rotation and the re-parameterization groups have been removed completely. In both these cases, and the ones shown in later examples, it is easy to see the improvement in resulting geodesic paths when the full shape-preserving group ($\Gamma_S \times SO(2)$) has been removed. The matching of points across the shapes is better and the deformation of shapes along the connecting geodesics is more natural. Take, for instance, the first case where one camel shape is being compared to another camel shape. In the first row, the two curves have been aligned rotationally and in placement of origins. Still the deformation from one to other does not seem natural, as the first leg almost disappears in the middle of the path. In the second row, the two curves lack rotational alignment, while in the last row the two curves are completely aligned in all variables. Consequently, the resulting geodesic path in the last seems most natural; the legs match the legs, the tail matches the tail, the head matches the head, etc.

Figure 6.26 shows some additional examples of geodesics between curves in $\mathscr{S}_2^c$. These curves have ticks placed on them to help visualize the speed of traversal along those curves. The ticks start with a uniform separation on the leftmost shape, denoting the arc-length parameterization in the first shape, but become nonuniform as we progress toward the target shape along the geodesic. This non-linear sampling in the second shape helps improve matching of similar features across the two shapes. Visually it appears as if the first curve is being stretched and/or compressed locally, in addition to being bent, to reach the second curve. For example, in the first row, the small tail of the tortoise is matched to a much larger tail of the dog; this matching of tails, in addition of a matching of their legs, becomes possible due to stretching and compressing of other parts along the tortoise curve.

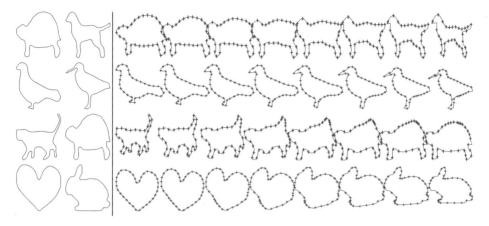

Fig. 6.26 Geodesic paths in $\mathscr{S}_2^c$ between the pair of curves shown to the left

Another way to analyze elastic geodesics is to study the way points are matched between two shapes, resulting from optimal rotations and re-parameterizations. Figure 6.27 shows elastic geodesics between pairs of shapes on the top and the corresponding elastic matching on the bottom. The success in matching similar parts across shapes results in more meaningful deformations in the geodesic process. Take the case of two hands in the last row, for example. The two hands (leftmost and the rightmost) differ in at least two ways: the middle finger is stretched and the last finger is bent. A study of the matching result shows us that despite these differences the corresponding parts on the two hands are matched well—the tips of the fingers in one hand match with the tips in the other, the thumb with the thumb, and so on. As a result, the geodesic path shows a deformation that seems natural—the middle finger stretches out from left to right, the last finger bends from left to right, etc. The algorithm is successful in using an optimal combination of bending and stretching to perform the matching and the deformation.

6.8 Examples of Elastic Geodesics

As further examples of geodesic computations, we look at the shapes of capital English letters under two fonts: Times New Roman and Comic Sans, as shown in Fig. 6.28. In order to compare the shapes of letters in these fonts, we compute the geodesic paths and geodesic distances between all the letters of one font with all the letters in the second font. Shown in the bottom part of this figure are two examples of geodesic paths and optimal matching between the corresponding curves. For the full comparison, we obtain a 26×26 distance matrix that is shown as a gray scale image in the right panel. As expected, the geodesic distances between the same letters are smaller and the image has darker pixels along the diagonals. Other darker spots result from shape similarities of other letters, e.g., M and W, D and O, etc.

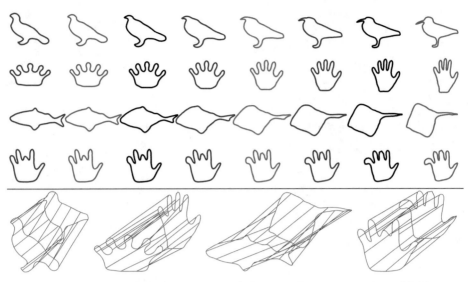

Fig. 6.27 *Top*: elastic geodesics between pairs of 2D curves. *Bottom*: optimal matching between points across given shapes

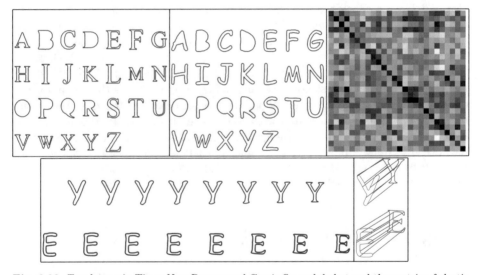

Fig. 6.28 *Top*: letters in Times New Roman and Comic Sans alphabet and the matrix of elastic shape distances between them. *Bottom*: examples of geodesic paths and the pairwise matching of parts between same letters of different fonts

Figure 6.29 shows an interesting example from biology where one is interested in classifying plant species according to the shape of their leaves. Several datasets have been constructed to study this classification. We show examples of leaves from 32 different species of plants, taken from the Flavia data. For some of these leaves, we display the elastic geodesic paths between their shapes in $\mathscr{S}_2^c$. The reader can see the effect of stretching boundary parts in matching sharper features across leaf boundaries.

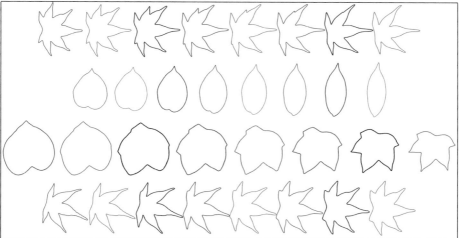

Fig. 6.29 *Top*: representative leaves from 32 different species of plants, taken from the Flavia dataset. *Bottom four rows*: examples of geodesic paths between some sample shapes computed in $\mathscr{S}_2^c$

6.8.1 Elastic Matching: Gradient Versus Dynamic Programming Algorithm

In the previous section we have presented a gradient-based approach for finding optimal re-parameterization of the second curve in order to minimize the geodesic distance between shapes in $\mathscr{C}_2^c$. As described in Appendix B, one can also use the dynamic programming algorithm (DPA) to solve this problem. Since DPA provides an optimal solution, albeit on a discrete grid in $[0, 1] \times [0, 1]$, what is the need for this gradient-based solution?

The gradient-based methods can sometimes be advantageous over DPA. First, it is a very general approach and works for more general objective functions. In contrast, DPA is applicable only when the objective functions is additive over

Table 6.1 Comparisons between DP and the gradient approach for elastic shape matching

	Elastic matching using dynamic programming	Elastic matching using gradient method
Solution type	Global solution for a fixed grid	Local solution
Cost function requirement	Requires additive cost function	Allows general cost function
Computational cost	Order $O(kT^2)$ (open curves) or $O(kT^3)$ (closed curves) T are the number of samples, $k << T$	Order $O(Tn)$, n is the number of basis elements of the re-parameterization algebra

the graph $(t, \gamma(t))$, e.g., the $\mathbb{L}^2$ norm. However, either this norm may not be the appropriate metric on the space of shape configurations or the cost function may take a more complicated form involving $\mathbb{L}^2$ norms. Secondly, the gradient approach may offer significantly lower computational complexity as it is $O(T)$ where T is the number of sample points on the curve. In contrast, the smallest computational cost in DPA for matching closed curves is generally $O(T^3)$ (some algorithms can do better and use $O(T^2 \log(T))$). Even if we search over the $\mathbb{S}^1$ component in Γ_S exhaustively and search the rest using the gradient approach, the computational cost remains low, on the order of $O(T)$.

Table 6.1 summarizes important differences between the two methods:

6.8.2 Fast Approximate Elastic Matching of Closed Curves

We have empirical evidence that the matching of points across curves is fast and optimal, albeit for the chosen grid, once the starting points on the two curves are determined. The difficulty comes in choosing optimal starting or seed points. Since the optimization over these seed points potentially has several local minima, in general, a more basic grid search can be efficient here. Note that one needs to search only on one curve, the other seed can be fixed arbitrarily. One can partition the domain $[0, 1]$ into coarse intervals, either uniform or nonuniform, and use the partition points as candidates for the seed during matching.

Let $\{t_i\} \subset \mathbb{S}^1$ represent the candidate seeds on curve 2. The seed on curve 1 is kept fixed. Then, for each i, we match the seed t_i on curve 2 with the fixed seed on curve 1, using the DPA for matching the remaining points. The seed with the least overall cost function is kept as the final solution.

Another possible approximation, in the interest of gaining speed, is to solve the geodesic problem on the larger space $\mathscr{S}_2$ and project the final path, element by element, on the space $\mathscr{S}_2^c$. Depending on the shapes involved, the projected geodesic in $\mathscr{S}_2^c$ may not be very different from the exact one computed using the techniques described earlier. In some application, this approximation will be acceptable, especially when it comes with a large gain in speed.

6.9 Elastic Versus Non-elastic Deformations

In this chapter we have presented two comprehensive frameworks for comparing shapes of planar, closed curves, termed non-elastic and elastic. In both cases we specify a representation of shapes, the groups that leave the shapes invariant,

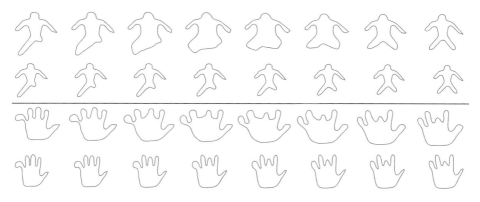

Fig. 6.30 Odd rows show non-elastic geodesic paths while even rows show elastic geodesics

the chosen Riemannian structures, and the algorithms for finding geodesics in the resulting shape spaces. It is natural to compare the two frameworks. We will do so using a number of examples, looking at the geodesic paths resulting from each method for the same pair of shapes.

As a first set of examples, Fig. 6.30 shows a comparison between the non-elastic geodesics and the elastic geodesics. For each pair, the top row is the non-elastic geodesic and the bottom row is the elastic geodesic. In the case of the non-elastic geodesic, we are allowed to deform one shape into another using only the bending process. No stretching or compressing is allowed. This is reflected in the results where, in order to match some salient features, the algorithm has to flatten a feature and create a new one at a new location. In the first example, the algorithm needs to grow the right leg in this human shape while keeping the hands matched. However, in view of the different relative locations of the legs (with respect to the hands), the algorithms ends up removing the left leg and growing two legs at new locations. This task is accomplished more naturally in the elastic framework where the features are matched first and the intermediate parts are stretched/compressed accordingly. As a result, the right leg grows smoothly along the geodesic path while the other parts (hands, leg) remain well matched and unchanged.

Figures 6.31 show a larger comparison set with geodesics in pre-shape and shape spaces for both non-elastic and elastic frameworks. The first figure deals with polygonal shapes while the second one shows more natural shapes.

Figure 6.32 shows examples of geodesic paths between closed curves with and without the registration step. A summary of the two approaches for shape analysis of closed curves is presented in Table 2.

6.10 Parallel Transport of Shape Deformations

We have previously discussed, in passing, an important tool that has many applications in shape analysis. This tool is parallel transport of tangent vectors from one point on a manifold to another, along a geodesic path between the two points. In the context of shape analysis, where the manifold of interest is a shape space, say $\mathscr{S}_2^c$, and a tangent vector represents a deformation of a shape, the notion of

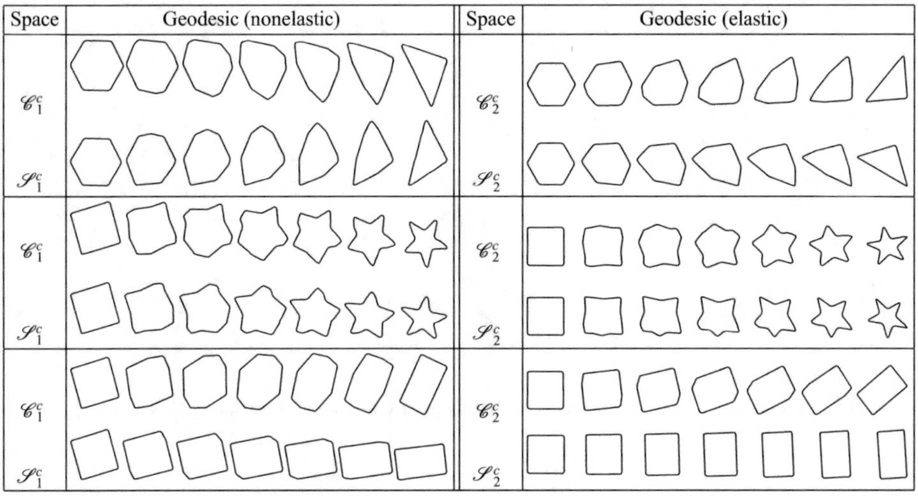

Space	Geodesic (nonelastic)	Space	Geodesic (elastic)
$\mathscr{C}_1^c$		$\mathscr{C}_2^c$	
$\mathscr{S}_1^c$		$\mathscr{S}_2^c$	
$\mathscr{C}_1^c$		$\mathscr{C}_2^c$	
$\mathscr{S}_1^c$		$\mathscr{S}_2^c$	
$\mathscr{C}_1^c$		$\mathscr{C}_2^c$	
$\mathscr{S}_1^c$		$\mathscr{S}_2^c$	

Fig. 6.31 Geodesic paths between polygonal shapes in pre-shape and shape spaces for both non-elastic and elastic frameworks

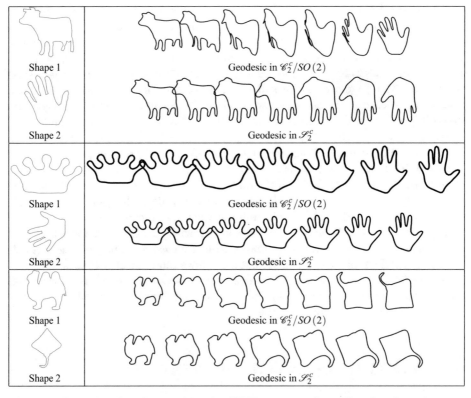

Fig. 6.32 Examples of geodesic paths under SRVF representation with and without the registration. In each case the top row shows the geodesic under a fixed arc-length parameterization for both curves, while the bottom row shows the geodesic under optimal re-parameterization of the second curve

Table 6.2 Summary of shape analysis of planar closed curves

Property	Bending-only planar closed curves	Elastic planar closed curves
Representation	Angle function θ (under arc-length param.)	Sqrt-velocity function q (arbitrary param.)
	$\dot{\beta}(s) = e^{i\theta(s)}$ with $|\dot{\beta}(s)| = 1$	$q(t) = \frac{\dot{\beta}(t)}{\sqrt{|\dot{\beta}(t)|}}$
Metric/larger space	$\mathbb{L}^2$ metric/$\mathbb{L}^2([0, 2\pi], \mathbb{R})$	$\mathbb{L}^2$ metric/$\mathbb{L}^2(\mathbb{S}^1, \mathbb{R}^2)$
Pre-shape space		
Pre-shape space	$\mathscr{C}_1^c = \{\theta \in \mathbb{L}^2([0, 2\pi], \mathbb{R})| \int_0^{2\pi} \theta(t)dt = \pi, \theta(2\pi) = \theta(0) + 2\pi, \int_0^{2\pi} \cos(\theta(t))dt = 0 \int_0^{2\pi} \sin(\theta(t))dt = 0\}$ Angle functions of closed curves of length 2π with fixed with orientation π	$\mathscr{C}_2^c = \{q \in \mathbb{L}^2(\mathbb{S}^1, \mathbb{R}^2)| \int_{\mathbb{S}^1} |q(t)|^2 dt = 1, \int_{\mathbb{S}^1} q(t)|q(t)|dt = 0\}$ Sqrt-velocity functions of closed curves of length 1
Normal space	$N_\theta(\mathscr{C}_1^c) = \text{span}\{1, \cos \circ \boldsymbol{\theta}, \sin \circ \boldsymbol{\theta}\}$	$N_q(\mathscr{C}_2^c) = \text{span}\{q, \{\left(\frac{q^i(t)}{|q(t)|}q(t) + |q(t)|e^i\right), i = 1, 2\}\}$
Tangent space	$T_\theta(\mathscr{C}_1^c) = \{v \in \mathbb{L}^2([0, 2\pi], \mathbb{R})|v \perp N_\theta(\mathscr{C}_1^c)\}$	$T_q(\mathscr{C}_2^c) = \{w \in \mathbb{L}^2(\mathbb{S}^1, \mathbb{R}^2)|v \perp N_q(\mathscr{C}_2^c)\}$
Geodesics	Computed using a shooting method	Computed using a path-straightening method
Shape space		
Shape-preserving transformations	Placement of origin: $\mathbb{S}^1$	Rotation $SO(2)$, $(O, q) = Oq$
		Re-parameterization Γ_S, $(q, \gamma) = (q \circ \gamma)\sqrt{\dot{\gamma}}$ (Γ_S: orientation-preserving diffeos of $\mathbb{S}^1$)
Equivalence class	$[\theta] = \{(u, \theta)|u \in \mathbb{S}^1\}$	$[q] = \{O(q \circ \gamma)\sqrt{\dot{\gamma}}|O \in SO(2), \gamma \in \Gamma_S\}$
Shape space	$\mathscr{S}_1^c = \mathscr{C}_1^c/\mathbb{S}^1 = \{[\theta]|\theta \in \mathscr{C}_1^c\}$	$\mathscr{S}_2^c = \mathscr{C}_2^c/(SO(2) \times \Gamma_S) = \{[q]|q \in \mathscr{C}_2^c\}$
Tangent space	$T_\theta(\mathscr{S}_1^c) = \{v \in T_\theta(\mathscr{C}_1^c)| \langle v, \dot{\theta}\rangle = 0\}$	$T_q(\mathscr{S}_2^c) = \{w \in T_q(\mathscr{C}_2^c)|w \perp T_q([q])\}$
		$T_q([q]) = \{A(\sqrt{\dot{\gamma}}(q \circ \gamma_{id})v + \frac{1}{2\sqrt{\dot{\gamma}_{id}}}q\dot{v})|v \in T_{\gamma_{id}}(\Gamma_S), A \in T_I(SO(2))\}$
Geodesics	Iterated over the optimal starting point on the second curve—u^*	Iterate over the two alignment steps until convergence:
		1. Find optimal rotation O^* in $SO(2)$ 2. Find optimal re-parameterization γ^* in Γ_S using a gradient process
	Set $\theta_2^* = (u^*, \theta_2)$ and compute geodesic between θ_1 and θ_2* in $\mathscr{C}_1^c$	Set $q_2^* = O^*(q_2 \circ \gamma^*)\sqrt{\dot{\gamma}^*}$ and compute geodesic between q_1 and q_2^* in $\mathscr{C}_2^c$.

parallel transport has a natural interpretation. This implies transferring the deformation from one shape to another, in such a way that represents least distortion in the deformation field during the transfer. There are many situations in which we want to transfer deformations from one shape to another. We motivate this application with two examples, although one can construct several more.

- **Prediction of Growth-Related Changes**: Consider the change in shape associated with the growth of a leaf over time. While it may also grow in size (or scale), let's focus primarily on the changes in its shape. These changes, over a certain interval $[T_0, T_1]$, can be treated as observation of an underlying continuous path $\alpha : [T_0, T_1] \to \mathscr{S}_2^c$. A short-term change in the shape at a time $\tau \in [T_0, T_1]$ is captured by the velocity vector $\dot{\alpha}(\tau)$. In other words, using these velocity vectors, also called *deformations*, one can represent and analyze instantaneous growth. Now, consider another leaf whose shape is given at time T_0 and we want to predict its shape at a time $\tau \in [T_0, T_1]$. One way to accomplish that is to *transfer* deformations from the previous leaf and apply those deformations to obtain predicted shapes for the new leaf.

- **Prediction of Silhouette Shapes from Novel Views**: One difficulty in using shapes for recognizing 3D objects is that their 2D appearances change according to the viewing angle. Since a large majority of imaging technology is oriented toward 2D images, there is a striking focus on planar shapes, their analysis and modeling, despite the viewing variability. Within this focus area, there is an interesting problem of predicting shapes of 3D objects from novel viewing angles. Our solution to the problem of shape prediction is the following. If we know how a known object deforms under a viewpoint change, perhaps we can apply the "same" deformation to a similar (yet novel) object and predict its deformation under the same viewpoint change.

These and similar applications require taking deformations from one shape and applying them to another. The basic technical issue in this procedure is that since shape spaces are nonlinear spaces, and the deformations are tangent vectors on these manifolds, the deformations at one point or shape cannot simply be applied to another. One would need to transport the required deformation from one shape to the other, before applying that deformation.

The mathematical statement of this problem is as follows: let $[q_1]$ and $[q_2]$ be two given shapes and let $v_1 \in T_{[q_1]}(\mathscr{S}_2^c)$ be a tangent vector, or a deformation, at the first shape. Our goal is to form a vector $v_2 \in T_{[q_2]}(\mathscr{S}_2^c)$ that is the parallel transport of v_1 along the geodesic connecting $[q_1]$ and $[q_2]$. We remind the reader that one transports a vector v_1 along a path α on a manifold by constructing a vector field $w : [0, 1] \to T\mathscr{S}_2^c$ along that path such that: (1) the initial value of the field is v_1, i.e., $w(0) = v_1$ and (2) the covariant derivative of w is zero along the whole path, i.e., $\frac{Dw}{d\tau} = 0$. Then, the end vector $w(1)$ is the desired v_2. See Chap. 3, Sect. 3.4 for details. Without any loss of generality, we assume that q_1 and q_2 are the nearest points on the two orbits $[q_1]$ and $[q_2]$ and let $\alpha : [0, 1] \to \mathscr{C}_2^c$ be the geodesic path in $\mathscr{C}_2^c$ that represents the corresponding geodesic in the shape space. That is, α is a path that is perpendicular to the rotation and re-parameterization orbits in $\mathscr{C}_2^c$. The algorithm for performing parallel transport has been studied earlier and is repeated here.

Algorithm 23 (Parallel Transport Along Geodesics in $\mathscr{S}_2^c$).

1. *Use Algorithm 22 to compute the geodesic α between q_1 and q_2 in $\mathscr{C}_2^c$. (Note that this geodesic represents the shortest path between $[q_1]$ and $[q_2]$ in $\mathscr{S}_2^c$.) This path is actually computed at uniformly spaced times with spacing $1/k$: $\alpha(0/k), \alpha(1/k), \ldots, \alpha(k/k)$.*
2. *Set $w(0) = v_1$. Let $l = \|v_1\|$ be the $\mathbb{L}^2$ norm of v_1.*
3. *For $\tau = 1, 2, \ldots, k$, perform the following steps:*

    a. *Project $w((\tau - 1)/k)$ into $T_{\alpha(\tau/k)}(M)$ to obtain $c(\tau/k)$.*
    b. *Set $w(\tau/k) = lc(\tau/k)/\|c(\tau/k)\|$.*

To implement this, we only require the mechanism for projecting an arbitrary element of $\mathbb{L}^2(\mathbb{S}^1, \mathbb{R}^2)$ into $T_q(\mathscr{C}_2^c)$. It must be kept in mind that since $v_1 \perp T_{q_1}([q_1])$, all resulting vectors $w(\tau/k)$ will also be perpendicular to the $T_{\alpha(\tau/k)}([\alpha(\tau/k)])$. The projection into the tangent $T_q(\mathscr{C}_2^c)$ is given in Eq. 6.36.

6.10.1 Prediction of Silhouettes from Novel Views

Now we illustrate the use of deformation transfer in predicting shapes of silhouettes of 3D objects from novel viewing angles.

The mathematical statement of this problem is as follows. Let $[q_1^a]$ and $[q_1^b]$ be the shapes of silhouettes a 3D object $\mathcal{O}^1$ when viewed from two viewing angles $\theta_a, \theta_b \in SO(3)$, respectively. The deformation in contours, in going from $[q_1^a]$ to $[q_1^b]$ depends on some physical factors: the geometry of $\mathcal{O}^1$ and the viewing angles involved. Consider another object $\mathcal{O}^2$ that is similar but not identical to $\mathcal{O}^1$ in geometry. Given the shape of its silhouette $[q_2^a]$ from the viewing angle θ_a, our goal is to predict the corresponding shape $[q_2^b]$ from the viewing angle θ_b. Our solution is based on taking the deformation that deforms $[q_1^a]$ to $[q_1^b]$ and applying it to $[q_2^a]$ after some adjustments. Here is the algorithm:

1. Let $\alpha_1(\tau)$ be a geodesic between $[q_1^a]$ and $[q_1^b]$ in $\mathscr{S}^c$ and $v_1 \equiv \dot{\alpha}_1(0) \in T_{[q_1^a]}(\mathscr{S}^c)$ be its initial velocity.
2. We need to *transport* v_1 to $[q_2^a]$; this is done using a forward parallel translation. Let $\alpha_{12}(\tau)$ be a geodesic from $[q_1^a]$ to $[q_2^a]$ in $\mathscr{S}^c$. Construct a vector field $w(t)$ such that $w(0) = v_1$ and $\frac{Dw}{d\tau} = 0$ for all points along α_{12}. This is accomplished in practice using Algorithm 23. Then, $v_2 \equiv w(1) \in T_{[q_2^a]}(\mathscr{S}^c)$ is a parallel translation of v_1.
3. Construct a geodesic starting from $[q_2^a]$ in the direction of v_2 with the path length equal to the norm of v_2.

Figure 6.33 shows two examples of this idea. In the top case, a hexagon ($[q_1^a]$) is deformed into a square ($[q_1^b]$) using an elastic geodesic; this deformation is then transported to a circle ($[q_2^a]$) and applied to it to result in the prediction $[q_2^b]$. A similar transport is carried out in the bottom example.

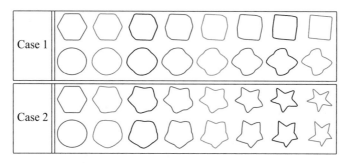

Fig. 6.33 In each case: a geodesic from the template shape (*hexagon*) to the training shape (*top*) and deformation of the test shape (*circle*) with the transported deformation (*bottom*)

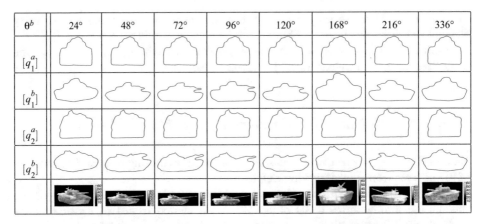

θ^b	24°	48°	72°	96°	120°	168°	216°	336°
$[q_1^a]$								
$[q_1^b]$								
$[q_2^a]$								
$[q_2^b]$								

Fig. 6.34 Shape predictions for novel pose. In each column, the first two are given shapes of the M60 from $\theta^a = 0$ and θ^b. The deformation between these two is used to deform the T72 shape in the third row and obtain a predicted shape in the fourth row. The accompanying pictures show the true shapes of the T72 at those views

We present another example using the M60 tank as $\mathscr{O}^1$ and the T72 as $\mathscr{O}^2$. Given shapes for different azimuthal pose (fixed elevation) of M60 and one azimuth for the T72, we would like to predict shapes for the T72 from the other azimuthal angles. Since both the objects are tanks, they have similar but not identical geometries. For instance, both have mounted guns but the T72 has a longer gun than the M60. In this experiment, we select $\theta^a = 0$ and predict the shape of the T72 for several θ^b The results are shown in Fig. 6.34. The first and the third rows show the shapes for $[q_1^a]$ and $[q_2^a]$, respectively, the shapes for the M60 and the T72 looking from head on. The second row shows $[q_1^b]$ for different θ^b given in the last column, while the fourth row shows the predicted shapes for the T72 from those θ^b.

6.10.2 Classification of 3D Objects Using Predicted Silhouettes

How can we evaluate the quality of these predictions? One application of prediction of silhouettes of 3D objects from novel views is in recognition. We perform a simply binary classification with and without the predicted shapes and compare results. Here is the experimental setup. We have 62 and 59 total azimuthal views of the M60 and the T72, respectively. Of these, we randomly select 31 views of M60 and one view of the T72 as the training data; the remaining 31 (58) views of the M60 (the T72) are used for testing. The classification results, using the nearest neighbor classifer and the elastic distance $d_{\mathscr{S}_2^c}$ (Eq. 6.42), are shown in the table below. While the classification for the M60 is perfect, as expected, the classification for the T72 is 46.55 %. (Actually, this number is somewhat higher than expected—we would expect a smaller performance with only one training shape.) Now we generate additional 31 shapes for the T72 using the prediction method described earlier. Using the 31 training shapes of the M60, we generate 31 corresponding shapes of the T72 using parallel transport. The θ^a used here was 90°. The classification result after including the 31 predicted shapes is found to

Table 6.3 Classification rate with (bold fonts) and without (normal fonts) use of predicted shapes for the T72

| Est./True | Experiment 1 ($\theta^a = 90°$) | | Experiment 2 ($\theta^a = 0°$) | |
	M60	T72	M60	T72
M60	100 % (**100 %**)	53.45 % (**39.66 %**)	100 % (**100 %**)	93.2 % (**82.8 %**)
T72	0 % (**0 %**)	46.55 % (**60.34 %**)	0 % (**0 %**)	6.8 % (**17.2 %**)

be 60.34 %, a 15 % increase in the performance when using shape predictions. We performed the same experiment for another azimuth, $\theta^a = 0°$, and the results are listed under experiment 2 in the table. In this case we improve the classification performance from 6.8 % to 17.2 %, an increase of almost 11 %, using the predicted shapes of the T72 (Table 6.3). While this experiment was performed with only one training shape, one can repeat this idea using multiple given shapes for the novel object and then perform prediction for a novel view using joint information from these views.

6.11 Symmetry Analysis of Planar Shapes

Reflection symmetry of objects is a property of interest in many applications. Several papers in computer vision and image analysis have been concerned with either evaluating the symmetry or exploiting its symmetric nature in its detection and classification. A 2D object is considered (reflection) symmetric around a line if its shape remains unchanged after reflection in that line. Thus, naturally, some quantification of differences between curves and their reflections will help quantify their symmetry. In this section, we present the use of elastic shape analysis in characterizing symmetry of planar curves.

Let H be a 2×2 reflection matrix (a reflection matrix satisfies the conditions $H = H^T$, $H^T H = H H^T = I_2$ and $\det(H) = -1$), and let $\mathscr{H}$ be the set of all such matrices. For any $v \in \mathbb{R}^2$ ($|v| \neq 0$), we can construct a reflection matrix using $H = I_2 - 2\frac{vv^T}{v^T v}$. For a point $x \in \mathbb{R}^2$, Hx is its reflection about the line orthogonal to v and passing through the origin. For a planar curve $\beta : D \to \mathbb{R}^2$ (where $D = [0, 1]$ or $\mathbb{S}^1$), the curve $\tilde{\beta}(t) = (H \cdot \beta)(t)$ is its reflection in the same line.

Remark 6.1. We remind the reader that our shape analysis requires a fixed direction of parameterization of curves throughout this textbook. However, a reflection changes the direction of parameterization. Therefore, along with the multiplication by H, one also has to reverse the direction of parameterization, in order to preserve the original direction and to perform shape analysis. With a slight abuse of notation, we will denote this direction-preserving reflection by H. In other words, for $\beta : [0, 1] \to \mathbb{R}^2$, $(H \cdot \beta)(t) \equiv H\beta(1 - t)$.

If the SRVF of β is given by q, then the SRVF of $H \cdot \beta$ is given by $-H \cdot q$ (proof is left as an exercise).

Recall that $\mathscr{C}_2$ is the set of SRVFs of absolutely continuous, unit-length curves (without the closure constraint). It is a Hilbert sphere and for any two points $q_1, q_2 \in \mathscr{C}_2$, such that $q_1 \neq -q_2$, there exists a unique, shortest (and constant speed) geodesic going from one to the other. The following result characterizes the midpoint of the geodesic between a curve and its reflection.

Lemma 6.4. *Let $q \in \mathscr{C}_2$ be the SRVF of a curve β and $\tilde{q} = -H \cdot q$ be the SRVF of $\tilde{\beta} = H \cdot \beta$, for any $H \in \mathscr{H}$. Then, the midpoint of the constant-speed geodesic path connecting them is a perfectly symmetric curve under H. That is, $(H \cdot q_0) = q_0$ where q_0 denotes the midpoint of the geodesic.*

Proof. Let $\alpha : [0, 1] \to \mathscr{C}_2$ denote the unique, constant-speed, shortest-geodesic path such that $\alpha(0) = q$ and $\alpha(1) = (H \cdot q)$. Form a new path $\tilde{\alpha}(\tau) \equiv (H \cdot \alpha(\tau))$ by reflecting each curve along α by the same H. Accordingly, we get $\tilde{\alpha}(0) = (H \cdot \alpha(0)) = (H \cdot q)$ and $\tilde{\alpha}(1) = (H \cdot \alpha(1)) = (H \cdot (H \cdot q)) = q$. Next, we claim that $\tilde{\alpha}$ is also a constant-speed geodesic path between the two SRVFs. (This claim is left as an exercise for the reader.) Since this geodesic was assumed to be unique, α and $\tilde{\alpha}$ are exactly the same paths but traversed in different direction, i.e., $\tilde{\alpha}(\tau) = \alpha(1 - \tau)$. Furthermore, since α is a constant-speed path, we conclude that $\tilde{\alpha}(\frac{1}{2}) = (H \cdot \alpha(\frac{1}{2})) = \alpha(\frac{1}{2})$ and that $\alpha(\frac{1}{2})$ is symmetric under H. $\square$

One can also use the length of that geodesic α, denoted by $d_{\mathscr{C}_2^c}(q, H \cdot q)$, as a measure of symmetry of β under that given H. Indeed, if the length of α is zero, then the original curve is perfectly symmetric under H to start with! A parameterized curve may be symmetric under one reflection and may not be symmetric under another. Thus, to assess reflection symmetry, one has to search over all possible reflections and find if there exists a reflection under which the curve is symmetric. In that case, one can use $\min_{H \in \mathscr{H}} d_c(q, H \cdot q)$ as a measure of asymmetry of the given curve with SRVF q. Note that:

$$
\min_{H \in \mathscr{H}} d_c(q, H \cdot q) = \min_{v \in \mathbb{R}^2} \left(\cos^{-1}(\int_0^1 \left\langle q(t), (I - \frac{vv^T}{v^T v})q(1 - t) \right\rangle dt) \right)
$$

$$
= \min_{O \in SO(2)} \left(\cos^{-1}(\int_0^1 \left\langle q(t), O(I - \frac{v_0 v_0^T}{v_0^T v_0})q(1 - t) \right\rangle dt) \right) ,
$$

for any $v_0 \in \mathbb{R}^2 (|v_0| \neq 0)$. The last equation uses the fact that any arbitrary reflection in $\mathbb{R}^2$ can be replaced by an arbitrary (fixed) reflection and an appropriate rotation! This suggests taking an arbitrary reflection of q, using a $v_o \in \mathbb{R}^2$, and then computing geodesic paths between the $SO(2)$ orbit of q and its arbitrary reflection.

One can take this discussion a step further and compare the symmetry of *shapes* of curves, rather than the curves themselves. In this situation, one compares shapes of a curve and its reflection, using a shape metric:

$$
\min_{O \in SO(2), \gamma \in \Gamma_I} \left(\cos^{-1}(\int_0^1 \left\langle q(\gamma(t))\sqrt{\dot{\gamma}(t)}, O(I - \frac{v_0 v_0^T}{v_0^T v_0})q(1 - t) \right\rangle dt) \right) . \quad (6.50)
$$

This is exactly $d_s([q], [(H \cdot q)])$ for any $H \in \mathscr{H}$, where d_s is the distance between shapes of curve in $\mathscr{S}_2$. This concept can be extended to study symmetry of *shapes* of closed curves using the geodesic distance d_s defined on $\mathscr{S}_2^c$. The only difference is that the re-parameterization group Γ_I in this equation is replaced by Γ_S. As described earlier, using the identification between Γ_S and $\mathbb{S}^1 \times \Gamma_I$, this implies repeated use of Eq. 6.50 for each shift of the origin on the second curve.

We illustrate this idea using a couple of examples. Shown in the top-left panel of Fig. 6.35 is a curve and the next panel shows its reflection around an arbitrary line (passing through the origin). The next panel shows the minimum value obtained

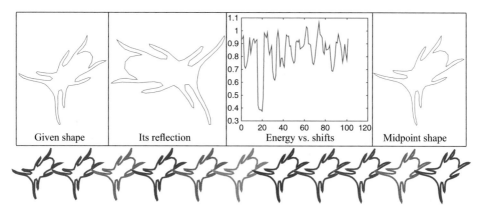

Fig. 6.35 Symmetrizing arbitrary shapes using elastic deformations

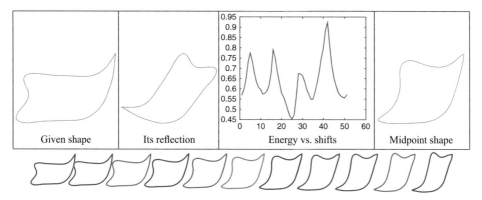

Fig. 6.36 Symmetrizing arbitrary shapes using elastic deformations

in Eq. 6.50 versus different placements of origin on the second curve. The last plot in the first row shows the midpoint of elastic geodesic, while the geodesic itself is shown in the second row. Figure 6.36 shows another example of this idea, using a different curve.

We summarize the results on symmetry analysis using elastic geodesics between a curve β and its arbitrary reflection $H\beta$:

1. The geodesic elastic distance, $d_{\mathscr{S}_2^c}$ in $\mathscr{S}_2^c$, can be used to quantify the level of asymmetry in a curve. In other words, the shape distance $d_{\mathscr{S}_2^c}$ between β and $H\beta$ is a quantification of asymmetry of β.
2. For any given curve, the halfway point along the geodesic between that curve and its reflection provides the nearest symmetric shape, under the distance $d_{\mathscr{S}_2^c}$.
3. This analysis provides the most efficient way to deform any given closed curve into a symmetric curve. We sketch a proof of this statement here. Note that for an arbitrary reflection matrix H about a line of symmetry L, the map $H : \mathbb{L}^2(\mathbb{S}^1, \mathbb{R}^2) \to \mathbb{L}^2(\mathbb{S}^1, \mathbb{R}^2)$ is an "involution", i.e., it has the property that $H \cdot (H \cdot q) = q$ for all $q \in \mathbb{L}^2(\mathbb{S}^1, \mathbb{R}^2)$. As a result, we obtain a direct sum decomposition $\mathbb{L}^2(\mathbb{S}^1, \mathbb{R}^2) = S \oplus A$, where S is the $+1$-eigen space of H and A is the -1-eigen space. Note that the subspace S consists precisely of SRVFs of curves that are symmetric with respect to L. Given $q \in \mathbb{L}^2(\mathbb{S}^1, \mathbb{R}^2)$, we can

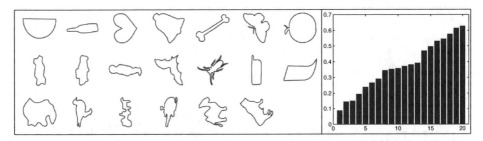

Fig. 6.37 Left panel shows a set of 20 curves and the right panel provides a quantification of their symmetries using geodesic lengths. The curves have been arranged in the order of increasing asymmetry

uniquely represent it under this direct sum decomposition by $q = s + a$, where $s \in S$ and $a \in A$; then, $H \cdot q = s - a$. Note that s represents the orthogonal projection of q onto the subspace S. Hence, s represents the closest curve to q that is symmetric about L. Furthermore, s is the midpoint of the line from q to $H \cdot q$. Since the space of all reflection matrices is compact, we can choose the particular H that minimizes $\|a\|$. This choice of H will minimize the distance from q to $H \cdot q$ and will produce the shortest deformation from q to a symmetric curve: this is simply the straight line from q to s, where s is defined using the minimizing choice of H.

4. It also provides the line of symmetry in any symmetric object. If v is the original reflection and O^* is the optimal rotation obtained during geodesic computation, then O^*v defines the axis of reflection or the line of symmetry.

Figure 6.37 shows a set of 20 curves (left panel) and the quantification of their level of asymmetry in the bar chart on the right. The curves have been displayed in the order of increasing asymmetry, starting with the half moon (most symmetric) and ending with a cow silhouette (least symmetric).

6.12 Exercises

6.12.1 Theoretical Exercises

1. Show that there is a natural bijection between Γ_S and $\Gamma_I \times \mathbb{S}^1$. Hint: first identify Γ_I with the subgroup of Γ_S that fixes $1 \in \mathbb{S}^1$ (thinking of $\mathbb{S}^1$ as the set of unit complex numbers). Then, associate to $(\gamma, z) \in \Gamma_I \times \mathbb{S}^1$ the diffeo $\tilde{\gamma} \in \Gamma_S$ that first applies γ and then rotates by z.
2. Derive expressions for angle functions of the following curves, ignoring those points in the domain where the derivative is not continuous:

   a. A unit circle
   b. An equilateral triangle with one side aligned to the x axis and a side length $2\pi/3$
   c. An ellipse defined by the condition: $\frac{\beta_x^2}{a^2} + \frac{\beta_y^2}{b^2} = 1$ for $a = 1$ and $b = 2$. Rescale the curve to be of length 2π.

Plot them with the starting point corresponding to $\theta(0) = 0$. Evaluate and display geodesic paths between these curves in $\mathscr{C}_1$.

3. Show that the space $\mathscr{D}$ defined in Eq. 6.7 is an affine space.

4. Show that for all $\theta \in \mathscr{D}$, the functions $\{1, \sin(\theta(s)), \cos(\theta(s))\}$ are linearly independent and lie in $\mathscr{B}$. Furthermore, verify that $d\Phi_1 : \mathscr{B} \to \mathbb{R}^3$ is surjective at every $\theta \in \Phi_1^{-1}(\pi, 0, 0)$.

5. For a parameterized curve $\beta : [0, 1] \to \mathbb{R}^2$ with the SRVF $q : [0, 1] \to \mathbb{R}^2$, show that $\beta(0) = \beta(1)$ implies that $\int_0^1 q(t)|q(t)|dt = 0$. Also, verify that $L[\beta] = \|q\|$.

6. Show that the closure constraint, i.e., the constraint that a curve is closed, is a nonlinear constraint in both the angle function and SRVF representation. That is, if $\theta_1, \theta_2 \in \mathscr{C}_1^c$, then $a\theta_1 + b\theta_2$ may not be in $\mathscr{C}_1^c$ for some $a, b \in \mathbb{R}$, and repeat it for the SRVF representation in $\mathscr{C}_2^c$ space.

7. Show that for any $u \in \mathbb{R}$, we have $\int_0^{2\pi} v(s-u)w(s-u)\,ds = \int_0^{2\pi} v(s)w(s)\,ds$ if $v, w : \mathbb{R} \to \mathbb{R}$ are both periodic with a period of 2π.

8. Prove Theorem 6.3: The action of $\mathbb{S}^1$ on $\mathscr{C}_1^c$ given by Eq. 6.14 is an action by isometries.

9. For the following paths α with their given parameterizations, calculate the path energy $E[\alpha] = \int_0^1 \langle \dot{\alpha}(\tau), \dot{\alpha}(\tau) \rangle \, dt$ in $\mathbb{R}^2$ under the Euclidean metric. Set $p_1 = (-1, 0)$ and $p_2 = (1, 0)$.

   - A straight line between p_1 and p_2
   - A unit half circle from p_1 to p_2 with its center at $(0, 0)$
   - The shorter part of ellipsoid $x_1^2 + \frac{x_2^2}{4} = 1$ that passes through p_1 and p_2

10. Denote elements of $\mathbb{S}^1$ by unit complex numbers and define a map $G_w : \mathbb{S}^1 \to \mathbb{S}^1$ as $G_w(z) = wz$ where w is also a unit complex number. Show that G_w is a diffeomorphism for all w and the set of all G_w's is a subgroup of Γ_S.

11. Prove that the mapping given in Eq. 6.44, Moebius transformation, is a diffeomorphism from $\mathbb{S}^1$ to itself. Also, show that the set of all such mappings forms a group. Verify the statement that the functions $\mathbf{1}$, $\cos(t)$ and $\sin(t)$ span the tangent space of this manifold at the identity γ_{id}.

12. Show that the basis given in Eq. 6.45 forms an orthonormal basis of $T_{\gamma_{id}}(\Gamma_S)$ under the Palais metric given in Eq. 6.29.

13. Show that $H = I_2 - 2\frac{vv^T}{v^Tv}$, where $v \in \mathbb{R}^2$ is a column vector with $|v| \neq 0$, is a reflection matrix.

14. For any 2×2 reflection matrix $H_0 \in \mathscr{H}$, show that the coset $[H_0] = \{OH_0|O \in SO(2)\}$ equals the set $\mathscr{H}$. That is, for any $H \in \mathscr{H}$, there exists a rotation $O \in SO(2)$ such that $H = OH_0$.

15. If the SRVF of $\beta : [0, 1] \to \mathbb{R}^2$ is given by q, then show that the SRVF of $\tilde{\beta} \equiv H \cdot \beta$ is given by $-H \cdot q$. Recall that $(H \cdot \beta)(t) = H\beta(1-t)$.

16. If $\alpha : [0, 1] \to \mathscr{C}_2$ is a geodesic path then show that its reflection $(H \cdot \alpha(\tau))$ is also a geodesic path in $\mathscr{C}_2$. If α is a constant-speed path, then so is its reflection.

6.12.2 Computational Exercises

1. Write a program to compute the angle function of a given parameterized curve. (Ignore those points where the derivative of a curve is not continuous.) Run this

program on the following curves: (1) a unit circle, (2) an equilateral triangle of side length $2\pi/3$ and one side parallel to the x axis, and (3) a general ellipse of the type $t \mapsto (a\cos(2\pi t), b\sin(2\pi t))$ (rescale it to have length 2π and re-parameterize it to have constant speed). Compare your results with the analytical expressions obtained in Problem 2.

2. Provide an example to illustrate that even if θ_1, θ_2 satisfy Eq. 6.2, their linear combination $c_1\theta_1 + c_2\theta_2$, for some $c_1, c_2 \in \mathbb{R}$, may not satisfy it.

3. Provide an example to illustrate that even if q_1, q_2 satisfy Eq. 6.3, their linear combination $c_1 q_1 + c_2 q2$, for some $c_1, c_2 \in \mathbb{R}$, may not satisfy it.

4. Implement a function to compute an orthonormal basis of the normal space $N_\theta(\mathscr{C}_1^c)$ for a given $\theta \in \mathscr{C}_1^c$. Then, use this function to project an arbitrary element of $\mathbb{L}^2([0, 2\pi], \mathbb{R})$ in $T_\theta(\mathscr{C}_1^c)$.

5. Implement a function to compute an orthonormal basis of the normal space $N_q(\mathscr{C}_2^c)$ for a given $q \in \mathscr{C}_2^c$. Then, use this function to project an arbitrary element of $\mathbb{L}^2(\mathbb{S}^1, \mathbb{R}^2)$ in $T_q(\mathscr{C}_2^c)$.

6. Implement a function to project an arbitrary element $\theta \in \mathbb{L}^2([0, 2\pi], \mathbb{R})$ in the set $\mathscr{C}_1^c$ using Algorithm 8. Similarly, implement a function to project an arbitrary element of $\mathbb{L}^2(\mathbb{S}^1, \mathbb{R}^2)$ into the set $\mathscr{C}_2^c$ using Algorithm 18.

7. Write a program to implement the shooting algorithm (Algorithm 10) for computing geodesics between any two closed curves in $\mathscr{C}_1^c$.

8. Implement a numerical procedure for computing a covariant derivative of a vector field v along a smooth path α on $\mathbb{S}^2$ under the Euclidean Riemannian metric. Test your program on the following setup: Let α be a half circle, forming a part of the equator on $\mathbb{S}^2$. Define a vector field w on it as follows: $w(\tau)$ at any point $\alpha(\tau)$ is the unit vector shooting from $\alpha(\tau)$ to the north pole. Compute and display the covariant derivative of w.

9. Use the previous program to compute the covariant derivative of a velocity vector field along the great circle between any two non-antipodal points.

10. Write a program that adapts the general path-straightening algorithm (Algorithm 17) for computing geodesics between any two points on $\mathbb{S}^2$. Test this program for any two non-antipodal points on $\mathbb{S}^2$ and compare with the analytical expressions for geodesics on $\mathbb{S}^2$ given earlier.

11. Write a program to implement the path-straightening algorithm (Algorithm 19) for computing geodesics between any two curves in $\mathscr{C}_2^c$ using their SRVF representation.

12. Write a program to compute the length of an arbitrary path $\alpha : [0, 1] \to \mathscr{C}_2^c$ under the elastic metric. Use this program to compare the lengths of the following two paths. For any two closed curves in $\mathscr{C}_2^c$: (1) compute a geodesic path between them in $\mathscr{C}_2$ (i.e., the pre-shape space without the closure constraint) and (2) compute a geodesic between them in $\mathscr{C}_2^c$ (i.e., the pre shape space with the closure constraint).

13. We will compare resulting deformations between shapes of closed curves using different strategies. For any given pair of closed curves: (1) compute a uniform-speed geodesic path between them in $\mathscr{S}_2^c$ using Algorithm 22 and (2) compute a uniform-speed geodesic path between them in $\mathscr{S}_2$, i.e., without the closure constraint, as described in Sect. 5.7, and project each point of this geodesic in $\mathscr{S}_2^c$ using Algorithm 18. Approximate both these paths by the same number of time points and compare the corresponding shapes along these two paths. You can also use the previous program to compute the lengths of these two curves and study how close they are!

14. Implement the parallel transport procedure given in Algorithm 23 to transport deformations from one shape to another in $\mathscr{S}_2^c$.

6.13 Bibliographic Notes

Klassen et al. [54] were the first to compute geodesics on shape spaces of closed curves, using angle function representation. The identification of the space of planar closed curves, using the complex square-root representation, with a Grassmann manifold, was discussed in [126]. The general elastic metric for planar curves was studied in [79] and using SRVF representation in [42, 43, 106]. The path-straightening algorithm for finding geodesics between closed curves was derived in [53], while the shooting method was presented in [54]. The use of parallel transport of tangent vectors for transferring deformations was covered in [106]. Samir et al. [99] studied symmetry analysis of planar closed curves. Applications of shape analysis of planar curves include 3D face recognition [97], silhouette analysis [1], and analysis of segmented contours [58].

Chapter 7
Statistical Modeling on Nonlinear Manifolds

It has been emphasized frequently in the earlier chapters that the representation spaces of our interest are both nonlinear and infinite dimensional. It is therefore expected that some standard ideas in statistics, where the domains are usually finite-dimensional vector spaces, will not be applicable here directly. Since these ideas involve addition, subtraction, and linear combinations of vectors, they will expectedly need some modifications before application to our representation spaces. In fact, one needs to use Riemannian geometry and related analytical or computational tools, developed for representation spaces earlier in Chaps. 4, 5, and 6, to define and compute statistical quantities. Before we consider statistical modeling of functional and shape data in later chapters, we use this chapter to introduce basic ideas for some familiar nonlinear manifolds, such as the unit spheres $\mathbb{S}^n$ for $n = 1$ and 2, $\mathbb{S}_\infty$ and some quotient spaces of the type M/G. We will restrict to a setup where G is a group that acts on a Riemannian manifold M in such a way that the action is by isometries under the metric of M.

7.1 Goals and Challenges

The main goals for this chapter are:

1. To establish the notions of means and covariances for Riemannian manifolds and their quotient spaces and, in case of sample data, to estimate these parameters using sample statistics. Although there is a possibility of viewing these statistics as estimators of population parameters, we will primarily treat them as tools for generating statistical summaries of the datasets that take values on the underlying manifolds. The challenge here is to define and compute these quantities in spaces where addition, subtraction, and multiplication are no longer valid and need to be replaced by geometric operations.
2. To extend the notion of a multivariate normal density to nonlinear Riemannian manifolds, such as the unit sphere $\mathbb{S}^n$, and their quotient spaces. The main challenges here come from nonlinearity and compactness of these spaces. Several choices will be discussed and one of them, involving truncated wrapped-normal densities, will be considered in detail.
3. To establish some basic tools for statistical inferences using these probability models. For instance, the truncated wrapped-normal densities can be used for

© Springer-Verlag New York 2016
A. Srivastava, E.P. Klassen, *Functional and Shape Data Analysis*,
Springer Series in Statistics, DOI 10.1007/978-1-4939-4020-2_7

statistical inferences in different ways. They can be used to form prior densities for Bayesian inferences, or they can be used to describe in-class variability for use in statistical decision theory. They can serve as generative models for the data in the sense that one can sample from them. These samples can then be useful, for instance, in Monte Carlo-type methods for generating inferences.

7.2 Basic Setup

We start with a basic question about statistical analysis on nonlinear manifolds. What are the main challenges in applying classical statistics if the underlying domain is nonlinear? Take the case of the most basis statistics in a Euclidean case, the sample mean, for a sample set $(x_1, x_2, \ldots, x_k)$ on $\mathbb{R}^n$:

$$\bar{x}_k = \frac{1}{k} \sum_{i=1}^{k} x_i, \qquad x_i \in \mathbb{R}^n \ . \tag{7.1}$$

Since $\bar{x}_k$ is a widely used and studied statistic, the pros and cons of using $\bar{x}_k$, as an estimate of the population mean, are well known. Assuming that $x_i \sim f(x)$, with $\mu = E_f[x]$, we know that $\bar{x}_k$ is an unbiased and efficient estimator of μ, but is susceptible to the outliers. Now what if the underlying space is not $\mathbb{R}^n$ but a nonlinear manifold, say $\mathbb{S}^n$, instead? In this situation, the summation in Eq. 7.1 is no longer a valid operation and that equation is not useful anymore. Even in the simple case of $\mathbb{S}^1$, if one tries to average points on $\mathbb{S}^1$ as real numbers, the results are likely to be wrong. Take, for example, the Euclidean average of two points with angles (relative to say X axis) given by ϵ and $2\pi - \epsilon$, for an $\epsilon > 0$ small. This value $\frac{2\pi - \epsilon + \epsilon}{2} = \pi$ is not a good representative of the original points and, thus, is not a useful summary. So, how can one define the sample mean for points on a nonlinear manifold, let alone higher moments, and obtain summaries that are representatives of the original data?

To answer this question, we consider an n-dimensional Riemannian manifold M. Let $d(p, q)$ be the length of the shortest geodesic between arbitrary points $p, q \in M$. To facilitate a discussion of different options available, we will assume that there exists an embedding $\mathcal{E} : M \to V$ where V is an m-dimensional Hilbert space ($n \leq m$). In this context, an embedding is an injective (one-to-one) map that preserves the Riemannian structure between M and V. (We clarify that this embedding is needed only to perform a certain type of statistical analysis, described later as an *extrinsic analysis*, and not for the main ideas presented in this chapter.) Stated differently, the Riemannian metric on M coincides with the standard inner product of V. We have chosen V to be a vector space so that we can perform more standard multivariate statistical analysis in V. The distance between any two elements $p, q \in M$, is the geodesic distance $d(p, q)$ when the geodesic is restricted to be in M and it is $|\mathcal{E}(p) - \mathcal{E}(q)|$, with $|\cdot|$ being the norm of V, when the geodesic is allowed to be in V. The latter distance, of course, depends on the choice of the embedding $\mathcal{E}$.

Assume that we have a probability density function f on M with respect to a chosen standard measure on M. This function $f : M \to \mathbb{R}_{\geq 0}$ satisfies the property that $\int_M f(p)dp = 1$, where dp denotes a base measure on M with respect to which the density f is defined. We can extend f to the larger set V by simply setting:

$$\tilde{f}(x) = \begin{cases} f(p) & \text{if } x = \mathcal{E}(p), \ p \in M \\ 0 & \text{if } x \notin \mathcal{E}(M) \end{cases} . \tag{7.2}$$

That is, $\tilde{f}(x) = f(\mathcal{E}^{-1}(x))1_x(\mathcal{E}(M))J_{\mathcal{E}}(x)$ where $1_x(A)$ denotes the indicator function for the set A and $J_{\mathcal{E}}(x)$ is the Jacobian of $\mathcal{E}$ at p. $\tilde{f}$ is called the extension of f from M to V and is unique by construction. Naturally, $\tilde{f}$ is a probability density function on V. Next, we will present some choices of f on a general M and will address the issue of defining means and covariances on M for some fs.

7.3 Probability Densities on Manifolds

Similar to the Euclidean domains, the two broad categories of probability densities on M are: parametric and nonparametric.

1. **Parametric densities**: A probability density function that is completely specified by a handful of parameters is said to take a parametric form. Denote the probability density by $f(p; a)$, where a is a set of parameters specifying the density function. In case of $M = \mathbb{R}^n$, a simple and important example of a parametric family is the multivariate normal density:

$$f(x; x_0, C) = \frac{1}{(2\pi)^{n/2} \det(C)^{1/2}} e^{-\frac{1}{2}(x-x_0)^T C^{-1}(x-x_0)} . \tag{7.3}$$

The mean $x_0 \in \mathbb{R}^n$ and the covariance $C \in \mathbb{R}^{n \times n}$ are the two parameters that completely specify a normal density. In case of nonlinear domains, a precise notion of a normal density may not exist. However, often there are some similar constructions that can be derived and used on a case-by-case basis.

In case of a nonlinear manifold M, at least three types of parametric densities are possible depending upon the geometry of M. Firstly, in cases where it applies, we will start with a uniform density. A uniform density is important for its role in sampling from other, more interesting densities. Secondly, we will study densities of the type $e^{-d(p,\mu)^2/\sigma^2}$, where d is the geodesic distance on M under the chosen Riemannian metric. Such a density is not feasible when M is infinite dimensional, but can be applicable to a finite-dimensional submanifold of M. Finally, we will study the truncated wrapped-normal densities, densities that are first defined on tangent spaces and then mapped to the manifold using the exponential or a similar map. To reach an explicit expression for the resulting density, one needs expressions for the Jacobians of these maps, and in case of $M = \mathbb{S}^n$, we shall present an explicit formula.

To estimate a parametric density from given data, one only has to estimate the relevant parameters, and there may be several estimators that are available for this task. For example, if we are given independent and identically distributed samples $p_1, p_2, \ldots, p_k$ on M, from a parametric density $f(p; a)$, then one can estimate a using the criterion:

$$\hat{a} = \underset{a}{\operatorname{argmax}} \prod_{i=1}^{k} f(p_i; a) = \underset{a}{\operatorname{argmax}} \sum_{i=1}^{k} \log(f(p_i; a)) . \tag{7.4}$$

This is called the **maximum-likelihood estimate** (MLE) of a and the estimated density is given by $f(p; \hat{a})$.

2. **Nonparametric densities**: In this case we do not make any prior assumption about the functional form of the density. The only constraints on f are the ones from its definition as a probability density—it is a non-negative function that integrates to one on its domain. Occasionally one imposes a smoothness constraint, e.g., assume that the first k derivatives of f are continuous, for the purpose of estimating f from the data. How can one estimate a nonparametric density function from the given sample data $\{p_1, p_2, \ldots, p_k\}$? Even though several types of estimators exist, the most common one is the kernel density estimator. This concept of kernel density estimator for nonlinear Riemannian domains has been developed by several researchers. For a point $p \in M$, we will use $\theta_p : M \to \mathbb{R}$ to denote the volume density function on M. This volume density function $\theta_q(p)$ is the determinant of the Jacobian of the exponential map $\exp_q : T_q(M) \to M$ evaluated at p in the neighborhood of mean μ. Then, for a collection of points $p_1, p_2, \ldots, p_k \in M$, one can define a Kernel density estimator to be:

$$\hat{f}_k(p) = \frac{1}{k} \sum_{i=1}^{k} \frac{1}{h_k^n \theta_{p_i}(p)} \mathcal{K}\left(\frac{d(p, p_i)}{h_k}\right) \qquad (7.5)$$

where $\mathcal{K}$ is a chosen kernel function with the desired smoothness properties, n is the dimension of M, h_k is the bandwidth for the sample size k, and $d(\cdot, \cdot)$ is the geodesic distance. The estimated density $\hat{f}_k$ is the superposition of kernels centered at different sample points p_i. If a certain point p is close to many samples p_i, then the estimated density function at that point will be high. If p is far away from the sample points, then the estimated density function will be low. By definition, $\hat{f}_k$ is non-negative and it integrates to one on M.

Now that we have some ideas for imposing a probability density f on M, we can return to the problem of defining and finding summary statistics for data sampled from f on M. These statistics can be viewed as estimates of the first few centralized moments under a probability density f on M, but we will motivate it from the perspective of generating statistical summaries. That is, given a set of points on M, we seek a method for generating data summary that is both intuitive and convenient. There are at least two possibilities for defining these summaries. The first one is to completely restrict to M and use its Riemannian structure to perform calculations. The second is to use the embedding $\mathcal{E}$ of M in V, perform the desired analysis on the extended pdf $\tilde{f}$ on V, using tools from multivariate statistics, and then project the final solution back to M. The former is called an *intrinsic analysis* and the latter is called an *extrinsic analysis*. Both have their advantages and disadvantages. We discuss these two ideas next.

7.4 Summary Statistics on Manifolds

7.4.1 Intrinsic Statistics

An intrinsic analysis results when the considerations are completely restricted to M. Once a Riemannian metric is defined and tools for computing geodesics, the exponential map and its inverse are available, there is no use of the ambient space V in this approach. In this context, we repeat a question we posed earlier: What is a suitable notion of mean of a probability density f, or a "sample mean," on a Riemannian manifold M? In other words, what is the counterpart of Eq. 7.1 on nonlinear manifolds? Of course, if M happens to be a vector space, with a given inner product, then we want this notion to coincide with the definition (Eq. 7.1) that we already know and use. A popular method for defining a mean on a manifold was proposed by Karcher who used the centroid of a density as its mean.

Definition 7.1 (Karcher Mean [46]). The Karcher mean μ_{kar} of a probability density function f on M is defined as a local minimizer of the cost function: $\rho : M \rightarrow \mathbb{R}_{\geq 0}$, where

$$\rho(p) = \int_M d(p, q)^2 f(q) \ dq \ . \tag{7.6}$$

Here, dq denotes the reference measure used in defining the probability density f on M. The value of ρ at the Karcher mean is called the **Karcher variance**. The existence and uniqueness of Karcher mean has to be studied on a case-by-case basis, depending upon the curvature of M and the nature of f. In case the minimum exists but is not unique, then the set of minimizers is said to be the Karcher mean on f. Instead of using a local minimum, one can also choose a global minimum in the definition. While this will be more satisfying from the perspective of a statistical formulation, the task of finding a global minimum, especially on complicated shape spaces, makes the practical use of this characterization somewhat limited. Thus, we content ourselves with having just a *local* minimum in the definition of Karcher mean. If we have multiple such means, we select the ones with the smallest value of ρ.

How does the definition of Karcher mean adapt to the sample set, i.e., a finite set of points drawn from an underlying probability density? Let $p_1, p_2, \ldots, p_k$ be independent random samples from the density f. Then, the sample Karcher mean of these points, denoted by $\hat{\mu}_{kar}$, is defined to be the local minimizer of the function:

$$\rho_k(p) = \frac{1}{k} \sum_{i=1}^{k} d(p, p_i)^2 \ . \tag{7.7}$$

Since the definition of $\hat{\mu}_{kar}$ involves solving an optimization problem, its computation requires some additional discussion. As $\hat{\mu}_{kar}$ is only a local minimum, an iterative gradient-based search of this point is usually sufficient. To derive a gradient algorithm, we use the following important fact from Karcher [46].

Theorem 7.1. *For any two points $p, q \in M$, assume that there exists a unique shortest geodesic between them in M. Then, the gradient of the function $-d(p, q)^2$, with respect to p, is given by the tangent vector $2v \in T_p(M)$ such that $\exp_p(v) = q$.*

Proof. This result can be proven more technically using Gauss's law (see, e.g., [46]) but we will establish it using some elementary arguments. We have assumed that there is precisely one shortest geodesic from p to q, and we wish to show that the direction of the gradient of $-d(p, q)^2$ at p is $v \in T_p(M)$ such that $\exp_p(v) = q$. We will argue this for the function $g(p) \doteq -d(p, q)$ rather than $-d(p, q)^2$. (This makes no difference in the direction of the gradient since the square is a monotone function). Now, recall the fact that the direction of the gradient is simply the direction in which the directional derivative of a function is the largest. Let ϵ be a

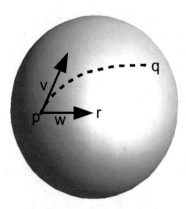

Fig. 7.1 The gradient of the distance function is along the geodesic

small positive number. Clearly, if we move a distance ϵ along the geodesic from p to q, the value of g will increase by precisely ϵ, i.e., it will be $g(p) + \epsilon$.

Now, let w be any other vector in the tangent space at p, as shown in Fig. 7.1. Suppose the directional derivative of g in the direction w was larger than the directional derivative along v. Then, if we travel a distance ϵ in the direction w (say, along that geodesic), the value of g will increase more than it increased when we moved along v. Now we will arrive at a point r such that $g(r) > g(p) + \epsilon$. In other words, $d(r, q) < d(p, q) - \epsilon$. However, this gives the following contradiction: Consider the path from p to q given first by the geodesic from p to r and then by a geodesic from r to q; its length is given by: $d(p, r) + d(r, q)$ or $\epsilon + d(r, q)$. We have shown that:

$$d(p, q) > d(r, q) + \epsilon = d(r, q) + d(p, r) .$$

This is a violation of the triangle inequality. Hence, the directional derivative of $d(p, q)$ was largest in the direction of v, proving that v is the direction of the gradient of g.

Now that we have the direction, what should be the magnitude of the gradient? It is easy to see that the magnitude of v should be the geodesic distance between p and q. Why? This is because an update (exponential map) from p in the direction of v reaches q exactly if the magnitude of v is $d(p, q)$. This proves the theorem. $\square$

Since the gradient of $-d(p, q)^2$ is $2v$, with $v = \exp_p^{-1}(q)$, the gradient of the cost function in Eq. 7.7 is simply the sum $\frac{2}{k} \sum_{i=1}^{k} v_i$, where $v_i = \exp_p^{-1}(q_i)$. Therefore, the next iterate in the estimation of Karcher mean is $\exp_p(\frac{\epsilon}{k} \sum_{i=1}^{k} v_i)$, where $\epsilon/2$ is the step size in the gradient direction.

An iterative algorithm for computing the sample Karcher mean is as follows.

Algorithm 24 (Karcher Mean on M). *Let μ_0 be an initial estimate of the Karcher mean. Set $j = 0$.*

1. *For each $i = 1, \ldots, k$, compute the tangent vector v_i such that $v_i = \exp_{\mu_j}^{-1}(q_i)$.*
2. *Compute the average direction $\bar{v} = \frac{1}{k} \sum_{i=1}^{k} v_i$.*
3. *If $\|\bar{v}\|$ is small, then stop. Else, update μ_j in the update direction using $\mu_{j+1} = \exp_{\mu_j}(\epsilon \bar{v})$, where $\epsilon > 0$ is small step size, typically 0.5.*
4. *Set $j = j + 1$ and return to Step 1.*

It can be shown that this algorithm converges to a local minimum of the cost function given in Eq. 7.7, which is the definition of the sample Karcher mean $\hat{\mu}_{kar}$. Depending upon the initial value μ_0 and the step size ϵ, it converges to the nearest local minimum.

Higher-Order Moments We now have a way of estimating the Karcher mean of a sample set on M. In a sense, we have tackled the first and most difficult challenge posed by the nonlinearity of M. This is because the definition of higher moments, under certain mild conditions, will be based on a linearization of M around this mean point $\hat{\mu}_{kar}$. This linearization is often based on the inverse of exponential map that will transform points in the neighborhood of $\hat{\mu}_{kar}$ to the corresponding points in the tangent space $T_{\hat{\mu}_{kar}}(M)$ that, being a vector space, allows standard multivariate statistics.

To illustrate this idea, we consider the covariance and higher-order (central) moments of f on M. The classical definition of the covariance of a density f on $\mathbb{R}^n$ is: $\int_{\mathbb{R}^n}(x - \mu_{kar})(x - \mu_{kar})^T f(x)dx$. Since M is not a vector space, the operation $p - \mu$ is no longer valid and we cannot use this expression. Still, we would still like to have some quantity that captures the second-order variation of shapes in a population. In particular, we would like a mechanism to study the directions of major variations in shapes, also termed *modes of variation*, similar to the dominant eigenvectors of covariance matrix in standard multivariate statistics. We exploit the fact that the tangent spaces of M are vector spaces and provide a natural domain for defining covariances. We can transfer the probability density f from M to a tangent space $T_p(M)$, using the inverse exponential map, and then use the standard definition of central moments in that vector space. The role of $x - p$ is played by $\exp_p^{-1}(x)$ and $x + p$ by $\exp_p(x)$.

Since M is nonlinear and $T_p(M)$ is linear, we will have some distortion in mapping points from M to $T_p(M)$. Specifically, when we take a set of points $\{p_1, \ldots, p_k\}$ in M and study how the distance between them in M changes when mapped to $T_p(M)$, it is clear that the distances are not preserved unless M is a flat Riemannian manifold. Let $v_i = \exp_p^{-1}(p_i)$, $i = 1, 2, \ldots, n$, and we compare the geodesic distance $d(p_i, p_j)$ in M with the Euclidean distance $\|v_i - v_j\|$. One way to measure this distortion is using the quantity:

$$D_p(p_1, p_2, \ldots, p_n) = \frac{\sum_{i,j}(d(p_i, p_j) - \|v_i - v_j\|)^2}{\sum_{i,j} d(p_i, p_j)^2} . \tag{7.8}$$

It seems intuitive that this distortion is minimum when the point p happens to be the Karcher mean $\hat{\mu}_{kar}$ of p_is. It will be good to establish this statement theoretically, but we simply leave it as an assertion based only on empirical evidence. Shown in Fig. 7.2 is an illustration of the idea. It shows a set of points on $\mathbb{S}^2$ that are mapped to tangent spaces at two different points using the inverse exponential map. The point p for the left case is the Karcher mean $\hat{\mu}_{kar}$, while p for the right case is a nearby point other than $\hat{\mu}_{kar}$. The figure shows the original points on $\mathbb{S}^2$ and their corresponding vectors $v_i \in T_p(\mathbb{S}^2)$ from two different viewing angles. It is noted that in this example, the distortion given in Eq. 7.8 is smaller for the left side than for the right side.

Choosing the Karcher mean μ_{kar} as the focal point for flattening the manifold, let, for any $p \in M$, $p \to \exp_\mu^{-1}(p)$ denote the inverse exponential map at μ from M to $T_\mu(M)$. The coordinate system on $T_{\mu_{kar}}(M)$ is chosen such that the point

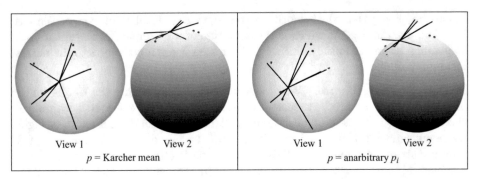

View 1	View 2	View 1	View 2
p = Karcher mean		p = anarbitrary p_i	

Fig. 7.2 Mapping a set of points from $\mathbb{S}^2$ to $T_p(\mathbb{S}^2)$ shows smaller distortion when p is the Karcher mean of the points. Distortion for the left case is $4.98e-04$ and distortion for the right case is $9.28e-04$

μ_{kar} maps to the origin $\mathbf{0} \in T_\mu(M)$. Now we can define the Karcher covariance matrix as:

$$K_{kar} = \int_{T_\mu(M)} vv^T f_v(v) dv, \quad \text{where} \quad v = \exp_{\mu_{kar}}^{-1}(q) \,,$$

and f_v is the probability density induced on $T_{\mu_{kar}}(M)$ by the original density f on M. Recall from Chap. 2 that if M is an n-dimensional manifold, then the tangent space $T_p(M)$ is a vector space of dimension n. Therefore, the matrix K_{kar} defined as the above is an $n \times n$ covariance matrix. For a finite sample set, the sample Karcher variance is given by

$$\hat{K}_{kar} = \frac{1}{k-1} \sum_{i=1}^{k} v_i v_i^T, \quad \text{where} \quad v_i = \exp_{\hat{\mu}_{kar}}^{-1}(q_i) \,. \tag{7.9}$$

Note that we need not subtract the mean here since it is zero under the chosen representation on the tangent space. $\hat{K}_{kar}$ is an $n \times n$ symmetric, positive-definite matrix that represents the observed covariance matrix on the vector space $T_{\hat{\mu}_{kar}}(M)$. Let

$$\hat{K}_{kar} = U\Sigma U^T = \sum_{i=1}^{n} \sigma_i^2 u_i u_i^T$$

denote the SVD of $\hat{K}_{kar}$. Here $U = [u_1, u_2, \ldots, u_n]$ are the singular vectors and $\{\sigma_i^2\}$ are the corresponding singular values. Assuming that $\sigma_1^2 \geq \sigma_2^2 \geq \sigma_3^2 \ldots$, the vectors u_1, u_2, etc, in that order, are the directions of largest variations in the population represented by f. One can depict these directions on M by mapping them back under the exponential map $\exp_{\hat{\mu}_{kar}}$. For instance, the geodesic starting at $\hat{\mu}_{kar}$ in the direction $u_1 \in T_{\hat{\mu}_{kar}}(M)$, i.e. $t \mapsto \exp_{\hat{\mu}_{kar}}(tu_1)$ is the first dominant path of f on M. It is also called the *first principal geodesic curve*. Similarly, the geodesic along u_2, i.e., $t \mapsto \exp_{\hat{\mu}_{kar}u}(tu_2)$ is the *second principal geodesic curve* and so on.

What happens if M is an infinite-dimensional Riemannian manifold or its quotient space? One can still apply the abovementioned framework for computing summary statistics, with the following caveat. Since the tangent space $T_{\hat{\mu}_{kar}}(M)$ is now infinite dimensional, an additional step of dimension reduction, as described in Sect. 4.3.1, becomes necessary. Treating elements of $T_{\hat{\mu}_{kar}}(M)$ as functions, we can apply an FPCA step to restrict our analysis to a relevant finite-dimensional subspace of $T_{\hat{\mu}_{kar}}(M)$. This process is called the *Tangent PCA* or TPCA and is related to FPCA covered earlier in Chap. 4.

7.4.2 Extrinsic Statistics

The other possibility for generating summary statistics of sample data on M is to use the vector-space structure of the embedding space V to simplify calculations. We remind the readers that $\mathcal{E} : M \to V$ is an embedding that preserves the Riemannian metric between M and V. Here one transfers the probability measure to V, computes pertinent statistical quantities in V and projects the final results back to M. Let $\Pi : V \to M$ be a projection map defined in such a way that

$$\Pi(v) = \underset{p \in M}{\mathrm{argmin}} \, \|v - \mathcal{E}(p)\|^2 \; . \tag{7.10}$$

The existence and the uniqueness of Π, of course, depend on the nature of M, p, and $\mathcal{E}$. Now, the extrinsic mean of a density f on M is defined as follows.

Definition 7.2 (Extrinsic Mean). The extrinsic mean of density f on M, specified with respect to an embedding $\mathcal{E}$ of M in a larger vector space V, is given by $\mu_{ext} = \Pi(\int_V v \tilde{f}(v) dv)$, where Π is the projection defined in Eq. 7.10, $\int_V v \tilde{f}(v) dv$ is the standard mean of $\tilde{f}$ in V, and $\tilde{f}$ is the unique extension of f from M to V (given by Eq. 7.2).

Once the embedding $\mathcal{E}$ has been chosen, and a mechanism for projection Π has been established, the rest of the process is quite straightforward. It requires computing the mean of $\tilde{f}$ in V and projecting it down to M. In case M is a Euclidean space, and the $\mathcal{E}$ is simply the identity map, the extrinsic mean coincides with the classical sample mean.

What about the covariance analysis in an extrinsic framework? An extrinsic covariance can be defined similarly to the extrinsic mean. Let $P : V \to T_\nu(M)$ be an orthogonal projection. Since it is a linear map, it can be written as a $n \times m$ matrix A so that $P(v) = Av$, where $n = \dim(V)$ and $m = \dim(M)$. Define the covariance $K_v = \int_V (v - \nu)(v - \nu)^t \tilde{f}(v) dv$, in the vector space V and project it using: $K_{ext} = A K_v A^T$. Similar to the intrinsic case, one can define principal geodesic curves in the extrinsic setup. Let $K_{ext} = U_{ext} \Sigma_{ext} U_{ext}^T = \sum_{i=1}^n \sigma_{ext,i}^2 u_{ext,i} u_{ext,i}^T$ denote the SVD of K_{ext}. Assuming that the singular values are labeled in a descending order, the corresponding directions given by $u_{ext,1}$, $u_{ext,2}$, etc, in that order, are the directions of largest variations in the population represented by f. The geodesic starting at μ_{ext} in the direction $u_{ext,1} \in T_{\mu_{ext}}(M)$, given by $t \mapsto \exp_{\mu_{ext}}(t u_{ext,1})$, is the *first extrinsic principal geodesic curve*. Similarly, the geodesic along $u_{ext,2}$, given by $t \mapsto \exp_{\mu_{ext}}(t u_{ext,2})$, is the *second extrinsic principal geodesic curve*, and so on. These geodesics are first constructed in the vector space V, using straight lines, and then projected on M using Π.

The pros and cons of using the extrinsic mean, as opposed to the Karcher mean, are quite straightforward. The main advantage is its computational simplicity. Once an embedding $\mathcal{E}$ is chosen, the rest of the analysis is quite standard and typically very efficient. In contrast, the computation of the Karcher mean requires repeated computations of the exponential and the inverse exponential maps on the manifold. The main disadvantage of the extrinsic analysis is that the result $\Pi(\nu)$ depends on the choice of embedding $\mathcal{E}$, which is quite arbitrary. Different embeddings will result in different solutions, and the projection Π itself may not be unique. Another limitation of the extrinsic analysis is that in some cases involving quotient manifolds—manifolds obtained as quotients of groups acting on certain other manifolds—it is difficult to find vector spaces in which we can embed the quotient spaces. This complicates the use of an extrinsic statistical analysis in quotient spaces. Since our shape spaces are quotient spaces of certain Hilbert manifolds, these arguments suggest that it is more natural to use intrinsic techniques for statistical analysis of shapes.

7.5 Examples on Some Useful Manifolds

7.5.1 Statistical Analysis on $\mathbb{S}^1$

In the next few sections, we will study these definitions and concepts for some well-known and relatively simple manifolds, such as $\mathbb{S}^1, \mathbb{S}^2, \mathbb{S}_\infty$ as a warmup for the functional spaces in the next two chapters. In each case we will define different representations of elements on these manifolds, choose Riemannian metrics, introduce some probability models, and eventually outline algorithms for computing sample statistics. The geometric and the algebraic properties of these manifolds have already been discussed in Chap. 3. Although this discussion on $\mathbb{S}^1$ and $\mathbb{S}^2$ is presented as preliminary to statistics on more complicated shape spaces, they are quite important in their own right.

Riemannian Structure As presented earlier in an example in Chap. 3, we will impose the standard Euclidean Riemannian metric on $\mathbb{S}^1$. Under this Riemannian structure, the geodesic between any two points (that do not form an antipodal pair) is given by the shorter arc on the circle and the distance is given by the length of that arc. To make it more concrete, let us represent an element of $\mathbb{S}^1$ by the angle $\theta \in [0, 2\pi]$, with the point 0 identified with the point 2π; with this identification each point of $\mathbb{S}^1$ is represented uniquely. Under the Euclidean Riemannian metric, the geodesic distance between any two points $\theta_1, \theta_2 \in [0, 2\pi]$ is given by:

$$d(\theta_1, \theta_2) = \min\{|\theta_1 - \theta_2|, |\theta_1 + 2\pi - \theta_2|, |\theta_1 - 2\pi - \theta_2|\} . \tag{7.11}$$

For any $\theta \in \mathbb{S}^1$ and $v \in T_\theta(\mathbb{S}^1) \equiv \mathbb{R}$, the exponential map $\exp_\theta : T_\theta(\mathbb{S}^1) \to \mathbb{S}^1$ is given by:

$$\exp_\theta(v) = (\theta + v)_{\mod 2\pi} . \tag{7.12}$$

The inverse of this map is given by: $\exp_{\theta_1}^{-1} : \mathbb{S}^1 \to T_{\theta_1}(\mathbb{S}^1)$,

$$\exp_{\theta_1}^{-1}(\theta_2) = \begin{cases} \theta_2 - \theta_1, & \text{if } d(\theta_1, \theta_2) = |\theta_2 - \theta_1| \\ \theta_2 + 2\pi - \theta_1, & \text{if } d(\theta_1, \theta_2) = |\theta_2 + 2\pi - \theta_1| \\ \theta_2 - 2\pi - \theta_1, & \text{if } d(\theta_1, \theta_2) = |\theta_2 - 2\pi - \theta_1| \end{cases} . \tag{7.13}$$

These quantities and maps will allow us to perform intrinsic analysis on $\mathbb{S}^1$.

For an extrinsic analysis, on the other hand, we have to look for ways to embed $\mathbb{S}^1$ in larger Euclidean spaces. A natural embedding comes from the mapping $\mathcal{E}_1(\theta) = [\cos(\theta) \ \sin(\theta)] \in \mathbb{R}^2$. Another embedding, this time in $\mathbb{R}^4$, is obtained using:

$$\theta \mapsto \mathcal{E}_2(\theta) = \begin{bmatrix} \cos(\theta) & -\sin(\theta) \\ \sin(\theta) & \cos(\theta) \end{bmatrix} .$$

The range of this mapping is actually the special orthogonal group $SO(2)$, which can be identified with $\mathbb{S}^1$ using precisely this mapping.

Under the embedding $\mathcal{E}_1$, the distance between any two points is given by:

$$|\mathcal{E}_1(\theta_1) - \mathcal{E}_1(\theta_2)| = \sqrt{2}\sqrt{1 - \cos(\theta_1 - \theta_2)} .$$

In case we choose the embedding in $SO(2)$, $\mathcal{E}_2(\theta)$, the distance between elements of $\mathbb{S}^1$ will be $2\sqrt{1 - \cos(\theta_1 - \theta_2)}$. Now that we have both a Riemannian structure and Euclidean embeddings of $\mathbb{S}^1$, we are ready to discuss possible stochastic models and statistical inferences on this space.

Probability Models To impose a probability model on $\mathbb{S}^1$, we can choose from both parametric and nonparametric families. Since a parametric family is often more efficient to work with, it becomes our first choice. Also, since $\mathbb{S}^1$ is a compact manifold, some of the traditional parametric families, such as the normal density or the exponential density functions, will not be applicable directly as these densities have infinite tail(s). So, what are the alternatives? Here are some suggestions:

1. **Uniform Density**: Of course, the simplest choice is the uniform density function: $f(\theta) = \frac{1}{2\pi}$. That is, all the elements of $\mathbb{S}^1$ occur with equal probability.
2. **von Mises Density**: Analogous to the normal random variable on $\mathbb{R}$, one would like a density such that: (i) it is unimodal, (ii) it is symmetric around the mode, and (iii) its spread is controlled by one parameter. One such density on $\mathbb{S}^1$ can be developed using the exponential of negative squared (extrinsic) distance, as follows. Consider a point on $\mathbb{S}^1$ as the mean μ and let θ be a generic point on $\mathbb{S}^1$. Using $\mathcal{E}_1$ as defined above, we can compute the extrinsic distance, *chord length* or *chordal distance*, between θ and μ: $\|\mathcal{E}(\theta) - \mathcal{E}(\mu)\| = \sqrt{2(1 - \cos(\theta - \mu))}$.

Then, we define a density function on $\mathbb{S}^1$ that is proportional to $e^{\frac{-d(\theta,\mu)^2}{\sigma^2}}$, for some value of the variance parameter σ^2. After dropping the constant terms, terms that do not depend on θ, and normalizing we obtain the probability density:

$$f(\theta; \mu, \sigma^2) = \frac{1}{I_0(\frac{1}{\sigma^2})} \exp(\frac{\cos(\theta - \mu)}{\sigma^2}) , \qquad (7.14)$$

where μ is the mean and σ^2 controls the variance. The value $\frac{1}{\sigma^2}$ is also called the dispersion. I_0 is the modified Bessel function of order zero and serves as the normalizing constant for f. This density is called one-dimensional **von Mises density**. Figure 7.3 shows some examples von Mises density for a fixed μ and decreasing values of σ^2. As σ^2 goes to infinity, the resulting density converges to a uniform density on $\mathbb{S}^1$.

We take a small detour to investigate this density. An important tool in generating statistical inference, especially using computational tools, is the Monte Carlo approach. Here one uses samples from the underlying density to estimate quantities (parameters) associated with that density. In case of statistical inferences from a von Mises density on $\mathbb{S}^1$, it will be useful to know how to

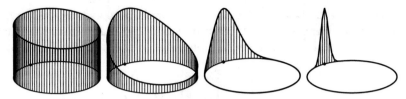

Fig. 7.3 Examples of von Mises density function on $\mathbb{S}^1$ for $\sigma^2 = 10, 1, 0.1$, and 0.01 from left to right

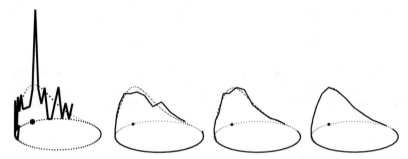

Fig. 7.4 Histograms of samples from von Mises density function on $\mathbb{S}^1$, with $\frac{1}{\sigma^2} = 5$, for increasing numbers of accepted samples from left to right

simulate from this density. In general there are two main techniques for directly simulating a random variable. Of these methods—inverse transform method and acceptance-rejection method—the second one is general enough to be useful here. Note that while the strength of an acceptance-rejection method is its wide applicability, its main weakness is the inefficiency resulting from rejection of proposed values. Let the von Mises density function have mean μ and dispersion $\frac{1}{\sigma^2}$, and let the candidate probability be the uniform density function. Then, the bounding coefficient is given by:

$$\frac{f(\theta; \mu, \sigma^2)}{1/2\pi} = \frac{2\pi}{I_0(\frac{1}{\sigma^2})} \exp(\frac{\cos(\theta - \mu)}{\sigma^2}) \leq \frac{2\pi}{I_0(\frac{1}{\sigma^2})} \exp(\frac{1}{\sigma^2}) \equiv c .$$

Hence, the ratio:

$$R(\theta) \equiv \frac{f(\theta; \mu, \sigma^2)}{c/2\pi} = \exp(\frac{(\cos(\theta - \mu) - 1)}{\sigma^2}) .$$

The acceptance-rejection sampling algorithm is given by:

Algorithm 25 (Acceptance-Rejection for von Mises Density on $\mathbb{S}^1$).

1. *Generate $U_1 \sim U[0, 1]$ and set $\phi = 2\pi U_1$.*
2. *Generate $U_2 \sim U[0, 1]$.*
3. *If $U_2 < R(\phi)$, then set $\theta = \phi$.*
Else, return to Step 1.

Shown in Fig. 7.4 are some examples of this procedure. Each panel shows histograms of samples generated using this algorithm; the number of samples increases from left to right. The solid line shows the observed histograms and the broken line shows the underlying von Mises density function. As the sample size gets larger, the reader can verify that the estimated density (histogram) converges to the underlying true density.

3. **Wrapped-Normal Density**: Since the previous item used an extrinsic distance on $\mathbb{S}^1$, it is natural to ask the question: what will happen if we use the intrinsic (Riemannian) distance, given in Eq. 7.11, instead. The distance between a point θ and the mean μ is the arc-length distance on $\mathbb{S}^1$, which is equivalent to the Euclidean distance in a tangent space at μ as long as one measures distances from μ. This approach has also been studied extensively for $\mathbb{S}^1$ and is defined as follows. Let $x \in \mathbb{R}$ be a normal random variable with mean $\mu = 0$ and variance σ^2. Let $\theta = \exp_\mu(x)$ to be the exponential map of x, as defined in Eq. 7.12. Since $\theta = (x + \mu)_{\mathrm{mod}\ 2\pi}$, we have $\theta = x + \mu + i2\pi$ for some integer i. This mapping from x to θ can be viewed as a *wrapping* of the real line around the unit circle. Therefore, we can write the probability density of θ as:

$$f(\theta; \mu, \sigma^2) = \sum_{i=-\infty}^{\infty} \frac{1}{\sqrt{2\pi}\sigma} e^{-\frac{1}{2\sigma^2}(x+\mu+i2\pi)^2} \ ,$$

where the summation accounts for all i associated with a given θ. The simulation of values from this wrapped-normal density is straightforward. Simply generate normal random variables and map them to $\mathbb{S}^1$ using the exponential map. An interesting observation here is that we are able to keep the full density on the tangent space $T_\mu(\mathbb{S}^1)$ and wrap it on $\mathbb{S}^1$ using infinite summation. In higher dimensional manifolds, this full wrapping is difficult and we resort to a truncated density in the tangent space before wrapping in on to the manifold.

4. **Truncated Wrapped-Normal (TWN) Density**: If we restrict the probability density on the real line to a proper subset $[-\lambda, \lambda]$ of $[-\pi, \pi]$, using truncation, then the exponential map from $\mathbb{R}$ to $\mathbb{S}^1$ is invertible. As a consequence, a wrapping of this density on the $\mathbb{S}^1$ avoids the infinite sum present in the previous item. Define a truncated normal density on $\mathbb{R}$, denoted by TN, to be:

$$f(x; \mu, \sigma^2, \lambda) = \frac{1}{Z_{\sigma,\lambda}} e^{-\frac{1}{2}\frac{(x-\mu)^2}{\sigma^2}} \mathbf{1}_{|x-\mu|<\lambda} \ , \quad Z_{\sigma,\lambda} = \int_{-\lambda}^{\lambda} e^{\frac{-x^2}{2\sigma^2}}\, dx \ . \tag{7.15}$$

Since the mapping from $\mathbb{S}^1$ to $\mathbb{R}$ is simply $\theta = x + \mu$, the corresponding density on $\mathbb{S}^1$ is exactly this expression.

Summary Statistics Now that we have a few choices for defining probability models on $\mathbb{S}^1$, we can turn our attention to the problem of defining and estimating central moments, such as mean and variance, using samples from a probability model. We will illustrate the ideas of extrinsic and intrinsic statistics on $\mathbb{S}^1$ using samples from a von Mises density.

1. **Intrinsic Statistics**: The first quantity that we want to estimate is the Karcher mean of f, which is defined using Definition 7.1. This is a purely intrinsic analysis and, thus, does not require any Euclidean embedding or extrinsic distances. Assume that we have independent samples $\theta_1, \theta_2, \ldots, \theta_n \sim f$. We can particularize Eq. 7.7 to define their Karcher mean on $\mathbb{S}^1$:

$$\hat{\mu}_{kar} = \underset{\theta \in \mathbb{S}^1}{\mathrm{argmin}} \sum_{i=1}^{n} d(\theta, \theta_i)^2 \ , \tag{7.16}$$

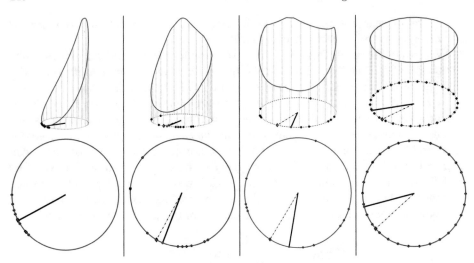

Fig. 7.5 Four examples of sample intrinsic and extrinsic means on a unit circle. The sample points show an increasing divergence from left to right. The intrinsic mean is connected to the origin using a *solid line*, while the extrinsic mean is connected using a *dotted line*

where d is given in Eq. 7.11. A local solution to this optimization problem can be found using Algorithm 24. The formulas for the exponential map and its inverse on $\mathbb{S}^1$ are given in Eqs. 7.12 and 7.13, respectively.

Let's look at some examples of this idea, displayed in Fig. 7.5. The top row shows four different scenarios—in each case we see some sample points, $\theta_1, \ldots, \theta_n$, on the unit circle and the cost function $\frac{1}{n} \sum_{i=1}^{n} d(\theta, \theta_i)^2$ drawn as a height function on the circle. The bottom row of this figure shows the estimated Karcher means for these cases. To highlight their positions on the circle, these Karcher means have been connected with the origin using solid lines. In the leftmost case where the sample points are clustered close to each other, the location of the Karcher mean falls somewhere in the middle of those points, as our intuition would suggest. Moving right in this figure, the sample points are scattered further apart and the Karcher mean tries to follow these points. In the third column, we see that the cost function has at least two local minimum and the algorithm selects one of them arbitrarily as an estimate for the Karcher mean. In the last case, the points are distributed uniformly around the circle and the cost function is nearly a constant. Any point on the circle is as good an estimate for the mean as any although the algorithm has been forced to select a point on the circle.

Once we have estimated the Karcher mean, we can estimate the sample variance as follows. For the estimated intrinsic mean $\hat{\mu}_{kar}$, let $v_i = \exp_{\hat{\mu}_{kar}}^{-1}(\theta_i)$ be the inverse exponential of a sample point θ_i, as defined in Eq. 7.13. Each v_i is an element of the tangent space $T_{\hat{\mu}_{kar}}(\mathbb{S}^1)$ and, by definition, their mean is zero. The sample variance of v_is is simply: $\frac{1}{n-1} \sum_{i=1}^{n} v_i^2$. It turns out that, except for the constant $\frac{1}{n-1}$, this is exactly the minimum value of the cost function given in Eq. 7.16. This becomes the sample Karcher variance of the dataset $\theta_1, \ldots, \theta_n$ on $\mathbb{S}^1$.

2. **Extrinsic Statistics**: The basic idea here is to embed the underlying space in a larger Euclidean space V and to compute statistics in that larger space. Since the computed statistic often does not lie on the manifold, one has to project the Euclidean solution back to the required space. We apply this procedure to $\mathbb{S}^1$ using some standard embeddings.

As the first example, we take $V = \mathbb{R}^2$ with the Euclidean metric and we embed $\mathbb{S}^1$ in $\mathbb{R}^2$ using the mapping $\mathcal{E}_1(\theta) = [\cos(\theta)\ \sin(\theta)]$. The sample mean on $\mathbb{R}^2$ is straightforward (Eq. 7.1) and the projection $\Pi : \mathbb{R}^2 \to \mathbb{S}^1$ is given by:

$$\Pi(v) = \operatorname*{argmin}_{\theta \in \mathbb{S}^1} \|v - \mathcal{E}_1(\theta)\|^2 = \tan^{-1}\left(\frac{v_2}{v_1}\right) . \tag{7.17}$$

To view this projection as an element of $\mathcal{E}_1(\mathbb{S}^1)$, we can use the concatenation:

$$\mathcal{E}_1(\Pi(v)) = \frac{v}{\|v\|} .$$

This projection is valid only if $\|v\| \neq 0$. In case $v = 0$, i.e., it coincides with the origin in $\mathbb{R}^2$, there is no preferred projection and any point of $\mathbb{S}^1$ is equally valid. For all other points in $\mathbb{R}^2$, this projection is well defined and unique.

Now, for a given set of points $\theta_1, \theta_2, \ldots, \theta_n \in \mathbb{S}^1$, we can derive an expression for its extrinsic mean in the chosen embedding. Define $x_i = \mathcal{E}_1(\theta_i) = [\cos(\theta_i)\ \sin(\theta_i)]$ and the sample mean $\bar{x}_n = \frac{1}{n}\sum_{i=1}^{n} x_i$ in $\mathbb{R}^2$. Finally, we project this mean from $\mathbb{R}^2$ to $\mathbb{S}^1$ using Eq. 7.17. This projected point $\hat{\mu}_n$ is called the sample *extrinsic mean* of the observed set. Figure 7.5 also shows some examples of the extrinsic means on the circle, this time connected to the origin by a dotted line. The bottom row simply shows the plots in the top row but from a different viewing angle. In the leftmost case, where the sample points are close together, we see that the two estimates of mean: extrinsic and intrinsic, coincide. However, as the points become further apart, the two means start being different. The rightmost case is the case of uniform distribution on $\mathbb{S}^1$ where any points on $\mathbb{S}^1$ is equally good as either intrinsic or extrinsic mean.

We re-emphasize the disadvantages of using extrinsic statistics using the example of $\mathbb{S}^1$. The main disadvantage is that the results depend on the choice of embedding. Different embeddings will result in different solutions. Earlier we have used the standard embeddings of $\mathbb{S}^1$ in $\mathbb{R}^2$, but many more non-standard choices are possible. Figure 7.6 a few illustrations of this issue. Consider an ellipsoidal embedding of $\mathbb{S}^1$ in $\mathbb{R}^2$ using:

$$\mathcal{E}_3(\theta) = [2\cos(\theta), \sin(\theta)] .$$

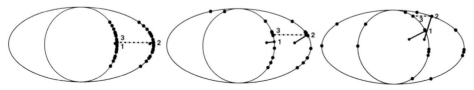

Fig. 7.6 Extrinsic means for different embeddings on $\mathbb{S}^1$ in $\mathbb{R}^2$. Each panel shows samples $\{\theta_i\}$ as points $\mathcal{E}(\theta_i)$, one extrinsic mean $\mathcal{E}(\mu_{ext})$ (labeled as 1), and another extrinsic mean $\mathcal{E}(\mu_{ext}^2)$ (labeled as 3)

Note that this embedding does not preserve the Riemannian metric on $\mathbb{S}^1$. For this embedding, we can define the projection $\Pi_2 : \mathbb{R}^2 \to \mathbb{S}^1$ as:

$$\Pi_2(v) = \underset{\theta \in \mathbb{S}^1}{\operatorname{argmin}} |v - \mathcal{E}_3(\theta)|^2 \ .$$

This projection is well defined and unique for all points other than those that fall on a line connecting the focal points. This sets up the computation of the extrinsic mean under this ellipsoidal embedding. We take the samples $\theta_1, \ldots, \theta_n$, compute their embeddings $\mathcal{E}_3(\theta_i)$, and then compute their sample mean in $\mathbb{R}^2$: $\frac{1}{n} \sum_{i=1}^n \mathcal{E}_3(\theta_i)$. This mean is projected back to $\mathcal{E}_3(\mathbb{S}^1)$ using Π_2 and that results in the extrinsic mean of the samples in this embedding. Call it $\hat{\mu}_{ext}^2$.

In order to compare this point in $\mathcal{E}_3(\mathbb{S}^1)$, with the extrinsic mean for $\mathcal{E}_1$, we can map it back to the circle using $\mathcal{E}_1(\hat{\mu}_{ext})$. Figure 7.6 shows several examples of this comparison. Each figure shows the following items: (i) the original sample points $\theta_1, \theta_2, \ldots, \theta_n$ displayed on the unit circle (using $\mathcal{E}_1(\theta_i)$) and the ellipse (using $\mathcal{E}_3(\theta_i)$), (ii) the extrinsic mean on the circle for the first embedding $\mathcal{E}_1(\mu_{ext})$, labeled as point 1, and (iii) the extrinsic mean on the ellipse for the second embedding $\mathcal{E}_3(\mu_{ext})$, labeled as point 2. To compare them on the same domain, we can map the point 2 from the ellipse to the unit circle using the inverse of $\mathcal{E}_3(\mathbb{S}^1)$, and that point is labeled 3. The differences between point 1 and 3 underscores the assertion that extrinsic means are critically dependent on the choice of embedding.

7.5.2 Statistical Analysis on $\mathbb{S}^2$

Although $\mathbb{S}^1$ is an interesting example to start with, the first real challenge comes from $\mathbb{S}^2$, the two-dimensional sphere. As in the previous section, we will look at some probability models on $\mathbb{S}^2$ and study the computation of both intrinsic and extrinsic summary statistics on $\mathbb{S}^2$.

Riemannian Structure In Chap. 3, we have discussed the common choice of Euclidean Riemannian metric and other elements of the differential geometry of $\mathbb{S}^2$. Given any two points p_1, $p_2 \in \mathbb{S}^2$, Eq. 3.5 provides an expression for the geodesic from p_1 to p_2. Similarly, for $p \in \mathbb{S}^2$ and a $v \in T_P(\mathbb{S}^2)$, Eq. 3.8 provides an expression for the geodesic starting from p in the direction v. In particular, these two equations provides expressions for computing the exponential map and its inverse. Therefore, we have the basic tools needed for performing intrinsic statistical analysis on $\mathbb{S}^2$. We can use the spherical coordinates to index points on a unit sphere. These coordinates are (ϕ_1, ϕ_2) with $\phi_1 \in [0, 2\pi]$ changing along the latitude and $\phi_2 \in [0, \pi]$ changing along the altitude and with $\phi_2 = 0$ being the north pole. This representation is not unique since, for example, for $\phi_2 = 0$, all the values of ϕ_1 map to the same point on $\mathbb{S}^2$. Still, it is useful to refer points on $\mathbb{S}^2$ by their spherical coordinates. The geodesic distance between points p_1 and p_2 is given by: $d(p_1, p_2) = \cos^{-1}(\langle p_1, p_2 \rangle)$.

For the extrinsic analysis, the elements of $\mathbb{S}^2$ can be represented as points in $\mathbb{R}^3$ with the north pole identified with $(0, 0, 1)$ and the south pole with $(0, 0, -1)$. This parameterization is given by:

$$\epsilon(p) \equiv \epsilon(\phi_1, \phi_2) = (\sin(\phi_1)\sin(\phi_2), \cos(\phi_1)\sin(\phi_2), \cos(\phi_2)) \in \mathbb{R}^3 \ .$$

The extrinsic, or chord-length, distance between any two points p_1 and p_2 is given by:

$$\|\epsilon(p^{(1)})-\epsilon(p^{(2)})\| = \sqrt{-2(\sin(\phi_2^{(1)})\sin(\phi_2^{(2)})\cos(\phi_1^{(1)} - \phi_1^{(2)}) + \cos(\phi_2^{(1)})\cos(\phi_2^{(2)}))} \, .$$

Probability Models What are the different choices of probability densities on $\mathbb{S}^2$? We discuss the same three classes as we studied for $\mathbb{S}^1$:

1. **Uniform Density**: The simplest case is the uniform density $f(p) = \frac{1}{4\pi}$ for all $p \in \mathbb{S}^2$. How can we generate samples from f? While this task was trivial in the case of $\mathbb{S}^1$, it requires some more effort for $\mathbb{S}^2$. We exploit the fact that the infinitesimal area element on $\mathbb{S}^2$ is given by $\sin(\phi_2)d\phi_2 d\phi_1$, where (ϕ_1, ϕ_2) are the spherical coordinates. As earlier, $\phi_1 \in [0, 2\pi]$ changes along the latitude and $\phi_2 \in [0, \pi]$ changes along the altitude, with $\phi_2 = 0$ being the north pole. So if we generate ϕ_1 uniformly in $[0, 2\pi]$ and ϕ_2 according to the density $\sin(\phi_2)$ using the inverse transform method, then we will have a uniform sample on $\mathbb{S}^2$. The following algorithm implements this idea.

   Algorithm 26 (Uniform Sampling on $\mathbb{S}^2$).

   a. *Generate a uniform random variable u_1 and set $\phi_1 = 2\pi u_1$.*
   b. *Generate a uniform random variable u_2 and set $\phi_2 = \cos^{-1}(u_2)$.*
   c. *Generate a uniform random variable u_3. If $u_3 < 0.5$, then set $\phi_2 = \pi - \phi_2$.*

   The last step is needed to offset the domain of the $\cos^{-1}$ function in the standard computer libraries from $[-\pi/2, \pi/2]$ to $[0, \pi]$. Some examples of samples generated by Algorithm 26 are shown in Fig. 7.7. The two rows show independent samples with increasing number of points from left to right, with the same number of samples in each column.

2. **Fisher's Density Function**: Next we study a density that is an extension of the von Mises density to $\mathbb{S}^2$. Let $\mu \in \mathbb{R}^3$ such that $\|\mu\| = 1$ and $\kappa > 0$ be positive number. Representing points on $\mathbb{S}^2$ with their Euclidean coordinates $x \in \mathbb{R}^3$, define a probability density function in $\mathbb{R}^3$ by

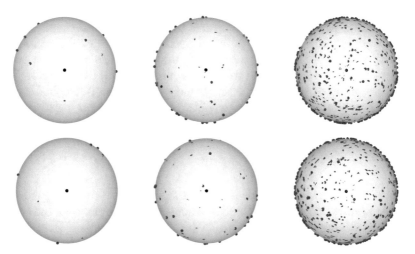

Fig. 7.7 Random samples from a uniform density on $\mathbb{S}^2$ with increasing number of samples from left to right. The two rows show independent samples

$$f(x) = \begin{cases} \frac{1}{Z(\kappa)} e^{\kappa \mu^T x} & x \in \mathbb{S}^2 \\ 0 & x \notin \mathbb{S}^2 \end{cases} . \tag{7.18}$$

Here $Z(\kappa)$ is the normalizing constant given by: $Z(\kappa) = \int_{\mathbb{S}^2} e^{\kappa \mu^T x} dx = \frac{\kappa^{0.5}}{(2\pi)^{1.5} I_{0.5}(\kappa)}$, where $I_{0.5}$ is the modified Bessel function of first kind and order 0.5. By definition, this density is restricted to $\mathbb{S}^2 \subset \mathbb{R}^3$. The mode and the mean of f are identical and are given by the unit vector μ. Its variance is inversely proportional to κ; κ is also called the *concentration parameter* . This f is known as the *Fisher density* and is the analog of the von Mises density on $\mathbb{S}^1$ (Eq. 7.14).

Now we consider the problem of simulating points from the Fisher density. Similar to the case of von Mises density on $\mathbb{S}^1$, we will use the acceptance-rejection idea. As a reference density, we choose the uniform density $g(x) = \frac{1}{4\pi}$ on $\mathbb{S}^2$ since we already know how to sample from it. We can bound the ratio of these two densities according to:

$$\frac{f(x; \mu, \kappa)}{g(x)} = \frac{4\pi e^{\kappa \mu^T x}}{Z(\kappa)} \leq \frac{4\pi e^{\kappa}}{Z(\kappa)} \equiv c .$$

Hence, the desired constant for comparing with uniform random variable is:

$$R(x) \equiv \frac{f(x; \mu, \kappa)}{g(x)c} = e^{\kappa(\mu^T x - 1)} . \tag{7.19}$$

The desired acceptance-rejection sampling algorithm is given by:

Algorithm 27 (Acceptance-Rejection for Fisher's Density on $\mathbb{S}^2$).

1. *Generate* $x \sim U(\mathbb{S}^2)$ *the uniform density on* $\mathbb{S}^2$.
2. *Generate* $U_2 \sim U[0, 1]$.
3. *If* $U_2 < R(x)$, *with* $R(x)$ *as given in Eq. 7.19, then set* $y = x$.

Else, return to Step 1.

This algorithm generates a sample y from the Fisher distribution $f(x; \mu, \kappa)$ on $\mathbb{S}^2$. Shown in Fig. 7.8 are some examples of samples generated from the Fisher density using Algorithm 7.19. From left to right, we see the same number of samples with the same mode μ but a decreasing value of the concentration parameter κ. The values of κ are 50, 10, and 1 in these cases. The two rows show two independent samples from the same density in each column.

An extension of this Fisher density that allows a non-isotropic distribution of mass around the mean is called a *Fisher-Bingham* or *Kent* density. It defines a five-parameter family given by:

$$f(x) = \frac{1}{c(\kappa, \beta)} \exp(\kappa \gamma_1 \cdot x + \beta[(\gamma_2 \cdot x)^2 - (\gamma_3 \cdot x)^2]) ,$$

where $x \in \mathbb{S}^2 \subset \mathbb{R}^3$ is viewed as a 3-vector, $c(\kappa, \beta)$ is a normalizing constant, κ is the concentration parameter, and β is a measure of ellipticity of the levels sets of f. The vectors γ_1, γ_2, and γ_3 are orthogonal to each other. Here γ_1 defines the mean direction, while γ_2 and γ_3 define the major and minor axes of the level sets of f.

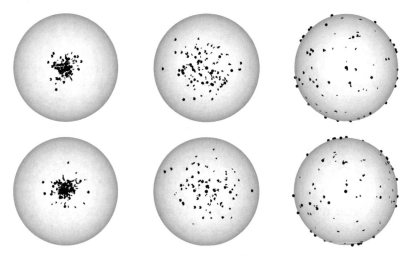

Fig. 7.8 Random samples from a Fisher density on $\mathbb{S}^2$ with increasing variance (decreasing κ) from left to right

3. **Truncated Wrapped-Normal (TWN) Density**: Another possibility that allows a non-isotropic density, i.e., different variances in different directions, akin to the multivariate normal density in $\mathbb{R}^n$, is the wrapped normal density function. The basic idea is to use the tangent space at the mean, $T_\mu(\mathbb{S}^2)$, a vector space, to define a bivariate normal density and then transfer it to the sphere. This transfer is often performed using the exponential map. Since the exponential map can be seen as wrapping the tangent space onto the sphere, we will call this density the *wrapped-normal density*. However, this case is different from that of $\mathbb{S}^1$ since the exponential map is a nonlinear Map, and its Jacobian appears in the final expression for the density resulting from wrapping. Furthermore, the magnitude of this Jacobian goes to infinity as we go away from the mean μ in the tangent space $T_\mu(\mathbb{S}^2)$. This is easy to see because a non-compact set $\mathbb{R}^2$ is being mapped (wrapped) to a compact set $\mathbb{S}^2$, and the differential of this mapping is going to be unbounded at some point. A simple solution is to restrict the density in the tangent space using truncation and then map it to the sphere. This way, a compact set is being mapped to a compact set and the Jacobian remains stable.

The details are as follows. The exponential map from a tangent space $T_\mu(\mathbb{S}^2)$ to $\mathbb{S}^2$ (using Eq. 3.8) is: $\exp_\mu(v) = \cos(\|v\|)\mu + \sin(\|v\|)\frac{v}{\|v\|}$. In order to make the exponential map invertible, we need to restrict the density to a subset of $T_\mu(\mathbb{S}^2)$ using truncation. While choosing a disc of radius π around the origin will be sufficient for this purpose, this will lead to a singularity in the resulting density at the point antipodal to the mean. To avoid that we will truncate the density at a smaller distance, say $\pi/2$, from the origin in the tangent space. Thus, $\exp_\mu$ becomes invertible in this domain and its inverse is given by: for $p \in \mathbb{S}^2$, $\exp_\mu^{-1}(p) = (\theta/\sin(\theta))(p - \mu\cos(\theta))$, where $\theta = \cos^{-1}(\langle p, \mu \rangle)$ and where p and μ are viewed as vectors in $\mathbb{R}^3$.

Let w_1, w_2 form an orthonormal basis of $T_\mu(\mathbb{S}^2)$. The set $\{\mu, w_1, w_2\}$ forms an orthogonal basis for $\mathbb{R}^3$. Using the basis $\{w_1, w_2\}$, we can identify any element v of $T_\mu(\mathbb{S}^2)$ with its coordinates $x = (x_1, x_2) \in \mathbb{R}^2$ such that $v = x_1 w_1 + x_2 w_2$. Define a truncated bivariate normal (TBN) density on $T_\mu(\mathbb{S}^2)$ using its identification with $\mathbb{R}^2$:

$$f(x; K, \lambda) = \frac{1}{Z_{K,\lambda}} e^{(-\frac{1}{2}x^T K^{-1} x)} \mathbf{1}_{\|x\| \leq \lambda} ,$$

where $K \in \mathbb{R}^{2 \times 2}$ is a covariance matrix and $Z_{K,\lambda}$ is simply the normalizing constant. Next we map this density, for $\lambda = \pi/2$, onto the two sphere $\mathbb{S}^2$ using the exponential map so that the origin of the tangent space coincides with the mean μ. For a $\theta \in \mathbb{R}$, and the point θw_1 in $T_\mu(\mathbb{S}^2)$, the exponential map equation becomes $\theta w_1 \mapsto \exp_\mu(\theta w_1) \equiv \cos(\theta)\mu + \sin(\theta)w_1$. Let this point on $\mathbb{S}^2$ be called p. We need to establish an orthogonal basis for the tangent space $T_p(\mathbb{S}^2)$ and the vectors $b_1 = (-\sin(\theta)\mu + \cos(\theta)w_1)$, and $b_2 = w_2$ provide a convenient orthonormal basis. To derive the differential of the map $\exp_\mu$, denoted $\exp_{\mu*}$, we take each of the basis elements of $T_\mu(\mathbb{S}^2)$ and map them to $T_p(\mathbb{S}^2)$ under $\exp_{\mu*}$. In order to induce the truncated Gaussian density on a sphere, using the exponential map, we need to compute the determinant of the Jacobian of the exponential map. In this case, the resulting determinant of the Jacobian matrix turns out to be $(\sin(\theta)/\theta)$.

Now we can write the expression for the induced density. For a point $p \in \mathbb{S}^2$, the local coordinates of the inverse map in $T_\mu(\mathbb{S}^2)$ are: for $i = 1, 2, x_i = \langle w_i, \exp_\mu^{-1}(p) \rangle = (\theta/\sin(\theta)) \langle w_i, p \rangle$, where $\theta = \cos^{-1}(\langle p, \mu \rangle)$. Then, the induced truncated normal density on $\mathbb{S}^2$ is given by

$$f(p; \mu, K) = \frac{1}{Z_2} \frac{\theta}{\sin(\theta)} \exp\{-\frac{1}{2}(p^T K_w^{-1} p)\} \mathbf{1}_{\theta \leq \pi/2} , \qquad (7.20)$$

with $K_w^{-1} = W K^{-1} W^T$ and where $W = [w_1 \ w_2] \in \mathbb{R}^{3 \times 2}$. Sometimes it is more convenient to express the density in a polar coordinate system. Let (ϕ_1, ϕ_2) denote polar coordinates on $\mathbb{S}^2$ such that ϕ_2 is the depression angle from μ ($\phi_2 = 0$ denotes the mean μ, the north pole) and ϕ_1 is the azimuth, i.e., $p = [\sin(\phi_2)\sin(\phi_1), \sin(\phi_2)\cos(\phi_1), \cos(\phi_2)]^T$, and $\mu = [0\ 0\ 1]^T$. For convenience, we choose the basis $\{w_1 = [1\ 0\ 0]^T, w_2 = [0\ 1\ 0]^T\}$ and the induced density becomes:

$$f(\phi_1, \phi_2; \mu, K) = \frac{\phi_2}{Z_2 \sin(\phi_2)} e^{-\frac{\phi_2^2}{2}(K_{11}^{-1}\sin(\phi_1)^2 + 2K_{12}^{-1}\cos(\phi_1)\sin(\phi_1) + K_{22}^{-1}\cos(\phi_1)^2)} \mathbf{1}_{\phi_2 \leq \pi/2} .$$
$$\qquad (7.21)$$

In the case of $K = \sigma^2 I_2$, we get

$$f(\phi_1, \phi_2; \mu, \sigma^2) = \frac{\phi_2}{Z_2 \sin(\phi_2)} e^{\{-\phi_2^2/2\sigma^2\}} \mathbf{1}_{\phi_2 \leq \pi/2} , \qquad (7.22)$$

and the normalizing constant takes the form $Z_2 = 2\pi\sigma^2(1 - e^{-(\pi^2/8\sigma^2)})$. Three examples of a truncated wrapped-normal density are shown in Fig. 7.9. In each case we show the original truncated normal density in $\mathbb{R}^2$ (top row) and the corresponding wrapped densities on $\mathbb{S}^2$, both as intensity functions (middle row) and the level curves (bottom row).

Summary Statistics We have discussed three examples of density functions on $\mathbb{S}^2$ and have also discussed techniques for sampling from them. Now we return to the problem of computing some simple statistics given a sample set on $\mathbb{S}^2$. Let $p_1, \ldots, p_n$ be independent samples from a density f, and we want to derive

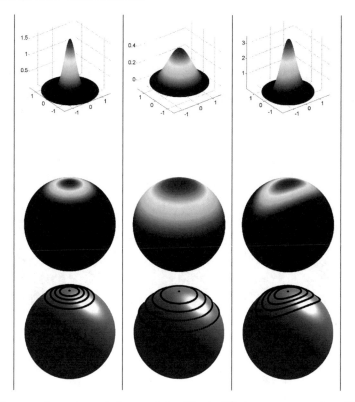

Fig. 7.9 Top row shows truncated normal densities in $\mathbb{R}^2$ that are wrapped on $\mathbb{S}^2$ using the exponential map. The next two rows display the resulting densities on $\mathbb{S}^2$ using an intensity map (*middle*) and level curves (*bottom*)

expressions for intrinsic and extrinsic statistics of this sample. For the extrinsic analysis, we choose the most natural embedding of $\mathbb{S}^2$ in $\mathbb{R}^3$. Designate a point on $\mathbb{S}^2$ as the north pole $(0, 0, 1)$, the diametrically opposite point as the south pole $(0, 0, -1)$, and an arbitrary point on the equator as $(1, 0, 0)$. This fixes the mapping of all points on $\mathbb{S}^2$ into $\mathbb{R}^3$; we will call the mapping of p_i into $\mathbb{R}^3$ as $\mathcal{E} : \mathbb{S}^2 \to \mathbb{R}^3$, $\mathcal{E}(p_i) = p_i$. Then, the extrinsic mean of the given set can be computed as:

$$\mu_{ext} = \frac{\bar{p}}{\|\bar{p}\|}, \quad \text{where} \quad \bar{p} = \frac{1}{n} \sum_{i=1}^{n} p_i \ . \tag{7.23}$$

For the intrinsic analysis, we first want to compute the intrinsic mean under the standard Riemannian metric. We simply particularize Algorithm 24 to $\mathbb{S}^2$ to obtain the following steps:

Algorithm 28 (Karcher Mean on $\mathbb{S}^2$). *Let μ_0 be an initial estimate of the Karcher mean of f. One can use μ_{ext} for this purpose. Set $j = 0$.*

1. For each $i = 1, \ldots, k$, compute

$$v_i = \frac{\theta_i}{\sin(\theta_i)} (p_i - \mu \cos(\theta_i)), \quad \text{where} \quad \theta_i = \cos^{-1}(\langle p_i, \mu \rangle) \ .$$

2. Compute the average direction $\bar{v} = \frac{1}{k} \sum_{i=1}^{k} v_i$.

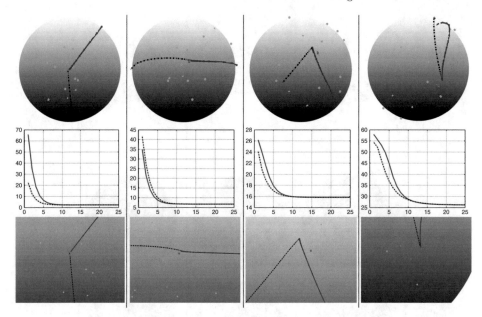

Fig. 7.10 Examples of sample mean on $\mathbb{S}^2$: the top row shows sample points and convergence of Algorithm 28 for two different initializations. Intrinsic mean is where these two points meet. Also, for comparison, we show the extrinsic mean as the yellow point. The middle row shows the evolution of Karcher variance for two paths, and the bottom row shows a zoom in to highlight differences between the means

3. If $\|\bar{v}\|$ is small, then stop. Else, update μ_j in the update direction using

$$\mu_{j+1} = \cos(\epsilon\|\bar{v}\|)\mu + \sin(\epsilon\|\bar{v}\|)\frac{\bar{v}}{\|\bar{v}\|} \; ,$$

where $\epsilon > 0$ is small step size, typically 0.5.
4. Set $j = j + 1$ and return to Step 1.

Let us look at some results obtained by this algorithm in Fig. 7.10. In each column we show a different sets of points sampled from a Fisher's density on $\mathbb{S}^2$ and compute their intrinsic and extrinsic means. In each case, we run Algorithm 28 twice with two random initial conditions and show their evolutions using marked lines on $\mathbb{S}^2$. The corresponding evolution of the Karcher variance function, defined in Eq. 7.6, for these two runs are shown in the middle row. In the bottom row, we show a magnified view of $\mathbb{S}^2$ near the point of convergence. Note that despite different initial conditions, the two runs of the algorithm converge to the same point. Shown in a thick point is the extrinsic mean of the same dataset. In cases where the sample points are close to each other, i.e., the variability is small, both the intrinsic and extrinsic means fall at the same point (first two columns). However, when the data exhibits larger variability, the two means are actually different (last two columns). As emphasized earlier, these extrinsic means were computed using a specific embedding $\mathcal{E}$ and will change with the embedding. The intrinsic means are however dependent only on the Riemannian structure of $\mathbb{S}^2$ and do not need any Euclidean embedding.

7.5.3 Space of Probability Density Functions

The next space we consider is the set of all probability density functions on the unit interval $[0, 1]$, i.e., $\mathcal{P} = \{g : [0, 1] \mapsto \mathbb{R}_{\geq 0} | \int_0^1 g(x)dx = 1\}$. The overall geometry of $\mathcal{P}$ was presented earlier in Sect. 4.11.2. Specifically, we discussed the construction of geodesic paths and geodesic distances between elements of $\mathcal{P}$ under the Fisher-Rao Riemannian metric. We start here with a brief recap of that discussion.

Riemannian Structure As described in Sect. 4.11.2, an important step in analyzing elements of $\mathcal{P}$ is using the half-density representation, $q(t) = \sqrt{g(t)}$, to represent any $g \in \mathcal{P}$. Under this representation, the Fisher-Rao metric becomes the standard $\mathbb{L}^2$ metric (Lemma 4.9) and the set $\mathcal{P}$ can be identified with $\mathcal{Q}$, the positive orthant of the Hilbert sphere $\mathbb{S}_\infty$:

$$\mathcal{Q} = \{q : [0, 1] \to \mathbb{R}_{\geq 0} \mid \|q\| = 1\} \ ,$$

where $\| \cdot \|$ denotes the $\mathbb{L}^2$ norm as usual. This is illustrated pictorially in Fig. 7.11. Consequently, the geodesic distance between any two densities is given by (Eq. 4.35): $d(g_1, g_2) = \cos^{-1} \left(\int_0^1 \sqrt{g_1(t)}\sqrt{g_2(t)}dt \right)$ and, since the set $\mathcal{Q}$ is geodesically convex, as a subset of $\mathbb{S}_\infty$, the geodesic path between them can also be computed easily using Eq. 4.9. Also, we can consider $\mathcal{Q}$ directly for performing statistical analysis since such an analysis on $\mathcal{Q}$ using the $\mathbb{L}^2$ metric is equivalent to (and simpler than) an analysis on $\mathcal{P}$ under the Fisher-Rao metric. For any $q \in \mathcal{Q}$, the tangent space $T_q(\mathcal{Q})$ is the set of all functions on $[0, 1]$ that are perpendicular to q under the $\mathbb{L}^2$ metric, i.e., $\langle v, q \rangle = 0$ implies $v \in T_q(\mathcal{Q})$. For any $v \in T_q(\mathcal{Q})$, the exponential map $\exp_p(v) = \cos(\|v\|)p + \sin(\|v\|)\frac{v}{\|v\|}$, where $\| \cdot \|$ denotes the $\mathbb{L}^2$ norm. Similarly, for any $q_1, q_2 \in \mathbb{S}_\infty$, and $q_1 \neq -q_2$, the inverse exponential map is given by $\exp_{q_1}^{-1}(q_2) = (\theta / \sin(\theta))(q_2 - q_1 \cos(\theta))$, where $\theta = \cos^{-1}(\langle q_1, q_2 \rangle)$.

Probability Models Next we take up some probability models that can potentially be used to model observed elements of $\mathcal{P}$. While there is a large literature on non-informative priors the can be used in this situation, see, e.g., Bayesian nonparametric methods, our interest lies in more efficient parametric expressions, similar to the forms that were studied for $\mathbb{S}^2$ in the previous section. We will use the identification of $\mathcal{P}$ with $\mathcal{Q} \subset \mathbb{S}_\infty$ to attach a spherical structure to this space, and then extend ideas of the previous section, from $\mathbb{S}^2$ to $\mathbb{S}^k$, for some large k, that replaces $\mathbb{S}_\infty$ as the representation space in practical situations. A finite-dimensional approximation of the elements of $\mathcal{Q}$ is warranted in practice for several reasons:

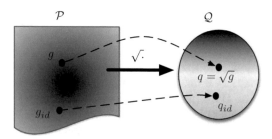

Fig. 7.11 Illustration of the square-root mapping from $\mathcal{P}$ to $\mathcal{Q}$

the observations are mostly finite-dimensional, the number of observations itself is finite, and so on.

1. **Uniform Density**: It is not possible to define a uniform density on an infinite-dimensional space but may be possible on a finite-dimensional, compact subset of Q.
2. **Squared-Distance Density**: The second option explored for $\mathbb{S}^2$ was of the form e^{-d^2}, up to a normalization constant, where d is a distance on underlying manifold. Can we use a similar expression for imposing density on Q? The answer is no and the reason is the infinite dimensionality of this space. Let us explain this further.

   Consider the space of square-integrable functions on $[0,1]$, denoted by $\mathbb{L}^2([0,1],\mathbb{R})$. The Euclidean distance between any two points is given by the $\mathbb{L}^2$ norm of their difference: $\|f_1 - f_2\| = \sqrt{\int_0^1 (f_1(t) - f_2(t))^2 dt}$. For any $f_0 \in \mathbb{L}^2([0,1],\mathbb{R})$, we can try to form the squared-distance density using the form $\frac{1}{Z(f_0,\sigma)} e^{-\|f-f_0\|^2/2\sigma^2}$, for a positive constant σ. Here Z is the normalization constant and is given by the integral:

$$Z(f_0,\sigma) = \int_{\mathbb{L}^2([0,1],\mathbb{R})} e^{-\|f-f_0\|^2/2\sigma^2} \, df \ ,$$

with df being the reference measure on $\mathbb{L}^2([0,1],\mathbb{R})$. For a fixed σ, what will be the value of Z? First, we try to compute this integral along a one-dimensional subspace of $\mathbb{L}^2([0,1],\mathbb{R})$. For a fixed $g \in \mathbb{L}^2([0,1],\mathbb{R})$, with $\|g\| = 1$, define a subspace: $\{f_0 + gt | t \in \mathbb{R}\}$ and then:

$$\int_{-\infty}^{\infty} e^{-\|f_0-f_0-gt\|^2/2\sigma^2} \, dt = \int_{-\infty}^{\infty} e^{-t^2\|g\|^2/2\sigma^2} \, dt = \sqrt{2\pi}\frac{\sigma}{\|g\|} = \sqrt{2\pi}\sigma \ .$$

So, integration along a one-dimensional subspace of $\mathbb{L}^2([0,1],\mathbb{R})$, passing through f_0, results in a term that is linear in σ.

Now consider a two-dimensional subspace of $\mathbb{L}^2([0,1],\mathbb{R})$, spanned by f_0, g_1, and g_2, for some square-integrable functions g_1 and g_2 with unit norm $\|g_1\| = \|g_2\| = 1$ and $\langle g_1, g_2 \rangle = 0$. The integral for evaluating the normalizing constant gives:

$$\int_{-\infty}^{\infty} \int_{-\infty}^{\infty} e^{-\|f_0-f_0-g_1t_1-g_2t_2\|^2/2\sigma^2} \, dt_1 \, dt_2$$
$$= \int_{-\infty}^{\infty} e^{-t_1^2\|g_1\|^2/2\sigma^2} \, dt_1 \cdot \int_{-\infty}^{\infty} e^{-t_2^2\|g_2\|^2/2\sigma^2} \, dt_2$$
$$= 2\pi\sigma^2 \ .$$

This term is quadratic in σ. As we include more directions in the model, the normalizing constant is a polynomial in σ of higher order. What will be the result if we integrate the squared-distance density over the whole space? Since there are infinite number of independent (orthogonal) directions in $\mathbb{L}^2([0,1],\mathbb{R})$, this integral will be infinite! Thus, it is not possible to define a probability model of the type e^{-d^2} on the entire space Q. The main reason for this is that this model is designed to be isotropic in infinitely many directions. To avoid this situation, there are two possibilities: one is to use only a finite-dimensional submanifold of $\mathcal{Q}$ to model the variability and other is to use a more complicated form that distributes its variance in only a finite number of dimensions. The truncated wrapped-normal distribution provides the latter possibility, and, thus, we will focus on that probability model for representing variability in $\mathcal{Q}$.

3. **Truncated Wrapped-Normal Density**: The basic idea here is to: (1) select a finite-dimensional subspace of a tangent space $T_q(\mathcal{Q})$, for a certain central point q, (2) define a truncated multivariate normal density on that vector space, and (3) use the exponential map to transfer this density onto $\mathcal{Q} \subset \mathbb{S}_\infty$. We will demonstrate the idea using a disk region centered at q.

For a point $q \in \mathcal{Q}$, let $\{b_i\}$ denote a complete, orthonormal basis for the tangent space $T_q(\mathcal{Q})$. That is, we can write any $v \in T_q(\mathcal{Q})$ as $v = \sum_{i=1}^{\infty} x_i b_i$, $x_i \in \mathbb{R}$. For any positive integer k, define a subspace of the tangent space using:

$$V_k \equiv \text{span}\{b_1, b_2, \ldots, b_k\} \subset T_q(\mathcal{Q}) .$$

Elements of V_k can be mapped onto $\mathcal{Q}$ using the exponential map as $\exp_q$. Let $D \subset V_k$ be a compact disk, i.e., $D = \{\sum_{i=1}^{k} x_i b_i \in T_q(\mathcal{Q}) | \sum_{i=1}^{k} x_i^2 \leq \pi/2\}$. Then, we can impose a truncated normal density on D and map to the relevant subset of $\mathcal{Q}$ using the exponential map:

$$\{x_i\} \in \mathbb{R}^k \mapsto \exp_q(\sum_{i=1}^{k} x_i b_i) = \cos(|x|)q + \sin(|x|)\frac{\sum_{i=1}^{k} x_i b_i}{|x|} . \tag{7.24}$$

Note that the range of D under the exponential map can be identified with a subset of the finite-dimensional sphere $\mathbb{S}^k$.

Now we will derive a truncated multivariate normal density on $\mathbb{S}^k$ and will use it to impose a probability density on the subset $\exp_q(D) \in \mathcal{Q}$. This is a simple extension of the derivation presented in the previous section on $\mathbb{S}^2$. A truncated normal density in $\mathbb{R}^k$ is given by: $f_x(x) = \frac{1}{Z_k} e^{(-\frac{1}{2}x^T K^{-1} x)} \mathbf{1}_{|x| \leq \pi/2}$, where K is a $k \times k$ covariance matrix and Z_k is the normalizing constant.

Using Eq. 7.24, we will perform a change of variable from $x \in \mathbb{R}^k$ to $p \in \mathcal{Q}$ and induce a density on a subset of $\mathcal{Q}$ by pushing f_x forward. The resulting expression is given by:

$$f_p(p; \mu, K) = \frac{1}{Z_k} \left(\frac{\theta}{\sin(\theta)}\right)^{(k-1)} e^{(-\frac{1}{2}x^T K^{-1} x)} \mathbf{1}_{\theta \leq \pi/2} ,$$
$$\text{where } \theta = |x| = \cos^{-1}(\langle p, q \rangle \, , x_i = \langle b_i, \exp_q^{-1}(p) \rangle .$$

If $K = \sigma^2 I_k$, then the induced density reduces to $\frac{1}{Z_k}(\theta/\sin(\theta))^{(k-1)} e^{-(\theta^2/2\sigma^2)} \mathbf{1}_{\theta \leq \pi/2}$.

Summary Statistics Given a set of probability density functions, we seek to compute their summary statistics so that we can capture the variability of the given functions using a smaller number of descriptors. Let $g_1, g_2, \ldots, g_n \in \mathcal{P}$ be the given set of probability density functions and let $q_i = \sqrt{g_i} \in \mathbb{S}_\infty$ be the corresponding half-density functions. We can compute the Karcher mean of $\{q_i\}$ using Algorithm 28 since all the expressions needed in that algorithm are identical for $\mathbb{S}^k$, for all k including infinity. Figure 7.12 shows some examples of computing Karcher means of unimodal densities where the location, height, and breadth of

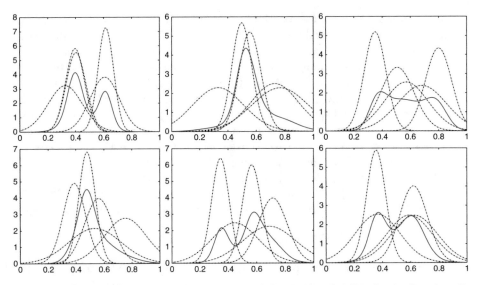

Fig. 7.12 Examples of computing Karcher mean of the given probability density functions. In each panel, we show five density functions using *broken lines* and their Karcher mean using the *solid line*

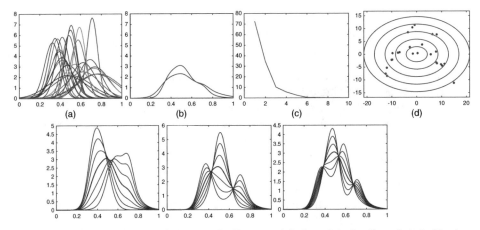

Fig. 7.13 Tangent PCA of pdf data on $\mathcal{Q}$. *Top row*: (**a**) the original pdfs and their Karcher mean, (**b**) the Karcher mean and extrinsic mean, (**c**) the singular values of the tangent covariance matrix, and (**d**) the scatter plot of the observed principal coefficients. *Bottom row*: three most dominant eigen directions mapped back in the pdf space $\mathcal{P}$

the mode is different for different densities. We note that in cases where the modes of different densities are located close to each other, the Karcher mean also has a single mode. However, when the peaks of the individual densities are scattered apart, the Karcher mean can become multimodal.

Once we have an estimate for the Karcher mean, say $\hat{\mu}$, we can estimate the covariance matrix in a straightforward manner. Simply take the observed half densities $\{q_i\}$ and map them onto the tangent space $T_{\hat{\mu}}(\mathcal{Q})$ using the inverse exponential map; call these points $\{v_i\}$. Then, form the sample covariance matrix K of the set $\{v_i\}$ to estimate the underlying covariance. A dimension reduction technique, such as FPCA, can be used to reduce dimension of the observed data and to capture the essential modes of variations in the data. Figure 7.13 shows

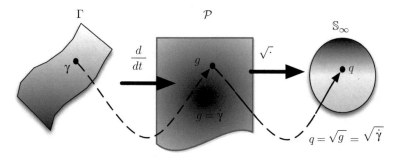

Fig. **7.14** Illustration of the square-root mapping from Γ to $\mathcal{P}$ to $\mathcal{Q}$

an example of this idea where the top-left panel shows the original pdfs and their Karcher mean. The next panel compares the Karcher mean with the extrinsic mean $(\frac{1}{n}\sum_{i=1} f_i)$. The singular values of the sample covariance matrix are plotted in the third panel, and the observed principal scores (in the first two dominant directions) are shown in the rightmost panel. The PCA of shooting vectors results the dominant modes of variability, shown in the bottom panel. From left to right, we see top three modes of variability—from $-0.6\sigma_i$ to $+0.6\sigma_i$—mapped back to $\mathcal{P}$.

7.5.4 Space of Warping Functions

Earlier, in Sect. 4.10.2, we have studied the geometry of Γ_I using the natural Fisher-Rao metric and have demonstrated the computation of geodesic paths between arbitrary warping functions. These constructions are based on the following idea. For a $\gamma \in \Gamma_I$, the derivative $\dot{\gamma}$ is an element of $\mathcal{P}$ defined in the previous section. Therefore, we have a natural mapping $\gamma \to \dot{\gamma} \to \sqrt{\dot{\gamma}}$ from Γ to $\mathcal{P}$ to $\mathcal{Q}$. This is illustrated in Fig. 7.14. Once again, the Fisher-Rao metric becomes the $\mathbb{L}^2$ metric on $\mathcal{Q}$, which is a positive orthant of $\mathbb{S}_\infty$ and the geodesics are simply arcs on great circles (Eq. 4.9). Each point on this geodesic in $\mathcal{Q}$ can be mapped back to Γ_I to obtain the desired geodesic.

Now we can use this representation to define and compute some elementary statistics such as Karcher means and covariance of a set of warping functions. Let $\gamma_1, \gamma_2, \ldots, \gamma_n \in \Gamma_I$ be a set of observed warping functions. Our goal is to develop a probability model on Γ_I that can be estimated from the data directly. There are two problems in doing this is in a standard way: (1) Γ_I is not a vector space although it has affine structure and (2) it is infinite dimensional. In fact, Γ_I has a nonlinear structure under the metric of interest, and this nonlinearity is handled using a convenient transformation, which coincidentally is similar to the definition of SRVF. The issue of infinite dimensionality is handled using dimension reduction, e.g., FPCA. We are going to represent an element $\gamma \in \Gamma_I$ by its SRVF $\psi = \sqrt{\dot{\gamma}}$. The identity map γ_{id} maps to a constant function with value $\psi_{id}(t) = 1$. As described in Sect. 4.3.3, the square-root representation simplifies the complicated geometry of Γ_I to a unit sphere. The Fisher-Rao distance between any two warping functions is found to be the arc length between the corresponding SRVFs $d_{FR}(\gamma_1, \gamma_2) = d_\psi(\psi_1, \psi_2) \equiv \cos^{-1}(\int_0^1 \psi_1(t)\psi_2(t)dt)$. Now we can define the Karcher mean of a

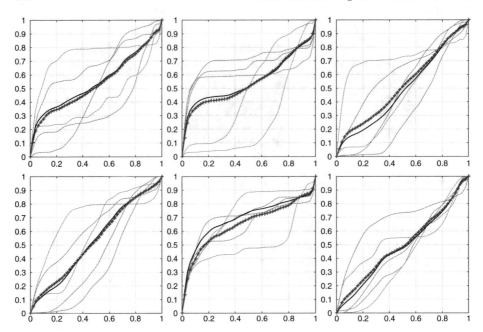

Fig. 7.15 Karcher mean of warping functions under Fisher-Rao distance on Γ_I. Each panel shows five random warping functions (*dotted curves*), their Karcher mean (*solid curve*), and their cross-sectional mean (*spotted curve*)

set of warping functions using the Karcher mean on $\mathcal{Q}$. For a given set of warping functions $\gamma_1, \gamma_2, \ldots, \gamma_n \in \Gamma_I$, their Karcher mean is $\gamma(t) \equiv \int_0^t \psi(s)^2 ds$ where ψ is the Karcher mean of $\sqrt{\dot{\gamma}_1}, \sqrt{\dot{\gamma}_2}, \ldots, \sqrt{\dot{\gamma}_n}$ in $\mathcal{Q}$. (Computation of Karcher mean of a finite set of points in $\mathcal{Q}$ is discussed in the previous section and is not repeated here.) We present some examples of computing the Karcher mean of warping functions in Fig. 7.15. Each panel shows five random warping functions (dotted curves), their Karcher mean (solid curve), and their cross-sectional mean (dashed curve). In most cases the cross-sectional mean is quite similar to the Karcher mean under the Fisher-Rao distance, although they are not equal.

To apply functional PCA to the warping data, we once again utilize the SRVF representation. Let $\{\gamma_i\}$ be the given warping functions and $\psi_i = \sqrt{\dot{\gamma}_i} \in \mathcal{Q}$ be the corresponding SRVFs. Let μ_ψ denote the Karcher mean of $\{\psi_i\}$ on $\mathcal{Q}$ under the standard $\mathbb{L}^2$ metric. Then, the shooting vector $v_i = \exp_{\mu_\psi}^{-1}(\phi_i)$ is the mapping of ψ_i to the tangent space $T_{\mu_\psi}(\mathcal{Q})$. At this stage, we assume that each ψ_i is represented as a vector using a uniform partition of $[0, 1]$ with T elements. Then, the sample covariance K of $\{v_i\}$ is a $T \times T$ covariance matrix and let its SVD be given by $K = U_\psi \Sigma_\psi U_\psi^T$. Assuming that the singular values are arranged in a non-increasing order from top left to bottom right, the first few columns of U_ψ denote the principle modes of variations. Note that it is easy to map elements from $T_{\mu_\psi}(\mathcal{Q})$ to Γ_I using the following steps:

$$v \in T_{\mu_\psi}(\mathcal{Q}) \quad \mapsto \quad \psi = \cos(\|v\|)\mu_\psi + \frac{\sin(\|v\|)}{\|v\|}v \in \mathcal{Q} \quad \mapsto \quad \gamma(t) = \int_0^t \psi(s)^2 ds \in \Gamma_I \, .$$
$$(7.25)$$

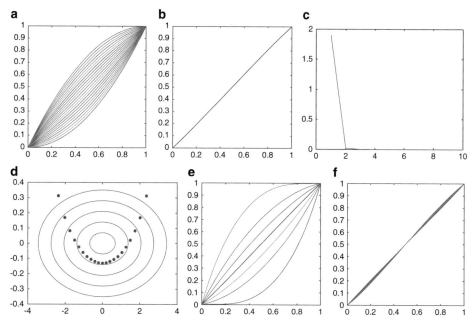

Fig. 7.16 (**a**) A set of warping functions, (**b**) their Karcher mean, (**c**) singular values of the sample covariance matrix, (**d**) top-two principal coefficients of observed functions, (**e**) first principal model of variation, and (**f**) second principal mode of variation

Thus, one can study the principal modes of variation in Γ_I by converting columns of U_ψ into the corresponding warping functions. We illustrate this idea using an example shown in Fig. 7.16. The top-leftmost panel shows a collection of warping functions that form the set $\{\gamma_i\}$. In this example, $\gamma_i(t) = t + a_i t(1-t)$, for $a_i = (2(i-1)-n)/n$, for $i = 1, 2, \ldots, n$ and $n = 20$. The top-right panel shows a plot of the singular values, diagonal elements of the matrix Σ. Since the data are taken from a one-parameter family, only the first singular value is significant; the rest are quite small. The bottom middle and right plots show the two main principal modes of variation in the data by taking $xU_\psi(j)$, where x goes from $-2\sigma_j$ to $+2\sigma_j$ and $U_\psi(j)$ is the jth column of U_ψ, into the warping space using Eq. 7.25. It can be seen that the first principal mode nicely captures the variability in the original dataset.

7.6 Statistical Analysis on a Quotient Space M/G

So far the discussion has been focused on Riemannian manifolds, with $\mathbb{S}^1$, $\mathbb{S}^2$, and $\mathcal{Q}$ as examples, but we realize that the shape spaces of interest are actually not manifolds by themselves but are quotient spaces of manifolds. Therefore, in preparation for shape spaces, we should study the techniques for defining probability models, computing summary statistics, and estimating probabilities on some quotient spaces. Here is a general setup of interest. Let M be a Riemannian manifold and let G act on M in such a way that: (1) the action of G on M is by isometries and (2) the orbits of G in M are closed. We are interested in statistical analysis on the quotient space M/G as defined in Definition 2. We can divide the discussion into two possibilities, (1) where the quotient space can be identified with an orthogonal section (see Definition 3.19) and the statistical analysis can be completely shifted to this section, and (2) such an identification is not possible and one has to deal with the space M/G in general terms.

7.6.1 Quotient Space as Orthogonal Section

In case M/G can be identified with an orthogonal section S of M, then the desired analysis can be performed on S, as described in Sect. 3.7. Since S is a submanifold of M, and a Riemannian manifold by itself, the earlier discussion about statistical analysis on a general Riemannian manifold applies directly. Therefore, instead of repeating that framework, we will simply illustrate the ideas using a couple of simple examples.

Example 7.1. 1. **Landmark Shapes Modulo Scaling (But Not Rotation)**: Consider the set of k landmarks in $\mathbb{R}^n$, given by $M = \mathbb{R}^{n \times k} - \{0\}$ and the action of the scaling group $G = \mathbb{R}^{\times}$ on M. According to Example 3.14, the orthogonal section of M under the action of G is given by:

$$S = \{X \in \mathbb{R}^{n \times k} | \sum_{j=1}^{k} \sum_{i=1}^{n} X_{i,j}^2 = 1\} .$$

Since this S is $\mathbb{S}^{nk-1}$, a unit sphere in $\mathbb{R}^{n \times k}$, one can use directly the techniques for a unit sphere on S, as described in Sect. 7.5.2. We refer the reader to that section for details.

2. **Rescaled Functions**: In this case, let M be the vector space of square-integrable, real-valued functions on the interval $[0,1]$, i.e., $M = \mathbb{L}^2([0,1], \mathbb{R})$. The scaling group $G = \mathbb{R}^{\times}$ acts on M with the action $(a, f) = af$. The orbit of a function is given by $[f] = \{af | a \in \mathbb{R}^{\times}\}$. Similar to the previous example, it can be shown that an orthogonal section of M under the action of G is given by:

$$S = \{f \in \mathbb{L}^2([0,1], \mathbb{R}) | \int_0^1 |f(t)|^2 dt = 1\} .$$

S is nothing but a unit Hilbert sphere $\mathbb{S}_{\infty}$ in $\mathbb{L}^2([0,1], \mathbb{R})$ and a statistical analysis on the quotient space M/G is performed using the geometry of this sphere, as described in Sect. 7.5.3. To illustrate this point further, take the case of computing the Karcher mean in the quotient space. Given a set of functions $\{f_i \in \mathbb{L}^2([0,1], \mathbb{R}) | i = 1, 2, \ldots, k\}$, one can compute the Karcher means of their orbits under G as follows. Project each function in its orthogonal section by simply scaling it: $f_i \to \tilde{f}_i = \frac{f_i}{\|f_i\|} \in S$. Now, since these projected points are elements of a unit sphere, use Algorithm 28 to compute their Karcher mean in S. The orbit of the resulting mean represents the sample Karcher mean of this given orbits in the quotient space M/G.

7.6.2 General Case: Without Using Sections

In situations where the quotient space M/G cannot be identified directly with an orthogonal section, then the analysis is more involved. The basic steps for a statistical analysis on M/G remain similar, at least conceptually, to the corresponding steps on M, except one has to define the central ingredients, such as the geodesics, geodesic distances, exponential map and its inverse, etc, accordingly.

As the first step, we define these central items for a quotient space and then discuss procedures for a statistical analysis on that space. Consider a point $p \in M$ and let $[p] = \{(g, p)|g \in G\}$ be the orbit of p under G. We will assume that $[p]$ is a manifold in itself, and, therefore, one can talk about $T_p([p])$, the space tangent to $[p]$ at $p \in M$.

1. **Tangent Space**: Let $T_p(M)$ and $T_p([p])$ denote the tangent spaces at p to the manifold M and the orbit $[p]$, respectively. Clearly, $T_p([p]) \subset T_p(M)$. Let $N_p(M)$ be the set of vectors that are perpendicular to $T_p([p])$ in $T_p(M)$. That is, the elements of $T_p([p])$ are tangent to the orbit $[p]$ at p and the elements of $N_p(M)$ are perpendicular to the orbit at p. Their direct sum:

$$T_p([p]) \oplus N_p(M) = T_p(M) ,$$

is the full tangent space at p. We can identify the perpendicular space $N_p(M)$ with the tangent space on the quotient set $T_{[p]}(M/G)$. That is, for every element $v \in T_{[p]}(M/G)$, there is a corresponding element $w \in N_p(M)$, and vice versa. We will use this identification to specify tangent vectors in the quotient space.

Example 7.2. a. Let M be the representation space of k landmarks in $\mathbb{R}^2$ or $\mathbb{C}$, $M = \mathbb{C}^k$, with the standard Euclidean Riemannian metric, and let $G = SO(2)$ or $\mathbb{S}^1$ or $U(1)$ be the rotation group acting on it. We already know that there is no orthogonal section to represent the quotient space M/G. For any $z \in \mathbb{C}^k$, its orbit is given by: $[z] = \{e^{i\theta}z|\theta \in \mathbb{S}^1\}$, with $i = \sqrt{-1}$, and the space tangent to the orbit at z is given by: $T_z([z]) = \{ixz|x \in \mathbb{R}\}$. Therefore, the space perpendicular to the orbit is given by:

$$N_z(\mathbb{C}^k) = \{v \in \mathbb{C}^k| \langle v, iz \rangle = 0\} .$$

This also specifies the tangent space in the quotient set $T_{[z]}(\mathbb{C}^k/\mathbb{S}^1) = N_z(\mathbb{C}^k)$.

2. **Geodesic and Geodesic Distance**: According to Definition 3.17, the inherited distance between elements in the quotient space is given by:

$$d_{M/G}([p_1], [p_2]) = d_M(p_1, (g^*, p_2)) ,$$

where $g^* = \operatorname{argmin}_{g \in G} d_M(p_1, (g, p_2))$. Since this distance is a geodesic distance, one would like to know the corresponding geodesic path in M/G, connecting $[p_1]$ and $[p_2]$, whose length achieves this distance. This geodesic is realized as follows. Let $\psi : [0, 1] \to M$ be the geodesic between the points p_1 and $p_2^* = (g^*, p_2)$ in the original manifold M. Then, $[\psi(\tau)]$, indexed by $\tau \in [0, 1]$, forms the desired geodesic path between $[p_1]$ and $[p_2]$ in M/G. Here, $[\psi(\tau)]$ denotes the orbit of the point $\psi(\tau)$ in M under the action of G.

This geodesic, by definition, is perpendicular to each orbit it meets. In other words, the velocity vector $\frac{d\psi(\tau)}{d\tau}$ is in the normal space $N_{\psi(\tau)}(M)$. Specifically, the shooting vector for the geodesic $v = \frac{d\psi(\tau)}{d\tau}|_{\tau=0}$ is in $N_{p_1}(M)$ or equivalently in $\tau_{[p_1]}(M/G)$.

With this notation, one can define the exponential map (and its inverse) in the quotient space M/G as follows.

3. **Exponential Map**: According to Definition 3.6, the exponential map is obtained using a unit-speed geodesic in the given direction. Applying this idea in this situation, we can obtain the exponential map as follows. For a point $[p] \in M/G$, and a tangent vector $v \in \tau_{[p]}(M/G)$ (i.e., in $N_p(M)$), we need to construct a constant-speed geodesic in M/G whose shooting direction is v. As mentioned in the previous item, a geodesic in M/G is realized using a corresponding geodesic in M that is perpendicular to all the orbits it meets. Let us construct a constant-speed geodesic $\psi(\tau)$ from p in the direction of v (existence of ψ is shown in Theorem 3.1). Then, since $\dot{\psi}(0) \perp \tau_p([p])$, it can be shown that $\dot{\psi}(\tau) \perp \tau_{\psi(\tau)}([\psi(\tau)])$ for all τ, due to the isometry condition. Therefore, $[\psi(\tau)]$ is the desired constant-speed geodesic in M/G and the exponential map $\exp : \tau_{[p]}(M/G) \to M/G$ is given by: $\exp_{[p]}(v) = [\psi(1)]$.

4. **Inverse Exponential Map**: The inverse of an exponential map takes a point on the quotient space M/G and maps it to an element (or multiple elements) of the tangent space $T_{[p]}(M/G)$. A vector $v \in T_{[p]}(M/G)$ is said to be the inverse exponential of $[q] \in M/G$ at p if $\exp_{[p]}(v) = [q]$. It is denoted by $v = \exp_{[p]}^{-1}([q])$ and is often not a unique point. That is, the inverse may be set-valued.

Example 7.3. **Landmark Shapes**: As described in Sect. 2.2.3, the shapes formed by a set of planar landmarks are conveniently represented as elements of the quotient space $\mathcal{C}/U(1) = \{[\mathbf{z}]|\mathbf{z} \in \mathcal{C}\}$ and where $\mathcal{C} = \{\mathbf{z} \in \mathbb{C}^n|\frac{1}{k}\sum_{i=1}^{k} z_i = 0, \|\mathbf{z}\| = 1\}$. Recall that $U(1) = SO(2) = \mathbb{S}^1$. The two constraints on the elements of $\mathcal{C}$ denote the removal of the translation and the scaling group. Each orbit represents all possible rotations of a configuration, $[\mathbf{z}] = \{e^{j\phi}\mathbf{z}|\phi \in \mathbb{S}^1\} \subset \mathcal{C}$. One can define the Karcher mean shape of several configurations $\mathbf{z}_1, \mathbf{z}_2, \ldots, \mathbf{z}_n$ as the configuration that minimizes the sum of squares of distances:

$$\hat{\mu} = \underset{\mathbf{z}\in\mathcal{C}}{\operatorname{argmin}} \left(\sum_{i=1}^{k} d_s([\mathbf{z}], [\mathbf{z}_i])^2 \right)$$

$$= \underset{\mathbf{z}\in\mathcal{C}}{\operatorname{argmin}} \left(\sum_{i=1}^{n} \left(\min_{\phi_i}(\cos^{-1}(\langle \mathbf{z}, \mathbf{z}_i e^{j\phi_i} \rangle)) \right) \right) = \underset{\mathbf{z}\in\mathcal{C}}{\operatorname{argmin}} \left(\sum_{i=1}^{n} \cos^{-1}(|\langle \mathbf{z}, \mathbf{z}_i \rangle|) \right) .$$

Algorithm 29 (Karcher Mean of Landmark Shapes). *Initialize $\mu_0 \in \mathcal{C}$ and set $t = 0$.*

1. *For each $i = 1, \ldots, k$, compute θ_i and r_i where $\langle \mu_t, \mathbf{z}_i \rangle = r_i e^{j\theta_i}$. Set: $v_i = \frac{\theta_i}{\sin(\theta_i)}(\mathbf{z}_i - \mu\cos(\theta_i))$.*
2. *Compute the average direction $\bar{v} = \frac{1}{k}\sum_{i=1}^{k} v_i$.*
3. *If $\|\bar{v}\|$ is small, then stop. Else, update μ_t in the update direction using*

$$\mu_{t+1} = \cos(\epsilon\|\bar{v}\|)\mu_t + \sin(\epsilon\|\bar{v}\|)\frac{\bar{v}}{\|\bar{v}\|} ,$$

where $\epsilon > 0$ is small step size, typically 0.5.

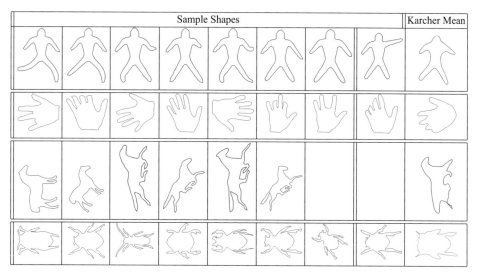

Sample Shapes									Karcher Mean

Fig. 7.17 Karcher mean shapes for given sets under Kendall's shape analysis

4. Set $t = t + 1$ and return to Step 1.

Figure 7.17 shows a number of examples of Karcher means of several sets of shapes, obtained using this algorithm. It is observed that several shape features present in the original curves have been blurred to some extent in the mean shapes. Take, for example, the fingers in the hand shape shown in the second row; fingers in the mean shape are not as explicit as the fingers in the original sample. This is due to a lack of alignment of features along different curves and the smoothing of features when averaged together with straighter parts of other curves. The issue of nonalignment of similar geometric features was explained previously in Sect. 2.2.4 with a simple example. Here we see some consequences of this nonalignment.

7.7 Exercises

7.7.1 Theoretical Exercises

1. Let $x_i \sim N(\mu, \sigma^2)$, for $i = 1, 2, \ldots, k$, be independent samples. Find the maximum likelihood estimate of μ given x_1, x_2, ..., x_k using Eq. 7.4.
2. Show that if $\hat{f}$ is a Kernel density estimator (Eqn. 7.5), then $\int \hat{f}(x)dx = 1$.
3. Let $M = \mathbb{S}^2$ with the standard Euclidean metric and viewed as a subset of $\mathbb{R}^3$ with the standard embedding. Let $q = (0, 0, 1)$ denote the north pole and $p = (0, 1, 0)$ represent a point on the equator. Evaluate the gradient of $d(p, q)^2$, the geodesic distance squared, with respect to p on M, expressed as an element of $T_p(\mathbb{S}^2)$.
4. Verify that the projection of a point v in $\mathbb{R}^2$ to the circle $\mathbb{S}^1$ is, according to Eq. 7.17, $\tan^{-1}(\frac{v_2}{v_1})$, as long as $\|v\| \neq 0$.
5. Show that the determinant of the Jacobian of the exponential map from $T_\mu(\mathbb{S}^2)$ to $\mathbb{S}^2$ is given by $\frac{\sin(\theta)}{\theta}$.

6. Let $\mathbb{S}_\infty = \{q : [0,1] \to \mathbb{R} | \int_0^1 |q(t)|^2 dt = 1\}$. For any $q \in \mathbb{S}_\infty$, let $\{b_i\}$ form an orthonormal basis of $T_q(\mathbb{S}_\infty)$. Show that the set $\exp_q(D)$, where $D = \{\sum_{i=1}^{k+1} \alpha_i b_i | \sum_{i=1}^{k} \alpha_i^2 \leq \pi/2\}$, can be identified with a subset of $\mathbb{S}_k$.

7. Check that if $\gamma \in \Gamma_I$, i.e., γ is a diffeomorphism from $[0,1]$ to itself with boundary points preserved, then $\dot{\gamma}$ is probability density function on $[0,1]$. Furthermore, show that $\sqrt{\dot{\gamma}}$ is an element of the unit Hilbert sphere $\mathbb{S}_\infty$.

7.7.2 Computational Exercises

1. Implement Algorithm 26 to generate samples from the uniform density on $\mathbb{S}^2$.
2. Implement Algorithm 27 to generate samples from a Fisher's density on $\mathbb{S}^2$.
3. Implement an algorithm for generating samples from a truncated normal density on $\mathbb{S}^2$ given in Eq. 7.21.
4. Write a program to estimate a probability density function f on $\mathbb{S}^2$ given its samples $\{x_i \in \mathbb{S}^2 | i = 1, 2, \ldots, n\}$ using the kernel method. Use a von-Mises kernel and study the results for different values of the kernel bandwidth h.
5. Write a program to compute the Karcher mean of a set of points $\{x_i \in \mathbb{S}^2 | i = 1, 2, \ldots, n\}$ using Algorithm 28. Then, using the inverse exponential map of the observed points into the tangent space at the Karcher mean, compute the sample Karcher covariance matrix.
6. Assume that we are given a set of probability density functions on $[0,1]$: $\{g_1, g_2, \ldots, g_k | g_k \in \mathcal{P}\}$. Each function is available numerically in form of T uniform samples on the interval $[0,1]$.

   a. Write a program to compute the Karcher mean of points in this set using the discussion in Sect. 7.5.3. Note that this computation is easier to perform after square-root mapping to the space $\mathbb{S}_\infty$.

   b. Use the inverse exponential map at the Karcher mean and compute the sample Karcher covariance matrix of this data.

   c. Using SVD of this covariance matrix, compute the first k_0 principal directions of variations. Display these principal directions using exponential maps in both positive and negative directions.

7. Implement an algorithm to generate a set of random warping functions as follows.

   a. For a large N, form smooth functions using a Fourier basis according to:

$$v(t) = \sum_{n=1}^{N} \left(a_i \cos(2\pi nt) + b_i \sin(2\pi nt) \right),$$

   where $a_i, b_i \sim N(0, \sigma^2/n)$.

   b. Project v into $T_{\psi_{id}}(\mathbb{S}_\infty)$ using $v \mapsto v - \int_0^1 v(t)dt$.

   c. Map the resulting ψ onto the sphere using $v \mapsto \psi = \exp_{\psi_{id}}(v)$.

   d. Integrate according to $\gamma(t) = \int_0^t \psi(s)^2 ds$ to obtain a random warping function.

8. Implement an algorithm compute the Karcher mean and sample Karcher variance of a set of random warping functions.

7.8 Bibliographic Notes

The notion of intrinsic Frèchet or Karcher mean has been covered by many authors, including [46, 88, 17]. Specifically, [88] outlined the concepts of intrinsic and extrinsic means on nonlinear Riemannian manifolds. Grenander et al. [35] used the Hilbert-Schmidt norm to develop extrinsic estimators on some matrix Lie groups. Similar ideas are used for some quotient spaces of matrix Lie groups in [103]. The special case of a circle was discussed in [48].

The classic density estimation for Euclidean domains is well covered in [101], while that for nonlinear domains is studied in [89]. Ghosh [31] provides a nice treatment of Bayesian nonparametric techniques.

Statistical methods for unit vectors in three-dimensional space have been studied extensively in the field of directional statistics [75]. In the landmark-based shape analysis of objects [28, 44, 56], where 2-D objects are represented by configurations of salient points or landmarks, the set of all such configurations, after removing translation and scale is a real sphere $\mathbb{S}^{2n-3}$ (for configurations with n landmarks).

Chapter 8
Statistical Modeling of Functional Data

In this chapter we will look at the problem of developing statistical models that capture the essential modes of variability in a given functional dataset. To keep the discussion simple, we will assume that all the functions are defined on a fixed domain, say $[0, 1]$. A piece of the puzzle, dealing with pairwise alignment of functions, was introduced earlier in Chap. 4. Now we face a bigger task—to develop generative models for function variables where model parameters can be estimated from past (training) data. We shall break this task down into several smaller pieces, seek convenient mathematical representations of functions, choose task-appropriate metrics, and develop statistical models that fit naturally to these choices. One important piece in this process is to align the given set of functions using nonlinear time warping. This step, also known as *phase-amplitude separation*, helps decompose the more complex variability of general functional data into two separate sets of relatively simpler variabilities—phase variability and amplitude variability. Another important challenge is to deal with the infinite-dimensional nature of functional data. The two components in functional data, phase and amplitude, are still, despite separation and simplification, elements of infinite-dimensional spaces. One needs a dimension-reduction tool to approximate and represent them as elements of finite-dimensional spaces. Finally, once they are projected onto a finite-dimensional space, one needs probability models, either parametric or nonparametric, to model the projected values.

The first task—alignment of multiple functions or group-wise registration of functional data—is an extension of the problem studied in Sect. 4.4. In that section, we developed a principled, metric-based framework for pairwise registration of functions; this metric was as an extension of the *nonparametric* Fisher-Rao Riemannian metric, originally defined for functions with positive derivatives, to a broader class of real-valued functions. While the original form of the metric is complicated, a remarkable computational efficiency results from the use of a novel representation, termed *square-root slope function* (SRSF), so that the objective function becomes the $\mathbb{L}^2$ distance between SRSFs of the functions being registered. This objective function has a special property that it is invariant to identical warping of functions and, consequently, the registration solution is inverse symmetric. Now, we consider the group-wise registration problem where we try to register peaks and valleys of several functions at the same time. Furthermore, we develop techniques for discovering principal modes of variations in functional data

© Springer-Verlag New York 2016

A. Srivastava, E.P. Klassen, *Functional and Shape Data Analysis*,
Springer Series in Statistics, DOI 10.1007/978-1-4939-4020-2_8

269

and to use these modes in imposing tractable statistical models on function spaces. These models can be used in drawing statistical inference, pattern recognition, and general data analysis using functional data.

8.1 Goals and Challenges

We start with a listing of goals and challenges. Some goals listed here are simple extensions of those studied in Chap. 4, but there are some additional, more ambitious goals.

1. **Alignment of Multiple Functions** or **Phase-Amplitude Separation**: We want a formal framework and an efficient algorithm for aligning peaks and valleys of *several* real-valued functions, defined on a common domain, using nonlinear time warpings. More precisely, given a set of functions $\{f_i \in \mathscr{F} | i = 1, 2, \ldots, n\}$, where $\mathscr{F}$ is the set of absolutely continuous functions on $[0, 1]$, our goal is to find a set of warping functions $\{\gamma_i \in \Gamma_I\}$ such that, for any $t \in [0, 1]$, the values $\{f_i(\gamma_i(t))\}$ are said to be registered or aligned. The challenge, of course, is in defining a mathematical framework, equipped with a proper objective function, that formalizes this problem. This problem is also called the *phase-amplitude separation*, where $\{\gamma_i\}$ are called the *phases* and $\{f_i \circ \gamma_i\}$ are treated as representatives of the equivalence classes representing their *amplitudes*.

2. **Joint Modeling of Phase-Amplitude Components**: In situations where one is interested in developing statistical models for capturing variability in given functional data, the phase-amplitude separation provides a novel perspective to the problem. Instead of modeling the original functions, one can model the two components individually, but not necessarily independently. As mentioned above, there are two general challenges in modeling the phase and amplitude components. The first is *infinite dimensionality* and is handled using a truncated basis and, thus, a *finite* number of coefficients to represent a function. These basis elements can come from a predetermined family or from an empirical analysis such as FPCA. The second challenge, more difficult to handle, comes from the *nonlinearity* of the phase components. As mentioned in Chap. 4 Sect. 4.10.2, the geometry of Γ_I is nonlinear and the direct use of FPCA is not possible. We will use a special geometric structure, obtained by transforming the warping functions, to help overcome that challenge.

3. **Joint FPCA and Registration of Functional Data**: Another task, critical in exploring and modeling functional data, is to account for the phase variability of given functions while, at the same time, discovering the principal modes in the amplitude data. This problem is slightly different from the last one, where the phase-amplitude separation was performed first and then followed by FPCA of each component. Now, we want to solve for the principal modes while *simultaneously* registering the data. The challenge is to formulate a joint optimization problem that can perform both these tasks in one solution.

In the following sections, we develop solutions to achieve these three goals.

8.2 Template-Based Alignment and $\mathbb{L}^2$ Metric

We start with the problem of nonlinear alignment of multiple functions. A natural extension of the pairwise solution to the group-wise alignment is a *template-based* registration. Here one designs a template function in a certain way and aligns each of the given functions to this template using the pairwise solution. Furthermore, these two steps can be performed iteratively to improve the quality of the template and to improve overall alignment. Thus, each iteration of multiple alignment involves:

1. **Averaging to form template**: Take the current versions of the functions $\{f_i | i = 1, 2, \ldots, n\}$ and average them under a proper metric to form the template μ.
2. **Pairwise alignment to the template**: For each $i = 1, 2, \ldots, n$, align the given function f_i to the template μ in a pairwise fashion. Call the optimal warping function γ_i, and update $f_i \leftarrow f_i \circ \gamma_i$.

The first item requires a distance on the function space $\mathscr{F}$ in order to define and compute the mean function μ. The second item requires an objective function for pairwise registration, a topic that was discussed in detail in Sect. 4.4. Furthermore, these two items cannot be solved in isolation. It is important for the metric, used for registration in the second step, to be compatible with the one used for defining mean in the first step. We will use the extended Fisher-Rao distance, discussed in Sect. 4.10, for both these items.

Before we introduce our approach, we present and evaluate a seemingly natural idea involving the $\mathbb{L}^2$ metric and demonstrate its limitations. The two main shortcomings of $\mathbb{L}^2$ norm for pairwise registration, namely, the *pinching effect* and the *lack of inverse symmetry*, are already described in Sect. 4.4. We want to emphasize that the pinching effect is still present in the multiple alignment problem, even though the need of symmetry is not there due to the presence of multiple functions. One can use a penalty term to avoid the pinching effect, but the results are not satisfactory, as shown below. So, let's use the $\mathbb{L}^2$ metric in the iterative framework suggested above and study the result. The iteration consists of the following two steps:

Algorithm 30 (Averaging Under Penalized $\mathbb{L}^2$ Norm).

1. **Form a template**: *Take the current versions of the aligned functions $\{f_i | i = 1, 2, \ldots, n\}$ and average them according to $\mu(t) = \frac{1}{n} \sum_{i=1}^{n} f_i(t)$.*
2. **Align to the template**: *For each $i = 1, 2, \ldots, n$, solve the problem:*

$$\gamma_i = \arg\inf_{\gamma} \left(\|\mu - f_i \circ \gamma\|^2 + \lambda \mathscr{R}(\gamma) \right),$$

and update $f_i \leftarrow f_i \circ \gamma_i$. Here, $\mathscr{R}$ represents a chosen roughness penalty on the warping function γ and $\lambda > 0$ is a scalar that controls the influence of $\mathscr{R}$ on the solution.
3. *Test for a stopping criterion. Return to Step 1, if convergence is not achieved.*

This process, of course, requires making a sensible choice for the parameter λ. In case $\lambda = 0$, these two steps can be merged together in the form of the optimization problem:

$$\min_{\mu \in \mathbb{L}^2} \left(\sum_{i=1}^{n} \inf_{\gamma_i} \|\mu - f_i \circ \gamma_i\|^2 \right). \tag{8.1}$$

In Algorithm 30, the first step solves the outer optimization (over μ) while fixing the warping functions $\{\gamma_i\}$, and the second step solves the inner optimization (over γ_i) while fixing the mean μ. To evaluate the results of these choices, we take a case study. For comparisons, we also present results for the case $\lambda > 0$ and where the roughness penalty is based on the first derivative of γ, i.e., $\mathcal{R}(\gamma) = \int_0^1 (\dot{\gamma}(t))^2 dt$.

Example 8.1. Figure 8.1 shows a set of bimodal functions $\{f_i\}$ that differ slightly in locations and heights of their peaks (see the top row). These functions were generated as follows; the individual functions are given by: $y_i(t) = z_{i,1} e^{-(t-1.5)^2/2} + z_{i,2} e^{-(t+1.5)^2/2}$, $i = 1, 2, \ldots, 21$, where $z_{i,1}$ and $z_{i,2}$ are *i.i.d* normal with mean one

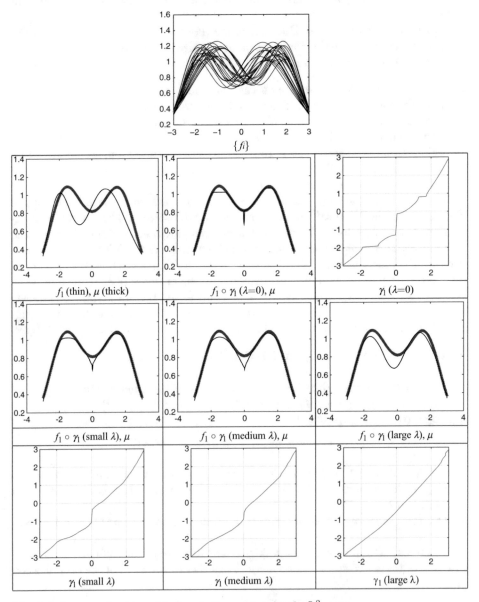

Fig. 8.1 An example of template-based alignment under the $\mathbb{L}^2$ norm

and standard deviation 0.25. Each of these functions is then warped according to: $\gamma_i(t) = 6(\frac{e^{a_i(t+3)/6}-1}{e^{a_i}-1}) - 3$ if $a_i \neq 0$; otherwise $\gamma_i = \gamma_{id}$, where a_i are equally spaced between -1 and 1, and the functions of interest are computed using $f_i(t) = y_i(\gamma_i(t))$. A set of 21 such functions forms the original data and is shown in the top panel of this figure. We apply Algorithm 30 on these functions and initialize the template with their cross-sectional mean μ, i.e., sample mean under the $\mathbb{L}^2$ metric. The mean function (drawn with a thick line) is also a similar bimodal function, as can be seen in this panel. The next two rows display results from pairwise alignment of one of the functions—say f_1—to this mean μ for different values of λ. The second row shows the alignment of f_1 to the mean μ when $\lambda = 0$, where pinching and flattening of parts of f_1 are clearly visible. This can also be seen in the warping function plotted in the rightmost panel. As we increase λ, the pinching effect is reduced but so is the overall warping, thus reducing the alignment level itself. This is shown in the third row. The bottom row shows the corresponding warping functions getting smoother but also getting closer to γ_{id}, as λ increases. These results highlight the main problem in using the penalized $\mathbb{L}^2$-distance for function alignment: the choice of λ. One can avoid the pinching effect using an external penalty term, but how much weight this penalty should carry is not obvious. There is no simple way to select a λ automatically, and the results can be very sensitive to this choice.

8.3 Elastic Phase-Amplitude Separation

As an alternative to the $\mathbb{L}^2$ metric, let us try the amplitude distance d_a, as given in Eq. 4.19, for multiple alignment. We shall use it to establish and compute the Karcher mean of functions, which , in turn, will serve as template for multiple alignment. Since this distance is actually defined on the quotient space $\mathscr{A} \equiv \mathscr{F}/\tilde{\Gamma}_I$, the Karcher mean will be an amplitude or an orbit rather than an individual function. Therefore, we will need to select a special element of this mean orbit, termed the *center of the orbit* (see Sect. 3.10), as the template μ. More specifically, we will define a template $\mu \in \mathbb{L}^2$ whose amplitude and relative phase denote the respective sample means of amplitudes and relative phases of the given functions.

8.3.1 Karcher Mean of Amplitudes

Continue representing each absolutely continuous function f_i by its SRSF $q_i(t) = \text{sign}(\dot{f}_i(t))\sqrt{|\dot{f}_i(t)|}$. The amplitude of f in the SRSF space is given by the orbit:

$$[q] = \text{closure}\{(q, \gamma) = (q \circ \gamma)\sqrt{\dot{\gamma}}|\gamma \in \Gamma_I\} \ .$$

This allows for defining an energy function for pairwise registration, between functions f_1 and f_2 (Definition 4.7)—$\inf_{\gamma \in \Gamma_I} \|q_1 - (q_2, \gamma)\|$. If a $\gamma^* \in \Gamma_I$ is (approximately) a minimizer of this energy term, then $f_1(t)$ is said to be optimally aligned to $f_2(\gamma^*(t))$. Furthermore, as described in Sect. 4.10, this quantity also approximates a proper distance between amplitudes (or orbits) $[q_1]$ and $[q_2]$ considered as elements of the quotient space $\mathscr{A} = \mathscr{F}/\tilde{\Gamma}_I$. With this distance, we have enough tools to define the Karcher mean of the amplitudes of a set of functions in $\mathscr{F}$.

Definition 8.1. Define the Karcher mean $[\mu_q]$ of the amplitudes $\{[q_i]\}$ of given functions $f_1, f_2, \ldots, f_n$, with SRSFs $q_1, q_2, \ldots, q_n$, as a local minimum of the sum of squares of distances:

$$[\mu_q] = \operatorname*{arginf}_{[q] \in \mathscr{A}} \sum_{i=1}^{n} d_a([q], [q_i])^2 \;, \tag{8.2}$$

where the definition of d_a is as given in Eq. 4.19.

We reiterate that $[\mu_q]$ is an orbit and not just a single element of $\mathbb{L}^2$. To form a template, we will select a specific element of this orbit, as described later. The full algorithm for computing the Karcher mean of amplitudes is as follows.

Algorithm 31 (Karcher Mean in $\mathscr{A}$).

1. *Initialization Step: Select $\mu_q = q_j$, where j is any index in $\operatorname{argmin}_{1 \leq i \leq n} \|q_i - \frac{1}{n} \sum_{k=1}^{n} q_k\|$.*
2. *For each q_i find γ_i by solving: $\gamma_i = \operatorname{arginf}_{\gamma \in \Gamma_I} \|\mu_q - (q_i \circ \gamma)\sqrt{\dot{\gamma}}\|$. This is the pairwise alignment studied in Sect. 4.6 where we used DPA to solve the optimization problem.*
3. *Compute the aligned SRSFs using $\tilde{q}_i \mapsto (q_i \circ \gamma_i)\sqrt{\dot{\gamma}_i}$.*
4. *If the increment $\|\frac{1}{n} \sum_{i=1}^{n} \tilde{q}_i - \mu_q\|$ is small, then stop. Else, update the mean using $\mu_q \mapsto \frac{1}{n} \sum_{i=1}^{n} \tilde{q}_i$ and return to Step 2.*

The iterative update in Steps 2–4 is based on the gradient of the cost function given in Eq. 8.2. (Due to the local nature of the iterative optimization, the convergence of this algorithm to a global solution is not guaranteed.) The mean amplitude is then given by the orbit $[\mu_q]$. Any element of this orbit, say μ_q, can be converted into the corresponding element of $\mathscr{F}$ by integration: $\mu_f(t) = \int_0^t \mu_q(s)|\mu_q(s)|ds$.

An example of this computation is shown in Fig. 8.2 where we start with $f_i(t) = \sin(2\pi t^{a_i})$, $a_i \sim \mathcal{N}(1, 0.09)$, $i = 1, 2, \ldots, 10$, as shown in the top-left panel. Then, we use Algorithm 31 to compute the mean amplitude of these given functions. The algorithm generates optimal warping functions $\{\gamma_i\}$ (second panel), the aligned functions $\{f_i \circ \gamma_i\}$ (third panel), and the mean amplitude function μ_f (rightmost panel). What happens if we warp all the f_is by the same $\gamma \in \Gamma_I$? How will it affect the amplitude mean? We see that in the second row where we further warp each f_i by $\gamma(t) = t^{1.2}$ and run the algorithm again. As expected the warping functions do not change but the aligned functions and μ_f reflect this extra warping. This is a special example where all the functions involved have the same amplitude and the result belongs to the same amplitude class. However, since Algorithm 31 generates an element of this class, we see differences in results between top and bottom rows.

8.3.2 Template: Center of the Mean Orbit

Now that we have defined the mean amplitude $[\mu_f]$ of given functions, the next task is to find a particular element of this mean orbit. This function, an element

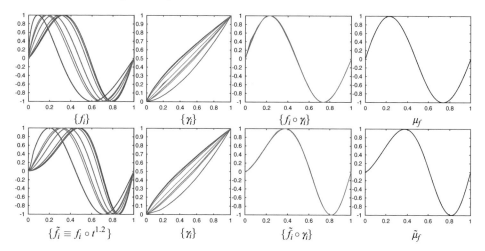

Fig. 8.2 Computation of mean amplitude. In each row we plot the given functions f_is, the optimal warping functions $\{\gamma_i\}$, the aligned functions $\{f_i \circ \gamma_i\}$, and the mean function μ_f. The input functions in the bottom row are obtained by applying an additional warping $t^{1.2}$ to the input functions in the top row

of $\mathscr{F}$, can then be used as a template to align the given functions. Toward this purpose, we will define the notion of a center of an orbit with respect to some given functions.

Definition 8.2. (Center of an Orbit) For a given set of SRSFs $q_1, q_2, \ldots, q_n$ and q, define an element $\tilde{q}$ of $[q]$ as the center of $[q]$ with respect to the set $\{q_i\}$ if the relative phases $\{\gamma_i\}$, where $\gamma_i = \mathrm{arginf}_{\gamma \in \Gamma} \|\tilde{q} - (q_i, \gamma)\|$, have their sample Karcher mean as identity, $\gamma_{id}(t) = t$.

(See Sect. 3.10 for a general definition.) In other words, the relative phases of all q_i's with respect to the template function $\tilde{q}$ should average γ_{id} under the chosen metric. Assume that q_i's are such that the action of Γ_I on each q_i is free.

To make sense of this definition, we will of course need a metric on the set of warping functions. A simple idea is to obtain the desired mean under the $\mathbb{L}^2$ norm, leading to the cross-sectional mean $\frac{1}{n} \sum_{i=1}^{n} \gamma_i(t)$. It is interesting to note that this mean is a valid warping function, i.e., it is an element of the set Γ_I, despite the fact that Γ_I is not a vector space. While this definition may be sufficient for the current purposes, we also bring up another possibility. Recall that the concept of sample Karcher mean of warping functions, under the Fisher-Rao metric, has already been presented in Sect. 7.5.4. We refer the reader to that section for the definition and an algorithm for computing this quantity. Either one of these two definitions, the $\mathbb{L}^2$-mean or the Fisher-Rao mean, will work, as long as we are consistent in our choice. Let $\bar{\gamma}$ denote the chosen mean of $\{\gamma_1, \gamma_2, \ldots, \gamma_n\}$. Now, we can use this mean to find the center of an orbit.

The algorithm for computing the center of the orbit $[q]$ is rather simple. It is depicted pictorially in Fig. 8.3. Take any element, say q, of the orbit $[q]$, compute the warping functions $\{\gamma_i\}$ that align $\{q_i\}$ to this q, compute their sample mean $\bar{\gamma}$, and apply the inverse of $\bar{\gamma}$ to q.

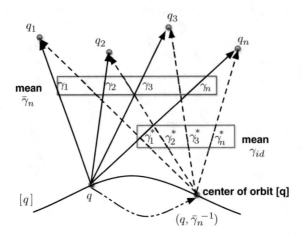

Fig. 8.3 Finding center of the orbit $[q]$ with respect to the set $\{q_i\}$

Algorithm 32 (Finding Center of an Orbit). *Let q be any element of the orbit $[\mu_q]$.*

1. *For each q_i find γ_i by solving: $\gamma_i = \mathrm{arginf}_{\gamma \in \Gamma_I} \left(\|\mu_q - (q_i \circ \gamma)\sqrt{\dot{\gamma}}\| \right)$.*
2. *Compute the mean $\bar{\gamma}$ of all $\{\gamma_i\}$ as described in Sect. 7.5.4. The center of $[\mu_q]$ wrt $\{q_i\}$ is given by $\tilde{q} = (\mu_q, \bar{\gamma}^{-1})$.*

We can check that $\tilde{q} \in [\mu_q]$ resulting from Algorithm 32 satisfies the mean condition in Definition 8.2. We know that γ_i is chosen to minimize $\|\mu_q - (q_i, \gamma)\|$, and also that

$$\|\tilde{q} - (q_i, \gamma)\| = \|(\mu_q, \bar{\gamma}^{-1}) - (q_i, \gamma)\| = \|\mu_q - (q_i, \gamma \circ \bar{\gamma})\| .$$

Therefore, $\gamma_i^* = \gamma_i \circ \bar{\gamma}^{-1}$ minimizes $\|\tilde{q} - (q_i, \gamma)\|$. That is, γ_i^* is a warping that aligns q_i to $\tilde{q}$. To verify the Karcher mean of γ_i^* (under the Fisher-Rao metric), we compute the sum of squared distances $\sum_{i=1}^n d_{FR}(\gamma, \gamma_i^*)^2 = \sum_{i=1}^n d_{FR}(\gamma, \gamma_i \circ \bar{\gamma}^{-1})^2 = \sum_{i=1}^n d_{FR}(\gamma \circ \bar{\gamma}, \gamma_i)^2$. As $\bar{\gamma}$ is already the mean of γ_i, this sum of squares is minimized when $\gamma = \gamma_{id}$. That is, the mean of γ_i^* is γ_{id}.

Some examples of this idea are shown in Fig. 8.4. In the top-left panel, we plot a set of warping functions $\{\gamma_i\}$, and their Karcher mean $\bar{\gamma}$, the middle panel shows the mean $\bar{\gamma}$ and its inverse $\bar{\gamma}^{-1}$, and the right panel shows the centered relative phases $\{\gamma_i \circ \bar{\gamma}^{-1}\}$. The bottom row shows another example along the same lines. This time the relative phases are further away from identity and the effect of centering is, therefore, more pronounced. Once again, the idea of finding the center of an orbit is essentially to warp all the given functions so that their relative phases, with respect to the chosen element of the mean amplitude, are also centered.

8.3.3 Phase-Amplitude Separation Algorithm

Now we can combine these pieces together to form an algorithm for separation of phase and amplitude components in a given set of functional data.

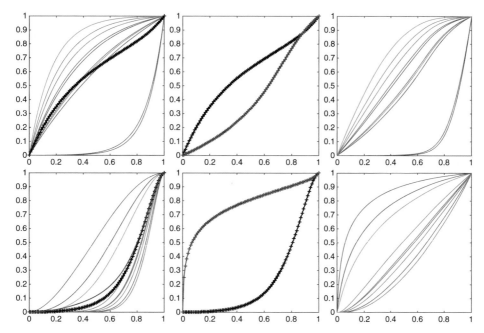

Fig. 8.4 In each row: (1) the left panel shows a number of warping functions $\{\gamma_i\}$ and their Karcher mean $\bar\gamma$ under the elastic metric, (2) the middle panel shows the mean $\bar\gamma$ and its inverse $\bar\gamma^{-1}$, and (3) the right panel shows the centered set $\{\gamma_i \circ \bar\gamma^{-1}\}$

Algorithm 33 (Phase-Amplitude Separation). *Given a set of functions $f_1, f_2, \ldots f_n$ on $[0,1]$, let $q_1, q_2, \ldots, q_n$ denote their SRSFs, respectively.*

1. *Compute the Karcher mean of amplitudes $[q_1], [q_2], \ldots, [q_n]$ in $\mathscr{A} = \mathscr{F}/\bar\Gamma_I$ using Algorithm 31. Denote it by $[\mu_q]$.*
2. *Find the center of the orbit $[\mu_q]$ with respect to the set $\{q_i\}$ using Algorithm 32; call it $\tilde{q}$. (Note that this algorithm, in turn, requires a tool for computing the Karcher mean of warping functions).*
3. *For $i = 1, 2, \ldots, n$, find γ_i by solving: $\gamma_i = \operatorname{arginf}_{\gamma \in \Gamma_I} \|\tilde{q} - (q_i, \gamma)\|$. This minimization is approximated using the DPA, as described in Appendix B.*
4. *Compute the aligned SRSFs $\tilde{q}_i = (q_i, \gamma_i)$ and aligned functions $\tilde{f}_i = f_i \circ \gamma_i$.*
5. *Return the template $\tilde{q}$, the warping functions $\{\gamma_i^*\}$, and the aligned functions $\{\tilde{f}_i\}$.*

To illustrate this method, we use a number of simulated and real datasets. Although this SRSF-based framework is developed for functions on $[0,1]$, it can easily be adapted to an arbitrary interval using an affine transformation.

MSE-Based Quantification of Alignment To quantify the extent of separation of original functions into amplitude and phase components, we can use the mean squared error (MSE) of these components. The MSE of any set of functions $\{g_i\}$ is given by $\sum_{i=1}^{n} \|g_i - \bar{g}\|^2$, where $\bar{g}$ is the cross-sectional mean of these functions and $\|\cdot\|$ denotes the $\mathbb{L}^2$ norm as usual. For our quantification, we can choose to compute the MSE in the function space $\mathscr{F}$ or the SRSF space $\mathbb{L}^2$. Since these numbers are simply for illustration purposes and not used in any optimization,

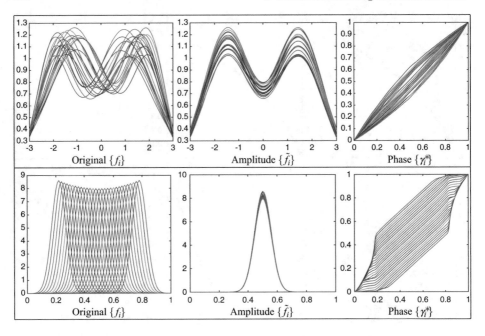

Fig. 8.5 Phase-amplitude separation of two simulated datasets, top, and bottom

we will go with MSE computation in $\mathscr{F}$, despite the known problems associated with using $\mathbb{L}^2$ norm in the function space. For quantifying the amplitude components, we will use the MSE of $\{\tilde{f}_i\}$, the aligned functions as elements of $\mathscr{F}$, and for quantifying the phase components, we will use the MSE of $\{\bar{\tilde{f}} \circ \gamma_i\}$, where $\bar{\tilde{f}}$ is the (cross-sectional) mean of the aligned functions $\{\tilde{f}_i\}$ and $\{\gamma_i\}$ are the warping functions.

Example 8.2. 1. As the first example, we study a set of simulated functions presented previously in Fig. 8.1. The top-left panel of Fig. 8.5 shows these functions again, and the remaining two panels in that row show the results of Algorithm 33. The middle panel presents the resulting aligned functions $\{\tilde{f}_i\}$, and the rightmost panel plots the corresponding warping functions $\{\gamma_i^*\}$. The plot of $\{\tilde{f}_i\}$ shows a tighter alignment of functions with sharper peaks and valleys. The two peaks are at ± 1.5, which is exactly what we expect. This means that the effects of warping generated by the γ_is have been completely removed and only the amplitude variability remains. To evaluate the extent of separation, we note that the MSE of the amplitude components is 0.41, the MSE of the phase components is 1.26, and the MSE of the original data is 1.66.

2. As the second example, we take the another simulated dataset. This time we study a family of Gaussian kernel functions with the same shape but with significant phase variability, in the form of horizontal shifts, and minor amplitude (height) variability. Figure 8.5 (bottom row) shows the original 29 functions $\{f_i\}$ (left), the aligned functions $\{\tilde{f}_i\}$ (middle), and the warping functions $\{\gamma_i^*\}$ (right). Once again we notice a tighter alignment of functions with only minor variability left in $\{\tilde{f}_i\}$ reflecting the differing heights in the original data. In this experiment, the MSE of the original data is 125.59, MSE of the amplitude components is 0.12, and the MSE of the phase components is 139.33.

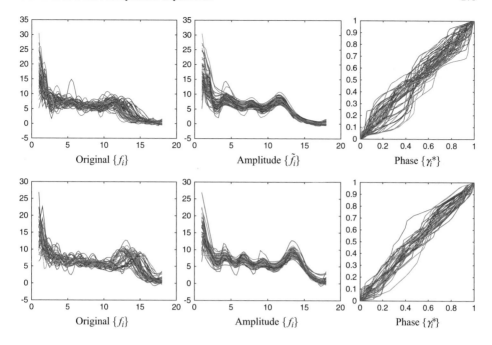

Fig. 8.6 Results of Algorithm 33 on biological growth data. *Top*: female data. *Bottom*: male data

3. **Berkeley Growth Data**: Next we consider the Berkeley growth dataset for 54 female and 39 male subjects. For better illustrations, we have used the first derivatives of the growth curves as the functions $\{f_i\}$ in our analysis. (In this case, since SRSF is based on the first derivative of f, we actually end up using the second derivatives of the growth functions.)

The results from Algorithm 33 on the female growth curves are shown in Fig. 8.6 (top row). The left panel shows the original data, which shows that while the growth spurts for different individuals occurs at slightly different times, there are some underlying patterns to be discovered. In the second panel, we see the aligned functions $\{\tilde{f}_i\}$ and they exhibit a much tighter alignment of the functions and, in turn, an enhancement of peaks and valleys in the aligned mean. In fact, this mean function suggests the presence of several growth spurts, one between 3 and 4 years, one between 5 and 7 years, and another between 10 and 12 years, on average. The MSE of the amplitude component is 2.09×10^3, the MSE of the phase component is 2.15×10^3, and the MSE of the original data is 2.22×10^3.

A similar analysis is performed on the male growth curves, and the bottom row shows the results: the original data (consisting of 39 derivatives of the original growth functions) and the aligned functions $\tilde{f}_i(t)$. The cross-sectional mean functions also show a much tighter alignment of the functions and, in turn, an enhancement of peaks and valleys in the aligned mean. This alignment suggests the presence of several growth spurts, between 3 and 5, 6 and 7, and 13 and 14 years, on average. The MSE of the amplitude component is 0.99×10^3, the MSE of the phase component is 1.44×10^3, and the MSE of the original data is 1.70×10^3.

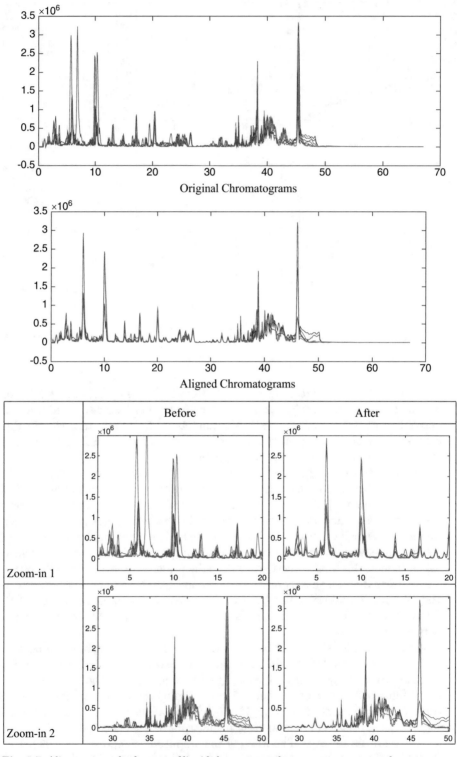

Fig. 8.7 Alignment results for a set of liquid chromatography-mass spectrometry chromatograms

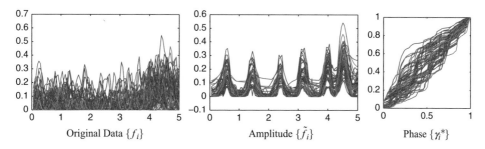

Fig. 8.8 Alignment of spike train data. The left panel shows the original functions, the middle panel shows the aligned functions of amplitudes, and the right panel displays their phase components

4. **Mass-Spectrometry Data**: Next, we study a set of eight chromatograms shown in the top panel on Fig. 8.7. In this data, some of the major peaks are not aligned. The outcome of Algorithm 33 on this data is shown in the second row where the peaks appear to be sharply aligned throughout the spectrum. To emphasize the quality of alignment, we look at a couple of smaller intervals in the spectrum more carefully. The zoom ins of these smaller regions are shown in the last two row of the figure. In each of the last two rows, we show the before and after alignment spectra for these two domains: 0–20 and 28–50. It can be seen in these zoom ins that the algorithm aligns both the major and minor peaks remarkably well. Note that this procedure does not require any prior peak detection or matching to reach this alignment.

5. **Neuronal Spike Train Data**: Statistical analysis of spike trains is one of the central problems in neural coding and can be pursued in several ways. While one option is to assume parametric or semi-parametric models, such as the Poisson model, for spike trains and use them in decoding spike trains, the other is a nonparametric option based on metrics for comparing the numbers and the placements of spikes in different trains. Here we demonstrate the use of phase-amplitude separation as a tool for aligning spike trains and estimating underlying neural signals. In order to enable this processing, one pre-processes discrete spike trains with Gaussian smoothing kernels. An example set of such smoothed functions is shown in the left panel of Fig. 8.8. The result of phase-amplitude separation algorithm on this dataset is shown in the next two panels.

8.4 Alternate Interpretation as Estimation of Model Parameters

The algorithm we have specified for separating phase and amplitude components of functional data can also be interpreted as estimation of parameters of a model. This model first transforms the given functions into their SRSFs and then assumes the SRSFs to be observations of a mean signal contaminated by white Gaussian observation noise. More precisely,

$$q_i = \text{SRSF}(f_i) ,$$
$$(q_i, \gamma_i) = \mu_q + \epsilon_i . \tag{8.3}$$

Under the same assumptions as in FPCA analysis of points $\mathbb{L}^2$ (see Eq. 4.7), we get the optimization function for estimating model parameters:

$$(\hat{\mu}_q, \{\hat{\gamma}_i\}) = \underset{\mu_q, \{\gamma_i\}}{\operatorname{argmin}} \left(\sum_{i=1}^{n} \|(q_i, \gamma_i) - \mu_q\|^2 \right).$$

Algorithm 31 is essentially a coordinate-descent procedure for minimizing this objective function. Additionally, if we impose the condition that the sample Karcher mean of $\{\gamma_i\}$ is identity, this results in Algorithm 33. The discussion on convergence properties of these estimators have been left out for the future.

8.5 Phase-Amplitude Separation After Transformation

We see that Algorithm 33 is quite successful in aligning functional data, but it can sometimes be more useful indirectly. For instance, instead of applying the algorithm to the given data, one can transform the given functions into a more convenient form and then apply the separation algorithm. We illustrate this idea using several examples.

1. **Absolute Value Transform**: This transform is useful in situations where the signs of peaks (and valleys) do not matter in the analysis. In other words, the positive peaks can be matched with negative peaks and zero crossings with zero crossings. Consider the set of functions shown in the top-left panel of Fig. 8.9. The top row shows the original data, its alignment using Algorithm 33 and the corresponding phase components $\{\gamma_i\}$. One can see that the warped functions are well aligned, peaks with peaks and valleys with valleys. However, if we perform FPCA of this aligned data, we will find that there are more than one significant singular values associated with the covariance operator of this data.

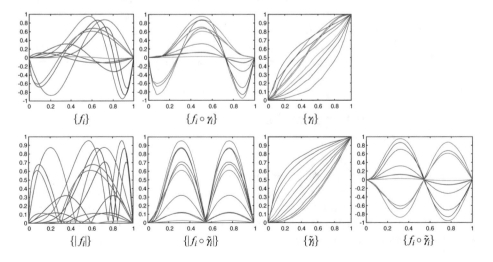

Fig. 8.9 *Top row*: phase separation of the original functional data using Algorithm 33. *Bottom row*: results of Algorithm 33 after taking absolute values of the functions

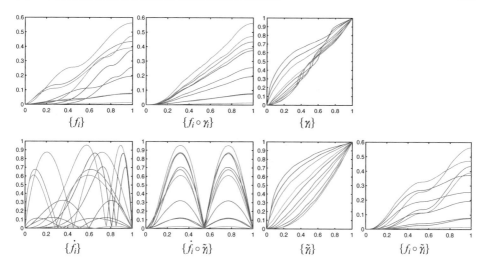

$\{f_i\}$ $\{f_i \circ \gamma_i\}$ $\{\gamma_i\}$

$\{\dot{f_i}\}$ $\{\dot{f_i} \circ \tilde{\gamma_i}\}$ $\{\tilde{\gamma_i}\}$ $\{f_i \circ \tilde{\gamma_i}\}$

Fig. 8.10 *Top row*: phase separation of the original functional data using Algorithm 33. *Bottom row*: results of Algorithm 33 after taking derivatives of the functions

In other words, these functions do not lie in a one-dimensional subspace of $\mathscr{F}$. This is not a problem in itself but a different result can be achieved if we first transform the given functions using the absolute value. This time when we apply Algorithm 33 to the absolute transformed data, we obtain results shown in the second row of this figure. Furthermore, if we take the resulting warping functions and apply them to the original functions, we get warped functions that are shown in the rightmost panel of this row. Imagine that the goal of this exercise was to form an efficient PCA basis of the given data, rather than just matching peaks to peaks and valleys to valleys. One can see that the aligned functions obtained using the absolute transformation provide a more succinct data for extracting an FPCA basis. In fact, a single principal component is now sufficient to describe the given functions!

2. **Derivative Transform**: The need for this transform arises when dealing with monotonically increasing or decreasing functions. Examples include survival functions in biostatistics and height functions in human growth analysis, The top row in Fig. 8.10 shows an example of such data and its alignment using Algorithm 33 directly on that data. One can see that although these functions are aligned, some of the features (bumps) have been smoothed out in this alignment. The algorithm seems to have straightened the functions into lines and thus the alignment is good. (Actually, one can show that for monotonic functional data, the aligned functions are scalar multiples of each other.) If, instead of aligning original functions, we take first derivatives $\{\dot{f_i}\}$ and then extract phase components $\{\tilde{\gamma_i}\}$ using Algorithm 33, the results look much better. When we visualize $\{f_i \circ \tilde{\gamma_i}\}$, we find that this approach is better in preserving data features while performing alignment at the same time.

The other potential transformations of data include log, higher-order derivatives, square root, and so on, depending on the needs of the problem.

8.6 Penalized Function Alignment

In some situations, it may be desirable to control the amount of warping, or the roughness level of the phase components. This is sometimes necessitated by the need to improve immunity against noise in the data or sometimes simply due to some contextual knowledge. One way to do this, of course, is to smooth the phase components after they have been separated. Another approach is to add a roughness penalty during the separation process itself. This penalization is analogous to Eq. 4.16 in Chap. 4 where one adds a roughness penalty on γ, controlled by a weight parameter λ, during optimization on γ_I.

Define a new optimization problem according to

$$\inf_{\gamma \in \Gamma_I} E_\lambda(q_1, (q_2, \gamma)), \quad \text{where} \quad E_\lambda(q_1, (q_2, \gamma)) = \left(\|q_1 - (q_2, \gamma)\|^2 + \lambda \mathcal{R}(\gamma) \right)^{1/2} ,$$

(8.4)

where $\mathcal{R}(\gamma)$ is a quantity that measures the roughness of a given warping γ and $\lambda > 0$ is a constant. As earlier, q_1 and q_2 are the SRSFs of the given functions to be aligned. The choices for $\mathcal{R}$ include functionals of first- or higher-order derivatives of γ and distances between γ and γ_{id} in the set Γ_I:

1. A term that includes the first derivative of γ is: $\mathcal{R}(\gamma) = \int_0^1 (\dot{\gamma}(t))^2 dt$. This penalty favors constant γ functions. Since γ is constrained to have $\int \dot{\gamma}(t) dt = 1$, this is minimized when $\gamma = \gamma_{id}$.
2. Similarly, the second-order roughness penalty involves the second derivative of γ according to: $\mathcal{R}(\gamma) = \int_0^1 (\ddot{\gamma}(t))^2 dt$. This penalty favors smooth functions and is minimized when γ is γ_{id}.
3. There are several choices of distances, in different associated spaces, that one can used as penalty functions:

    a. The squared $\mathbb{L}^2$ norm between γ and γ_{id}: $\mathcal{R}(\gamma) = \|\gamma - \gamma_{id}\|^2$.
    b. The squared $\mathbb{L}^2$ norm between their derivatives: $\mathcal{R}(\gamma) = \|\dot{\gamma} - \mathbf{1}\|^2$, where $\mathbf{1}$ is the constant function with value one. This is same as the first item above, up to a constant.
    c. The squared $\mathbb{L}^2$ norm between their SRSFs: $\mathcal{R}(\gamma) = \|\sqrt{\dot{\gamma}} - \mathbf{1}\|^2$, where $\mathbf{1}$ is the constant function with value one.
    d. The arc length between their SRSFs on the unit sphere: $\mathcal{R}(\gamma) = \cos^{-1}(\langle \sqrt{\dot{\gamma}}, \mathbf{1} \rangle)$. As described in Sect. 4.10.2, this is exactly the Fisher-Rao distance between γ and γ_{id}.

In the following, we will look at some experimental results based on the first-order penalty $\mathcal{R}(\gamma) = \|1 - \sqrt{\dot{\gamma}}\|^2$. The resulting E_λ serves as a criterion for pairwise alignment of functions:

$$E_\lambda(q_1, (q_2, \gamma)) = \left(\|q_1 - (q_2, \gamma)\|^2 + \lambda \|1 - \sqrt{\dot{\gamma}}\|^2 \right)^{1/2} ,$$

The parameter λ can be used to control the level of smoothness of relative phase components. One can solve this pairwise registration problem using DPA. For multiple alignment, one first computes a template (e.g., as the cross-section mean of given functions $\{f_i, i = 1, 2, \ldots, n\}$) and then aligns the individual functions to this template. Since this procedure is defined for $\mathbb{L}^2$ directly, rather than its quotient space, there is no need to perform the centering step described earlier (although it might still be interesting to study its effects).

Algorithm 34 (Penalized Alignment of Functions).

1. *Initialization Step: Select* $\mu_q = q_j$, *where* j *is any index in* $\operatorname{argmin}_{1 \le i \le n} E_0(q_i, \frac{1}{n} \sum_{k=1}^{n} q_k)$.
2. *For each* q_i *find* γ_i *by solving:* $\gamma_i = \operatorname{arginf}_{\gamma \in \Gamma_I} E_\lambda(\mu_q, (q_i, \gamma))$. *This is the pairwise alignment solved using a modification of the DPA presented in Algorithm 3.*
3. *Compute the aligned SRSFs using* $\tilde{q}_i \mapsto (q_i \circ \gamma_i)\sqrt{\dot{\gamma}_i}$.
4. *If the increment* $\|\frac{1}{n} \sum_{i=1}^{n} \tilde{q}_i - \mu_q\|$ *is small, then stop. Else, update the mean using* $\mu_q \mapsto \frac{1}{n} \sum_{i=1}^{n} \tilde{q}_i$ *and return to Step 2.*

Shown in Fig. 8.11 is an illustration of this idea on a simulated dataset. In the top-left panel, we show the given functions, with the resulting alignment results in the next four panels under $\lambda = 0, 75, 150$, and 300, respectively. In the second row, we show the corresponding (column-wise) time-warping functions. The results show that the warping functions stay closer to γ_{id} as λ is increased, at the loss of alignment level relative to the case for $\lambda = 0$. To quantify the extent of separation of original functions into amplitude and phase components, we compute the mean squared error (MSE) of these components. As stated earlier, the MSE of any set of functions $\{g_i\}$ is given by $\sum_{i=1}^{n} \|g_i - \bar{g}\|^2$. Thus, for quantifying the amplitude

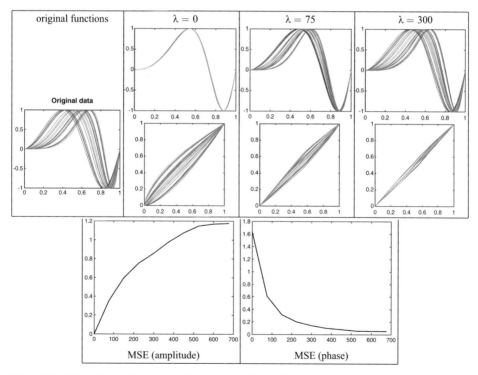

Fig. 8.11 Illustration of phase-amplitude separation with penalty term. *First row*: the original data (*left*) and the aligned functions with the four λ values. *Second row*: the corresponding time-warping functions. *Bottom row*: MSE of the amplitude data (aligned functions) and MSE of phase data ($\{\gamma_i\}$ applied to mean of aligned functions) versus λ

components, we compute MSE of $\{\tilde{f}_i\}$, the aligned functions as elements of $\mathscr{F}$, and for phase components, we compute MSE of $\{\bar{\tilde{f}} \circ \gamma_i\}$, where $\bar{\tilde{f}}$ is the mean of the aligned functions and $\{\gamma_i\}$ are the warping functions. The last row shows the evolution of these two MSEs—for amplitude and for phase—as we increase λ.

We point out that one loses the formal notion of amplitude and phase in this setup, especially the earlier identities involving d_a and d_p. One can still refer to the resulting warping functions as relative phases, but they will not satisfy the properties listed in Sect. 4.7 for the non-penalized ($\lambda = 0$) case.

8.7 Function Components, Alignment, and Modeling

An overarching goal in this chapter is to develop faithful statistical models for function variables. The essential steps in this process are to discover a suitable finite representation of functional data, e.g., using FPCA, and to impose models on that representation. The use of FPCA to extract principal modes of variation in functional data was described earlier in Sect. 4.3.1. To remind the reader of the setup, the basic idea is to find orthonormal functions $\phi_1, \phi_2, \ldots, \phi_J$, for some predetermined J, for a given dataset $\{f_1, f_2, \ldots, f_n\}$, with $f_i \in \mathbb{L}^2$, that minimize the squared residual error. The computation of FPCA basis solves the following optimization problem:

$$\{\hat{\phi}_1, \ldots, \hat{\phi}_J\} = \underset{\substack{\{\phi_j \in \mathbb{L}^2([0,1], \mathbb{R}) | j = 1, 2, \ldots, J\}, \\ \{c_{i,j} \in \mathbb{R} | i = 1, 2, \ldots, n; j = 1, 2, \ldots, J\}}}{\text{arginf}} \left(\sum_{i=1}^{n} \left\| f_i - \sum_{j=1}^{J} c_{i,j} \phi_j \right\|^2 \right),$$

$$(8.5)$$

and the actual minimization is based on SVD of the sample covariance operator estimated from the given data (see Sect. 4.3.1).

Now consider a situation where the functions f_is are observed under unknown time warpings or arbitrary phase variability. That is, instead of observing f_is directly, we observe $f_i \circ \gamma_i$ instead, with $\gamma_i \in \Gamma_I$, the group of boundary-preserving diffeomorphisms of $[0, 1]$. What happens if we perform FPCA of this dataset? Figure 8.12 shows an illustration of this idea using three simple examples. In the following, we use $\mathscr{N}(\mu, \sigma^2)$ to denote the normal pdf with mean μ and variance σ^2.

1. *Case 1—Row 1*: We construct the data using $f_i(t) = c_i \sin(2\pi t)$, with $c_i \sim \mathscr{N}(0, 5)$. The f_is are shown in the top-left panel of this figure. Note that these f_is happen to lie a one-dimensional space—span($\sin(2\pi t)$). Also, note that while the zero crossings of the derivatives coincide, the peaks and valleys are not well registered across functions here. The second panel in this row shows the top three principle basis functions (eigenfunctions) obtained using FPCA, and the third panel shows the reconstruction of f_1 using $\hat{f} = \langle f_1, \phi_1 \rangle \phi_1$, drawn over f_1. The top principle basis function happens to be $\phi_1(t) = -\sin(2\pi t)$ in this example, and the reconstruction $\hat{f}_1$ is very similar to f_1. The last panel plots the total residual error (minimum value of the object function in Eq. 8.5) versus J, and it confirms the one-dimensional nature of the data.

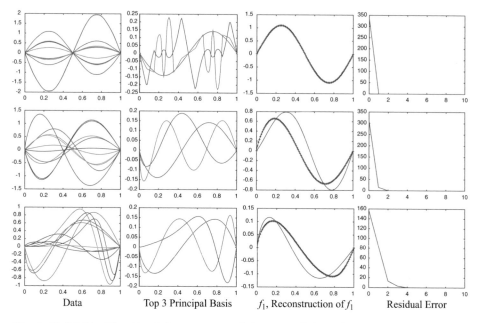

Fig. 8.12 FPCA of functional data constructed from a single basis function but with different levels of phase variability. In each row, the left panel shows original data, the second panel shows the three principle basis functions, the third panel shows f_1 and its reconstruction using the first principle direction, and the last panel shows the evolution of of the residual error plotted against the number of basis elements

2. *Case 2—Row 2*: Now we construct the data using $f_i(t) = c_i \sin(2\pi\gamma_i(t))$, with $c_i \sim \mathcal{N}(0,5)$, where $\gamma_i(t) = t^{a_i}$ and $a_i = 0.75 + b_i$, with $b_i \sim \exp(0.25)$. As a result, for the $\{f_i\}$, shown in the left panel of the second row, the zero crossings of derivatives are no longer aligned. They also show some variability in the locations of peaks and valleys. When we apply FPCA to this data, we obtain results shown in the remaining panels of the second row. As we can see in the third panel, a single basis function is no longer sufficient to reconstruct f_1 perfectly. The presence of phase variability has created the need for an additional basis element. Thus, one needs two basis functions to reach a perfect reconstruction, as shown by the residual error plot in the last panel.

3. *Case 3—Row 3*: Next, we first increase amount of warpings by using $\gamma_i(t) = t^{a_i}$ and $a_i = 0.4 + b_i$, with $b_i \sim \exp(0.6)$. The resulting f_is, shown in the bottom-left panel, display as increased phase variability. Consequently, it now takes four principal basis elements to bring the residual error closer to zero.

Another example, albeit more complex, of the FPCA is shown in Fig. 8.13. In this data, the function f_is are more complex than the previous example, and the extent of phase variability is not obvious just by observing the data. One can see that the peaks are not aligned but the matching of peaks, i.e., which peak in a function matches with which peak in another function, is also not as obvious as in the last example. In this data, it takes eight basis elements to bring the residual error down.

As this discussion illustrates, an increase in phase variability of functional data implies increases variance in the functional data, requiring a larger number of

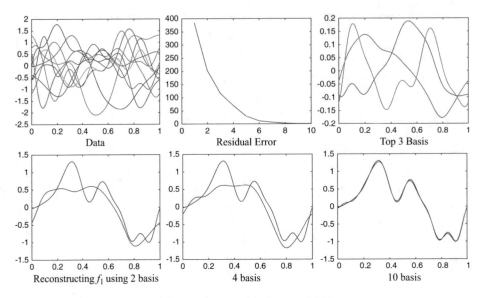

Fig. 8.13 FPCA of functional data with potential phase variability

principle components to model and represent the given functions. What is the alternative to a direct FPCA on functions in situations where they contain some phase variability? Here are some possibilities:

1. **Sequential Approach**: Separate the functional data into phase and amplitude components, perform individual FPCA on these components, and impose statistical models on the resulting coefficients.
2. **Joint Approach Using $\mathbb{L}^2$ Metric**: Perform phase-amplitude separation and FPCA jointly using an objective function involving $\mathbb{L}^2$ metric on the function space, and then impose statistical models on the coefficients.
3. **Joint Approach Using Elastic Metric**: Perform phase-amplitude separation and FPCA jointly using an objective function based on the elastic metric, and then impose statistical models on the coefficients.

We discuss these three ideas in the next few sections.

8.8 Sequential Approach

In this approach we use Algorithm 33 to first separate the phase and amplitude components and then develop generative models for the two components. Focusing on the latter task, one way to accomplish this task is to derive an FPCA basis of these functions individually and then impose a joint statistical model on principle coefficients. FPCA is simply viewed as a dimension-reduction step here, used in order to bring the representation to a finite dimension and to enable standard multivariate modeling.

8.8.1 FPCA of Amplitude Functions: A-FPCA

Once the given functions have been aligned using Algorithm 33, we can use the aligned SRSFs as representatives of their respective amplitudes and perform cross-sectional computations. Recall that cross-sectional analysis implies the use of the $\mathbb{L}^2$ metric in the SRSF space, which, in turn, denotes the use of the elastic metric in the original function space. One can compute their principal directions and use coefficients as low-dimensional representations. Since we are focused on the amplitude variability in this section, we will call this analysis A-FPCA.

Let $f_1, \cdots, f_n$ be a given set of functions, and $q_1, \cdots, q_n$ be the corresponding SRSFs. Also, let μ_q be the center of their mean orbit, and let $\tilde{q}_i$s be the corresponding aligned SRSFs obtained using Algorithm 33. In performing A-FPCA, one should not forget about the variability associated with the initial values, i.e., $\{f_i(0)\}$, of the given functions. Since representing functions by their SRSFs loses this initial value, this information is represented separately. That is, a functional variable f is analyzed using the pair $(q, f(0))$ rather than just q. In this way the mapping from the function space $\mathscr{F}$ to $\mathbb{L}^2 \times \mathbb{R}$ becomes a bijection. We will use the product metric on the space $\mathbb{L}^2 \times \mathbb{R}$ to analyze data in that space.

The underlying framework for computing principal directions of functional data has been laid out in Sect. 4.3.1. Thus, we proceed directly to computations using discrete time samples of the SRSFs. An aligned SRSF $\tilde{q}$ is represented using a finite partition of $[0, 1]$, say with cardinality T, so that $\mathbf{q}_j = \tilde{q}(j/T)$, $j = 1, 2, \ldots, T$. The combined vector $\mathbf{h}_i = [\mathbf{q}_i \quad f_i(0)]$ has dimension $(T + 1)$. We can define a sample covariance operator for the aligned combined vector as:

$$K_h = \frac{1}{n-1} \sum_{i=1}^{n} \left((\mathbf{h}_i - \mu_h)(\mathbf{h}_i - \mu_h)^T \right) \quad \in \mathbb{R}^{(T+1)\times(T+1)} , \qquad (8.6)$$

where $\mu_h = [\mu_q \quad \bar{f}(0)]$. Here $\bar{f}(0)$ is the sample mean of the function values at 0. Taking the SVD, $K_h = U_h \Sigma_h V_h^T$ we can calculate the directions of principle variability in the given SRSFs using the first $p \leq (T + 1)$ columns of U_h and can be converted back to $\mathscr{F}$, via integration, for displaying the principal components in the original function space. Moreover, we can calculate the observed principal coefficients as $\langle \mathbf{h}_i, U_{h,j} \rangle$.

Figure 8.14 shows the results of A-FPCA on the simulated dataset from the top row of Fig. 8.5. It shows the aligned functions $\{\tilde{q}_i\}$, the three principal directions, and the singular values. The first three singular values for the data are $0.0481, 0.0307$, and 0.0055, the remaining being negligibly small. We also show the corresponding pairwise scatter plots of the observed principle coefficients in the bottom row of this figure. In some cases these scatter plots can be used to evaluate statistical models associated with functional data. For instance, the scatter plots reveal a strong pairwise dependence between the three coefficients.

A similar experiment involving the female growth velocity functions is shown in Fig. 8.15. Here the top row shows the original data in the SRSF space $\{q_i\}$, the aligned data $\{\tilde{q}_i\}$, the top three singular vectors (or principal directions of amplitude variations), and the singular values. The scatter plots of the observed coefficients are shown in the bottom row. These plots show only a weak dependence between the coefficients, and some support for Gaussian models for the underlying random variables.

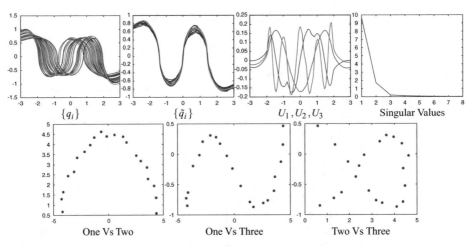

Fig. 8.14 FPCA of the amplitude components of a simulated data

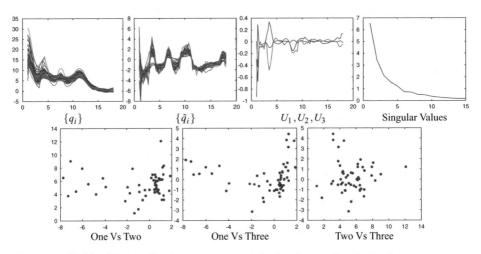

Fig. 8.15 FPCA of the amplitude components of the female growth velocity data

Formally, one can test if the observed principal coefficients follow a normal distribution. There are several tests possible, one of them is the Kolmogorov-Smirnov test. Since the principal coefficients are uncorrelated, they can be tested independently for Gaussian behavior. In the case of female growth velocity data, we find that the null hypothesis of having a Gaussian distribution is rejected for the top-two principal coefficients at 5 % confidence level, but this hypothesis is not rejected for the third principal coefficient.

8.8.2 FPCA of Phase Functions: P-FPCA

Here we want to model the phase components of the functional data, available to us in the form of the warping functions $\{\gamma_i^*\}$ outputted by Algorithm 33. An explicit statistical modeling of the warping functions can be of interest to an analyst since they represent an important part of the original data. This is the exact problem studied in Sect. 7.5.4 where we used the SRSF representation of warping functions to reach a Hilbert sphere structure. Here we illustrate those ideas using some experimental results.

Let $\gamma_1, \gamma_2, \ldots, \gamma_n \in \Gamma_I$ be a set of observed warping functions. Our goal is to develop a probability model on Γ_I, albeit implicitly, that can be estimated from the data directly. There are two problems in doing this in a standard way: (1) Γ_I is not a vector space and (2) it is infinite dimensional. The issue of nonlinearity is handled using a convenient transformation, which is similar to the definition of SRSF, and the issue of infinite dimensionality is handled using dimension reduction, e.g., FPCA, which we will call phase-FPCA or *P-FPCA*. We are going to represent an element $\gamma \in \Gamma_I$ by the square root of its derivative $\psi = \sqrt{\dot{\gamma}}$. Note that this is the same as the SRSF defined earlier for f_is and takes this form since $\dot{\gamma} > 0$. Refer to Fig. 7.14 for an illustration of this mapping. The identity map γ_{id} maps to a constant function with value $\psi_{id}(t) = 1$. Since $\gamma(0) = 0$, the mapping from γ to ψ is a bijection and one can reconstruct γ from ψ using $\gamma(t) = \int_0^t \psi(s)^2 ds$. An important advantage of this transformation is that since $\|\psi\|^2 = \int_0^1 \psi(t)^2 dt = \int_0^1 \dot{\gamma}(t) dt = \gamma(1) - \gamma(0) = 1$, the set of all such ψs forms the positive orthant of Hilbert sphere $\mathbb{S}_\infty$, the unit sphere in the Hilbert space $\mathbb{L}^2$. In other words, the SRSF representation simplifies the complicated geometry of Γ_I to a unit sphere. The distance between any two warping functions is exactly the arc length between their corresponding SRSFs on the unit sphere $\mathbb{S}_\infty$: $d_\psi(\psi_1, \psi_2) \equiv \cos^{-1}\left(\int_0^1 \psi_1(t)\psi_2(t) dt\right)$. The definition of a distance on $\mathbb{S}_\infty$ helps define a Karcher mean of sample points on $\mathbb{S}_\infty$, as described in Sect. 7.5.4.

Definition 8.3. For a given set of points $\psi_1, \psi_2, \ldots, \psi_n \in \mathbb{S}_\infty$, their Karcher mean in $\mathbb{S}_\infty$ is defined to be a local minimum of the cost function $\psi \mapsto \sum_{i=1}^n d_\psi(\psi, \psi_i)^2$.

Now we can define the Karcher mean of a set of warping functions using the Karcher mean in $\mathbb{S}_\infty$. For a given set of warping functions $\gamma_1, \gamma_2, \ldots, \gamma_n \in \Gamma_I$, their Karcher mean in Γ_I is $\bar{\gamma}(t) \equiv \int_0^t \mu_\psi(s)^2 ds$ where μ_ψ is the Karcher mean of $\sqrt{\dot{\gamma}_1}, \sqrt{\dot{\gamma}_2}, \ldots, \sqrt{\dot{\gamma}_n}$ in $\mathbb{S}_\infty$. The search for this minimum is performed as described in Sect. 7.5.4.

Since $\mathbb{S}_\infty$ is a nonlinear space (a sphere), one cannot perform principal component analysis on it directly. Instead, we choose a vector-space tangent to the space, at a certain fixed point, for analysis. This idea is also called *Tangent PCA* or simply TPCA. The tangent space at any point $\psi \in \mathbb{S}_\infty$ is given by: $T_\psi(\mathbb{S}_\infty) = \{v \in \mathbb{L}^2 | \int_0^1 v(t)\psi(t) dt = 0\}$. In the following, we will use the tangent space at μ_ψ to perform analysis. Note that the outcomes of the mean computation include the Karcher mean μ_ψ and the tangent vectors $\{v_i\} \in T_{\mu_\psi}(\mathbb{S}_\infty)$. These tangent vectors, also called the *shooting vectors*, are the mappings of ψ_is into the tangent space $T_{\mu_\psi}(\mathbb{S}_\infty)$. In this tangent space we can define a sample covariance function: $(t_1, t_2) \mapsto \frac{1}{n-1} \sum_{i=1}^n v_i(t_1)v_i(t_2)$. In practice, this covariance is computed using a finite number of points, say T, on these functions and one obtains a $T \times T$ sample covariance matrix instead, denoted by K_ψ. The SVD of $K_\psi = U_\psi \Sigma_\psi V_\psi^T$ provides the estimated principal components of $\{\psi_i\}$: the principal directions $U_{\psi,j}$ and the observed principal coefficients $c_{i,j} = \langle v_i, U_{\psi,j} \rangle$.

As an example, we compute the Karcher mean of a set of random warping functions. These warping functions and their sample Karcher mean are shown in the left panel of Fig. 8.16. Using the $\{v_i\}$'s that result from the mean computation,

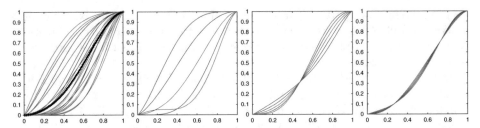

Fig. 8.16 *Left panel*: observed warping functions and their Karcher mean. *Next three panels*: principal directions of the observed data

we form their covariance matrix K_ψ and take its SVD. The first three columns of U_ψ are used to visualize the principal geodesic paths, in the last three panels of this figure.

8.8.3 Joint Modeling of Principle Coefficients

To develop statistical models for capturing the phase and amplitude variability, there are several possibilities. Once we have obtained the FPCA coefficients for these components, we can impose probability models on the coefficients (while keeping the basis functions fixed) and induce a distribution on the function space $\mathscr{F}$. In this section, we explore two such possibilities: a joint Gaussian model and a nonparametric model, and demonstrate them with experiments.

Let $c_a \in \mathbb{R}^{k_1}$ and $c_p \in \mathbb{R}^{k_2}$ be the dominant principal coefficients of the amplitude and phase components, respectively, as described in the previous two sections. Recall that $c_{a,j} = \langle h, U_{h,j} \rangle$ and $c_{p,j} = \langle v, U_{\psi,j} \rangle$. We can reconstruct the amplitude component using $q = \mu_q + \sum_{j=1}^{k_1} c_{a,j} U_{h,j}$ and $f(t) = f(0) + \int_0^t q(s)|q(s)|ds$. Here, $f(0)$ is a random initial value. Similarly, we can reconstruct the phase component (a warping function) using $v = \sum_{j=1}^{k_2} c_{p,j} U_{\psi,j}$ and then using $\psi = \cos(\|v\|)\mu_\psi + \sin(\|v\|)\frac{v}{\|v\|}$, and $\gamma(t) = \int_0^t \psi(s)^2 ds$. Combining the two random quantities, we obtain a random function $f \circ \gamma$.

1. **Gaussian Models on FPCA Coefficients** In this setup the model specification reduces to the choice of models for $f(0)$, c_p and c_a. We are going to model them as multivariate normal random variables. The mean of $f(0)$ is $\bar{f}(0)$, while the means of c_a and c_p are zero vectors. Their joint covariance matrix is of the type: $\begin{bmatrix} \sigma_0^2 & L_1 & L_2 \\ L_1^T & \Sigma_h & S \\ L_2^T & S & \Sigma_\psi \end{bmatrix} \in \mathbb{R}^{(k_1+k_2+1) \times (k_1+k_2+1)}$. Here, $L_1 \in \mathbb{R}^{1 \times k_1}$ is the sample covariance between $f(0)$ and c_a, $L_2 \in \mathbb{R}^{1 \times k_2}$ between $f(0)$ and c_p, and $S \in \mathbb{R}^{k_1 \times k_2}$ between c_a and c_p. As discussed in the previous sections, $\Sigma_h \in \mathbb{R}^{k_1 \times k_1}$ and $\Sigma_\psi \in \mathbb{R}^{k_2 \times k_2}$ are diagonal matrices and are estimated directly from the data.

2. **Nonparametric Models on FPCA Coefficients** An alternative to the parametric model suggested above is the nonparametric approach, where the density function for each coefficient is allowed to take an arbitrary form. One can esti-

mate the densities of all the variables of interest: $f(0)$, each of the k_1 components of c_a, and the k_2 components of c_p, using kernel density estimation:

$$\frac{1}{nb}\sum_{i=1}^{n}\mathscr{K}\left(\frac{x-x_i}{b}\right),$$

where $\mathscr{K}(\cdot)$ is the kernel function (a valid kernel function is a symmetric function that integrates to 1) and $b > 0$ is the smoothing parameter or bandwidth. A range of kernel functions can be used, but the most common choice is the Gaussian kernel. Note that separate estimation of densities for each coefficient implies independence of the coefficients, a condition that will be difficult to satisfy in practice. Still, from a modeling perspective, one can try this model on real data and evaluate its effectiveness.

We illustrate the use of these models via random sampling. That is, we first compute sample statistics (means and the covariances) from the given data, estimate the model parameters, and then generate random samples from these estimated models. We demonstrate results on the same two simulated datasets used in the previous examples. For the first simulated dataset, shown in Fig. 8.5 (top-left), we randomly generate 35 samples from the amplitude model, 35 domain-warping functions from the phase model, and then apply group action to generate random functions. These random samples are shown in Fig. 8.17, where the first panel is a set of random warping functions $\{\gamma_i\}$, the second panel is a set of corresponding amplitude functions $\{\tilde{f}_i\}$, and the third panel shows their compositions $\{\tilde{f}_i \circ \gamma_i\}$. Comparing them with the original datasets (Fig. 8.5 top-left), we conclude that the random samples are very similar to the original data and the proposed model is successful in capturing the variability in the given data. Furthermore, if we compare these sampling results to the FPCA-based Gaussian model directly on f (without separating the phase and amplitude components) in the last panel, we notice that this model is more consistent with the original data. A good portion of the samples from the non-separated model just contain three peaks or have a higher variation than the original data and poorly represent the original data. Beyond visual validation of the composition model, one can actually select among the two competing models using a proper likelihood ratio test. The basic idea is to evaluate the likelihood of the training data under the competing model and select the one providing the larger value.

For the second simulated dataset, we use the data shown in Fig. 8.5 (bottom left) and perform A-FPCA, P-FPCA. As before, we randomly generate 35

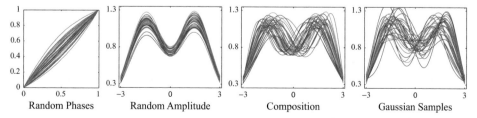

Fig. 8.17 Random samples from jointly Gaussian models on FPCA coefficients of γ (*left*) and f (*middle*) and their combinations $f \circ \gamma$ (*right*) for Simulated Data 1. The last plot are random samples if a Gaussian model is imposed on f directly without any phase and amplitude separation

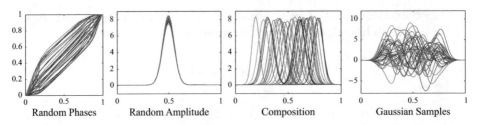

Fig. 8.18 Random samples from jointly Gaussian models on FPCA coefficients of γ (*left*) and f (*middle*) and their combinations $f \circ \gamma$ (*right*) using parameters estimated from the training data. The last panel shows the random samples resulting from a Gaussian model imposed on f directly, without separating phase and amplitude components

functions from the amplitude model and 35 domain-warping functions from the phase model and then combine them to generate random functions. The corresponding results are shown row of Fig. 8.18, where the first panel is a set of random warping functions, the second panel is a set of corresponding amplitude functions, and the last panel shows their compositions. Under a visual inspection, the proposed models seem successful in capturing the variability in the given data. One can validate these models more formally using likelihood-based tests. In this example, performing FPCA directly on the function space does not correctly capture the data and fails to generate any single unimodal function, as shown in the last panel.

8.9 Joint Approach: Elastic FPCA

In case one is interested in low-dimensional representations of functional data, especially for data containing phase variability, one can also formulate this problem directly, without performing phase-amplitude separation as a preliminary step. In some cases it may seem more natural to perform these two steps—dimension-reduction and phase-amplitude separation—simultaneously. The resulting low-dimensional representations can then be used in statistical modeling, pattern recognition, or other applications. One can also view this formulation as a model-based approach where low-dimensional representations become parameters in a model. Then, the search for representations becomes a problem of estimation or optimization under an appropriate objective function. Depending on the choice of model, or the objective function, there are several possibilities. Here we look at two possibilities, involving the $\mathbb{L}^2$ metric on two different spaces, one on the original function space $\mathscr{F}$ and the other on the SRSF space $\mathbb{L}^2$.

8.9.1 Model-Based Elastic FPCA in Function Space $\mathscr{F}$

We start with the case where each functional observation f_i is described in terms of its components: amplitude g_i and phase γ_i. We caution the reader that in this model-based approach, we will develop a completely new notion of phase and amplitude, different from the definitions provided in Sect. 4.7. So, there is no connection between the usage of these terms *phase* and *amplitude* here and that in Sect. 4.7. We model the amplitude variable g_i to be an element of a low-dimensional subspace, and the phase γ_i is left to be completely arbitrary. Thus, the amplitude can be expressed using an orthonormal basis set $\{b_j\}$ so that it is simply a linear combination of these basis elements.

$$f_i(t) = g_i(\gamma_i^{-1}(t)), \quad g_i(t) = \mu_f(t) + \sum_{j=1}^{J} c_{i,j} b_j(t) + \epsilon_i(t) \ , \ i = 1, 2, \ldots, n,$$

where

- μ_f is the mean of the amplitude,
- $\{f_i \in \mathbb{L}^2 | i = 1, 2, \ldots, n\}$ are the given observations,
- $\{\gamma_i \in \Gamma_I | i = 1, 2, \ldots, n\}$ represent arbitrary phases of the observations,
- $\{b_j \in \mathbb{L}^2 | j = 1, 2, \ldots, J\}$ is a chosen basis set,
- $\{c_{i,j} \in \mathbb{R} | i = 1, 2, \ldots, n, j = 1, 2, \ldots, J\}$ are the low-dimensional representations of interest, and
- $\{\epsilon_i \in \mathbb{L}^2 | i = 1, 2, \ldots, n\}$ is observation noise modeled as a white Gaussian noise process.

(This is a simple extension of the model presented in Eq. 4.7, which did not involve any time warping.) Another way to express this model is:

$$f_i(\gamma_i(t)) = \mu_f(t) + \sum_{j=1}^{J} c_{i,j} b_j(t) + \epsilon_i(t) \ , \ i = 1, 2, \ldots, n \ . \tag{8.7}$$

Having selected the model, we now focus our attention on the estimation problem: Given the observations $\{f_i\}$, the main goal is to estimate the coefficients $\{c_{i,j}\}$. However, since the phases $\{\gamma_i\}$ are unknown, and they play the role of nuisance parameters in this estimation, one has to estimate them too from the data. Furthermore, in case we do not want to assume any predetermined basis, then the basis set $\{b_j\}$ also becomes a parameter of the model and needs to be estimated. Under the white Gaussian noise model, the full estimation problem is:

$$\left(\hat{\mu}_f, \hat{c}, \hat{b}, \hat{\gamma}\right) = \operatorname*{arginf}_{\mu_f, \{\gamma_i\}, \{c_{i,j}\}, \{b_j\}} \left(\sum_{i=1}^{n} \operatorname*{arginf}_{\gamma_i \in \Gamma_I} \left(\| f_i \circ \gamma_i - \mu_f - \sum_{j=1}^{J} c_{i,j} b_j \|^2 \right) \right) \ . \tag{8.8}$$

One can solve for these unknowns using a suitable optimization approach. A particularly simple solution is to use coordinate descent, where one solves for one set while fixing the others. These conditional solutions can be evaluated as follows:

1. **Updating Phase Components**: For fixed $\{c_{i,j}\}$ and $\{b_j\}$, the optimization over relative phases $\{\gamma_i\}$ is given by:

$$\hat{\gamma}_i = \operatorname*{argmin}_{\gamma \in \Gamma_I} \left(\| f_i \circ \gamma - \mu_f - \sum_{j=1}^{J} c_{i,j} b_j \|^2 \right) \ , \ i = 1, 2, \ldots, n \ . \tag{8.9}$$

Notice that computations of $\hat{\gamma}_i$, for different i, are independent of each other. As discussed earlier in this textbook, these problems can be solved using the DPA, one for each i, and the solution can actually be implemented in parallel due to this independence across i's.

2. **Updating Mean Function**: We can estimate the mean function μ_f using the cross-sectional average:

$$\hat{\mu}_f(t) = \frac{1}{n} \sum_{i=1}^{n} f_i(\hat{\gamma}_i(t)) \ .$$

3. **Updating Basis Set Using PCA**: For a fixed set of warping functions $\{\hat{\gamma}_i\}$, the problem of estimating the basis set is given by:

$$\hat{b} = \underset{\substack{\{b_j \in \mathbb{L}^2([0,1],\mathbb{R})|j=1,2,\ldots,J, \\ b_j \perp b_i, i \neq j \\ \|b_j\| = 1, j = 1,2,\ldots,J\}}}{\operatorname{arginf}} \left(\sum_{i=1}^{n} \underset{\gamma_i \in \Gamma}{\operatorname{arginf}} \|f_i \circ \hat{\gamma}_i - \hat{\mu}_f - \sum_{j=1}^{J} c_{i,j} b_j\|^2 \right) \ .$$

(8.10)

It is easy to see that this is exactly the problem of finding FPCA of given function data $\{f_i \circ \hat{\gamma}_i\}$ (see Sect. 4.3.1). FPCA has been used several times in this chapter already.

4. **Updating Coefficients**: Given the phase components $\{\gamma_i\}$ and an orthonormal basis $\{b_j\}$, the coefficients $\{c_{i,j}\}$ can be computed using simple projections:

$$c_{i,j} = \langle f_i \circ \hat{\gamma}_i, b_j \rangle, \ \ i = 1, 2, \ldots, n, \ \ j = 1, \ldots, J \ .$$

(8.11)

An iterative algorithm for implementing this estimation procedure is as follows.

Algorithm 35 (Joint FPCA Under Phase Variability).

1. *Initialize the warping functions to be identity $\hat{\gamma}_i = \gamma_{id}$, $i = 1, 2, \ldots, n$.*
2. *$\boldsymbol{FPCA}$: Perform FPCA on the warped data $\{f_i \circ \hat{\gamma}_i\}$ and select the basis elements for the top J principal components. This results in an orthonormal basis $\{b_j, j = 1, 2, \ldots, J\}$.*
3. *$\boldsymbol{Coefficients}$: Given the basis elements $\{b_j\}$, compute $\{c_{i,j}\}$ using Eq. 8.11.*
4. *$\boldsymbol{Alignment}$: Given the basis $\{b_j\}$ and the coefficients, update the warping functions $\{\hat{\gamma}_i\}$ according to Eq. 8.9 using the DPA, for each i.*
5. *Check for convergence. If not converged, return to Step 2.*

The convergence properties of this approach have been left out of the discussion here. Generally speaking, the cost function has to be convex in all the variables being optimized to guarantee convergence of coordinate descent to a global solution.

We study this approach empirically using a couple of simulated examples. In Fig. 8.19, we show results on a dataset of 20 bimodal functions used earlier in phase-amplitude separation experiments. The figure shows the original functions, the estimated warping functions $\{\hat{\gamma}_i\}$, the warped functions $\{f_i \circ \hat{\gamma}_i\}$, and the evolution of the objective function during optimization. In this experiment we use $J = 3$. In the results we can clearly see the presence a pinching effect. The warping

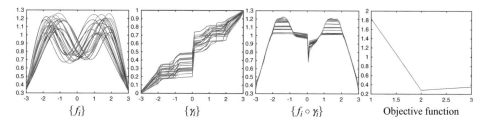

$\{f_i\}$ $\{\gamma_i\}$ $\{f_i \circ \gamma_i\}$ Objective function

Fig. 8.19 FPCA with data warping using Algorithm 35. Here we use $J = 3$ to perform FPCA. The results shows a pinching of functions during joint alignment, a shortcoming of the $\mathbb{L}^2$ norm used in this algorithm

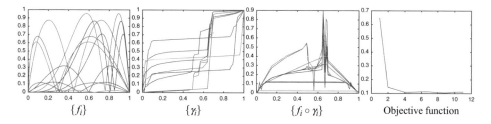

$\{f_i\}$ $\{\gamma_i\}$ $\{f_i \circ \gamma_i\}$ Objective function

Fig. 8.20 FPCA with data warping using Algorithm 35. Here we use $J = 2$ to perform FPCA. The pinching of aligned functions is clearly visible and is a drawback of this framework

functions show near-vertical jumps that result in pinching of the corresponding f_i's, in order to minimize the objective function.

We take another example, this time using data relating to the absolute, warped sine functions, shown in the top-left panel of Fig. 8.20. We apply Algorithm 35 on this dataset for $J = 3$. Once again, we observe a large pinching effect in these results.

Thus, we conclude that the main problem with this formulation is *pinching*, similar to the discussion presented in Sect. 4.5. In the absence of any additional constraint on $\{\gamma_i\}$, we observe sharp jumps in these warping functions resulting in a pinching of the given functions. This suggests that we reformulate this problem using the Fisher-Rao distance, which, in turn, motivates the use of SRSFs under the $\mathbb{L}^2$ norm.

8.9.2 Elastic FPCA Using SRSF Representation

In order to avoid the pinching effect, and to avoid the use of external roughness penalties (since they come with a difficult choice of relative weights), we suggest another model to study the same problem. Similar to a generalized linear model, we first transform the given functions into their SRSFs and then model them using a linear model as earlier. Let q_is denote the SRSFs of the given f_is, and impose the model:

$$q_i(t) = (g_i, \gamma_i^{-1})(t), \quad g_i(t) = \mu(t) + \sum_{j=1}^{J} c_{i,j} b_j(t) + \epsilon_i(t) , \; i = 1, 2, \ldots, n,$$

where the quantities in the model have the same roles as in Eq. 8.7. The difference from Eq. 8.7 is that the current model is posed in the SRSF space, and, thus, the warping group action is different. Once again, we can rearrange terms in this model to get:

$$(q_i, \gamma_i)(t) = \mu(t) + \sum_{j=1}^{J} c_{i,j} b_j(t) + \epsilon_i(t) \ , \ i = 1, 2, \ldots, n \ . \tag{8.12}$$

Note that for $J = 0$, this equation reduces to Eqn. 8.3. Under the assumption of white Gaussian noise, the full estimation problem becomes:

$$\left(\hat{\mu}, \hat{c}, \hat{b}, \hat{\gamma} \right) = \underset{\mu, \{\gamma_i\}, \{c_{i,j}\}, \{b_j\}}{\mathrm{arginf}} \left(\sum_{i=1}^{n} \underset{\gamma_i \in \Gamma_I}{\mathrm{arginf}} \left\| (q_i, \gamma_i) - \mu - \sum_{j=1}^{J} c_{i,j} b_j \right\|^2 \right) \ . \tag{8.13}$$

Compared to Eq. 8.8, this formulation is more stable since the action of Γ_I is norm preserving here and prevents pinching of the SRSFs being aligned. Similar to the previous section, we can apply a coordinate-descent technique and the resulting algorithm can be summarized in the following.

Algorithm 36 (Elastic Function Principal Component Analysis).

1. *Initialization: Set $\gamma_i^* = \gamma_{id}$ for all i.*
2. *Form the warped SRSFs $\tilde{q}_i = (q_i, \gamma_i^*)$ for all i, and estimate μ using $\hat{\mu} = \frac{1}{n} \sum_{i=1}^{n} \tilde{q}_i$.*
3. *Compute the covariance matrix*

$$K_q(s, t) = \frac{1}{n-1} \sum_{i=1}^{n} (\tilde{q}_i(s) - \hat{\mu}(s))(\tilde{q}_i(t) - \hat{\mu}(t)) \ .$$

4. *Take the SVD of K_q, and set b_js to be the first J eigenvectors of K_q.*
5. *Compute the coefficients $c_{i,j} = \langle \tilde{q}_i, b_j \rangle$ for $i = 1, \ldots, n$ and $j = 1, \ldots, J$.*
6. *For each i, solve the optimization problem using the DPA:*

$$\gamma_i^* = \underset{\gamma \in \Gamma_I}{\mathrm{arginf}} \left(\left\| (q_i, \gamma) - \hat{\mu} - \sum_{j=1}^{J} c_{i,j} b_j \right\|^2 \right) \ .$$

7. *Check for convergence. If not converged, return to Step 2.*

The convergence can be tested using the decrease in the objective function from one iteration to the other.

We apply this framework to the same datasets as in the previous approach. Figure 8.21 shows the original functions, their SRSFs, the warped SRSFs, and the estimated μ in the top row. The bottom row shows the warping functions $\{\gamma_i\}$ and the aligned functions $\{f_i \circ \gamma_i\}$. In the rightmost panel, we also see the singular values of the functional data before and after the alignment. Based on these results, several remarks are in order. Firstly, it is easy to see that the aligned data does not show any signs of pinching effects. The use of the SRSF avoids the pinching problem as hypothesized earlier in the formulation. In the original data, we have four significant singular values, but after alignment, there is no

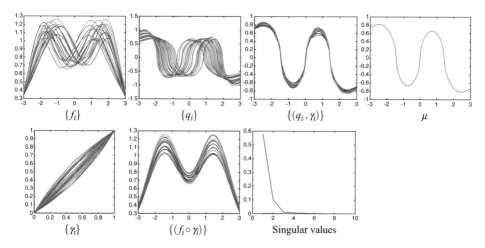

Fig. 8.21 FPCA with data warping in the SRSF space using Algorithm 36 using $J = 2$ to perform FPCA. Compare these results with those in Fig. 8.19

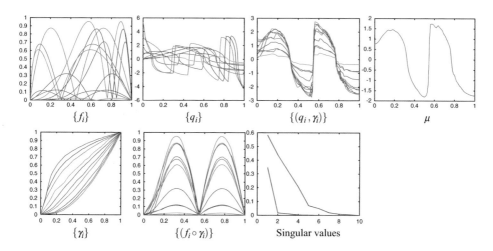

Fig. 8.22 FPCA with data warping in the SRSF space using Algorithm 36 using $J = 3$ to estimate model parameters. Compare these results with those in Fig. 8.20

significant variability left in the data. Note that we used $J = 2$ in this experiment. Compare the shape of the aligned functions with those shown in the bottom row of Fig. 8.5. The aligned functions are unimodal in both cases, but the modal shape is much broader in this case than the previous alignment. It is difficult to justify these $\{f_i \circ \gamma_i\}$ as simple alignments of the original functions; there is an additional distortion in these functions, due to the nature of the underlying model being used to estimate the warping functions.

Another example, involving sine functions, is shown in Fig. 8.22.

Alternative Definition of Phase and Amplitude This **model-based** estimation of $\{\gamma_i\}$ can be viewed as an alternative definition of phase (and thus amplitude) of a functional variable. One can compare this definition with the

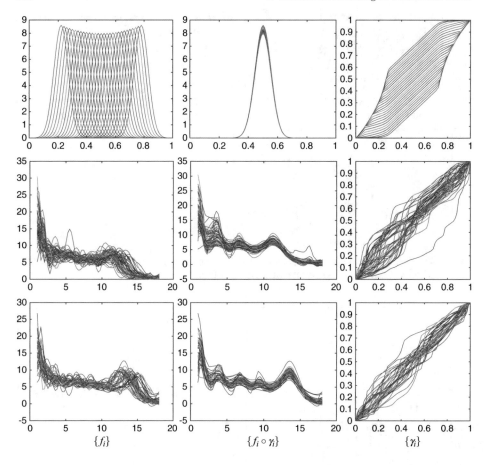

Fig. 8.23 Model-based separation of phase and amplitude in some functional datasets. Even though these functions are not always on $[0, 1]$, the warping functions are scaled for display as elements of Γ_I

metric-based definition presented in Sect. 4.7. We make some remarks about these two competing definitions (Fig. 8.23).

1. Firstly, the phase-amplitude separation based on the model in $\mathscr{F}$, given in Eq. 8.7, is plagued by the pinching effect and destroys the structure of the given data. Even if one can impose a regularization term, or restrict to only simple warping functions, in order to tame the pinching effect, these choices are difficult and often require a manual selection to yield interesting results. Consequently, this model does not yield a satisfactory definition of phase and amplitude in general and we are left only with SRSF models (e.g., Eq. 8.12) for a model-based approach.

2. This model-based approach is primarily for multiple function alignment and does not provide useful results for pairwise alignment. Recall from Sect. 4.4 that it is desirable for a pairwise solution to satisfy certain properties: *invariance to simultaneous warping, robustness to random warping,* and *inverse symmetry.* The alignment obtained from a model-based approach will not necessarily satisfy these properties. The metric-based approach, on the other hand, han-

dles both the pairwise and multiple function alignment in a consistent, elegant manner.

3. In the metric-based approach, the definition of *amplitude* is absolute—it consists of a whole equivalence class (all possible warpings) of a function. Once the set of allowed warping functions are chosen, this definition is fixed—independent of choice of metric or representation. Only the definition of phase depends on the metric in this approach. This characterization seems more natural and satisfactory than the model-based approach where all components (phase and amplitude) depend on model choices—representation space, noise model, number of basis elements, and so on.

4. Relative to the metric-based approach, the model-based ideas seem better placed for statistical analysis, since we are able to specify an underlying model and pose the phase-amplitude separation as estimation. In this setup, one can potentially develop asymptotic analyses relating to the efficiency and consistency of the estimators of phase and amplitude. To accomplish the same task in the metric-based approach, one would need to specify an underlying model that is compatible with all the algebraic definitions laid out in that approach.

5. Interestingly, the alignment results are quite similar under the two approaches—metric-based and model-based—for the datasets studied here.

In summary, we conclude that the metric-based approach is mathematically more fundamental and elegant for phase-amplitude separation in functional data, while the model-based approach is more amenable to a statistical analysis.

8.10 Exercises

8.10.1 Theoretical Exercises

1. For the functions $f_1, f_2, \ldots, f_n \in \mathbb{L}^2([0,1], \mathbb{R})$, show that their Karcher mean under the $\mathbb{L}^2$ norm is the same as the cross-sectional mean. That is,

$$\mu = \underset{f \in \mathbb{L}^2([0,1],\mathbb{R})}{\mathrm{arginf}} \sum_{i=1}^{n} \|f - f_i\|^2 = \frac{1}{n} \sum_{i=1}^{n} f_i \ .$$

2. Let $\gamma_1, \gamma_2, \ldots, \gamma_n \in \Gamma_I$ be a set of warping functions. Show that:

   a. Their sample average under the $\mathbb{L}^2$ norm is simply $t \mapsto \frac{1}{n} \sum_{i=1}^{n} \gamma_i(t)$.
   b. This average is an element of Γ_I, i.e., it is a diffeomorphism, invertible, and its inverse is a diffeomorphism.

3. Provide an example of a situation where:

$$\frac{1}{n} \sum_{i=1}^{n} f(\gamma_i(t)) \neq f(\frac{1}{n} \sum_{i=1}^{n} \gamma_i(t)) \ .$$

   Repeat the problem when the cross-sectional average of γ_i's is replaced by their sample Karcher mean.

4. Show that the cost function $\sum_{i=1}^{n} d_a([\mu_q], [q_i])^2$ does not increase from one iteration to the next in Algorithm 31.

5. What happens in the Definition 8.2 when $n = 1$? That is, for an arbitrary $q, q_1 \in \mathbb{L}^2([0,1], \mathbb{R})$, what is the center of the orbit $[q]$ with respect to the singleton set $\{q_1\}$?

6. **Averaging within an orbit**: Let $q \in \mathbb{L}^2([0,1], \mathbb{R})$ (assume that $q \neq 0$ *a.e.*) and let $[q]$ be its orbit under $\tilde{\Gamma}_I$. Let $q_1, q_2, \ldots q_n$ be elements of this orbit. In other words, there exist $\gamma_1, \gamma_2, \ldots, \gamma_n \in \tilde{\Gamma}_I$ such that $q_i = (q, \gamma_i)$ for all $i = 1, 2, \ldots, n$.

   - $\mathbb{L}^2$ **Average**: If we define a cross-sectional mean $\hat{q}(t) = \frac{1}{n} \sum_{i=1}^{n} q_i(t)$, will the average in general lie in the orbit $[q]$?
   - d_a **Average**: Now, if we define a sample Karcher mean $[\mu_q]$ of the amplitudes $[q_1], [q_2], \ldots, [q_n]$ according to Definition 8.1, how will the mean amplitude relate to $[q]$?

7. Once again, let $q_1, q_2, \ldots q_n$ be elements of the same orbit $[q]$, i.e., $q_i = (q, \gamma_i)$ for some $\gamma_i \in \tilde{\Gamma}_I$, and let $[\mu_q]$ be the sample Karcher mean of the amplitudes $[q_1], [q_2], \ldots, [q_n]$ under d_a. Let $q_0 \in [\mu_q]$ be the *center* of this orbit with respect to $q_1, q_2, \ldots, q_n$ (according to Definition 8.2). Show that the calculation of q_0 boils down to averaging of the corresponding γ_is in $\tilde{\Gamma}_I$ (under the Fisher-Rao distance).

8.10.2 Computational Exercises

1. Write a program to compute the cross-sectional mean of a set of functions $f_1, f_2, \ldots, f_n$ on $[0,1]$. Assume that the function values are given to you at arbitrary sample points in $[0,1]$. Thus, you will need a function for *interpolation* on the full domain. Display the original functions and the results. Test this program with the functions $f_i(t) = \sin(2\pi i t)$.

2. Write a program to compute the SRSF q of a given function f and another program to recover the function back from its SRSF (up to a vertical translation).

3. Implement the dynamic-programming algorithm (DPA) to align any two given functions f_1 and f_2 by minimizing the objective function:

$$\underset{\gamma \in \Gamma_I}{\arg\inf} \left(\|f_1 - f_2 \circ \gamma\|^2 + \lambda \mathscr{R}(\gamma) \right) ,$$

where $\mathscr{R}(\gamma)$ can be: (1) $\int_0^1 \dot{\gamma}(t)^2 dt$, or (2) $\int_0^1 \ddot{\gamma}(t)^2 dt$. Try the program with $f_1(t) = \sin(2\pi t)$ and $f_2(t) = 2\sin(2\pi t)$ and different values of $\lambda > 0$.

4. Implement Algorithm 31 for computing means of amplitudes of a given set of functions. Test this program on the following sets of functions:

   a. $f_i(t) = \sin(2\pi \gamma_i(t))$, $\gamma_i(t) = t^{2i/n}$, for $i = 1, 2, \ldots, n$ and where n is an even number.
   b. $f_i(t) = \sin(2\pi t - (i - \frac{n}{2})\pi/n)$, for $i = 1, 2, \ldots, n$, and where n is an even number.

5. Write a program to compute the center of an orbit $[q]$ with respect to the functions $q_1, q_2, \ldots, q_n \in \mathbb{L}^2([0,1], \mathbb{R})$, according to Algorithm 32.

6. Implement Algorithm 33 to perform phase-amplitude separation of a given set of functions. Test your program on simulated data created as follows: let $y_i(t) = z_{i,1}e^{-(t-1.5)^2/2} + z_{i,2}e^{-(t+1.5)^2/2}$, $i = 1, 2, \ldots, 21$, where $z_{i,1}$ and $z_{i,2}$ are $i.i.d$ normal with mean one and standard deviation 0.25. Each of these functions is then warped according to: $\gamma_i(t) = 6\left(\frac{e^{a_i(t+3)/6}-1}{e^{a_i}-1}\right) - 3$ if $a_i \neq 0$; otherwise, $\gamma_i = \gamma_{id}$, where $\{a_i\}$ are equally spaced between -1 and 1, and the functions of interest are computed using $f_i(t) = y_i(\gamma_i(t))$.

7. Implement the penalized alignment presented in Algorithm 34, and test it on the data generated in the previous example for different values of λ.

8.11 Bibliographic Notes

The problem of alignment of multiple functions has been studied in several places, including [71, 113, 67, 90, 91, 55]. Most of these methods rely on using the $\mathbb{L}^2$ norm on functional data directly and suffer from the limitations described in this chapter. An introduction to the problem of phase-amplitude separation can be found in [76]. The framework for phase-amplitude separation using elastic Riemannian metric was first proposed in [108]. Modeling of phase and amplitude components using FPCA was discussed in [117].

Several case studies involving real datasets and phase amplitude have been presented in [122, 116, 59, 123].

Chapter 9
Statistical Modeling of Planar Shapes

As emphasized in the introduction chapter (Sect. 1.3), one of the main goals in this textbook is to develop statistical models of shapes of curves. Within that broad topic, there are several tools that we wish to develop:

1. **Clustering of Shapes**: A basic task in any pattern analysis of objects is their clustering into homogeneous groups. Since we are focused on the shape of objects in this textbook, we seek a method for clustering curves according to their shapes. The basic idea is to divide given objects into subsets such that the shape variability is small within the subsets, and as large as possible across subsets. The quantification of shape variability is based on shape metrics derived in the earlier chapters.

2. **Summary Statistics of Shapes**: An important tool in shape analysis is the computation of summary statistics. One is given a set of observations of shapes, from the same class or different classes, and the goal is to replace them with a short summary that characterizes the observed set. In our context, one may be interested in finding a representative shape for a given collection of shapes (see item 2 in Sect. 1.3). This can be obtained using sample mean, sample truncated mean, or sample median shape. Once a representative shape is obtained, the next item of interest is covariance, a quantity that encodes the principal or dominant modes of variability in shape data. These summaries may also provide estimates for corresponding parameters under certain probability distributions, but that aspect is not developed in this textbook.

3. **Stochastic Modeling of Shapes**: If we treat shapes as random quantities, then we would like to specify the probabilistic rules that govern these random quantities. For instance, we may want to model the shapes of the hippocampi in the human brain, the silhouettes of ducks floating in a pond, the shapes of tanks in a battlefield, or the shapes of human silhouettes under specific poses. What this means is that we want to define probability density functions on shape spaces, a different density for each shape class, that take higher values in regions where more shapes are observed or expected and lower values where fewer are observed. One can view this as a problem of estimating density functions—parametric or nonparametric—using past observations of shapes. In computer science, this problem of estimating densities is often termed as that of *learning*. A natural option is to use sample means and covariances and impose densities on shape spaces that are analogous to Gaussian densities but are adapted to

© Springer-Verlag New York 2016
A. Srivastava, E.P. Klassen, *Functional and Shape Data Analysis*,
Springer Series in Statistics, DOI 10.1007/978-1-4939-4020-2_9

shape spaces (or rather their finite-dimensional subsets). In case the densities belong to a parametric family, the task of learning reduces to that of estimating parameters.

4. **Testing/Classification of Shapes**: A common problem in shape analysis is the classification of given shapes into predetermined shape classes. For example, given a contour of an object in an image, we can attempt to classify it as a vehicle, a building, or a human silhouette. As another example, we may be interested in studying the shape of a ventricle in a human brain, in order to classify patients as healthy or unhealthy. From a statistical point of view, this problem can be formulated as a problem in hypothesis testing (see item 4 in Sect. 1.3 of the introduction chapter). Similarly, the problem of detecting an object can also be studied as that of a binary hypothesis testing, to determine whether an object is present in given data or not. Solutions to such testing problems require statistical models for candidate shape classes.

9.1 Goals and Challenges

We set up the following specific goals for this chapter:

1. Given a set of planar curves, closed or open, we want to cluster them or subdivide them into a given number of subsets so as to optimize a prescribed objective function. This objective function is typically constructed using shape metrics derived in the previous chapters.
2. Extend ideas of Chap. 7 and define statistical summaries—sample means, truncated means, and medians—of given sets of shapes associated with specific objects and shape classes.
3. Define parametric probability densities on (dominant subsets of) shape spaces of general and closed curves. In particular, we will adapt the notion of a *truncated wrapped-normal* (TWN) density to shape spaces of curves. Specifically, our modeling approach will be to: (1) Select a representative shape $\mu \in \mathcal{S}_2$ (or $\mathcal{S}_2^c$) as the central tendency of the population. (2) Select a finite-dimensional subspace of a tangent space $T_{[\mu]}(\mathcal{S}_2)$ (or $T_{[\mu]}(\mathcal{S}_2^c)$). This will give us a finite-dimensional representation of shapes. (3) Define a probability density, such as the truncated multivariate normal density, on that vector space. (4) Use the exponential map to transfer this density onto a finite-dimensional subset of $\mathcal{S}_2$ (or $\mathcal{S}_2^c$). The previous item on computing summaries can be related to estimating parameters of truncated wrapped-normal densities using observed data.
4. Develop algorithms for generating random samples from shape models. This tool will aid us in applying Monte Carlo-type methods for generating statistical inferences, especially in Bayesian frameworks. For the TWN densities, this task becomes quite simple with the use of exponential maps.
5. Formulate hypothesis tests for shape classifications and define likelihood ratios, and other classification tools.

Statistical models of shapes can have many further uses, including Bayesian shape estimation, shape tracking in dynamical scenes, and pattern analysis of a collection of shapes, but we will focus mainly on the items mentioned above.

Challenges The challenges associated with accomplishing these goals are similar to those discussed in the previous chapters. While a metric-based clustering is relative easy, given a shape metric, the remaining goals are still difficult to achieve using standard multivariate statistics. To generate statistical summaries, we will need to deal with the nonlinear nature of shape spaces, using computational tools for calculating geodesics and geodesic distances between arbitrary shapes. In summary, the nonlinearity, infinite dimensionality, and complexity of shape spaces are the main challenges in reaching the goals laid out here.

9.2 Clustering in Shape Spaces

We start by clustering curves on the basis of their shapes. There is an important need in shape analysis to cluster similar shapes together using an automated procedure. The goal is to allocate n curves into k subsets $(k < n)$ such that the homogeneity within these subsets is maximized and across the subsets is minimized. This process is also called an *unsupervised learning* process. There are two sets of unknowns in a clustering solution: (1) the number of clusters k and (2) the $n \times k$ membership matrix M that describes the allocation of n objects to k clusters. Most of the elements of M are zero; each row has only one nonzero element, which is one. In this chapter, we will assume that k is given and we focus on the search for an optimal M. We will treat shapes as elements of $\mathcal{S}_2^c$, the elastic shape space of closed planar curves as described in Chap. 6, and note that we have already developed algorithms for computing geodesic distances between elements of this space.

Traditional algorithms for clustering in Euclidean spaces are well studied researched and generally fall into two main categories: partitional and agglomerative. These techniques are also called *hierarchical clustering*. The partitional algorithms are top-down. They start with one big cluster of n shapes and successively divide the current clusters, while optimizing a chosen cost function, until the desired number of clusters is reached. The agglomerative algorithms, on the other hand, take a bottom-up approach. They start with each shape as a cluster and they iteratively merge sets until the number of clusters is reduced to k. The objective functions are based on metrics between clusters themselves, which, in turn, are often based on pairwise distances between shapes. For example, one objective function can be to minimize the sum of all pairwise distances between clusters, where the distance between any two clusters is defined to be the average of all distances between points across those sets.

The main difference between clustering in a Euclidean space and a shape space is the nonlinearity of the shape space. One can use the notion of the mean and the covariance on the shape space $\mathcal{S}_2^c$ and use it to perform clustering. However, we mostly restrict to ideas that rely only on pairwise geodesic distances between shapes. This avoids the computation of means and covariances for the purpose of clustering.

9.2.1 Hierarchical Clustering

Consider the problem of clustering n shapes into k clusters. Let a configuration C consist of clusters denoted by $C_1, C_2, \ldots, C_k$, with cluster sizes $n_1, n_2, \ldots, n_k$, respectively. A hierarchical clustering iteratively either merges small clusters or

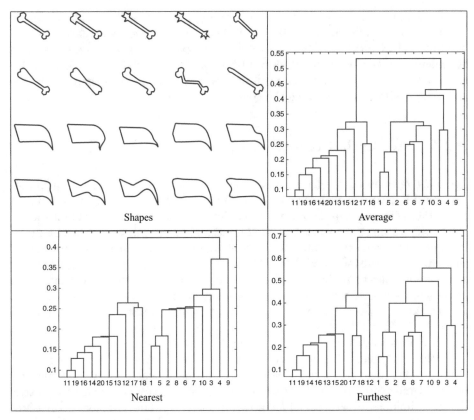

Fig. 9.1 A set of 20 shapes of the left have been clustered using different linkage criterion: average (*top-right*), nearest distance (*bottom left*), and compete or furthest distance (*bottom right*)

divides large clusters until the required number of clusters is reached. Some computational softwares have built-in routines for clustering data points according to a chosen metric between the sets $\{C_i\}$. For instance, one can define linkage using the nearest-neighbor criterion, $d(C_i, C_j) = \min_{[q_1]\in C_i,[q_2]\in C_2} d_s([q_1], [q_2])$, or the furthest $d(C_i, C_j) = \max_{[q_1]\in C_i,[q_2]\in C_2} d_s([q_1], [q_2])$. In case of average linkage, the distance is given by $d(C_i, C_j) = \frac{1}{|C_i||C_j|} \sum_{[q_1]\in C_i,[q_2]\in C_2} d_s([q_1], [q_2])$. One can also use the median distance or a weighted distance, or other combinations of pairwise shape distances.

An example of this idea is shown in Fig. 9.1 where we use the 20 contours displayed in the left panel. In the remaining three panels, we show dendrograms resulting from three different kinds of linkages in the clustered sets: average, nearest distance, and furthest distance. Since the shapes here are very well separated into two sets, the clustering result is quite stable with respect to different linkage criteria.

We show another example in Fig. 9.2 where we have four shape classes and the shapes are not as different as in the previous example. In this example, there are some changes, albeit small, as we change the criterion for clustering. Specifically, shapes 13 and 15 change clusters based on the definition of distance between

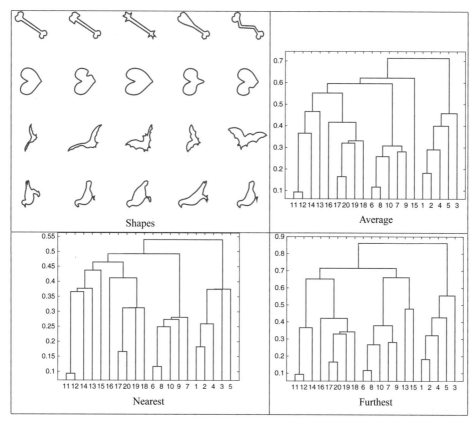

Fig. 9.2 Same as Fig. 9.1

clusters. In this example, the nearest-neighbor clustering provides the best result as the four classes are all clustered separately.

9.2.2 A Minimum-Dispersion Clustering

Consider the problem of clustering n shapes (in $\mathcal{S}_2^c$) into k clusters. A general approach is to form clusters in such a way that they minimize total within-cluster variance, which relates to a scaled sum of distances (squared). Let a configuration C consist of clusters denoted by $C_1, C_2, \ldots, C_k$, with sizes $n_1, n_2, \ldots, n_k$, and define a cost function:

$$Q(C) = \sum_{i=1}^{k} \frac{2}{n_i} \left(\sum_{q_a \in C_i} \sum_{b < a, q_b \in C_i} d([q_a], [q_b])^2 \right). \tag{9.1}$$

We seek configurations that minimize Q, i.e., $C^* = \operatorname{argmin} Q(C)$.

An important question in any clustering problem is: How to choose k? In some cases, such as those involving shape retrieval, the answer comes from the context. Here, k determines a balance between the retrieval speed and performance, and a wide range of k can be tried. In the worst case, one can set $k = n/2$ at every level of hierarchy and still obtain $O(\log(n))$ retrieval speeds. (This assumes that the shapes are uniformly distributed in the clusters.) However, the choice of k is much more important in the case of learning. Probability models estimated from the clustered shapes are sensitive to the clustering performance. To obtain a possible k automatically, one option is to study the variation of $Q(C^*)$ for different values of k and select a k that provides the largest decrease in $Q(C^*)$ from its value at $k-1$. Another possibility is to use human supervision in selecting k. A third possibility is to use a Bayesian approach, under a Dirichlet process prior, for searching over different values of k [127].

Clustering Algorithm We will take a stochastic simulated annealing approach to solve for C^*. (For the reader that are not familiar with simulated annealing, it is an optimization approach that searches for solutions in a larger space and iteratively reduces the search space as it gets closer to a solution. This search domain is controlled by altering the profile of energy landscape using a temperature parameter—high temperature implies flatter profile and larger search space, and vice versa.) We will minimize the clustering cost using a Markov chain search process on the configuration space. Please refer to [93] for a version of Markov chain simulated annealing algorithm. The basic idea is to start with a configuration of k clusters and to reduce Q by rearranging shapes among the clusters. The rearrangement is performed in a stochastic fashion using two kinds of moves. These moves are performed with probability proportional to the negative exponential of the Q-value of the resulting configuration. The two types of moves are:

1. **Move a shape**: Here we select a shape randomly and reassign it to another cluster. Let $Q_j^{(i)}$ be the clustering cost when a shape q_j is reassigned to the cluster C_i keeping all other clusters fixed. If q_j is not a singleton, i.e., not the only element in its cluster, then the transfer of q_j to cluster C_i is performed with probability: $P_M(j, i; T) = \frac{\exp(-Q_j^{(i)}/T)}{\sum_{i=1}^{k} \exp(-Q_j^{(i)}/T)}$, $i = 1, 2, \ldots, k$. Here T plays a role similar to temperature in simulated annealing. If q_j is a singleton, then moving it is not allowed in order to fix the number of clusters at k.

2. **Swap two shapes**: Here we select two shapes randomly from two different clusters and swap them. Let $Q^{(1)}$ and $Q^{(2)}$ be the Q-values of the original configuration (before swapping) and the new configuration (after swapping), respectively. Then, swapping is performed with probability: $P_S(T) = \frac{\exp(-Q^{(2)}/T)}{\sum_{i=1}^{2} \exp(-Q^{(i)}/T)}$.

Additional moves can also be designed to improve the efficiency of search over the configuration space, although their computational cost becomes a factor too. In view of the computational simplicity of moving a shape and swapping two shapes, we have restricted our algorithm to these two moves.

In order to seek global optimization, we have adopted a simulated annealing approach. That is, we start with a high value of T and reduce it slowly as the algorithm searches for configurations with smaller dispersions. Additionally, the moves are performed according to an acceptance-rejection procedure that is a variant of more conventional simulated annealing. Here, the candidates are proposed randomly and accepted according to certain probabilities (P_M and P_S are defined above). Although simulated annealing and the random nature of the search help in avoiding local minima, the convergence to a global minimum is difficult to establish. The output of this algorithm is a Markov chain that is neither homogeneous nor convergent to a stationary chain. If the temperature T is decreased slowly, then the chain is guaranteed to converge to a global minimum [93]. However, it is difficult to make explicit the required rate of decrease in T and instead we rely on empirical studies to justify this algorithm. First, we state the algorithm and then describe some experimental results.

Algorithm 37 (Clustering Using Simulated Annealing). *For n shapes and k clusters, initialize by randomly distributing n shapes among k clusters. Set a high initial temperature T.*

1. *Compute pairwise geodesic distances between all n shapes. This requires $n(n - 1)/2$ geodesic computations.*
2. *With equal probabilities, pick one of the two moves:*
   - ***Move a shape****: Choose a shape q_j randomly. If it is not a singleton in its cluster, then compute $Q_j^{(i)}$ for all $i = 1, 2, \ldots, k$. Compute the probability $P_M(j, i; T)$ for all $i = 1, \ldots, k$ and reassign q_j to a cluster chosen according to the probability P_M.*
   - ***Swap two shapes****: Select two clusters randomly and select a shape from each. Compute the probability $P_S(T)$ and swap the two shapes according to that probability.*
3. *Update the temperature using $T = T/c$ and return to Step 2. We suggest using $c = 1.0001$.*

It is important to note that once the pairwise distances are computed, they are not computed again in the iterations. Secondly, unlike k-mean clustering, the mean shapes are never calculated in this clustering. These factors make Algorithm 37 efficient and effective in clustering diverse shapes.

Figure 9.3 shows two examples of clustering MPEG7 shapes into a predetermined number ($k = 6$) of clusters using Algorithm 37. In each row, the left panel shows the given shapes arranged into clusters (rows), while the evolution of Q versus iterations is displayed in the corresponding right panel.

9.3 A Finite Representation of Planar Shapes

The main challenges in shape modeling stem from infinite dimensionality and nonlinearity of shape spaces. We suggest a solution, naturally an approximation, that reduces the representation space to a finite-dimensional vector space.

9.3.1 Shape Representation: A Brief Review

Recall shape representations of planar curves, developed in Chap. 5, where we have multiple choices of representations/metrics: a non-elastic representation that uses the angle functions of arc-length curves and an elastic representation that uses SRVFs of parameterized curves. Since the modeling procedure is similar for the

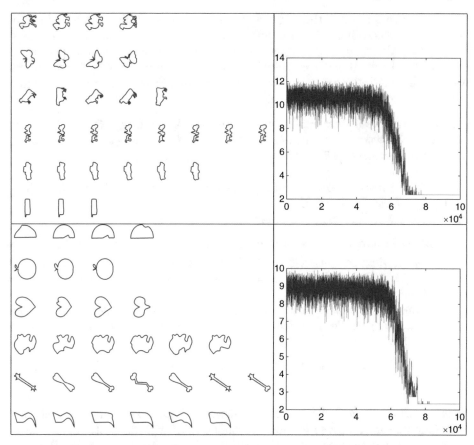

Fig. 9.3 Two examples of clustering of given shapes using Algorithm 37. In each case, we see the resulting clusters (row-wise) and the corresponding evolution of Q during the algorithm

two representations, we shall cover only one of them in detail, namely, the shape analysis of elastic curves under the SRVF representation as elements of $\mathcal{S}_2$ or $\mathcal{S}_2^c$.

We remind the reader of the underlying framework for an elastic shape analysis of planar curves introduced in Chap. 5. A parameterized curve $\beta : [0,1] \to \mathbb{R}^2$ is represented by its SRVF q, where $q(t) = \frac{\dot{\beta}(t)}{\sqrt{|\dot{\beta}(t)|}}$. In case $\dot{\beta}(t) = 0$ for some t, we simply set $q(t) = 0$ for that t. According to Eq. 5.6, the set of SRVFs of unit-length curves forms a Hilbert sphere: $\mathcal{C}_2 = \{q : [0,1] \to \mathbb{R}^2 | \int_0^1 |q(t)|^2 dt = 1\}$. For any $q \in \mathcal{C}_2$, the tangent space $T_q(\mathcal{C}_2) = \{v \in \mathbb{L}^2([0,1], \mathbb{R}^2) | \langle v, q \rangle = 0\}$. The shape space is the quotient space of $\mathcal{C}_2$ modulo rotation group $SO(2)$ and reparameterization group $\tilde{\Gamma}_I$, i.e., $\mathcal{S}_2 = \mathcal{C}_2/(SO(2) \times \tilde{\Gamma}_I)$. The elements of $\mathcal{S}_2$ are given by the orbits: $[q] = \text{closure}\{O(q, \gamma) | (O, \gamma) \in (SO(2) \times \Gamma_I)\}$. Since $\mathcal{C}_2$ is a unit sphere in the Hilbert space $\mathbb{L}^2 \equiv \mathbb{L}^2([0,1], \mathbb{R}^2)$, with a natural embedding, both the extrinsic distance (chord length) and the intrinsic distance (arc length on the sphere) are easily computable. Also, one can easily project an element of $\mathbb{L}^2$ to $\mathcal{C}_2$ by rescaling. Therefore, it becomes easier to perform statistical analysis in $\mathcal{C}_2$ using both extrinsic and intrinsic approaches (see Chap. 7). However, we will restrict ourselves to only the intrinsic approach for performing statistical analysis on the shape space $\mathcal{S}_2$. As described in Chap. 5 (Sect. 5.5), certain computational tools are available for analysis on $\mathcal{S}_2$ and $\mathcal{S}^c$. For any two curves β_1 and β_2, with the SRVFs $q_1, q_2 \in \mathcal{C}_2$, we can compute the following:

1. **Registration**: An optimal registration between SRVFs of the two curves using a combination of Procrustes method (for optimal rotation) and dynamic programming (for optimal re-parameterization).
2. **Geodesic path and geodesic distance**: Using an explicit expression, given in Eq. 5.29, we can compute a constant-speed geodesic α between the two shapes $[q_1]$ and $[q_2]$ such that $\alpha(0) = [q_1]$ and $\alpha(1) = [q_2]$. We use $d_s([q_1], [q_2]) = L[\alpha]$ to denote the geodesic distance or length of α.
3. **Exponential and its inverse**: Given any q_1 and q_2, we can obtain the shooting vector from the first shape to the second using $v = \dot{\alpha}(0) \in T_{[q_1]}(\mathcal{S}_2)$. This v is also termed the inverse exponential map $v = \exp_{[q_1]}^{-1}([q_2])$. Similarly, we can also compute the exponential map $exp_{[q]} : T_{[q]}(\mathcal{S}_2) \to \mathcal{S}_2$, using:

$$\exp_{[q]}(v) = \left[\left(\cos(\|v\|)q + \frac{\sin(\|v\|)}{\|v\|} v \right) \right]. \tag{9.2}$$

In the above equation, if $\|v\| = 0$, then the value of the exponential map is simply $[q]$.

Planar Closed Curves If we restrict ourselves to closed curves, we obtain the pre-shape space $\mathcal{C}_2^c$, see Eq. 6.5, and the shape space $\mathcal{S}_2^c$, the set of all orbits under the actions of $SO(2)) \times \tilde{\Gamma}_S$ on $\mathcal{C}_2^c$. The only difference from representation for general curves comes from the closure constraint: SRVFs representing closed curves have to satisfy an additional condition: $\int_{\mathbb{S}^1} q(t)|q(t)|dt = 0$. (Notice that the domain of representation changes from $[0, 1]$ to $\mathbb{S}^1$, both for curves and re-parameterization functions.) This is actually a two-dimensional vector equation implying two scalar constraints on q. Due to this constraint, the pre-shape space $\mathcal{C}_2^c$ becomes a proper subset of the Hilbert sphere $\mathcal{C}_2$. Consequently, the tangent space is reduced by having two additional perpendicular elements:

$$T_\mu(\mathcal{C}_2^c) = \left\{ v \in T_\mu(\mathcal{C}_2) | v \perp \left(\frac{q^1(t)}{|q(t)|} q(t) + |q(t)|\mathbf{e}^1 \right) \text{ and } v \perp \left(\frac{q^2(t)}{|q(t)|} q(t) + |q(t)|\mathbf{e}^2 \right) \right..$$

(We remind the reader that we set $\frac{q^i(t)}{|q(t)|} q(t) = 0$ whenever $q(t) = 0$.) Here $\mathbf{e}^1$ and $\mathbf{e}^2$ are canonical basis elements for $\mathbb{R}^2$. The expressions for geodesic paths and geodesic distances are no longer available explicitly. They are computed using a numerical path-straightening algorithm (see Sect. 6.6). This algorithm also provides, with obvious modifications, tools for computing the exponential map and its inverse.

9.3.2 Finite Shape Representation: Planar Curves

Next we seek a finite-dimensional, vector-space representation of planar shapes. Since the shape space is nonlinear and infinite dimensional, we need to locally flatten this set into a vector space and then restrict to a finite-dimensional subspace of this vector space. However, we recall that the shape space does not have a manifold structure even though the pre-shape space is a manifold. Due to this situation, we can only view the shape space as a set with a distance inherited from the pre-shape space. However, in the context of deriving statistical models, and reaching finite-dimensional representations of shapes, we will utilize ideas such as the tangent spaces of orbits and the tangent spaces of the quotient space to help out conceptually.

Our approach is to take an orthonormal basis of the full tangent space and find a relevant finite-dimensional subspace. While there are several choices of bases for a function space, we will start with a basic Fourier basis. Later, in the next section, we will advocate the use of training data, if available, to find an appropriate basis using a statistical procedure, say principal component analysis. The structure of $T_{[\mu]}(\mathcal{S}_2)$ is described in Sect. 5.8, specifically in Eq. 5.36, where we use its identification with $N_\mu([\mu])$, which, in turn, is defined by:

$$T_\mu([\mu]) \oplus N_\mu([\mu]) = T_\mu(\mathcal{C}_2) \quad \subset \quad \mathbb{L}^2([0,1], \mathbb{R}^2) \ ,$$

where $T_\mu([\mu]) = T_\mu([\mu]_{SO(2)}) \oplus T_\mu([\mu]_{\Gamma_I})$. Figure 9.4 illustrates this decomposition of tangent spaces into their constituents. Here $T_\mu([\mu]_{SO(2)})$ and $T_\mu([\mu]_{\Gamma_I})$ are spaces tangent to the rotation and re-parameterization orbits of μ, respectively. Recall that, if a $w \in \mathbb{L}^2([0,1], \mathbb{R}^2)$ is perpendicular to μ, then it is in $T_\mu(\mathcal{C}_2)$ (since $\mathcal{C}_2$ is a unit sphere). Additionally:

$$T_\mu([\mu]_{SO(2)}) = \text{span}\{u\}, \quad u(t) = \begin{bmatrix} 0 & -1 \\ 1 & 0 \end{bmatrix} \mu(t), \text{ and}$$

$$T_\mu([\mu]_{\Gamma_I}) = \text{span}\{d\phi_{\gamma_{id}}(v_i), i = 1, 2, \dots\} \ .$$

Recall that $\phi : \Gamma_I \mapsto [\mu]$ is essentially the group action on μ according to $\phi(\gamma) = (\mu, \gamma)$, and the differential of ϕ at γ_{id}, $d\phi_{\gamma_{id}}$ is repeated below in Eq. 9.4 for convenience. In summary, an element w of $T_{[\mu]}(\mathcal{S}_2)$ is perpendicular to (i) μ to respect the unit-length constraint, (ii) $d\phi_{id}(v_i)$s to be orthogonal to the re-parameterization orbit, and (iii) u to be orthogonal to the rotation orbit.

We can use this structure to construct elements of the tangent space $T_{[\mu]}(\mathcal{S}_2)$. We will start with the larger space $\mathbb{L}^2([0,1], \mathbb{R}^2)$ and remove these smaller subspaces to reach the desired set.

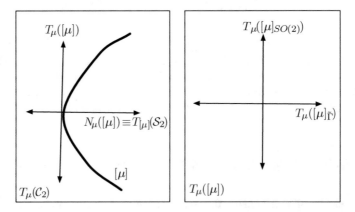

Fig. 9.4 *Left*: Decomposition of the tangent space at a pre-shape $\mu \in \mathcal{C}_2$ into spaces parallel and perpendicular to the orbit $[\mu]$. *Right*: Decomposition of the tangent space parallel to the orbit into spaces parallel to $SO(2)$ and $\tilde{\Gamma}_I$ orbits

- **Basis for the full space** $\mathbb{L}^2([0,1], \mathbb{R}^2)$: We know that:

$$\mathcal{B}_n = \text{span}\{\frac{1}{\sqrt{2}}\cos(2\pi it), \frac{1}{\sqrt{2}}\sin(2\pi it); i = 0, 1, 2, \dots, n\}, \quad (9.3)$$

form a $2n+1$-dimensional orthonormal subspace of $\mathbb{L}^2([0,1], \mathbb{R})$. Then, $\mathcal{B}_n \times \mathcal{B}_n$ forms a convenient orthonormal subspace of $\mathbb{L}^2([0,1], \mathbb{R}^2)$. An element of this product space can be written as:

$$\begin{bmatrix} \sum_{i=0}^n (\frac{a_{i,1}}{\sqrt{2}}\cos(2\pi it) + \frac{b_{i,1}}{\sqrt{2}}\sin(2\pi it)) \\ \sum_{i=0}^n (\frac{a_{i,2}}{\sqrt{2}}\cos(2\pi it) + \frac{b_{i,2}}{\sqrt{2}}\sin(2\pi it)) \end{bmatrix}, a_{i,1}, a_{i,2}, b_{i,1}, b_{i,2} \in \mathbb{R} .$$

- **Basis for the tangent space** $T_\mu(\mathcal{C}_2)$: Since $\mathcal{C}_2$ is simply a unit sphere inside $\mathbb{L}^2([0,1], \mathbb{R}^2)$, the space normal to $T_\mu(\mathcal{C}_2)$, inside $\mathbb{L}^2([0,1], \mathbb{R}^2)$, is a one-dimensional space with the basis element given by μ. Thus, one can take the elements of $\mathcal{B}_n \times \mathcal{B}_n$ and make them orthogonal to μ, followed by the Gram-Schmidt algorithm to reach the desired basis.
- **Basis for the space tangent to** $\tilde{\Gamma}_I$ **at** γ_{id}: The elements of this space are smooth functions on $[0,1]$ that are zero on the boundaries. One orthonormal basis of $T_{\gamma_{id}}(\Gamma_I)$ is given by:

$$\{\frac{sin(2\pi it)}{\sqrt{2}i\pi}, \frac{(cos(2\pi it)-1)}{\sqrt{2}i\pi}, i = 1, 2, 3, ..\} .$$

(See Sect. 4.10.2 for a discussion on possible bases for $T_{\gamma_{id}}(\Gamma_I)$.)
- **Space Tangent to the Orbit** $T_\mu([\mu])$. Here $[\mu]$ is the joint re-parameterization and rotation orbit of μ in $\mathcal{C}_2$. Since the group Γ_I is dense in the set $\tilde{\Gamma}_I$, we will actually use a basis for the $T_\mu([\mu]_{\Gamma_I})$ instead.

Algorithm 38. *Computation of orthonormal basis of $T_\mu([\mu])$*

1. *Start with a finite orthogonal basis $\{v_1, v_2, \dots, v_n\}$ of a subspace of $T_{\gamma_{id}}(\Gamma_I)$ for a large value n. Since, in a computer implementation, one works with a finite partition of the domain $[0,1]$, the number n is typically smaller than the size of this partition.*
2. *Using Eq. 5.35, we can compute the basis vectors $d\phi_{id}(v_i)$ of the re-parameterization orbit:*

$$\tilde{v}_i \equiv (d\phi_{id}(v_i))(s) = \dot{\mu}(s)v_i(s) + \frac{1}{2}\mu(s)\dot{v}_i(s), \quad s \in [0,1] . \quad (9.4)$$

3. *Compute the function $u(t) = \begin{bmatrix} 0 & -1 \\ 1 & 0 \end{bmatrix}\mu(t)$.*
4. *Using Gram-Schmidt (under the standard $\mathbb{L}^2$ metric), construct an orthonormal basis for the span of the set $\{\tilde{v}_1, \tilde{v}_2, \dots, \tilde{v}_n, u\}$. Call these elements $\{c_1, c_2, \dots, c_k\}$ where $k \le (n+1)$; each one is a function from $[0,1]$ to $\mathbb{R}^2$.*

- **Basis for** $T_{[\mu]}(\mathcal{S}_2)$: For each element w in the larger basis, $\mathcal{B}_n \times \mathcal{B}_n$, perform two projections. First, use $w = w - \langle w, \mu \rangle$ to project w into the tangent space $T_\mu(\mathcal{C}_2)$. Then, use $w \mapsto \tilde{w} = w - \sum_{i=1}^k \langle w, c_i \rangle c_i$ to project the resulting w to the tangent space $T_{[\mu]}(\mathcal{S}_2)$. We apply the Gram-Schmidt algorithm to these basis elements to reach an orthonormal basis of a finite-dimensional subspace of $T_{[\mu]}(\mathcal{S}_2)$. We will call this basis set $W = \{\tilde{w}_i, i = 1, 2, \ldots, m\}$, where $m = (2n+1)^2 - (k+1)$.

Using a set of coefficients with respect to this basis set W, i.e., $v = \sum_{i=1}^{m} x_i \tilde{w}_i \in T_{[\mu]}(\mathcal{S}_2)$, we obtain a parametric representation for tangent vectors to the shape space of planar curves. Under the exponential map at the point μ, the span of W projects to a finite-dimensional sphere $\mathbb{S}_m \subset \mathcal{S}_2$. Once again, note that since the elements of W are perpendicular to $T_\mu([\mu])$, we can conveniently identify the exponential map of the span of W as a subset of $\mathcal{S}_2$.

$$\text{Represent}: [q] \in \mathcal{S}_2 \xrightarrow{\exp_{[\mu]}^{-1}([q])} v \in T_{[\mu]}(\mathcal{S}_2) \xrightarrow{\text{Basis } W} x \in \mathbb{R}^m$$

$$\text{Reconstruct}: x \in \mathbb{R}^m \xrightarrow{\text{Basis } W} v \in T_{[\mu]}(\mathcal{S}_2) \xrightarrow{\exp_{[\mu]}(v)} [q] \in \mathcal{S}_2$$

9.3.3 Finite Representation: Planar Closed Curves

Here we make the following modification from the previous section. The domain of a function changes from $[0,1]$ to $\mathbb{S}^1$. We use $s \in [0,1] \mapsto t \equiv (\cos(2\pi s), \sin(2\pi s)) \in \mathbb{S}^1$ to identify the two domains and compose the basis elements in the previous section with this map to reach basis elements for the new domain.

- **Basis for the full space** $\mathbb{L}^2(\mathbb{S}^1, \mathbb{R}^2)$: Since the domain of a closed curve is $\mathbb{S}^1$, we adjust the basis elements of the full space accordingly. Let $\mathcal{B}_n = \text{span}\{\frac{1}{\sqrt{\pi}}\cos(it), \frac{1}{\sqrt{\pi}}\sin(it); i = 0, 1, \ldots, n\}$, a $2n+1$-dimensional orthogonal subspace of $\mathbb{L}^2(\mathbb{S}^1, \mathbb{R})$. Then, $\mathcal{B}_n \times \mathcal{B}_n$ forms a convenient orthogonal basis for a subspace of $\mathbb{L}^2(\mathbb{S}^1, \mathbb{R}^2)$.
- **Basis for the tangent space** $T_\mu(\mathcal{C}_2^c)$: This time, the normal space is actually a three-dimensional space spanned by:

$$\left\{\mu, \left(\frac{\mu^1(t)}{|\mu(t)|}\mu(t) + |\mu(t)|\mathbf{e}^1\right), \left(\frac{\mu^2(t)}{|\mu(t)|}\mu(t) + |\mu(t)|\mathbf{e}^2\right)\right\}. \tag{9.5}$$

- **Space Tangent to Γ_S at γ_{id}**: Now we build a finite basis for the space tangent to the rotation and re-parameterization orbits of μ. Let $\text{span}\{v_1, v_2, \ldots, v_n\}$ be a subspace of $T_{\gamma_{id}}(\Gamma_S)$. The basis functions v_i's essentially span the set of smooth functions on $\mathbb{S}^1$ with no boundary conditions. An orthonormal basis of $T_{\gamma_{id}}(\Gamma_S)$, under the first-order Palais metric, is given by:

$$\left\{\frac{\sin(it)}{\sqrt{\pi i}}, \frac{\cos(it)}{\sqrt{\pi i}}, i = 1, 2, 3, ..\right\}.$$

- **Space Tangent to the Orbit** $T_\mu([\mu])$: This part remains identical to the previous section and Algorithm 38 applies directly with the above definitions. This step results in a basis $\{c_1, c_2, \ldots, c_k\}$ of the $T_\mu([\mu])$.
- **Basis for** $T_{[\mu]}(\mathcal{S}_2^c)$: For each element w in the larger basis, $\mathcal{B}_n \times \mathcal{B}_n$, perform two projections. First, form an orthonormal basis $\{g_1, g_2, g_3\}$ for the three-dimensional space spanned by the elements of the set given in Eq. 9.5. Remove these elements from w using $\tilde{w} = w - \sum_{i=1}^{3} \langle w, g_i \rangle g_i$. Then, remove the basis element of the space tangent to the orbit $[\mu]$ according to

$\tilde{w} \mapsto \tilde{w} = \tilde{w} - \sum_{i=1}^{k} \langle \tilde{w}, c_i \rangle c_i$ to obtain a basis element of the tangent space $T_\mu(\mathcal{S}_2^c)$. These basis elements are rescaled so that they form an orthonormal basis of a finite-dimensional subspace of $T_\mu(\mathcal{S}_2^c)$. We will call this basis set $W = \{\tilde{w}_i, i = 1, 2, \ldots, m\}$, where $m = (2n+1)^2 - (k+3)$.

9.4 Models for Planar Curves as Elements of $\mathcal{S}_2$

Having obtained a (truncated) finite-dimensional representation space for planar shapes, we can now develop statistical models to capture their variability.

9.4.1 Truncated Wrapped-Normal (TWN) Model

What kinds of probability models can be imposed on the shape space $\mathcal{S}_2$? We will closely follow the discussion in Chap. 7 where we studied stochastic models for $\mathcal{P}$, the space of all probability density functions on $[0, 1]$; Γ_I, the group of all warping functions; and $\mathbb{L}^2/\Gamma_I$, the quotient of square-integrable functions modulo the group of warping functions. As discussed there, there is no canonical way to impose a uniform density on full $\mathcal{S}_2$, except when we restrict to a finite-dimensional and bounded subset. Similarly, it is not possible to use an isotropic density on $\mathcal{S}_2$ as it has infinitely many directions and we will not be able to integrate it to reach the normalization constant. Thus, we go directly to imposing a truncated wrapped-normal (TWN) density on $\mathcal{S}_2$, induced via a corresponding density on a Euclidean space (say $\mathbb{R}^m$).

We will impose a truncated normal model on the coefficients $\{x_i\}$ and reach a parametric model on the shape space of planar curves. Similar to Sect. 7.5.3, we are using $TWN(x; K, \lambda)$ to denote the density:

$$TWN(x; K_x, \lambda) \equiv \frac{1}{Z_m} e^{-\frac{1}{2}x^T K_x^{-1} x} \mathbf{1}_{|x| \leq \lambda} , \quad \lambda < \pi .$$

One can add a mean to this model also, if needed. In addition to the threshold λ, the normalization constant also depends on the dimension m. Throughout this chapter, we truncate the density at $\lambda = \pi/2$ although one can allow a larger value, up to π, and still keep the exponential map invertible. A random element of the space spanned by W can be mapped to the shape space $\mathcal{S}_2$ using the exponential map given in Eq. 9.2. The range space of this mapping is a subset of the m-dimensional unit sphere $\mathbb{S}^m$. An explicit expression for the resulting density function on $\mathbb{S}^m$ is available:

$$f(x; \mu, K_x) = \frac{1}{Z_m} \left(\frac{\theta}{\sin(\theta)} \right)^{(m-1)} e^{(-\frac{1}{2}x^T K_x^{-1} x)} \mathbf{1}_{\theta \leq \pi/2} , \quad \theta = \cos^{-1}(\langle \sum_{i=1}^{m} x_i \tilde{w}_i, \mu \rangle).$$
$$(9.6)$$

This lays out a mechanism for constructing a truncated wrapped-normal (TWN) probability model on shapes of planar curves.

Shown in Fig. 9.5 are some examples of random samples from a wrapped-normal model. The first shape in each row is the mean shape $[\mu]$ and the remain-

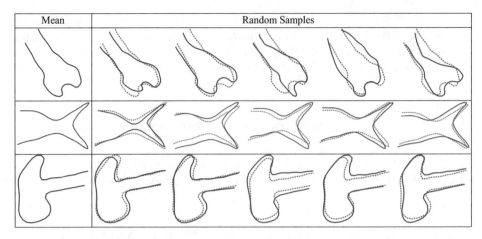

Mean	Random Samples

Fig. 9.5 In each row, the leftmost panel shows a mean shape μ and the other panels show random samples of a truncated wrapped-normal density with mean μ

ing shapes in that row are random samples generated from the wrapped-normal density on the $\mathcal{S}_2$ with that mean. For the first row we used $m = 9$ with the coefficients $x_i \sim TN(0, 0.2e^{-0.3i}, \pi/2)$, for the second row we had $m = 19$ with $x_i \sim TN(0, 0.1e^{-0.3i}, \pi/2)$, and for the third row $m = 39$ with $x_i \sim TN(0, 0.1e^{-0.3i}, \pi/2)$.

9.4.2 Learning TWN Model from Training Shapes in $\mathcal{S}_2$

So far we have used a basic Fourier basis as a starting point for reaching a parametric representation of shapes. However, in the situations where we are given a set of training exemplars, and our goal is to develop a model to capture the shape variability in the given set, we may be better off extracting a basis from the data itself. More generally, we may be interested in estimating the TWN model from the given data or, in other words, estimating the parameters of this model from the data. There are two parameters to be estimated: the shape mean μ and the covariance K_x in the tangent space at Karcher mean $T_\mu(\mathcal{S}_2)$. A spectral representation of this covariance operator also provides a convenient basis for parametric representation of shapes.

Remark 9.1. We clarify that we do not develop a formal asymptotic theory for estimating these mean and covariance parameters of TWN model. Instead, we use the notion of Karcher mean and variance introduced in the previous chapter to estimate these parameters. A rigorous analysis of these sample estimates, especially their convergence to the population parameters under asymptotic situations, remains to be performed.

While the pre-shape space $\mathcal{C}_2$ can be embedded in the Hilbert space $\mathbb{L}^2([0, 1], \mathbb{R}^2)$ naturally, it does not seem natural to embed the shape space $\mathcal{S}_2$ inside a Hilbert space. Therefore, we do not pursue the idea of computing extrinsic means and covariances for shapes of curves. For the purpose of comparison, we will define and compute the extrinsic mean in $\mathcal{C}_2$ and will discuss it in relation to the intrinsic definition of mean on $\mathcal{S}_2$. Let $\{\beta_1, \beta_2, \ldots, \beta_n\}$ be a given collection of curves, with shape representations $\{[q_1], [q_2], \ldots, [q_n]\}$. Considering the SRVFs as elements of the space $\mathbb{L}^2([0,1], \mathbb{R}^2)$, we can compute their average:

$$\mu_{ext}(t) = \frac{1}{n}\sum_{i=1}^{n} q_i(t) \quad \text{and project using} \quad \mu_{ext} \rightarrow \frac{\mu_{ext}}{\|\mu_{ext}\|} \in \mathcal{C}_2 . \tag{9.7}$$

The resulting $\mu_{ext} \in \mathcal{C}_2$ is the extrinsic mean of curves in the pre-shape space.

Next we look at the problem of computing the sample Karcher mean as specified in Definition 7.1. For the given SRVFs, $\{q_1, q_2, \ldots, q_n\}$, the sample Karcher mean is given by:

$$\bar{\mu}_n = \underset{[q] \in \mathcal{S}_2}{\text{argmin}} \sum_{i=1}^{n} d_s([q], [q_i])^2 , \tag{9.8}$$

where d_s denotes the geodesic distance in the shape space $\mathcal{S}_2$ under the elastic metric (Eq. 5.28). A gradient-based approach for finding a Karcher mean on a general Riemannian manifold is given in Algorithm 24. Its modification to include the truncation and its adaptation to $\mathcal{S}_2$ is presented here.

Algorithm 39 (Truncated Karcher Mean on $\mathcal{S}_2$). *Let μ_0 be an initial estimate of the Karcher mean. Since the $q_i s$ are elements of a hypersphere $\mathcal{C}_2$, it seems natural to use their extrinsic mean (Eq. 9.7) to initialize the gradient algorithm. Other choices may also be available in specific contexts. Set $j = 0$.*

1. *For each $i = 1, \ldots, n$, compute the tangent vector v_i such that $v_i = \exp_{[\mu_j]}^{-1}([q_i])$.*
   This can be done as follows. First find the optimal rotation and the optimal re-parameterization of q_i to best match μ_j; call it $\tilde{q}_i$. Then, compute the inverse exponential (using Eq. 5.18):

$$v_i = \frac{\theta_i}{\sin(\theta_i)}(\tilde{q}_i - \cos(\theta_i)\mu_j), \quad \theta_i = \cos^{-1}(\langle \tilde{q}_i, \mu_j \rangle) .$$

2. *Discard those vectors whose magnitude is larger than $\pi/2$ (to facilitate truncation). Compute the average of the remaining vectors: $\bar{v} = \frac{\sum_{i=1}^{n} v_i \mathbf{1}_{\|v_i\| < \pi/2}}{\sum_{i=1}^{n} \mathbf{1}_{\|v_i\| < \pi/2}}.$*
3. *If $\|\bar{v}\|$ is small, then stop. Else, update μ_j in the direction $\bar{v}$ using (Eq. 5.17):*

$$\mu_{j+1} = \cos(\epsilon\|\bar{v}\|)\mu_j + \sin(\epsilon\|\bar{v}\|)\frac{\bar{v}}{\|\bar{v}\|}$$

   where $\epsilon > 0$ is small step size, typically 0.5.
4. *Set $j = j + 1$ and return to Step 1.*

We present an example to illustrate this computation. First, take the set of nine shapes of glasses formed by the open curves shown in Fig. 9.6. Figure 9.7 shows the results of Algorithm 39 for four different initial conditions—once with the extrinsic mean (Eq. 9.7) shown in the top row, twice with individual shapes from the original sets (the bottom two rows), and once with a completely arbitrary shape (second row). The final estimate of the mean shape is similar in all cases. If we look at the variance function that is being minimized here, the final value is close to 0.001, which was the stopping criterion for the iteration.

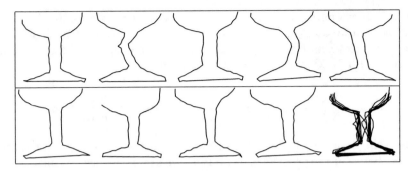

Fig. 9.6 A sample set of nine curves, drawn individually and all together (*bottom right*)

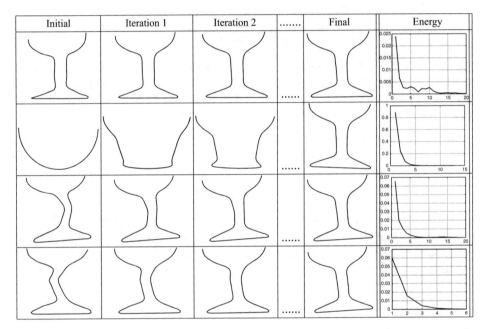

Fig. 9.7 Estimation of Karcher mean of the nine shapes shown in Fig. 9.6. Each row corresponds to a different run for Algorithm 39 with different initial condition. The left panel shows the initial shape and the subsequent panels show some iterations along the final estimate. The last panel shows the change in the cost function versus iteration index

Data	Mean	Data	Mean

Fig. 9.8 Examples of sample Karcher means of 10 handwritten signatures each, for two persons

Figure 9.8 shows an example of computing the truncated sample mean, this time of signature curves for two persons. For each pair, the left panel shows the sample shapes and the right panel shows the corresponding truncated mean shapes.

Once the sample Karcher mean has been computed, the evaluation of the Karcher covariance follows. Let $v_i = \exp_{[\mu]}^{-1}([\tilde{q}_i])$, $i = 1, 2, \ldots, n$ be the shooting vectors generated by the above algorithm. If their magnitude is greater than

$\pi/2$, then they are automatically discarded. The covariance kernel can be defined as a function $K : [0,1] \times [0,1] \to \mathbb{R}$ given by:

$$K(s,t) = \frac{\sum_{i=1}^n v_i(s)v_i(t)\mathbf{1}_{\|v_i\|<\pi/2}}{\sum_{i=1}^n \mathbf{1}_{\|v_i\|<\pi/2}} .$$

Note that the trace of this covariance function is proportional to the Karcher variance, i.e.:

$$\int_0^1 K(t,t)dt = \frac{\sum_{i=1}^n d_s([\mu],[q_i])^2 \mathbf{1}_{d_s([\mu],[q_i])<\pi/2}}{\sum_{i=1}^n \mathbf{1}_{\|v_i\|<\pi/2}} .$$

In practice, since the curves have to be sampled with a finite number of points, say T, the resulting covariance matrices are finite dimensional. Often the observation size n is much less than T and, consequently, n controls the degree of variability in the stochastic model. In other words, n dictates m, the dimension of the subspace (of $T_{[\mu]}(\mathcal{S}_2)$) on which the Gaussian model is eventually imposed.

Principal Component Basis for Shape Representation In the previous section, we used a generic basis for the tangent space $T_{[\mu]}(\mathcal{S}_2)$; we started with the Fourier basis for $\mathbb{L}^2([0,1],\mathbb{R}^2)$ and made its components orthogonal to the subspace $T_\mu([\mu])$. However, in the case of learning shape models from the observations, one can reach a more efficient basis for $T_{[\mu]}(\mathcal{S}_2)$ using the traditional principal component analysis, as described next.

With a slight abuse of notation, let $\mathbf{v}_i \in \mathbb{R}^{2T}$ denote the tangent vector function $v_i(t)$ sampled at T points and the components concatenated to form a tall vector. Then, the sample covariance matrix $\hat{K} = \frac{1}{n-1}\sum_{i=1}^n \mathbf{v}_i\mathbf{v}_i^T$ is a $2T \times 2T$ symmetric, non-negative matrix with the singular value decomposition $\hat{K} = U\Sigma U^T$. We will assume that the elements of Σ have been arranged in a non-increasing order from top left to bottom right. Then, a multivariate Gaussian model for the tangent vector $\mathbf{v} \in \mathbb{R}^{2T}$ is given by:

$$\mathbf{v} = \sum_{i=1}^n x_i U_i, \qquad (9.9)$$

where $x_i \sim TN(0,\Sigma_{ii})$ (independently) and U_i are columns of U matrix. One can rearrange the elements of $\mathbf{v} \in \mathbb{R}^{2T}$ (to undo the earlier concatenation) and form an approximation of a corresponding element of the tangent space $T_\mu(\mathcal{S}_2)$; call it v. This procedure leads to a basis $\{U_i\}$, which is typically more efficient than the earlier Fourier construction. Note that, by definition, these U_is are orthogonal to the subspace $T_\mu([\mu])$ and do not require any further orthogonalization. Also, these U_is are precisely in the directions where the actual variability of shapes lies. This random v can then be projected in the shape using the exponential map $v \mapsto \exp_\mu(v)$ (Eq. 9.2) to obtain a random shape. This provides a technique for sampling from the wrapped-Gaussian model on $\mathcal{S}_2$. The functional form of this density is the same as that given in Eq. 9.6, except the basis W used there is replaced by the basis U constructed here.

We illustrate this idea using the nine shapes shown in Fig. 9.6; we have already computed their Karcher mean shape earlier. Now we can project these nine shapes into the tangent space $T_\mu(\mathcal{S}_2)$ and compute the sample covariance matrix. In this implementation, these curves are sampled using $T = 100$ points each. Therefore, the sample covariance matrix is of size 200×200 although its rank is much lower

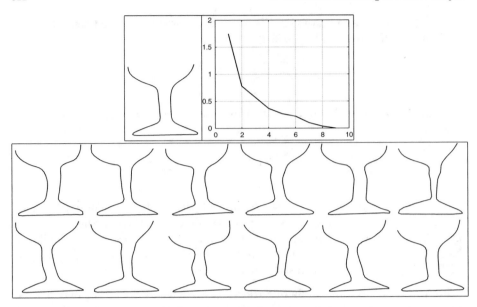

Fig. 9.9 *Top*: Karcher mean (*left*) and the singular values of the sample covariance matrix (*right*) for the nine wineglass shapes. *Bottom two rows*: Random samples from the wrapped-Gaussian density on $\mathcal{S}_2$ whose parameters (mean and covariance) were estimated from the given nine shapes

since we use only nine independent samples. Figure 9.9 top row shows a plot of the singular values Σ_{ii} in the decreasing order. Now, using these singular vectors and the singular values, we define a normal model on the tangent space $T_\mu(\mathcal{S}_2)$ according to Eq. 9.9. Wrapping these random vectors onto the shape space $\mathcal{S}_2$, using the exponential map, results in random shapes from this wrapped-Gaussian model. Some of these random shapes are shown in the bottom two rows of Fig. 9.9. A visual comparison of these random shapes with the original nine shapes in Fig. 9.6 tells us that the model is not only successful in capturing the variability in the original set; it generates realistic random shapes.

Figure 9.10 shows the variations of shapes along the principal geodesic curves around the mean shape μ. Each of the three rows show shapes along the paths $t \mapsto \exp_\mu(\pm 2t\sqrt{\Sigma_{ii}}U_i)$ for $i = 1, 2$, and 3. These plots start from $t = -1$ on the left and increase t until we reach $+1$ on the right. The middle shape in each row is the mean μ that corresponds to $t = 0$. The three rows capture the three main "directions" of deformations (from the mean) in the original nine curves.

9.5 Models for Planar Closed Curves

The next step is to study the statistics of shapes of planar-closed curves using the elastic representation. For comparisons, however, we will also provide results from the non-elastic representation, without providing the full details. We will focus on learning of TWN model and its parameters: mean, covariance, and the basis elements, from the training data itself. We will assume that we have some training curves in several shape classes and our goal is to develop and study some stochastic

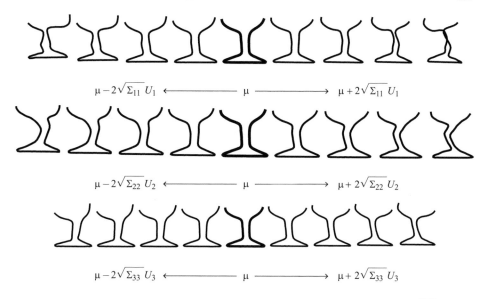

Fig. 9.10 Principal geodesic curves: the three rows show the paths $t \mapsto \exp_\mu(\pm 2\sqrt{\Sigma_{ii}} t U_i)$ for $i = 1, 2$, and 3, where U_is are the principal singular vectors of the sample covariance matrix

models that capture shapes in those training sets. Given a set of closed curves, we first want to compute its Karcher mean and then use that mean for defining the tangent space on which the covariance is computed.

1. **Sample Karcher Mean**: The algorithm for computing the Karcher mean of elastic closed curves is given below:

   Algorithm 40 (Sample Karcher Mean in $\mathcal{S}_2^c$).
   Set $k = 0$. Choose some time increment $\epsilon \leq \frac{1}{n}$. Choose a point $[\mu_0] \in \mathcal{S}_2$ as an initial guess of the mean. (For example, one could just take $\mu_0 = q_1$.)

   a. *For each $i = 1, \dots, n$, choose the tangent vector $v_i \in T_{[\mu_k]}(\mathcal{S}_2^c)$ that is tangent to the geodesic from $[\mu_k]$ to $[q_i]$, and whose norm is equal to the length of this shortest geodesic. This is accomplished using Algorithm 22 presented in Chap. 6.*
   b. *Compute the shooting vector $\bar{v} = \frac{1}{n} \sum_{i=1}^{n} v_i \in T_{[\mu_k]}(\mathcal{S}_2^c)$.*
   c. *Flow for time ϵ along the geodesic that starts at $[\mu_k]$ and has velocity vector $\bar{v}$. Call the point where you end up $[\mu_{k+1}]$, i.e., $[\mu_{k+1}] = \exp_{[\mu_k]}(\epsilon \bar{v})$. This is implemented numerically using Algorithm 20.*
   d. *Set $k = k + 1$ and go to Step 1.*

Figure 9.11 shows examples of Karcher means of some sets of shapes. For each example, shown in a separate row, the given shapes are shown in the left panel and their Karcher mean is shown in the right panel. A visual inspection of these mean shapes reveals that although the means appeared to have been smoothed out, they still retain the important geometric features characterizing the original sets. In the case of human running silhouettes, the average shape has distinct body parts such as head, legs, feet, etc. Similarly, the mean crown shape retains the five crown tips that are present in each of the given crown shapes; these tips have different sizes and locations in the individual shapes

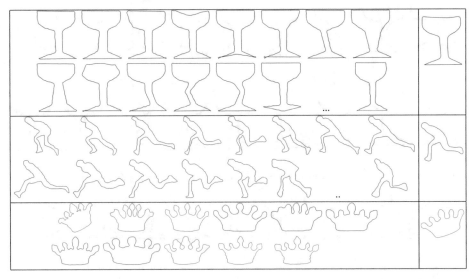

Fig. 9.11 Karcher mean of elastic shapes. The sample shapes are shown in the top three rows, while the Karcher mean and the evolution of the norm (gradient) are shown in the bottom row

but the process of elastic matching allows them to be matched and averaged together for computing the Karcher mean.

2. **Sample Karcher Median**: In situations where there is an outlier present in the sample set, it has the possibility of influencing the sample mean. Therefore, as an alternative, one can compute the sample median as a more robust estimate of the population mean shape. The algorithm for computing the Karcher median of elastic curves is given below:

Algorithm 41 (Sample Karcher Median in $\mathcal{S}_2^c$).

Set $k = 0$. Choose some time increment $\epsilon \leq \frac{1}{n}$. Choose a point $[\mu_0] \in \mathcal{S}_2$ as an initial guess of the median.

a. *For each $i = 1, \ldots, n$, find the shooting vector (i.e., tangent vector $v_i \in T_{[\mu_k]}(\mathcal{S}_2^c)$ that is tangent to the geodesic) from $[\mu_k]$ to $[q_i]$, and whose norm is equal to the length of this shortest geodesic. This is accomplished using Algorithm 22 presented in Chap. 6.*

b. *Let $d_i = \|v_i\| \in \mathbb{R}$ be the norm of the shooting vectors. Then, define the update direction to be $\bar{v} = \frac{\sum_{i=1}^{n} v_i/d_i}{\sum_{i=1}^{n} 1/d_i} \in T_{[\mu_k]}(\mathcal{S}_2^c)$.*

c. *Flow for time ϵ along the geodesic that starts at $[\mu_k]$ and has shooting vector $\bar{v}$. Call the point where you end up $[\mu_{k+1}]$, i.e., $[\mu_{k+1}] = \Psi([\mu_k], \epsilon, \bar{v})$. This is implemented numerically using Algorithm 20.*

d. *Set $k = k + 1$, and go to Step 1.*

In order to demonstrate the robustness of Karcher median shape to outliers, we perform the following experiment. We take 12 shapes from a set of car silhouettes (some examples are shown in the top row on Fig. 9.12) and estimate their sample means by changing the sample sets. Starting with 12 cars, we add replace 2 cars by 2 runner shapes (examples in the second row of Fig. 9.12) in successive experiment. The sample mean shapes are shown in the right column of Fig. 9.13. As more and more cars are replaced by runners, the effect on the sample mean starts growing and finally, in the last row, the average shape is

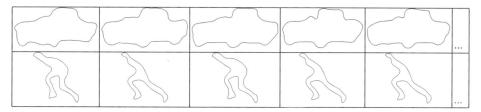

Fig. 9.12 Database of shapes used in comparing mean and median shapes

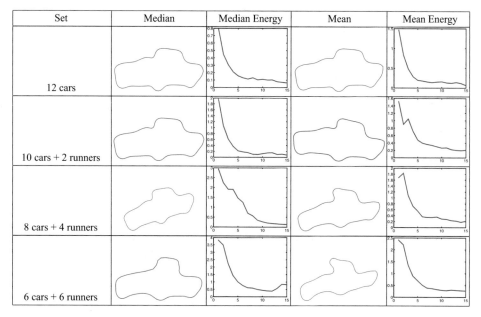

Fig. 9.13 Median and mean shapes obtain for different combinations of cars and runner shapes taken from the set shown in Fig. 9.12. Cars are the main set and runners are considered outliers

far from a car shape. In contrast, consider the corresponding median shapes shown in the left panel of this figure. Despite an increasing number of outliers, the median shape is found to be relatively more stable and better resembles a car silhouette.

3. **Covariance Estimation**: Once we have an estimate of the Karcher mean, the tangent space at that point is used to define and compute a covariance matrix for the given sample. Let $f_i = \exp_{[\mu]}^{-1}([q_i])$ be the projection of the ith shape in the tangent space $T_{[\mu]}(\mathcal{S}_2^c)$. This is actually implemented using the algorithm for finding geodesics between given shapes in $\mathcal{S}_2^c$. Also, since the implementation involves using discrete set of points on curves, the covariance matrices are of size $2T \times 2T$, where T is the number of points sampled on each curve. Let $\hat{K}$ be a covariance matrix for a shape class and let $\hat{K} = U\Sigma U^T$ be its singular value decomposition. Assuming that the singular values in Σ are arranged in a non-increasing order, the corresponding singular vectors $U_1, U_2, \ldots$, denote the principal directions of variations for that class. These directions can be mapped back to the shape space using the exponential map, and the resulting geodesic curves are called the principal geodesic curves. For instance, the

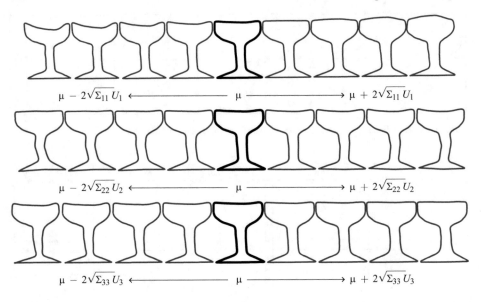

$$\mu - 2\sqrt{\Sigma_{11}}\,U_1 \longleftarrow \qquad \mu \qquad \longrightarrow \mu + 2\sqrt{\Sigma_{11}}\,U_1$$

$$\mu - 2\sqrt{\Sigma_{22}}\,U_2 \longleftarrow \qquad \mu \qquad \longrightarrow \mu + 2\sqrt{\Sigma_{22}}\,U_2$$

$$\mu - 2\sqrt{\Sigma_{33}}\,U_3 \longleftarrow \qquad \mu \qquad \longrightarrow \mu + 2\sqrt{\Sigma_{33}}\,U_3$$

Fig. 9.14 Display of shape variations around the mean in the tangent eigen directions. The first row shows the first principal direction, the second row shows the second direction, and so on

curve $t \mapsto \exp_{[\mu]}(t\sqrt{\Sigma_{11}}U_i)$ denotes the first principal geodesic curve of that class. Figure 9.14 shows the three top principal geodesic curves associated with the shape class containing wine glasses. As we move along the first principal geodesic, the main deformation in the shape is the bending up and down of the rim of the glass. The stem and the base of the glass remain unchanged. The second principal geodesic curve, on the other hand, changes the shape of the step—bending it left and right—while keeping the rim and the base unchanged. It is interesting to note how these principal directions (and the associated principal geodesics) capture certain dominant yet unrelated components of the deformations present in the original data.

As another example, Fig. 9.15 shows the principal geodesic curves associated with the given runner silhouettes. The deformations associated with the three principal directions make an interesting study. The first principal geodesic curve shows a deformation that bends and straightens the two legs—as if to capture a stride. The bending of the front leg is much more pronounced than that of the hind leg. In case of the second principal geodesic, the bending of the hind leg is much more significant than that of the front leg. The third principal geodesic keeps the legs fixed while changing the posture of the back. Since the movement of legs is much more than the back, the deformation along the third principal geodesic is relatively small.

4. **TWN Shape Model**: Now we have a simplified representation of shapes—each shape q is represented by a set of coefficients with respect to an orthonormal basis $\{U_1, U_2, \ldots, U_n\}$ in the tangent space $T_\mu(\mathcal{S}_2^c)$. In other words:

$$q \equiv \exp_\mu\left(\sum_{i=1}^{n} x_i U_i\right). \quad \text{(Using Algorithm 20.)} \qquad (9.10)$$

In order to specify a stochastic model on q, it is sufficient to specify a model on the coefficients $\mathbf{x} = \{x_1, x_2, \ldots, x_n\}$.

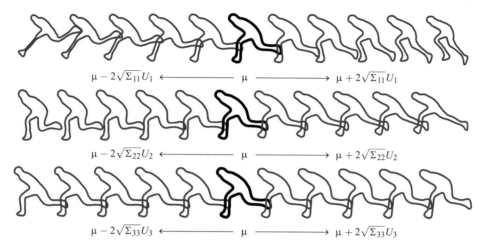

$$\mu - 2\sqrt{\Sigma_{11}}U_1 \longleftarrow \qquad \mu \qquad \longrightarrow \mu + 2\sqrt{\Sigma_{11}}U_1$$

$$\mu - 2\sqrt{\Sigma_{22}}U_2 \longleftarrow \qquad \mu \qquad \longrightarrow \mu + 2\sqrt{\Sigma_{22}}U_2$$

$$\mu - 2\sqrt{\Sigma_{33}}U_3 \longleftarrow \qquad \mu \qquad \longrightarrow \mu + 2\sqrt{\Sigma_{33}}U_3$$

Fig. 9.15 Display of shape variations around the mean in the tangent eigen directions. The first row shows the first principal direction, the second row shows the second direction, and so on

9.6 Beyond TWN Shape Models

In the previous sections, we have suggested the use of independent normal densities for finite-dimensional representations of shapes. Although this choice seems convenient for analysis and simulation, is it actually supported by the data? To answer this question, we study the observed statistics of the principal coefficients for some biological cell shapes. This application, coming from image cytometry, is concerned with measuring shapes of biological cell populations *in vitro* using electron microscopy images of blood samples. This technique is used in cancer studies, drug screening, and genomic research, among other applications.

Shown in the top panel of Fig. 9.16 are 54 examples (out of a set of 100 used in this experiment) of cell contours extracted automatically from electron microscopy images. These cells are characterized by elongated spindle-like structures with 2, 3, 4, or even more corners. Shown in the bottom row of that figure is the mean shape of these 100 samples—it shows similar elongated structure with two prominent corners and a third smaller corner. The right panel shows a plot of the eigenvalues of the covariance matrix. One can see that most of the variability can be captured using 20–25 principal components. It should be noted that these contours have 100 points on them.

Using the principal components of the covariance matrix, we obtain the observed values of the principal coefficients. Seeking a nonparametric model, we estimate the underlying probability density for each coefficient independently using a kernel density estimator. This assumption of independence may not be appropriate but the task of estimating the joint density nonparametrically is quite difficult. The results are presented in Fig. 9.17 where we show the estimated pdfs for the top ten principal coefficients. In this case, most of the estimated pdfs are unimodal and can pass the test for normality. One of them (the dominant one) has two modes, and some are heavily skewed in one direction. In general situations where TWN does not apply, we can use the following alternatives.

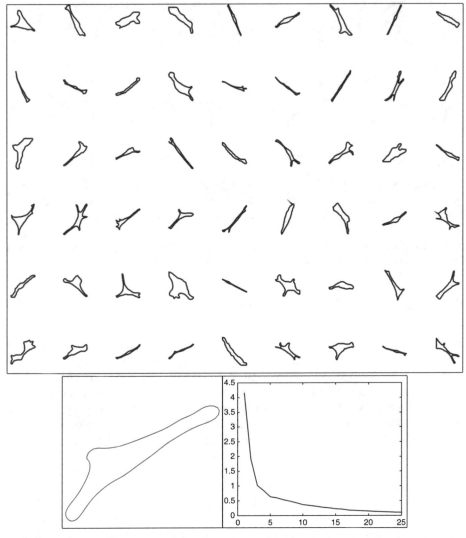

Fig. 9.16 Sample Karcher mean (*bottom left*) of a set of cell shapes shown at the top. The bottom right shows the evolution of the sum of squared distances during the mean computation

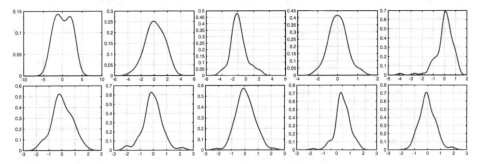

Fig. 9.17 Kernel density estimates for the top ten principal coefficients of the covariance matrix for NIH cell data

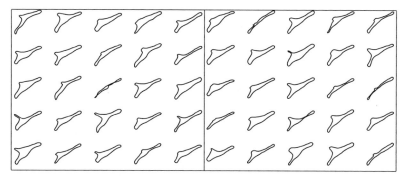

Fig. 9.18 Random samples from a wrapped-Gaussian model (*left*) and the independent nonparametric model (*right*) for cell shapes

1. **Nonparametric Model**: In this case, we assume that each coefficient is an independent random variable with the probability given by the nonparametric estimate of its pdf from the training data. Let the estimated pdfs be given by f_i and we model the coefficient $z_i \sim f_i$. The joint density function of the coefficients is given by $f(z_1, z_2, \ldots, z_n) = \prod_{i=1}^{n} f_i(z_i)$. The shape variable is generated using the mapping $(z_1, z_2, \ldots, z_n) \mapsto q \equiv \exp_\mu(\sum_{i=1}^{n} z_i U_i)$. In order to simulate from this model, we can simply simulate each coefficient z_i from its density function f_i and then reconstruct the resulting SRVF q using the above map. Then, we can integrate q to form coordinates of the random shapes: $\beta(t) = \int_0^t q(s)|q(s)|ds$. Figure 9.18 (right) shows some examples of the samples from this nonparametric model learned from the training shapes shown in Fig. 9.16.

2. **Mixture of Gaussian Model**: The other idea, to handle the non-Gaussian nature of the principle coefficients, is to model them as mixtures of Gaussians. As the plot for the estimated density of first principal coefficient, shown in Fig. 9.17, suggests, one should be able to model it using two components in the mixture. More generally, each coefficient is modeled independently and the parameters of the components (mean, variance, and proportion) can be different for different coefficients. Formally:

$$z_i \sim f_i \equiv w_{i,1} N(\mu_{i,1} \sigma_{i,1}^2) + w_{i,2} N(\mu_{i,2} \sigma_{i,2}^2) \ .$$

Estimation of parameters $\{\mu_{i,j}, \sigma_{i,j}, w_{i,j} | i = 1, 2, \ldots, n, \ j = 1, 2\}$ is performed using the standard expectation maximization (EM) algorithm.

9.7 Modeling Nuisance Variables

We have discussed statistical models for shapes of continuous curves in the previous three sections. In some problems, one also needs to model all the nuisance variables that were so meticulously removed in the shape analysis. An example of this situation is when we develop a model for the full curve, and not just its shape. The nuisance variables can in general be translation, rotation, scale, and re-parameterization. Since translation and scaling can be removed and parameterization reintroduced in a relatively simple manner, we restrict our discussion to modeling the two remaining variables. We start with the re-parameterization variability.

9.7.1 Modeling Re-Parameterization Function

We start by considering models for elements of Γ_I (one can treat Γ_S similarly) that can be used to re-parameterize curves. Since Γ_I is infinite dimensional, it is not possible to use standard probability models on this set and we will restrict ourselves to a finite-dimensional subset. The other issue, albeit a minor one, is that Γ_I is an affine space and one needs to take that into account. In fact, if we use SRVF representation for elements of Γ_I, as suggested in Sect. 7.5.4 earlier, then the resulting space is actually nonlinear (an orthant of a unit sphere) and we will need to use the geometry of a sphere to model a re-parameterization function.

Since, for any $\gamma \in \Gamma_I$, $\dot\gamma$ is a proper pdf, we can use this relation to implicitly impose a model on γ. Therefore, a natural finite-dimensional subset is to find a parametric family of pdfs on $[0,1]$, and to impose probability models on the corresponding parameters. Some examples are:

1. **Power Family**: A simple one-parameter family of pdfs on $[0,1]$ is $p_a(t) = at^{a-1}$, for $a > 0$, and the corresponding re-parameterization function $\gamma_a(t) = t^a$, $a > 0$. We can make γ_a random by assuming $a \sim \exp(\theta)$, exponential random variable with mean $\theta > 0$, or a scaled-gamma density. Figure 9.19 shows a set of 20 random samples of γ_a generated using a scaled-gamma density for a with the shape parameter set to 1.0 and the scale parameter being 10.0.

2. **Beta Family**: Another possibility, this time a two-parameter family, is the *beta density* $p_{a,b}(t) = \frac{t^a(1-t)^b}{B(a,b)}$ and the corresponding re-parameterization function $\gamma_{a,b}(t) = \int_0^t p_{a,b}(s)\, ds$. The pdfs in this family are all unimodal with the location of the mode decided by the mode parameters. When $b = 0$, the pdf reduces to the one in the previous item. We can now make $\gamma_{a,b}$ random by imposing probability models on a and b. Similar to the previous case, we can assume a and b to be random variables, e.g., exponential or scaled gamma. Figure 9.19 shows 20 random samples from Beta family with a, b being scaled gamma with shape and scale parameters being 1.0 and 10.0, respectively.

3. **Truncated Normal Distribution**: Since the normal distribution is defined over $\mathbb{R}$, we can restrict it to $[0,1]$ by truncation and obtain a desired pdf. The general expression for this pdf is given by, for $\mu \in (0,1)$, $\sigma > 0$, and $t \in [0,1]$,

$$p_{\mu,\sigma}(t) = \frac{\frac{1}{\sigma}\phi(\frac{t-\mu}{\sigma})}{\Phi(\frac{1-\mu}{\sigma}) - \Phi(\frac{-\mu}{\sigma})}\ ,\quad \phi(t) = \frac{1}{\sqrt{2\pi}}\exp(-\frac{1}{2}t^2),\quad \Phi(t) = \int_{-\infty}^{t}\phi(s)\, ds\ .$$

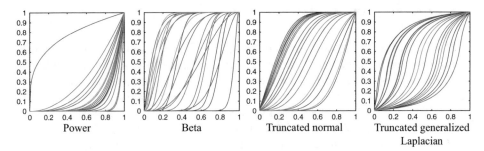

Fig. 9.19 Random simulations of γ from different parametric families

The resulting re-parameterization function is given by:

$$\gamma_{\mu,\sigma} : [0,1] \to [0,1], \quad \gamma_{\mu,\sigma}(t) = \frac{\Phi(\frac{t-\mu}{\sigma}) - \Phi(\frac{-\mu}{\sigma})}{\Phi(\frac{1-\mu}{\sigma}) - \Phi(\frac{-\mu}{\sigma})}.$$

For sharper peaks in the pdfs, one can also use the truncated Laplacian density and a special case of that is the truncated double exponential density.

4. **Other Families**: There are a number of other parametric families of pdfs on $[0,1]$ that can also be used for this purpose. The key is a choose family that allows different numbers, locations and, heights of the modes in pdfs, resulting in a rich family of re-parameterization functions. For instance, one can use a truncated mixture of normal densities. The difficulty in choosing such complex families stems from the challenge in estimating the warping parameters from the given data.

Distance-Based Model So far we have looked at the parametric families, but one may need a larger class of re-parameterization functions in practical situations. One can try for a nonparametric approach albeit with some probabilistic structure. One nonparametric idea is to impose a Gaussian-type distribution (on a finite dimensional subset of Γ) of the form:

$$p(\gamma|\gamma_0) \propto e^{-\frac{1}{2\sigma_s^2} d_\gamma(\gamma,\gamma_0)^2}, \tag{9.11}$$

where d_γ is the Fisher-Rao distance on Γ_I and γ_0 is the chosen central point. This, of course, is not a proper pdf since, in general, its integral may not be defined. Nevertheless, one can use this as an improper pdf, especially in situations where sampling from a pdf does not require the knowledge of the normalization constant. As stated in Sect. 4.10.2, the expression for the Fisher-Rao distance on Γ_I is simply $d_\gamma(\gamma_1, \gamma_2) = \cos^{-1}(\langle \sqrt{\dot{\gamma}_1}, \sqrt{\dot{\gamma}_2} \rangle)$. We discuss two possibilities for γ_0:

1. **Uniform Sampling**: The simplest possibility is to emphasize the samplings of a curve that are uniform with respect to its arc-length parameterization by choosing $\gamma_0(s) = s$.
2. **Curvature-Based Sampling**: Alternatively, γ_0 may depend on local geometrical properties; for example, the sampling density may increase with increasing curvature of the underlying curve β. Define $E(s) = \int_0^s \exp(|\kappa(s)|/\rho) \, ds'$, where $\kappa(s)$ is the curvature of β at arc-length parameter point s' and $\rho \in \mathbb{R}_+$ is a constant. The ratio $\gamma_I(s) = E(s)/E(1)$ is a diffeomorphism, from $[0,1]$ to itself, and the desired sampling for that curve is $\gamma = \gamma_I^{-1}$. The inverse of γ_I can be numerically estimated using a spline interpolation. To define a single γ_0 for each class, we use training curves, as follows. First we compute γ_q for each training curve, and then, using the techniques presented in Sect. 7.5.4, we compute their Karcher mean, which we use as γ_0. Here we use the Karcher variance for σ_s^2. For this computation, the training curves are aligned, something which is done automatically when geodesics are computed between the shapes of the curves. We now illustrate these ideas with some examples.

Remark 9.2. It is interesting to note that the re-parameterization variability is closely related to variability in the way curves can be discretized. A discretization requires two items: the number of points and their placements on the curve. It is

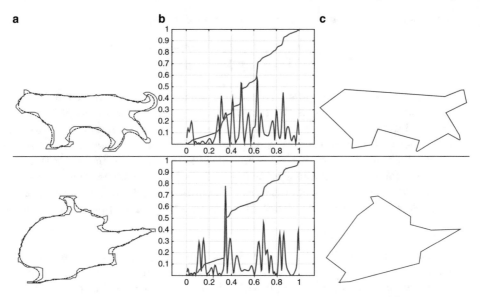

Fig. 9.20 Curvature-driven sampling: (**a**) a curve and its smoothed version, (**b**) the curvature function $\kappa(s)$ and the corresponding parameterization function $\gamma(s)$, and (**c**) curve samples using $\gamma(i/n)$

the placement part that can be controlled by re-parameterization, as follows. Fix the number of points to be n and start with a uniform partition of the $[0,1]$ in n subintervals. If we parameterize a unit-length curve β by its arc length, then these points $\{\beta(\frac{i}{n})|i = 0, 1, 2, \ldots, n\}$ form a uniform discretization of β. However, for any $\gamma \in \Gamma_I$, if we select the points $\{\beta(\gamma(\frac{i}{n}))|i = 0, 1, 2, \ldots, n\}$, this represents a different discretization of β. Thus, one can control discretization of curves using piecewise-linear re-parameterization functions.

Shown in the left column of Fig. 9.20, column (a), are two contours. We smooth these curves using Gaussian filters and their smoothed versions are drawn on top of them. For these smoothed curves, we compute the curvature function κ and then $E(s)$. These functions are displayed in (b), along with the resulting γ_I's. Column (c) shows the original curves sampled using the resulting γ_I. Figure 9.21 shows some examples of class-specific means of the γ_I's for two classes. By using these means as γ_0 for each class, we can form class-specific priors of the form given in Eq. 9.11.

To simulate random samples from distance-based models (of the form given in Eq. 9.11), we take the following steps: Define $\psi_0 = \sqrt{\gamma_0}$.

Algorithm 42 (Random Sampling of γ from a Gaussian-Type Model).

1. *Form an orthogonal basis of the tangent space $T_{\psi_0}(\mathbb{S}_\infty)$ as follows. We know that $T_{\psi_0}(\mathbb{S}_\infty) = \{v \in \mathbb{L}^2([0,1], \mathbb{R})| \langle v, \psi_0 \rangle = 0\}$. So, we can start with any basis of $\mathbb{L}^2([0,1], \mathbb{R})$ and make it orthogonal to ψ_0 to reach a basis $\{b_1, b_2, \ldots,\}$ for the tangent space $T_{\psi_0}(\mathbb{S}_\infty)$.*
2. *Generate a set of coefficients from i.i.d. normal distribution $\mathcal{N}(0, \sigma_s^2)$ and use the chosen basis to construct an element $f(t) = \sum_{i=1}^{l} c_i b_i(t) \in T_{\psi_0}(\mathbb{S}_\infty)$, for a large enough l.*

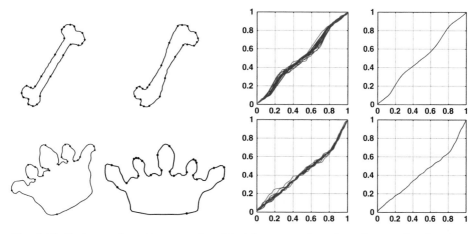

Fig. 9.21 Each row shows two examples of training curves in a class, the sampling functions for that class, and their Karcher means

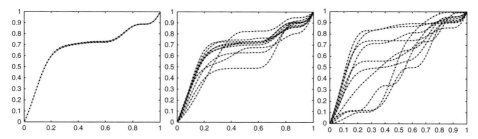

Fig. 9.22 Random samples generated using Algorithm 42 with σ_s^2 increasing from left to right

3. *Finally, use the exponential map on $\mathbb{S}_\infty$, according to $\psi = \cos(|c|)\psi_0 + \sin(|c|)f/|c|$.*
4. *The random re-parameterization function is then given by $\gamma(t) = \int_0^t \psi(s)^2 \, ds$.*

Figure 9.22 shows some examples of random simulations from such a class-specific prior density for increasing values of σ_s^2. If σ_s is too large, then many of the sampled points will lie outside the desired set since ψ can become negative. Such points may, however, simply be rejected from the samples, thus preserving the form of the density given above. For efficiency's sake, though, the proportion of such points should not be too large, and this implies a constraint on σ_s.

9.7.2 Modeling Shape Orientations

For a planar shape, its orientation in a fixed coordinate system is given by a rotation matrix $O \in SO(2)$. Since $SO(2) \equiv \mathbb{S}^1$, one can impose a probability model on $\mathbb{S}^1$ to capture the rotation variability of placing a shape in a scene. The are several choices available for imposing probability distributions on $\mathbb{S}^1$, and some of them are discussed in Sect. 7.5.1. The two main cases discussed there are the uniform density and the von Mises density. The parameters of the von Mises density and their estimation are described previously.

9.8 Classification of Shapes With Contour Data

One of the main reasons for developing statistical models for shapes and associated variables is their use in classification of previously unobserved shapes. The framework for shape representations and statistical models on shape spaces developed so far has important applications in decision theory. Statistical models of shapes can be used to make a variety of decisions such as: Does this shape belong to a given family of shapes? Do the given two families of shapes have similar means and/or variances? Given a test shape and several competing probability models, which one explains the test shape better? These questions can be addressed using TWN and other models within the framework of decision theory. Next, we are going to develop techniques for answering these questions. In this section, we will assume that the data is already available in the form of planar, closed curves. Therefore, we can directly use shape models developed for the shape space $\mathcal{S}_2^c$ and apply them in classification.

We shall use several examples to demonstrate our methods. One of the datasets we will use is that of several civilian vehicles viewed from different horizontal views. These vehicles—cars, vans, and pickup trucks—are rotated horizontally and are imaged from a fixed camera location. We extract their silhouettes from the resulting images and study the problem of classifying these vehicles by only using the shapes of these contours. There are 10 vehicles used in this experiment and labeled as Avalon, Camry, Jeep93, Jeep99, Maxima, MazdaMPV, Mitsubishi, Sentra, Tacoma, TaurusSE96, and Civic4dr. Some examples of their contours are shown in Fig. 9.23 where we display the silhouettes of each of the 10 vehicles

Fig. 9.23 Shapes of silhouettes of 10 vehicles from three different poses: 0°, 50°, 90°, and 180°. The vehicles are Avalon, Camry, Jeep93, Jeep99, Maxima, MazdaMPV, Mitsubishi, Sentra, Tacoma, TaurusSE96, and Civic4dr

from four viewing angles: $0°$ (sideways), $50°$, $90°$, and $180°$. For each vehicle, we obtain a total of 36 silhouettes, taken from viewing angles separated by $10°$. In the following, we will use the notation $i(j)$ to denote a profile where $i = 1, 2, \ldots, 10$ is the vehicle index and $j = 1, 2, \ldots, 36$ is the viewing angle index.

We establish some notation for the classification problem. Let C_i, $i = 1, 2, \ldots, 10$ denote the ten shape classes and let q_{test} be the SRVF of a test silhouette. We define a classifier to be a function that assigns a class C_i to the test silhouette. More precisely, define a classifier to be the mapping $\zeta : \mathcal{S}_2^c \to \{C_i | i = 1, 2, \ldots, 10\}$.

9.8.1 Nearest-Neighbor Classification

In cases where the feature space is a metric space, i.e., one can compute a distance between any two values of the chosen feature; one of the simplest classifiers to implement is the nearest-neighbor classifier. Here one takes the feature of the test object; compares with all the elements in the training set, using the distance function; and finds the nearest training objects. Then, the class associated with the nearest object is selected for classifying the test object.

We particularize this method to the classifying of contours according to their shapes and will use the distance function for elastic closed curves (in $\mathcal{S}_2^c$) for comparing shapes. Mathematically, the nearest-neighbor classifier is defined to be a mapping $\zeta_{near} : \mathcal{S}_2^c \to \{C_i | i = 1, 2, \ldots, 10\}$ according to:

$$\zeta_{near}([q_{test}]) = C_{\hat{i}}, \quad \text{where} \quad \hat{i} = \underset{i}{\operatorname{argmin}} \left(\min_j d_s([q_{test}], [q_{i(j)}]) \right) . \tag{9.12}$$

Here $q_{i(j)}$ denotes the SRVF of the jth silhouette of the ith vehicle. If, instead of finding the nearest shapes, we select k-nearest shapes, we obtain a k-nearest-neighbor (kNN) classifier.

We present an example of this idea in Fig. 9.24. In this experiment, we choose the even-numbered profiles $(2, 4, 6, \ldots, 36)$ of each vehicle as the training set and the odd-numbered profiles $(1, 3, 5, \ldots, 35)$ as the test set. For each test shape, we search over the even-numbered profiles for each object and select the five $(k = 5)$ closest shapes in the training set. In Fig. 9.24, we show several examples this classification. In each row, we first show the test shape and then show the five nearest shapes in the order of increasing distances.

With the same experimental setting, i.e., the same test and training data, we can evaluate the average classification performance of the nearest-neighbor classifier. In Fig. 9.24 (bottom), we present the change in the classification performance of the k-NN classifier as k changes from one to 20.

9.8.2 Probabilistic Classification

Another possibility for shape classification is to use a model-based classifier. For instance, we can define a probability distribution associated with each shape class and for a given test shape find the class that maximizes the value of the likelihood function. To understand this idea for shape analysis, let us start with a binary

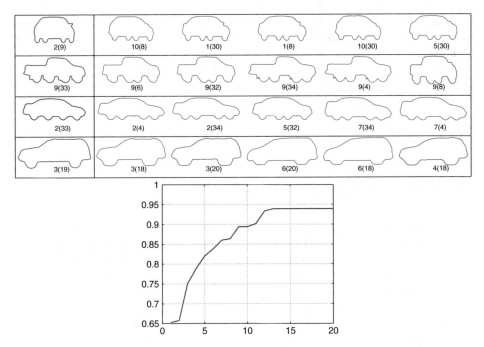

Fig. 9.24 *First four rows*: In each row, we first show a test shape and then its five nearest neighbors using the elastic distance in $\mathcal{S}_2^c$. The number below each shape denotes the vehicle index (pose index). *Bottom*: The classification performance of the k-NN classifier, versus k, for the 10 class vehicle classification problem

problem. The goal now is to use the statistical models of shape variations within two shape populations (classes) to classify a test shape into one of the two classes. Consider two shape families specified by their probability models: h_1 and h_2. In the following, we will use the TWN model defined earlier. In order to compare any two such shape models, there are two possibilities:

1. **Comparisons in Shape Space**: Here the probability densities are defined directly on the shape space $\mathcal{S}_2^c$ and are compared as functions on that space. Since TWN densities are defined only on finite-dimensional subsets of $\mathcal{S}_2^c$, we will require that the supports of both the densities include the regions where the test shapes are observed. Then, one can follow the process of computing likelihood ratios and standard decision theory modified to this specific domain.
2. **Comparison in a Tangent Space**: The other possibility is to compare the truncated normal densities, before the wrapping step, on a certain tangent space. Since the two shape models originate from a truncated normal density on a tangent space, this approach sounds natural, especially because it avoids the need for a complicated Jacobian term that comes from the wrapping step. However, the two models involve two *different* tangent spaces and, as earlier, one requires a common domain for comparing shape models. The solution is to parallel transport along geodesics the probability densities to a common tangent space. The candidate points for placing this common tangent space are the two original means or a new point that lies half way between the two means. In addition to the mean point, these shape models are completely specified using the principal components in the tangent space. This simplifies the task

of transporting a model as one has to only transport the principal directions of variations from the original mean to a new point.

3. **Comparing Likelihoods Computed in Different Tangent Spaces**: A third possibility is to compute the likelihood of a test shape under each competing model, in its own tangent space, and then to compare the resulting likelihoods. We will develop this option in more detail. For a given test shape, represented by its SRVF $q \in \mathcal{S}_2^c$, we are interested in selecting one of two following hypotheses:

$$H_0 : [q] \sim h_1$$
$$H_1 : [q] \sim h_2$$

Our classification, or a selection of a class for q, is based on finding the log-likelihood ratio:

$$l(q) \equiv \log\left(\frac{h_1(q)}{h_2(q)}\right) ,$$

and comparing it to a suitable threshold value. In case we use the TWN densities, we can derive the log-likelihood ratio in more detail. Recall that a TWN is essentially a truncated multivariate normal density in a finite-dimensional subspace of the tangent space at the mean and then projected on the shape space using the exponential map. Let $[\mu_1], [\mu_2] \in \mathcal{S}_2^c$ be the mean shapes associated with the two shape classes. Also, let $M_1^{\|}$ be the m-dimensional principal subspace of the tangent space $T_{[\mu_1]}(\mathcal{S}_2^c)$, obtained using the tangent principal component analysis (TPCA) of observed shapes from the first class, and let $M_1^{\perp}$ be the space orthogonal to $M_1^{\|}$ in $T_{[\mu_1]}(\mathcal{S}_2^c)$ so that $M_1^{\|} \oplus M_1^{\perp} = T_{[\mu_1]}(\mathcal{S}_2^c)$. Denote by K_1 the $m \times m$ covariance matrix of the tangent vector projected in $M_1^{\|}$. That is, if $[q] \sim h_1([\mu_1], K_1)$, then for $v_1 \equiv \exp_{[\mu_1]}^{-1}([q])$, the projection v_1 into $M_1^{\|}$, denoted by $v_1^{\|}$, we have that $v_1^{\|} \sim N(0, K_1)$. To complete a model on all of v_1, we assume that the components of $v_1^{\perp}$ are independent and normally distributed with mean zero and variance ϵ^2 for a small ϵ. (Note that this last item is meaningful only in practical situations where the shapes and tangent vectors are computed as finite-dimensional vectors for some large dimension, say T. If T is infinite, as it is in theory, this model for $v_1^{\perp}$ is not meaningful.) The probability density function of a shape $[q]$, under this shape class, is given by:

$$h_1([q]) = \frac{1}{\sqrt{(2\pi)^m \det(K_1)}} \exp(-\frac{1}{2}(v_1^{\|})^T K_1^{-1}(v_1^{\|})) \frac{1}{\sqrt{(2\pi\epsilon^2)^{(T-m)}}}$$
$$\exp(-\frac{1}{2}\frac{\|v_1^{\perp}\|^2}{2\epsilon^2}) J_1([\mu_1], q) , \tag{9.13}$$

where:

- $v_1 \equiv \exp_{[\mu_1]}^{-1}([q])$, and
- J_1 is the determinant of the Jacobian of the exponential map at $[\mu_1]$, when evaluated at $[q]$.

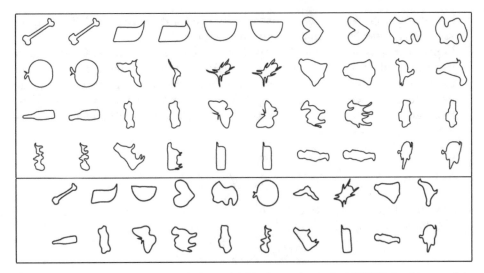

Fig. 9.25 *Top*: Two examples each from the 20 shape classes in MPEG7 database used in likelihood-based classification. *Bottom*: Mean shapes of each shape class

The probability density function under the second shape class, h_2, will have a similar expression. Note that we need to use the same dimension m for the principal subspace as in the case of h_1. Taking the log of their ratios, we obtain the log-likelihood of a shape q as:

$$L(q) \approx -(v_1^{\parallel})^T K_1^{-1} v_1^{\parallel} + (v_2^{\parallel})^T K_2^{-1} v_2^{\parallel} - (\|v_1^{\perp}\|^2 - \|v_2^{\perp}\|^2)/(2\epsilon^2) - \log(\det(K_1))$$
$$+ \log(\det(K_2)) . \tag{9.14}$$

Since the Jacobians J_1 and J_2 are hard to compute in practice, we make a simplifying approximation, albeit without further justification, that they cancel each other out and we get the expression given above. The likelihood ratio test reduces to:

$$L(q) \underset{h_2}{\overset{h_1}{\underset{<}{\gtrless}}} \nu , \tag{9.15}$$

where ν is a certain predetermined threshold. For the moment, we can take ν to be zero.

We describe an experiment that demonstrates the use of this framework. In this experiment, we take MPEG7 shape database that contains 20 shape classes with 10 exemplars each. Figure 9.26 (top) shows two examples for each of these 20 classes. For each of the shape classes C_i, $i = 1, 2, \ldots, 20$, we compute their mean $[\mu_i]$, m-dimensional principal subspace $M_i^{\parallel}$, and $m \times m$ tangent covariance K_i (with $m = 8$). The bottom part of Fig. 9.25 shows the mean shapes of each of these classes. Then, we take a random shape out of these 200 shapes, compute its SRVF q_{test}, and add noise according to:

$$q_{test}(t) = q_0(t) + \sigma\delta(t), \quad \delta(t) \sim \mathcal{N}(0, 1) ,$$

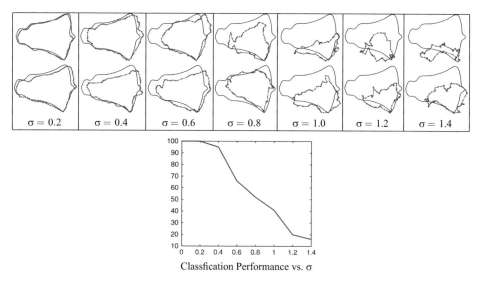

Fig. 9.26 *Top*: Examples of noisy curve data used in the classification experiment. *Bottom*: Decrease in classification rate as noise increases

where $\delta(t)$ is a random observation noise added independently at each time t. This provides a test shape for classification and evaluation. Shown in Fig. 9.26 (top two rows) are some examples of these random shapes for different values of σ. For small values of σ, the shapes are still recognizable visually, but as σ gets higher, the shapes are corrupted to the level that they are unrecognizable. We then evaluate the likelihood of this shape belonging to each of the classes and assign it to the class with the highest likelihood. If the highest likelihood belongs to the class from which the original shape q_0 came, then we call the classification successful.

In order to quantify the classification performance, we generate 100 test shapes, taken uniformly across the 20 shape classes, and calculate the percentage of correct classification. Shown in Fig. 9.26 (bottom panel) is a plot of the classification performance as a function of increasing σ. When σ is zero or small (~ 0.2), the classification performance remains perfect, but it starts deteriorating as σ decreases.

9.9 Detection/Classification of Shapes in Cluttered Point Clouds

In practical situations, the data available for detection and classification of shapes may be more complicated. Instead of observing contours directly, we may get data in the form of discrete, unordered points that are both noisy and cluttered. To be precise, we are given a point cloud $\mathbf{Y} = \{\mathbf{y}_i \in D , i = 1, 2, \dots, m\}$ in a domain $D \subset \mathbb{R}^2$ and we want to develop a statistical framework for deciding if there is a predetermined shape contained in this set. Only the shape is known but its location, orientation, and scale in the scene is unknown. One can extend this idea to detection of full shape classes, i.e., a set of shapes belonging to the same population, using statistical shape models. In order to analyze such point clouds,

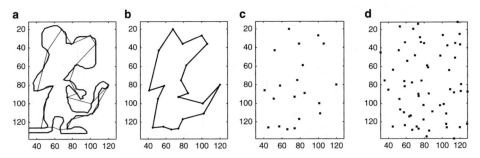

Fig. 9.27 A generative model for a cluttered point cloud containing a sampled shape. (**a**) Sampling a curve. (**b**) Sampled shape. (**c**) Samples points plus noise. (**d**) Cluttered point cloud

we will need generative models for the observed data. In this section, we develop one such model that governs generation of cluttered point clouds from arbitrary elements of our shape space $\mathcal{S}_2^c$.

Let $[q] \in \mathcal{S}_2^c$ denote an arbitrary planar shape. In order to reach a noisy point cloud, starting from $[q]$, we need to specify the steps illustrated pictorially in Fig. 9.27: (a) assign a position, orientation, and parameterization to the given shape and form a parameterized curve β; (b) sample the given curve according to a statistical model into a random number of points $\{\beta(t_i)\}$; (c) add observation noise to the given points $\{\beta(t_i) + \epsilon_i\}$; and (d) add clutter points to the set, $\{\mathbf{y}_i\} = \{\beta(t_i) + \epsilon_i\} \cup \{\mathbf{y}_j^c\}$, to result in a 2D point cloud with clutter. We will develop models for each of these steps.

There are at least two models for random selection of points along a parameterized curve:

1. **Point Process Model**: Use a point process for generation of discrete points along a curve. For instance, we can assume that the points are sampled from a Poisson process with a known intensity function along β. The number n is a Poisson random variable with mean given by the integral of the intensity function along the full curve, and the placement is dictated by the value of the intensity function.
2. **Random Re-parameterization Model**: Another possibility is to impose a statistical model on n, the number of points, and seek an independent model for their placement. As described in Remark 9.2, we can also control the placement variability via re-parameterization of β. Assume that the curve is arc-length parameterized and n points are placed at equal distances $\{0, 1/n, 2/n, \ldots, 1\}$ along the curve. Now, if we take a random re-parameterization $\gamma \in \Gamma_S$, then the selected points become $\{0, \gamma(1/n), \gamma(2/n), \ldots, 1\}$. This way one can control selection of points using random γ.

In this section, we will use the point process approach, developing a fully statistical model for generation of cluttered point clouds from a given parameterized curve.

Returning to the problem of shape detection and classification, we will start with the problem of binary hypothesis testing—the null hypothesis is that $\mathbf{Y}$ is simply clutter, i.e., the shape of interest is NOT present in $\mathbf{Y}$, and the alternate hypothesis is that $\mathbf{Y}$ is generated from that shape, i.e., a shape is present in $\mathbf{Y}$.

$$\mathrm{H}_0 : \text{Shape is absent, likelihood} \quad P(\mathbf{Y}|C)$$

$$H_1 : \text{Shape is present, likelihood} \quad P(\mathbf{Y}|S)$$

Here, C denotes the clutter and S denotes the shape of interest. The challenge, of course, is to develop appropriate probability models that will enable us to evaluate the two likelihoods.

9.9.1 Point Process Models for Cluttered Data

Shape is a characteristic that is invariant to similarity transformations, but when a shape occurs in a scene, it has a specific scale, position, and orientation. From the perspective of shape detection, these variables are considered nuisance variables that have to be either estimated or integrated out. In order to better explain the model description and a detection solution, we will start with a simpler problem where we seek a specific object, i.e., known shape, position, orientation, and scale. Let $\beta : \mathbb{S}^1 \to \mathbb{R}^2$ be a parameterized curve of interest. We will assume that the curves are parameterized by a constant-speed parameter in the 2D case, i.e., $\beta(s) \in \mathbb{R}^2$ such that $|\dot{\beta}(s)| = $ constant. To develop the data model, we make the following assumptions:

1. **Points belonging to** β: We assume that these points are realizations of a Poisson process on the parameterized object β. Let $g : \mathbb{S}^1 \to \mathbb{R}_{\geq 0}$ be the intensity function of the Poisson process along β. The number of points generated from any part of the object is a Poisson random variable with mean being the integral of g on that part. In particular, k, the total number of points belonging to the object, is a Poisson random variable with mean $G = \int_{\mathbb{S}^1} g(s)\, ds \in \mathbb{R}_{\geq 0}$. Let the points sampled from β be denoted by $\mathbf{X} = [\mathbf{x}_1, \mathbf{x}_2, \ldots, \mathbf{x}_k]$, $\mathbf{x}_j \in \mathbb{R}^n$. The actual observations $\mathbf{y}_j$ are assumed to be noisy versions of $\mathbf{x}_j$. For given $\mathbf{x}_1, \mathbf{x}_2, \ldots, \mathbf{x}_k$, the $\mathbf{y}_j$s are assumed to be independent of each other with the identical density $f(\mathbf{y}_j|\mathbf{x}_j)$. Under this model, the two hypotheses can be rewritten as:

$$\begin{aligned} H_0 : \quad & g = 0, \quad \text{Likelihood} \quad P(\mathbf{Y}|C) \\ H_1 : \quad & g > 0, \quad \text{Likelihood} \quad P(\mathbf{Y}|S) \end{aligned} \ .$$

2. **Points associated with clutter**: This subset of observations, independent of the first subset, comes from the clutter and we model them as realizations of a Poisson process with the intensity $l : D(\subset \mathbb{R}^2) \to \mathbb{R}_{\geq 0}$, where D is the region containing observed points, e.g., $D = [a, b]^2$. Let $L = \int_D l(\mathbf{y})d\mathbf{y} \in \mathbb{R}_{\geq 0}$.

The full observation $\mathbf{Y}$ can now be modeled as a Poisson process with the intensity function: $\xi(\mathbf{y}) = l(\mathbf{y}) + \int_{\mathbb{S}^1} f(\mathbf{y}|\beta(s))g(s)\, ds$. The probability density function of $\mathbf{Y}$, given β, g, l, and for a fixed m, is given by $P_m(\mathbf{Y}|\beta, g, l) = (\prod_{i=1}^m \xi(\mathbf{y}_i))e^{-L-G}$, where m is the total number of points in the data. The null hypothesis is that all the points belong to the Poisson clutter. In that case, the likelihood function is given by $Q_m(\mathbf{Y}|l) = e^{-L} \prod_{i=1}^m l(\mathbf{y}_i)$. The likelihoods for both the cases, H_0 and H_1, involve certain parameters that are generally not known beforehand. Thus, taking a simple likelihood ratio is not possible and we resort to the *generalized likelihood ratio test* (GLRT). This is based on maximum likelihood estimates (MLEs) of parameters, under the respective hypotheses, and uses the MLEs for evaluating the likelihood ratio. The generalized likelihood ratio is given by:

$$\frac{Q_m(\mathbf{Y}|C)}{P_m(\mathbf{Y}|S)} = \frac{\max_l Q_m(\mathbf{Y}|l)}{\max_{l,g} P_m(\mathbf{Y}|\beta, g, l)} = \frac{\max_l(e^{-L} \prod_{i=1}^m l(\mathbf{y}_i))}{\max_{l,g}(e^{-G-L}(\prod_{i=1}^m \xi(\mathbf{y}_i)))}. \qquad (9.16)$$

So far the unknown parameters are full functions and that involves tremendous computational complexity. We will simplify the evaluation of GLR in Eq. 9.16 by making the following additional assumptions:

1. The noise added to the points sampled from β is i.i.d. Gaussian with mean zero and variance $\sigma^2 I_{2\times 2}$. Therefore, the conditional density $f(\mathbf{y}|\mathbf{x})$ takes the form $\frac{1}{(2\pi)\sigma^2} e^{-\frac{1}{2\sigma^2} \|\mathbf{y}-\mathbf{x}\|^2}$ for $\mathbf{y}, \mathbf{x} \in \mathbb{R}^2$.
2. Both the Poisson intensities are constant, i.e., $l(\mathbf{y}) = l$ and $g(s) = g$ and we get $L = l \int_D d\mathbf{y}$ and $G = g \int_{\mathbb{S}^1} ds$. To simplify the discussion, we scale both the integrals to be one such that $L = l$ and $G = g$.

With these assumptions, the likelihood ratio simplifies to:

$$\frac{Q_m(\mathbf{Y}|C)}{P_m(\mathbf{Y}|S)} = \frac{\max_l(e^{-l} \prod_{i=1}^m l)}{\max_{l,g,\sigma}(e^{-g-l}(\prod_{i=1}^m (l + g\alpha_\sigma(\mathbf{y}_i))))} \ .$$

The numerator on maximization becomes $e^{-m}m^m$. The quantity $\alpha_\sigma : \mathbb{R}^n \to \mathbb{R}_+$ in the denominator is a scalar map given by $\alpha_\sigma(\mathbf{y}_i) = \frac{1}{(2\pi)\sigma^2} \int_D e^{-\frac{1}{2\sigma^2} \|\mathbf{y}_i - \beta(s)\|^2} ds$. Notice that $\alpha_\sigma(\mathbf{y}_i)$ is high if a point $\mathbf{y}_i$ is close to the object β, with the closeness being measured relative to the scale σ. Some illustrations of α_σ in $\mathbb{R}^2$ are shown as gray scale images in Fig. 9.28. The top row shows the case for different σs (from left to right: $\sigma = 0.01, 0.02, 0.03$) but a fixed curve. As σ increases, the region of high likelihood spreads further away from the curve. The bottom row shows α_σ maps for different curves but a fixed $\sigma = 0.02$.

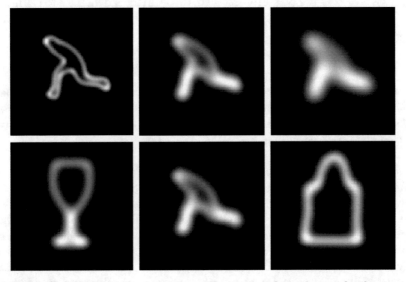

Fig. 9.28 Likelihood maps α_σ for some curves. The top row shows the case for the same curve but different σs and the bottom row shows different curves but with the same σ

9.9.2 Maximum Likelihood Estimation of Model Parameters

Define a function H to be the logarithm of $P_m(\mathbf{Y}|\beta, g, l)$. Let $\boldsymbol{\theta} = [g, l, \sigma] \in \mathbb{R}^3$ denote three unknown parameters associated with the shape. Then, the function $H : \mathbb{R}^3 \to \mathbb{R}_+$ is given by $H(\boldsymbol{\theta}) = -g - l + \sum_{i=1}^{m} \log(l + g\alpha_\sigma(\mathbf{y}_i))$ and let $\hat{\boldsymbol{\theta}} = \mathrm{argmax}_{\boldsymbol{\theta}} H(\boldsymbol{\theta})$ be maximizer (actually, $\hat{\boldsymbol{\theta}}$ is the MLE of $\boldsymbol{\theta}$ under the hypothesis H_1). We solve for the MLE of $\boldsymbol{\theta}$ using a gradient approach. Of the three components of $\boldsymbol{\theta}$, we search exhaustively for the parameter σ and use a gradient-based approach to search over the remaining two g and l. For each value of σ in a certain range, say $[\sigma_l, \sigma_u]$, we maximize H over the pair (g, l).

For a fixed σ, the function $H : \mathbb{R}^2 \to \mathbb{R}$, given by $H_\sigma(l, g) = -g - l + \sum_{j=1}^{m} \log(l + g\alpha_\sigma(\mathbf{y}_j))$, has the following properties:

1. Its derivatives with respect to l and g are given by:

$$\frac{\partial H_\sigma}{\partial l} = -1 + \sum_{i=1}^{m} \frac{1}{l + g\alpha_\sigma(\mathbf{y}_i)}, \quad \frac{\partial H_\sigma}{\partial g} = -1 + \sum_{i=1}^{m} \frac{\alpha_\sigma(\mathbf{y}_i)}{l + g\alpha_\sigma(\mathbf{y}_i)}. \quad (9.17)$$

2. Its Hessian matrix is given by:

$$\begin{pmatrix} \sum_{i=1}^{m} \frac{-(\alpha_\sigma(\mathbf{y}_i))^2}{(l+g\alpha_\sigma(\mathbf{y}_i))^2} & \sum_{i=1}^{m} \frac{-\alpha_\sigma(\mathbf{y}_i)}{(l+g\alpha_\sigma(\mathbf{y}_i))^2} \\ \sum_{i=1}^{m} \frac{-\alpha_\sigma(\mathbf{y}_i)}{(l+g\alpha_\sigma(\mathbf{y}_i))^2} & \sum_{i=1}^{m} \frac{-1}{(l+g\alpha_\sigma(\mathbf{y}_i))^2} \end{pmatrix}.$$

It is easy to show that the two eigenvalues of the Hessian matrix are non-positive, so that H_σ is a concave function in l and g.

Therefore, one can use the gradient search over l and g and reach a global optimizer. For σ, the situation is different and we use an exhaustive grid search over allowable values of σ to reach a global maximizer. This combined gradient and grid search algorithm is summarized below:

Algorithm 43 (MLE of $\boldsymbol{\theta}$).

- *For each $\sigma \in [\sigma_l, \sigma_u]$, perform the following:*

  1. *Set $t = 0$ and initialize the pair $[g_t, l_t]$ with random values in the range $[0, m]$.*
  2. *Update the estimates using $\begin{bmatrix} g_{t+1} \\ l_{t+1} \end{bmatrix} = \begin{bmatrix} g_t \\ l_t \end{bmatrix} + \delta \begin{bmatrix} \frac{\partial H_\sigma}{\partial g}(g_t, l_t) \\ \frac{\partial H_\sigma}{\partial l}(g_t, l_t) \end{bmatrix}$, for a small $\delta > 0$.*
  3. *If the norm of the gradient vector is small, then stop the loop. Else, set $t = t + 1$ and return to Step 2.*

- *Set the current values to be $(\hat{g}(\sigma), \hat{l}(\sigma))$.*
- *Define the MLE $\hat{\boldsymbol{\theta}}$ to be $(\hat{g}(\hat{\sigma}), \hat{l}(\hat{\sigma}), \hat{\sigma})$ where $\hat{\sigma} = \mathrm{argmax}_{\sigma \in [\sigma_l, \sigma_u]} H(\hat{g}(\sigma), \hat{l}(\sigma), \sigma)$.*

With the estimated parameters, the log-likelihood ratio (LLR) becomes:

$$R(\mathbf{Y}) = \log \frac{Q_m(\mathbf{Y}|C)}{P_m(\mathbf{Y}|S)} = -m + m\log(m) - H(\hat{\boldsymbol{\theta}}). \quad (9.18)$$

The generalized likelihood ratio test is given by:

$$R(\mathbf{Y}) \underset{S}{\overset{C}{\underset{<}{\overset{>}{\gtrless}}}} \nu.$$

Next, we will focus on how to choose the threshold ν.

Empirical Detection of Threshold In the binary test, the LLR $R(\mathbf{Y})$ is to be compared with a threshold ν to decide if a shape is present in the data or not. Ideally, this threshold is dictated by the probability distributions of $R(\mathbf{Y})$ under the null hypothesis. In practical situations, where it is difficult to ascertain these distributions, one uses either the asymptotic theory or an empirical approach to reach an optimal value of ν. To demonstrate the empirical approach, we estimate the pdf of $R(\mathbf{Y})$ under the null hypothesis. We generate 1000 realizations of $\mathbf{Y}$, each using $g = 0$ (null hypothesis = clutter) and a fixed m equal to the observed number of points in our data, and compute a histogram of $R(\mathbf{Y})$ values. Using this estimated density function, we can decide the threshold ν for a specific type I error rate, denoted by α. One can repeat this for different values of m to catalog distributions of $R(\mathbf{Y})$ for different ms.

The main advantage of a numerical evaluation of the threshold is that we need not assume any specific form for the underlying density, nor do we need to invoke any asymptotic argument. The disadvantage, however, is that we need to do this for every shape we are interested in, since we do not have an analytical expression. We point out that this computation is offline and can be performed for each of the shapes beforehand.

Consider the point cloud shown in the left column of Fig. 9.29. For this data, we apply Algorithm 43 for two curves—a runner and a wineglass, shown in the second column. The third column shows the estimated $\alpha_{\hat{\sigma}}$ map for each curve and the rightmost column shows the value of $\alpha_{\hat{\sigma}}(\mathbf{y}_i)$ for each of the data points using its thickness. Recall that $\alpha_{\hat{\sigma}}(\mathbf{y}_i)$ is large if the point $\mathbf{y}_i$ is close to the curve β. Since the observation $\mathbf{Y}$ here was generated from the runner curve, $\alpha_{\hat{\sigma}}(\mathbf{y}_i)$s have

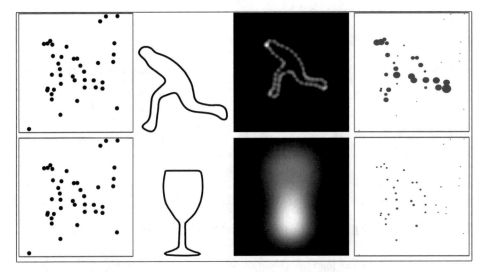

Fig. 9.29 A point cloud, two hypothesized curves, $\alpha_{\hat{\sigma}}$ profiles, and $\alpha_{\hat{\sigma}}$ at $\mathbf{y}_i$s for the two curves

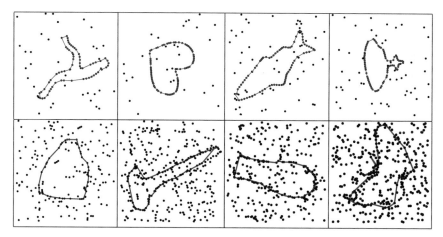

Fig. 9.30 Examples of shape estimation in cluttered point clouds (*black*, point clouds; *blue*, estimated shapes)

larger values for that shape and lower values for the wineglass shape. This points to a higher likelihood of the runner shape being present in the given point cloud.

Another interesting subproblem here, in addition to the detection of a given curve, is the estimation of that curve in the point cloud. For a given point cloud and a hypothesis shape β_0, we first detect if the shape is present in the data. If yes, using the estimates $\hat{g}$, $\hat{\sigma}$, and $\hat{\omega}$, we select $\hat{g}$ number of points in $\mathbf{Y}$ with the largest values of $\alpha_{\hat{\sigma}}$. Then, we connect them in the same order as their nearest neighbors on β_0. Some examples of this process are shown in Fig. 9.30. Note that in this fully statistical framework, one not only verifies the presence of a shape but also estimates the shape and provides the likelihood of this estimated shape being present in the given data.

9.10 Problems

9.10.1 Theoretical Problems

1. Let C_i and C_j be any two nonempty subsets of the set $\mathcal{B} = \{\beta_1, \beta_2, \ldots, \beta_n\}$, a set of given curves.

   a. Show that the quantity:

   $$D(C_i, C_j) = \min_{[q_1] \in C_i, [q_2] \in C_j} d_s([q_1], [q_2])$$

   is not a metric on the power set of $\mathcal{B}$.
   b. What happens if instead we use average of all pairwise distances across C_i and C_j?
   c. **Hausdorff**: Verify that the following quantity is a metric on the power set of $\mathcal{B}$:

   $$D(C_i, C_j) = \max\{\max_{[q_1] \in C_i} \min_{[q_2] \in C_j} d_s([q_1], [q_2]), \max_{[q_2] \in C_j} \min_{[q_1] \in C_i} d_s([q_1], [q_2])\} \ .$$

2. For any nontrivial $\mu \in \mathcal{C}_2$, show that the function $u(t) = \begin{bmatrix} 0 & -1 \\ 1 & 0 \end{bmatrix} \mu(t)$ spans the space $T_\mu([\mu]_{SO(2)})$, the tangent space to the rotation orbit of μ, at μ.

3. Recall that the Γ_I is the group of positive diffeomorphisms on $[0,1]$ and $\tilde{\Gamma}_I$ is the monoid of weakly increasing, absolutely continuous functions on $[0,1]$ (see Definition 4.3). We know that the set Γ_I is dense in $\tilde{\Gamma}_I$ using the $\mathbb{L}^2$ norm. Show that a set orthogonal functions that spans $T_\mu([\mu]_{\Gamma_I})$ also spans the set $T_\mu([\mu]_{\tilde{\Gamma}_I})$.

4. Show that the set $\mathcal{B}_n$, given in Eq. 9.3, forms an orthonormal basis for $\mathbb{L}^2([0,1], \mathbb{R})$ in the limit as n goes to infinity.

9.10.2 Computational Problems

1. Write a program to implement the simulated annealing search for optimal clustering, as given in Algorithm 37.

2. Write a program to generate the basis set $\mathcal{B}_n$ for a given n and the sample points $\{0, \frac{1}{T}, \frac{2}{T}, \ldots, 1\} \subset [0,1]$.

3. Similarly, write a program to generate a basis of $T_{\gamma_{id}}(\Gamma_I)$, as provided in this chapter.

4. Implement Algorithm 38 to generate an orthonormal basis of the tangent space $T_\mu([\mu])$ for $[\mu] \in \mathcal{S}_2$.

5. Write a program that takes (1) the SRVF of mean curve $\mu \in \mathbb{L}^2([0,1], \mathbb{R}^2)$, (2) an orthogonal basis of $T_\mu([\mu])$, and (3) SRVF of an arbitrary curve $q \in \mathbb{L}^2([0,1], \mathbb{R}^2)$; to find the finite-dimensional representation $x \in \mathbb{R}^m$ of $[q] \in \mathcal{S}_2$.

6. Write a program to generate a basis of $T_{\gamma_{id}}(\Gamma_S)$, as provided in this chapter.

7. Adapt Algorithm 38 to the case for closed curves, to generate an orthonormal basis of the tangent space $T_\mu([\mu])$ for $[\mu] \in \mathcal{S}_2^c$.

8. Write a program that takes (1) the SRVF of mean curve $\mu \in \mathbb{L}^2(\mathbb{S}^1, \mathbb{R}^2)$, (2) an orthogonal basis of $T_\mu([\mu])$, and (3) SRVF of an arbitrary curve $q \in \mathbb{L}^2(\mathbb{S}^1, \mathbb{R}^2)$; to find the finite-dimensional representation $x \in \mathbb{R}^m$ of $[q] \in \mathcal{S}_2^c$.

9. Write a program implementing Algorithm 39 to compute sample Karcher mean of a given set of curves in the shape space $\mathcal{S}_2$.

10. Write a program implementing Algorithm 40 to compute sample Karcher mean of a given set of closed curves in the shape space $\mathcal{S}_2^c$.

11. Write a program implementing Algorithm 41 to compute sample Karcher median of a given set of closed curves in the shape space $\mathcal{S}_2^c$.

12. Write a program to generate a basis W to provide a finite-dimensional basis of closed, planar shapes using the tangent space of a mean shape $[\mu] \in \mathcal{S}_2^c$.

13. Write a program that performs principal component analysis of shooting vectors $\{v_i\}$ obtained in Algorithm 40. Display the top three modes of variability in the given data using the wrapping (exponential map) of the tangent space geodesics on the shape space $\mathcal{S}_2^c$.

14. Write a program to generate random shapes using the TWN model with model parameters (mean and covariance) estimated using previous problems from a training dataset.

15. Write a program that takes in an arc-length parameterized curve and generates a 2D point cloud according to the Poisson model specified in Sect. 9.9.1.

16. Implement Algorithm 43 that takes in a given point cloud, and a parameterized curve, and generates MLE of the parameter θ.
17. Write a program to compute the likelihood ratio for a given test curve according to Eq. 9.14. (As suggested in this equation, ignore the terms involving Jacobians of the exponential map.)
18. Write a program to generate an empirical distribution of the test statistics, under the null hypothesis, for a given value of m and a given curve β.

9.11 Bibliographic Notes

The textbook by Jain and Dubes [40] is an excellent presentation of different ideas in clustering techniques. Several authors have explored the use of annealing in clustering problems, including soft clustering [95] and deterministic clustering [37]. An interesting idea presented in [37] is to solve an approximate problem, termed mean-field approximation, where Q is replaced by a function in which the roles of elements q_is are decoupled. The advantage is the resulting efficiency although it comes at the cost of error in approximation. Monte Carlo simulated annealing has been used in optimization in several places. See, for example, Algorithm A.20, pg. 200 [93] for a discussion. Srivastava et al. [105] studied the problems of clustering, modeling, and testing shapes of curves using the non-elastic framework.

The problem of modeling shapes and different appearance parameters (rotation, position, scale, and parameterization) of curves, for the purpose of estimating shapes in point clouds, has been tackled in [104]. Su et al. [110] discussed the problem detecting and estimating 2D and 3D shapes in cluttered point clouds using Poisson point process models.

Chapter 10
Shapes of Curves in Higher Dimensions

So far in this text we have considered only planar curves. Although the shapes of planar curves are quite important, especially for recognizing objects in images, the shapes of curves in higher-dimensional spaces also have an important role to play. While the shape analysis of curves in $\mathbb{R}^3$ and higher dimensions is directly useful for studying many important objects, such as the white matter fiber tracts, curves formed by sulcal folds, backbones in protein structures, level curves of geological terrains and oceans, and flight paths of migratory birds, they also have indirect uses in many situations. One example is in shape analysis of facial surfaces where each surface is represented by a collection of certain intrinsically defined curves. These curves are either open or closed curves in $\mathbb{R}^3$, and a shape comparison of two facial surfaces can be, in effect, performed by comparing shapes of their respective curves. Another interesting example is in shape analysis of augmented curves. An augmented curve is a parameterized curve in a certain space, say $\beta(t) \in \mathbb{R}^n$, which has a certain auxiliary function, say $f(t) \in \mathbb{R}^k$, associated with it. An example of this situation is in a color image: the parameterized curve comes from the boundary of an object in the image, while the auxiliary function represents the colors or textures along that boundary. Together the mapping $t \mapsto (\beta(t), f(t)) \in \mathbb{R}^{n+k}$ can be treated as a curve in a higher-dimensional space, albeit with some modifications to the general method.

10.1 Goals and Challenges

In this chapter, we will develop techniques for shape analysis of curves in $\mathbb{R}^n$, with n being three or more, and will demonstrate them with some applications. The constructions and techniques will be straightforward extensions of the corresponding ideas for the planar case in Chaps. 5 and 6. Our goals are to:

1. Develop mathematical representations, Riemannian metrics, and algorithms for geodesic computations for analyzing shapes of curves in $\mathbb{R}^n$ for $n \geq 3$. We will consider both open and closed curves.
2. Develop techniques for computing shape summaries and for modeling shapes of populations of curves.

© Springer-Verlag New York 2016
A. Srivastava, E.P. Klassen, *Functional and Shape Data Analysis*,
Springer Series in Statistics, DOI 10.1007/978-1-4939-4020-2_10

3. Demonstrate these ideas using applications involving open and closed curves in areas such as human biometrics, bioinformatics, medical image analysis, and geology.

10.2 Mathematical Representations of Curves

Previously we have utilized two representations for studying shapes of curves: the angle function and the square-root velocity function. The use of $\mathbb{L}^2$ metric in each space leads to a different framework, for finding optimal matching and deformations between curves—bending-only deformations in the first case and elastic deformations (bending-stretching combination) in the second. Although we will extend both these representations to curves in $\mathbb{R}^n$, we will treat the elastic framework in more detail, as there are distinct advantages to using it in many applications. Here are the mathematical representations for the two frameworks:

1. **Direction Function Representation**: Let $\beta : [0,1] \mapsto \mathbb{R}^n$ be an absolutely continuous curve of length one, parameterized by the arc length. For $\theta(s) \equiv \dot{\beta}(s) \in \mathbb{R}^n$, we have $|\theta(s)| = 1$ for all $s \in [0,1]$, due to the use of the arc-length parameterization. Here $|\cdot|$ denotes the Euclidean norm in $\mathbb{R}^n$. The function θ is called the *direction function* of β; it is the extension of the angle function θ, used in Chaps. 5 and 6 for studying planar curves, to curves in $\mathbb{R}^n$. Since $|\theta(s)| = 1$ for all s, the function θ itself can be viewed as a curve on the unit sphere $\mathbb{S}^{n-1}$, i.e., $\theta : [0,1] \mapsto \mathbb{S}^{n-1}$. Shown in Fig. 10.1 is an illustration of this idea where a closed curve β in $\mathbb{R}^3$ is represented by a curve θ in $\mathbb{S}^2$. We will use the direction function θ to represent and analyze the shape of the curve β. By the virtue of unit-speed parameterization and the domain size being one, the length of curves represented by a θ is also one. Note that we can reconstruct the curve from a given angle function θ, up to a translation, using the formula: $\beta(t) = \int_0^t \theta(s) ds$.

2. **Square-Root Velocity Function Representation**: The second representation is based on the function $q(t)$ that has the same direction as the velocity vector $\dot{\beta}(t)$ but its magnitude is given by $\sqrt{|\dot{\beta}(t)|}$. We will restrict to the curves β that are absolutely continuous. This implies that their q function is square integrable. Define a mapping $F : \mathbb{R}^n \to \mathbb{R}^n$ according to:

$$F(v) \equiv \begin{cases} v/\sqrt{|v|}, & \text{if } |v| \neq 0 \\ 0, & \text{otherwise} \end{cases} \tag{10.1}$$

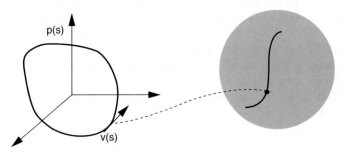

Fig. 10.1 A closed curve in $\mathbb{R}^n$ is denoted by a curve on $\mathbb{S}^{n-1}$

Here, $|\cdot|$ is the 2 norm in $\mathbb{R}^n$ and note that F is a continuous map. For the purpose of studying the shape of β, we will represent it using the square-root velocity function defined as $q : [0,1] \to \mathbb{R}^n$, where:

$$q(t) \equiv F(\dot{\beta}(t)) = \dot{\beta}(t)/\sqrt{|\dot{\beta}(t)|} \ . \tag{10.2}$$

Since the curve β is absolutely continuous, its square-root velocity function is an element of $\mathbb{L}^2([0,1],\mathbb{R}^n)$. This representation includes those curves whose parameterizations can become singular in the analysis. Also, for every $q \in \mathbb{L}^2([0,1],\mathbb{R}^n)$, there exists a curve β (unique up to a translation) such that the given q is the square-root velocity function of that β. One can reconstruct this curve from q, up to a translation, using the formula $\beta(t) = \int_0^t q(\tau)|q(\tau)|d\tau$. If the curve β is of unit length, the function q satisfies $\int_0^1 |q(t)|^2 dt = \int_0^1 |\dot{\beta}(t)|dt = 1$.

10.3 Elastic and Non-elastic Metrics

An interesting point about our earlier discussion on shape analysis of the planar curves was that the same $\mathbb{L}^2$ metric under different representations leads to two different quantifications of shape differences—both having their own physical interpretations. This point still holds when we extend to curves in higher dimensions. Similar to Sect. 5.6.1, we can motivate the two metrics for the higher-dimensional curves. Furthermore, we will make a tighter connection between the two metrics. More specifically, the bending-only metric will be shown to be a special case of the elastic metric when one restricts to curves with arc-length parameterizations.

Let $\beta : [0,1] \to \mathbb{R}^n$ be a smooth curve in $\mathbb{R}^n$. (In this section, we will assume additional properties, such as smoothness and non-singularity of parameterizations, to facilitate discussion.) Assume that for all $t \in [0,1]$, $\dot{\beta}(t) \neq 0$. We then define $\phi : [0,1] \to \mathbb{R}$ by $\phi(t) = \ln(|\dot{\beta}(t)|)$, and $\theta : [0,1] \to \mathbb{S}^{n-1}$ by $\theta(t) = \dot{\beta}(t)/|\dot{\beta}(t)|$. The difference from Sect. 5.6.1 is that this time $\theta(t)$ is an element of $\mathbb{S}^{n-1}$ rather than the unit circle $\mathbb{S}^1$. As earlier, ϕ and θ completely determine $\dot{\beta}$, since for all t, $\dot{\beta}(t) = e^{\phi(t)}\theta(t)$. Thus, we have defined a map from the space of parameterized curves in $\mathbb{R}^n$ to $\Phi \times \Theta$, where $\Phi = \{\phi : [0,1] \to \mathbb{R} \text{ smooth}\}$ and $\Theta = \{\theta : [0,1] \to \mathbb{S}^{n-1} \text{ smooth}\}$. From the point of physical interpretations, $\phi(t)$ is the instantaneous (log of the) speed, while $\theta(t)$ is the instantaneous direction of the curve at each time t.

In order to quantify the magnitudes of perturbations of β, we wish to impose a Riemannian metric on the space of curves that is invariant under rigid motions and re-parameterizations, and we will do this by putting a Riemannian metric on $\Phi \times \Theta$. The tangent of space of $\Phi \times \Theta$ at any point (ϕ, θ) is given by:

$$T_{(\phi,\theta)}(\Phi \times \Theta) = \{(u,v) : u \in \Phi \text{ and } v : [0,1] \to \mathbb{R}^n \text{ smooth and } v(t) \perp \theta(t), \forall t \in [0,1]\}$$

Definition 10.1 (Elastic Metric for Curves in $\mathbb{R}^n$). For any point $(\phi,\theta) \in (\Phi, \Theta)$, and a pair (u_1, v_1) and (u_2, v_2) in $T_{(\phi,\theta)}(\Phi \times \Theta)$, define an inner product by

$$\langle (u_1, v_1), (u_2, v_2) \rangle_{(\phi,\theta)} = a^2 \int_0^1 u_1(t) u_2(t) e^{\phi(t)} \, dt + b^2 \int_0^1 \langle v_1(t), v_2(t) \rangle \, e^{\phi(t)} \, dt \ , \tag{10.3}$$

where a and b are positive real numbers. This defines a Riemannian metric on the manifold $\Phi \times \Theta$, which is called the *elastic metric*.

The two terms measure the amount of stretching and bending, respectively, and the numbers a and b provide relative weights to the two measurements. Note that $\langle \cdot, \cdot \rangle$ in the second integral denotes the Euclidean inner product in $\mathbb{R}^n$.

Similar to the case of planar curves, under the elastic metric, the groups $SO(n)$ and Γ_I both act on the representation space by isometries. To elaborate on this, recall that $O \in SO(n)$ acts on a curve β by $(O, \beta)(t) = O\beta(t)$, and $\gamma \in \Gamma_I$ acts on β by $(\gamma, \beta)(t) = \beta(\gamma(t))$. Using our identification of the set of curves with the space $\Phi \times \Theta$ results in the following actions of these groups. $O \in SO(n)$ acts on (ϕ, θ) by $(O, (\phi, \theta)) = (\phi, O\theta)$. $\gamma \in \Gamma_I$ acts on (ϕ, θ) by $(\gamma, (\phi, \theta)) = (\phi \circ \gamma + \ln \circ \dot{\gamma}, \theta \circ \gamma)$. As stated, $SO(n)$ acts on $\Theta \times \Phi$ from the left and Γ_I acts from the right, making it difficult to form the joint action directly. Thus, if needed, one can switch the action of $SO(n)$ to the right using $((u, v), O) = (u, O^{-1}v)$. However, we will overlook this discrepancy here.

Now we look at the differentials of these group actions on the tangent spaces of $\Phi \times \Theta$. $SO(n)$ is easy; since each $O \in SO(n)$ acts by the restriction of a linear transformation on $\Phi \times L^2([0, 1], \mathbb{R}^n)$, it acts in exactly the same way on the tangent spaces $(O, (u, v)) = (u, Ov)$, where $(u, v) \in T_{(\phi, \theta)}(\Phi \times \Theta)$, and $(u, Ov) \in T_{(\phi, O\theta)}(\Phi \times \Theta)$. The action of $\gamma \in \Gamma_I$ given in the above formula is not linear, but affine linear, because of the additive term $\ln \circ \dot{\gamma}$. Hence, its action on the tangent space is the same, but without this additive term $(\gamma, (u, v)) = (u \circ \gamma, \theta \circ \gamma)$, where $(u, v) \in T_{(\phi, \theta)}(\Phi \times \Theta)$, and $(u \circ \gamma, \theta \circ \gamma) \in T_{(\gamma, (\phi, \theta))}(\Phi \times \Theta)$. Combining these actions of $SO(n)$ and Γ_I with the above inner product on $\Phi \times \Theta$, it is an easy verification that these actions are by isometries, i.e.:

$$\langle (O, (u_1, v_1)), (O, (u_2, v_2)) \rangle_{(O, (\phi, \theta))} = \langle (u_1, v_1), (u_2, v_2) \rangle_{(\phi, \theta)}$$
$$\langle (\gamma, (u_1, v_1)), (\gamma, (u_2, v_2)) \rangle_{(\gamma, (\phi, \theta))} = \langle (u_1, v_1), (u_2, v_2) \rangle_{(\phi, \theta)}. \qquad (10.4)$$

Since we have identified the space of curves with $\Phi \times \Theta$, we may identify the space of shapes with the quotient space $(\Phi \times \Theta)/(SO(n) \times \Gamma_I)$. Furthermore, since these group actions are by isometries with respect to all the metrics we introduced above, *no matter what values we assign to a and b*, we get a corresponding two-parameter family of metrics on the quotient space $(\Phi \times \Theta)/(SO(n) \times \Gamma_I)$. Given two shapes, we can find the geodesics between the corresponding elements of $(\Phi \times \Theta)$ with respect to any of these metrics (i.e., with respect to any choice of a and b).

Given a curve $\beta : [0, 1] \to \mathbb{R}^n$, its square-root velocity representation is the function $q : [0, 1] \to \mathbb{R}^n$ given by $q(t) = \frac{\dot{\beta}(t)}{\sqrt{|\dot{\beta}(t)|}}$ (Eq. 10.2). Relating this to the (ϕ, θ) representation of the curve gives $q(t) = e^{\frac{1}{2}\phi(t)}\theta(t)$. A couple of simple differentiations show that if $(u, v) \in T_{(\phi, \theta)}(\Phi \times \Theta)$, then the corresponding tangent vector to $L^2([0, 1], \mathbb{R}^n)$ at q is given by $f = \frac{1}{2}e^{\frac{1}{2}\phi}u\theta + e^{\frac{1}{2}\phi}v$.

Theorem 10.1. *The L^2 metric on the space of square-root velocity functions for curves in $\mathbb{R}^n$ corresponds to the elastic metric on $\Phi \times \Theta$ with $a = \frac{1}{2}$ and $b = 1$.*

Proof. Let (u_1, v_1) and (u_2, v_2) denote two elements of $T_{(\phi, \theta)}(\Phi \times \Theta)$ and let f_1 and f_2 denote the corresponding tangent vectors to $L^2([0, 1], \mathbb{R}^n)$ at $q = e^{\frac{1}{2}\phi}\theta$. Computing the L^2 inner product of f_1 and f_2 yields:

$$\langle f_1, f_2 \rangle = \int_0^1 \left\langle \frac{1}{2} e^{\frac{1}{2}\phi} u_1 \theta + e^{\frac{1}{2}\phi} v_1, \frac{1}{2} e^{\frac{1}{2}\phi} u_2 \theta + e^{\frac{1}{2}\phi} v_2 \right\rangle dt$$

$$= \int_0^1 \frac{1}{4} e^{\phi} u_1 u_2 + e^{\phi} \langle v_1, v_2 \rangle \, dt. \tag{10.5}$$

In this calculation, we have used the facts that $\langle \theta(t), \theta(t) \rangle = 1$, since $\theta(t)$ is an element of the unit sphere, and that $\langle \theta(t), v_i(t) \rangle = 0$, since each $v_i(t)$ is a tangent vector to the unit sphere at $\theta(t)$. Comparing the right side with the Eq. 10.3 give us the desired result. $\square$

This theorem shows that the square-root velocity representation with the $\mathbb{L}^2$ metric provides a framework for elastic shape analysis of curves.

The non-elastic metric is a special case of the elastic metric in the following sense. In case we restrict to curves that are parameterized by the arc length, then the log-speed $\phi(t) = 0$ for all curves and $\theta(t)$ is the only relevant function left. Restricting to that part of the elastic metric, we obtain the non-elastic metric.

Definition 10.2 (Non-elastic Metric). For any point $\theta \in \Theta$, and a pair v_1, v_2 in $T_\theta(\Theta)$, define an inner product by:

$$\langle v_1, v_2 \rangle = \int_0^1 \langle v_1(t), v_2(t) \rangle \, dt \ . \tag{10.6}$$

This defines a Riemannian metric on Θ and is called the *non-elastic metric*.

These two representations for analyzing shapes of curves in $\mathbb{R}^n$ are considered in the next two sections.

10.4 Shape Spaces of Curves in $\mathbb{R}^n$

In this section, we consider curves in $\mathbb{R}^n$ and describe elements of shape analysis of these curves using the two representations. In each case, we will describe the mathematical representation, the Riemannian metric, and the constructions of geodesic path between arbitrary shapes. The case of elastic curves is considered in more detail, where the computation of Karcher mean and the stochastic modeling of shapes is also considered. The ideas presented here are similar to those used in Chap. 5 for planar curves.

10.4.1 Direction Function Representation

We will develop a set of basic tools needed to perform shape analysis of curves using the direction function representation. We start with the pre-shape space, compute geodesics on it, and repeat these steps for the shape space.

Pre-shape Space In this case, the curves are rescaled to be of unit length and are parameterized by the arc length. Each curve is represented by its direction function and the pre-shape space of open curves is given by the set of all such direction functions:

$$\mathcal{C}_1 = \{\theta : [0, 1] \to \mathbb{S}^{n-1}\} \ . \tag{10.7}$$

Note that $\mathscr{C}_1 \subset \mathbb{L}^2([0,1], \mathbb{R}^n)$ since the $\mathbb{L}^2$ norm of any of its elements is one. Also note that we are using the same notation here for pre-shape and shape spaces as was used for planar curves. As mentioned in the previous section, we use the $\mathbb{L}^2$ metric to impose a Riemannian structure on the manifold $\mathscr{C}_1$. Similar to the planar case, this metric, with the directional function representation, is called the **bending-only** metric as the curves are fixed to be arc-length parameterized and are not allowed to stretch or compress in this representation. For any $\theta \in \mathscr{C}_1$, the tangent space $T_\theta(\mathscr{C}_1)$ is given by:

$$T_\theta(\mathscr{C}_1) = \{v : [0,1] \to \mathbb{R}^n | v(t) \perp \theta(t), \; \forall t \in [0,1]\} \; .$$

Geodesics in Pre-shape Space

How do we compute geodesic paths between curves in this representation? In the planar case, we saw that the geodesic paths are straight lines. In this case also, the geodesics are simple but with a small difference. Since each point $\theta(t)$ is an element of the unit sphere $\mathbb{S}^{n-1}$, this structure has to be maintained in the computation of the geodesics. Let θ_1, θ_2 be two elements of $\mathscr{C}_1$. Then, the uniform-speed geodesic connecting them in $\mathscr{C}_1$ is given by $\alpha(\tau)$, with $\tau \in [0,1]$, such that for all $t \in [0,1]$:

$$\alpha(\tau)(t) = \frac{1}{\sin(c(t))} \left[\sin((1-\tau)c(t))\theta_1(t) + \sin(\tau c(t))\theta_2(t)\right] \qquad (10.8)$$

where $c(t)$ is the angle between the points $\theta_1(t)$ and $\theta_2(t)$ on the sphere $\mathbb{S}^{n-1}$, i.e., $c(t) = \cos^{-1}(\langle\theta_1(t), \theta_2(t)\rangle)$. The geodesic distance between any two curves, represented by their direction functions θ_1 and θ_2 in $\mathscr{C}_1$, is given by:

$$d_c(\theta_1, \theta_2) = \int_0^1 c(t) \; dt = \int_0^1 \cos^{-1}(\langle\theta_1(t), \theta_2(t)\rangle) \; dt \; . \qquad (10.9)$$

Shape Space

Next we look at the shape-preserving transformations that we wish to remove in our shape analysis. The transformations resulting from translation and scaling have already been removed. The former due to the dependence of the direction function on $\dot{\beta}(t)$ and the latter due to the arc-length parameterization of the curve with the parameter taking values in $[0,1]$. However, the variability due to rotations of curves remains. In other words, the curve β and the curve $O\beta$, for $O \in SO(n)$ not equal to the identity, will have two distinct direction functions despite having the same shape. The action of $SO(n)$ on the pre-shape space $\mathscr{C}_1$ is given by:

$$SO(n) \times \mathscr{C}_1 \to \mathscr{C}_1, \qquad (O, \theta) = O\theta \; .$$

Here, $O\theta(t)$ denotes the multiplication of an $n \times n$ matrix O with the unit vector $\theta(t) \in \mathbb{R}^n$. With respect to the chosen Riemannian metric on $\mathscr{C}_1$, the group $SO(n)$ acts on $\mathscr{C}_1$ by isometries, and we can define the shape space as the quotient space:

$$\mathscr{S}_1 = \mathscr{C}_1/SO(n) \; .$$

Since the action of $SO(n)$ is by isometries, the metric on $\mathscr{C}_1$ descends to $\mathscr{S}_1$. The computation of geodesics in this shape space relies on the following optimization problem:

$$d_s([\theta_1],[\theta_2]) = \underset{O \in SO(n)}{\operatorname{argmin}} \ d(\theta_1, O\theta_2) = \underset{O \in SO(n)}{\operatorname{argmin}} \left(\int_0^1 \cos^{-1}(\langle \theta_1(t), O\theta_2(t) \rangle) dt \right)$$

$$(10.10)$$

We fix the element θ_1 of the orbit $[\theta_1]$ and search over the whole orbit $[\theta_2]$ to find a point nearest to θ_1 under the distance d_c. This nearest distance gives us $d_s([\theta_1],[\theta_2])$.

One way to solve this optimization is using the gradient method. As described in Example A.6, the tangent space $T_{I_n}(SO(n))$ is the set all $n \times n$ skew-symmetric matrices. Let E_i, $i = 1, 2, \ldots, n(n-1)/2$ be the standard orthonormal basis of that space; E_i is an $n \times n$ matrix that is zero everywhere except two symmetric places where it has values $\frac{1}{\sqrt{2}}$ and $-\frac{1}{\sqrt{2}}$. The directional derivative of the cost function, in the direction of E_i, is computed as following. Let $A(t)$ be an $n \times n$ matrix given by $\theta_2(t)\theta_1^T(t)$. Then:

$$\nabla_{E_i} d_s([\theta_1],[O\theta_2]) = \nabla_{E_i} \int_0^1 \cos^{-1}(\operatorname{trace}(OA(t))) dt$$

$$= \int_0^1 \frac{1}{\sqrt{1 - \operatorname{trace}(OA(t))^2}} \operatorname{trace}(OE_i A(t)) dt \ . \quad (10.11)$$

The full gradient of the cost function is given by $\sum_{i=1}^{n(n-1)/2} \nabla_{E_i} d_s([\theta_1],[O\theta_2])OE_i$, and the gradient-based iterative update of O becomes:

$$O \mapsto O \exp\left(-\delta \sum_{i=1}^{n(n-1)/2} \nabla_{E_i} d_s([\theta_1],[O\theta_2])E_i\right) ,$$

where $\delta > 0$ is the step size.

Algorithm 44. *Given two curves in $\mathbb{R}^n$, represented by their angle functions θ_1 and θ_2 in $\mathscr{C}_1$, our goal is to compute a geodesic between $[\theta_1]$ and $[\theta_2]$ in $\mathscr{S}_1$.*

1. *Compute $A(t) = \theta_1 \theta_2^T \in \mathbb{R}^{n \times n}$ for all $t \in [0,1]$.*
2. *Let $E_1, E_2, \ldots, E_{n(n-1)/2}$ be an orthonormal basis of all $n \times n$ skew-symmetric matrices. For each of these basis elements, compute the directional derivatives:*

$$\kappa_i = \int_0^1 \frac{1}{\sqrt{1 - trace(OA(t))^2}} trace(OE_i A(t)) dt \ .$$

3. *Compute the full gradient $E = \sum_{i=1}^{n(n-1)/2} \kappa_i E_i$ and update $\theta_2 \mapsto exp(-\delta E)\theta_2$ for a small step size $\delta > 0$.*

Some examples of geodesics resulting from this algorithm are shown in Fig. 10.2. Here we look at some cylindrical helical curves with different numbers and placements of loops. The deformations along these geodesics clearly show the bending nature of this representation. In order to go from one helix to another, the algorithm has to straighten some loops and create some new loops, all by bending the first curve. No stretching or compression of the curves is allowed. It is worth noting that the intermediate curves hardly preserve the helical structure of the original curves.

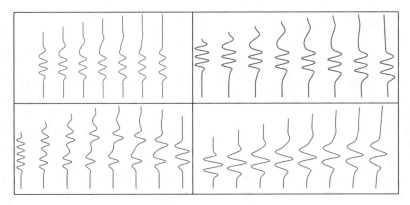

Fig. 10.2 Examples of bending-only geodesics between helical curves in the shape space $\mathscr{S}_1$

The steps for computing sample statistics in $\mathscr{S}_1$ and for imposing a stochastic model on non-elastic shapes are identical to those presented in Chap. 5. We refer the reader to that chapter for statistical analysis of shapes of non-elastic curves in $\mathbb{R}^n$.

10.4.2 Under SRVF Representation

The next representation to be studied is the elastic curve representation—the square-root velocity function under the $\mathbb{L}^2$ metric. As we have seen earlier, this representation allows for a combination of bending and stretching of curves to allow for a better matching of features than the previous case and, thus, is a more attractive framework for shape analysis of curves in general. To remove the scale variability, we fix the length of curves. Similar to the previous chapters, we fix the length to be one.

1. **Pre-shape Space**: The pre-shape space of all square-root velocity representations of curves in $\mathbb{R}^n$ is:

$$\mathscr{C}_2 = \{q : [0, 1] \to \mathbb{R}^n | \int_0^1 |q(t)|^2 dt = 1\} .$$

By definition, $\mathscr{C}_2$ is a hypersphere of radius one in $\mathbb{L}^2([0, 1], \mathbb{R}^n)$ with the standard metric:

$$\langle w_1, w_2 \rangle = \int_0^1 \langle w_1(t), w_2(t) \rangle \, dt , \qquad (10.12)$$

where the inner product inside the integral is the simple Euclidean product between vectors in $\mathbb{R}^n$. For any $q \in \mathscr{C}_2$, the tangent space $T_q(\mathscr{C}_2)$ is:

$$T_q(\mathscr{C}_2) = \{w : [0, 1] \to \mathbb{R}^n | \langle w, q \rangle = 0\} .$$

As described in the previous section, the use of the $\mathbb{L}^2$ metric on the space of square-root velocity representations generates a framework for elastic shape comparisons and the manifold $\mathscr{C}_2$ becomes a Riemannian manifold under this metric.

2. **Geodesics in Pre-shape Space**: With this Riemannian structure and the spherical geometry of the pre-shape space, one can write explicit forms for geodesics between curves: Given any two length 1, parameterized curves β_1 and β_2 in $\mathbb{R}^n$, we can compute geodesic paths between their representations q_1 and q_2 using the shorter arc on $\mathscr{C}_2$ (same as Eq. 5.14): $\alpha : [0,1] \to \mathscr{C}_2$:

$$\alpha(\tau) = \frac{1}{\sin(\theta)} \left[\sin(\theta(1-\tau))q_1 + \sin(\tau\theta)q_2 \right] \tag{10.13}$$

where $\theta = \cos^{-1}(\langle q_1, q_2 \rangle)$ and $q_1 \neq \pm q_2$. Even though these curves are in $\mathbb{R}^n$, the expressions for geodesics and geodesic distances remain the same as those for planar curves. The geodesic distance between any two points in $\mathscr{C}_2$ is given by (same as Eq. 5.15):

$$d_c(q_1, q_2) = \cos^{-1}(\langle q_1, q_2 \rangle) \ . \tag{10.14}$$

According to Eq. 5.17, the exponential map, $\exp : T_q(\mathscr{C}_2) \mapsto \mathscr{C}_2$ is given by:

$$\exp_q(w) = \cos(\|w\|)q + \sin(\|w\|)\frac{w}{\|w\|}. \tag{10.15}$$

For any $q_2 \in \mathscr{C}_2$, the inverse of the exponential map at $q_1 \in \mathscr{C}_2$, denoted by $\exp_{q_1}^{-1} : \mathscr{C}_2 \to T_{q_1}(\mathscr{C}_2)$, is computed as follows: $\exp_{q_1}^{-1}(q_2) = v$, where v is computed according to:

$$v = \frac{\theta}{\sin(\theta)}(q_2 - \cos(\theta)q_1), \quad \text{where} \quad \theta = \cos^{-1}(\langle q_1, q_2 \rangle). \tag{10.16}$$

3. **Shape Space**: Similar to the notation in Chap. 5, $\mathscr{C}_2$ is the pre-shape space since its elements do not represent the shape of a curve uniquely. A re-parameterization of β, using an element $\gamma \in \Gamma_I$, where Γ_I is the group of diffeomorphisms from $[0,1]$ to itself, results in a different square-root velocity function while preserving its shape. The action of Γ_I on $\mathscr{C}_2$ is given by $(q, \gamma) = \sqrt{\dot{\gamma}}(q \circ \gamma)$. Additionally, any rigid rotation of β changes q but not its shape. The action of the rotation group $SO(n)$ on $\mathscr{C}_2$ is given by $(O, q) = Oq$. It can be shown that the actions of these two groups on $\mathscr{C}_2$ commute, i.e., a re-parameterization of a curve followed by its rotation results in the same representation when the two operations are applied in a different order. Also, the two groups act on $\mathscr{C}_2$ by isometries.

Lemma 10.1. *The action of the product group $\Gamma_I \times SO(n)$ on $\mathscr{C}_2$ is by isometries with respect to the chosen metric.*

Proof. Same as Lemma 5.2.

For these reasons, we will call $\mathscr{C}_2$ the pre-shape space of open curves and we define the individual shapes as the orbits:

$$[q] = \text{closure}\{O\sqrt{\dot{\gamma}}(q \circ \gamma) | \gamma \in \Gamma_I, O \in SO(n)\} \ .$$

The set of all such orbits is defined as the shape space: $\mathscr{S}_2 = \{[q] | q \in \mathscr{C}_2\}$. The shape space $\mathscr{S}_2$ inherits a distance from the pre-shape space $\mathscr{C}_2$; the distance between any two orbits $[q_1]$ and $[q_2]$ is given by:

$$d_s([q_1], [q_2]) = \inf_{\gamma \in \Gamma_I, O \in SO(n)} d_c(q_1, \sqrt{\dot{\gamma}}O(q_2 \circ \gamma)) \ . \tag{10.17}$$

where d_c is as defined in Eq. 10.14. (In some cases where no optimal γ exists in Γ_I, one may be able to find a pair $(\gamma_1^*, \gamma_2^*) \in \tilde{\Gamma}_I \times \tilde{\Gamma}_I$ for optimal alignment, as discussed in Sects. 4.10 and 5.5.1.) The actual geodesic between $[q_1]$ and $[q_2]$ in $\mathscr{S}_2$ is given by $[\alpha(\tau)]$, where $\alpha(\tau)$ is the geodesic in $\mathscr{C}_2$ between q_1 and $\sqrt{\dot{\gamma}^*} O^*(q_2 \circ \gamma^*)$ as defined in Eq. 10.13. Here (O^*, γ^*) are the optimal transformations of q_2 that minimize the right side in Eq. 10.17. A closer look at that distance function reveals the following:

$$\underset{\gamma \in \Gamma_I, O \in SO(n)}{\arg\min} \cos^{-1}\left(\left\langle q_1, \sqrt{\dot{\gamma}} O(q_2 \circ \gamma)\right\rangle\right)$$

$$= \arg\inf_{\gamma \in \Gamma_I, O \in SO(n)} \|q_1 - \sqrt{\dot{\gamma}} O(q_2 \circ \gamma)\|^2$$

$$= \arg \sup_{\gamma \in \Gamma_I, O \in SO(n)} \left\langle q_1, O(q_2 \circ \gamma)\sqrt{\dot{\gamma}}\right\rangle$$

where the last norm is simply the $\mathbb{L}^2$ norm on the representation space. This equality says that minimizing the arc length on a unit sphere is the same as minimizing the chord length. If one is minimized, then so is the other. Therefore, for the purpose of finding the minimizer, we can use the $\mathbb{L}^2$ norm, which opens up the possibility of a computationally efficient solution.

4. **Geodesics in Shape Space**: Consider the problem:

$$(\gamma^*, O^*) = \arg\inf_{\gamma \in \Gamma_I, O \in SO(n)} \|q_1 - \sqrt{\dot{\gamma}} O(q_2 \circ \gamma)\|^2 . \tag{10.18}$$

This is a joint optimization problem with a well-known solution.

a. **Optimal Rotation**: For a fixed $\gamma \in \Gamma_I$, the optimization problem in Eq. 10.18 over $SO(n)$ is solved by $O^* = UV^T$, where USV^T is the singular valued decomposition of $A = \int_0^1 q_1(t)(\sqrt{\dot{\gamma}} q_2(\gamma(t)))^T dt$. In case the determinant of A is negative, one needs to modify V by making its last column negative (of it current value) before multiplying to U to obtain O^*. This is a known result from rigid alignment of objects when the points across objects are already registered.

b. **Optimal Registration**: For a fixed O, the optimization problem in Eq. 10.18 over Γ_I can be solved using the dynamic-programming (DP) algorithm or the gradient method described earlier in Chap. 5. Since the cost function is defined by the $\mathbb{L}^2$ norm and, thus, is additive over the path $(t, \gamma(t))$, the DP algorithm applies. As described in Appendix B, the DP algorithm forms a finite-dimensional grid in $[0,1]^2$ and searches over all the paths on that grid, satisfying the required constraints, to obtain an approximation to γ^*. WLOG, we will denote that estimated diffeomorphism by γ^*.

Once the optimal rotation and re-parameterization of q_2 are obtained, we can compute the geodesic path between the orbits $[q_1]$ and $[q_2]$, as mentioned above.

Algorithm 45. *Compute a geodesic path between two curves in $\mathbb{R}^n$ represented by their square-root velocity functions q_1 and q_2.*

a. *Initialize $\gamma = \gamma_{id}$.*
b. *Compute the $n \times n$ matrix $A = \int_0^1 q_1(t)(\sqrt{\dot{\gamma}} q_2(\gamma(t)))^T dt$. Perform SVD of $A = USV^T$ and compute the optimal rotation $O^* = UV^T$.*

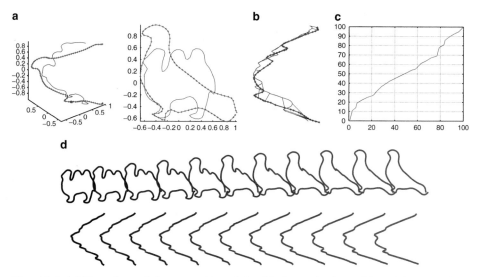

Fig. 10.3 (**a**) Two views of the curves β_1 and β_2, (**b**) the optimal matching γ^* between them and (**c**) the right panel shows the optimal γ^* estimated by the DP algorithm. (**d**) The geodesic path between the two shapes in $\mathscr{S}_2$ from two viewing angles

   c. Perform the gradient process (or DP algorithm) to find the optimal $\gamma^* \in \Gamma_I$.
   d. Update $q_2 \mapsto \sqrt{\dot{\gamma}^*}O^*(q_2 \circ \gamma^*)$.
   e. If not converged, go to step (b).

Shown in Fig. 10.3 is an example of computing geodesics between open curves in $\mathbb{R}^3$. In this toy example, we take two 2D contours and add a linear third coordinate to them to make them open curves in $\mathbb{R}^3$. In the top-left panels, we show the curves, β_1 and β_2, from two different angles. The procedure outlined above is used to find the optimal rotation and re-parameterization of q_2; these matching results are shown in the top-right panels. The bottom two rows show the resulting geodesic path in $\mathscr{S}_2$ from two viewing angles. Since the original curves came from 2D shapes, it is easier to verify the correct registration and the elastic nature of the resulting geodesic path.

   Shown in Fig. 10.4 are two examples of geodesics between some cylindrical helices. In each case, the panels (a) and (b) show the two helices, and (c) is the optimal matching between them obtained using the estimated γ function shown in panel (d). The resulting geodesic paths in $\mathscr{S}_2$ between these curves are shown in the bottom row. It is easy to see the combination of bending and stretching/compression that goes into deforming one shape into another. In the left example, where the turns are quite similar and the curves differ only in the placements of these turns along the curve, a simple stretching/compression is sufficient to deform one into another. However, in the right example, where the number of turns is different, the algorithm requires both bending and stretching to reach from the first to the second shape.

   We can use the example of helical curves to compare the nature of geodesics in the two shape spaces $\mathscr{S}_1$ and $\mathscr{S}_2$. Figure 10.5 shows an example of geodesics between the same two curves but in different spaces. The top example is for the non-elastic shape space and the bottom one is for the elastic shape space. It is

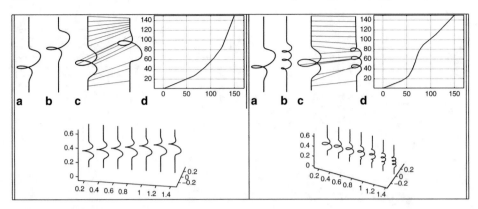

Fig. 10.4 Coil matching: Two example of computing geodesics between 3D curves containing spiral features. In each case (**a**) and (**b**) show the original curves, (**c**) shows the optimal registration between them, and (**d**) the optimal γ function. The lower panels show the corresponding geodesic paths between them

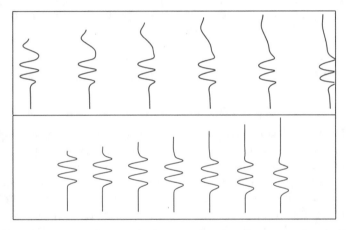

Fig. 10.5 Comparison of geodesic paths between the same two curves in $\mathscr{S}_1$ (*top*) and $\mathscr{S}_2$ (*bottom*). The bottom right shows the matching of points in the elastic matching of curves

easy to see the preservation of features (loops, vertical parts, etc.) in the elastic geodesic, as compared to the non-elastic one.

Finally, in Fig. 10.6, we present an example of comparing two protein backbones. In this experiment, we use two simple proteins—1CTF and 2JVD—that contain three and two α-helices, respectively. The top row of this figure shows depictions of the two backbones, while the bottom row shows the geodesic path between them in $\mathscr{S}$. The bottom left shows the registration of points between the curves and the bottom right shows the estimated γ function that results in that registration. These results on both simulated helices and real backbones suggest a role for elastic shape analysis in protein structure analysis.

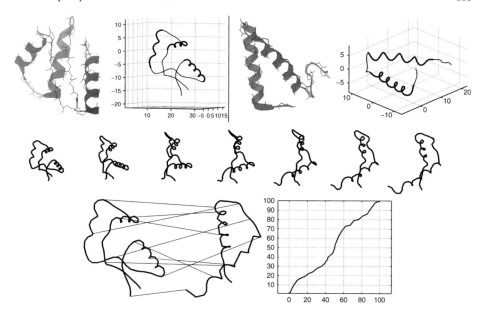

Fig. 10.6 Elastic deformations to compare shapes of two proteins: 1CTF and 2JVD (obtained from PDB). The top row shows the two proteins: 1CTF on the left and 2JVD on the right. The bottom row shows the elastic geodesic between them. The bottom left shows the optimal registration between the two curves and the bottom right shows the optimal γ function

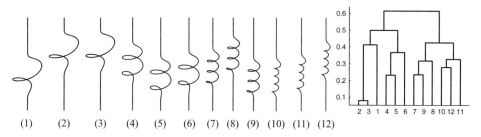

Fig. 10.7 A set of helices with different numbers and placements of spirals and their clustering using the elastic distance function

10.4.3 Hierarchical Clustering of Elastic Curves

Figure 10.7 shows an example of using the elastic distances between curves for clustering and classification. In this example, we experiment with 12 cylindrical helices that contain different numbers and placements of turns. The first three helices have only one turn, the next three have two turns, and so on. (We point out that the radii of these turns have some randomness that is difficult to see with the naked eye.) Using the elastic geodesic distances between them in $\mathscr{S}_2$, and the dendrogram clustering program in MATLAB, we obtain the clustering shown in the right panel. This clustering demonstrates the success of the proposed elastic metric in that helices with similar numbers of turns are clustered together.

10.4.4 Sample Statistics and Modeling of Elastic Curves in $\mathbb{R}^n$

Now that the tool for computing geodesic in $\mathscr{S}_2$ is available, we can discuss the computation of sample statistics and a probability model to capture shape variation in a population. We start with the sample statistics and Karcher mean.

Karcher Mean The very first step in stochastic modeling of shapes is to compute a statistical mean shape for a given collection of shapes. For this purpose, we use the notion of Karcher mean that has been used by many papers for finding means on nonlinear manifolds. For a given collection of curves $\{\beta_1, \beta_2, \ldots, \beta_n\}$, with shape representations, $\{q_1, q_2, \ldots, q_n\}$, the Karcher mean is defined as:

$$\bar{\mu}_n = \operatorname*{argmin}_{[q] \in \mathscr{S}_2} \sum_{i=1}^{n} d_s([q], [q_i])^2 . \tag{10.19}$$

Although a global mean would be desirable, a local minimum is considered sufficient in this definition. Therefore, a gradient-based approach is used for finding the Karcher mean, as follows.

Algorithm 46 (Karcher Mean on $\mathscr{S}_2$). *Let μ_0 be an initial estimate of the Karcher mean. Since the q_i's are elements of a hypersphere $\mathscr{C}_2$, it seems natural to use their extrinsic mean (Eq. 9.7) in $\mathbb{L}^2([0,1], \mathbb{R}^n)$ to initialize the gradient algorithm. Set $j = 0$.*

1. *For each $i = 1, \ldots, n$, compute the tangent vector $v_i \in T_{\mu_j}(\mathscr{S}_2)$ as follows: find the optimal (O_i^*, γ_i^*) in Eq. 10.18 for minimizing the distance between μ_j and $[q_i]$. Set $q_i^* = O_i^* \sqrt{\dot{\gamma}_i^*}(q_i \circ \gamma_i^*)$, and compute:*

$$v_i = \frac{\theta_i}{\sin(\theta_i)}(q_i^* - \cos(\theta_i)\mu_j), \quad \text{where} \quad \cos(\theta_i) = \langle \mu_j, q_i^* \rangle .$$

2. *Compute the average direction $\bar{v} = \frac{1}{n} \sum_{i=1}^{n} v_i$.*
3. *If $\|\bar{v}\|$ is small, then stop. Else, update μ_j in the direction $\bar{v}$ using (Eq. 5.17)*

$$\mu_{j+1} = \cos(\epsilon \|\bar{v}\|)\mu_j + \sin(\epsilon \|\bar{v}\|)\frac{\bar{v}}{\|\bar{v}\|}$$

where $\epsilon > 0$ is small step size, typically 0.5.
4. *Set $j = j + 1$ and return to Step 1.*

An example of computing the Karcher mean of some curves in $\mathbb{R}^3$ is shown in Fig. 10.8. The two left panels in (a) show five curves in $\mathbb{R}^3$ (from different viewing angles), panel (b) shows the estimated Karcher mean of their shapes, and panel (c) shows the evolution of the cost function (given in Eq. 10.19) during the gradient iterations.

Wrapped-Normal Density Given a mean shape $\mu \in \mathscr{S}_2$ and a covariance matrix on a subspace of $T_\mu(\mathscr{S}_2)$, we can define a wrapped-normal density on $\mathscr{S}_2$.

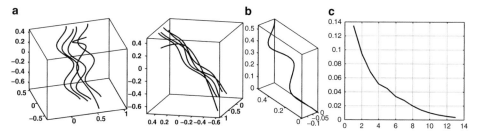

Fig. 10.8 (**a**) A collection of five curves (shown from two different viewpoints in the first two panels), (**b**) their Karcher mean (middle panel), and (**c**) the decrease in the cost function (Eq. 10.19) during gradient iterations

Let the covariance operator K admit an orthonormal basis $\{w_1, w_2, \ldots, w_j\}$ and variances $\{\sigma_1^2, \sigma_2^2, \ldots, \sigma_j^2\}$ such that:

$$K(s,t) = \sum_{i=1}^{j} \sigma_i^2 w_i(s) w_i(t) .$$

To generate samples from this wrapped-normal density, one simply generates coefficients $a_i \sim \mathcal{N}(0, \sigma_i^2)$ and forms a tangent function: $w(t) = \sum_{i=1}^{j} a_i w_i(t), \quad t \in [0, 1]$. Then, using the exponential map $w \mapsto \cos(\|w\|)\mu + \sin(\|w\|)\frac{w}{\|w\|}$, this tangent vector is mapped to the shape space.

Figure 10.9 shows an experimental illustration of this idea. The top panel (a) in this figure shows 20 spiral shapes that are used in this experiment. The first step is to compute a Karcher mean of these shapes as discussed above. The panel (b) shows the decrease in the variance function (the cost function in Eq. 9.8) during the estimation of the mean, panel (c) shows the estimated Karcher mean μ, and panel (d) plots the singular values σ_i^2 of the covariance matrix. Since the first two singular values capture 95 % of the total variations, we use $j = 2$ components in the wrapped-normal model. The first two singular vectors form the orthonormal basis elements w_1 and w_2. The bottom panel shows some random samples from this estimated wrapped-normal density.

There are several important applications of shape analysis of open curves in $\mathbb{R}^3$. Two common applications involve shapes of curves associated with the brain anatomy—neuronal fiber tracts (especially in human brain) and sulcal curves formed by folds of white matter. A sulcus is a depression or fissure in the surface of the brain that surrounds the gyri, creating the characteristic appearance of the brain in humans (see Fig. 10.10 top). The problem of automatic labeling of sulcal curves, i.e., putting an anatomical label on a sulcal curve using its different physical features, is a well-known problem. One feature that can be useful in this process is the shape, and thus, shape analysis of sulcal curves becomes important. Shown in Fig. 10.10 is an example of comparing shapes of two sulcal curves as elements of $\mathscr{S}_2$.

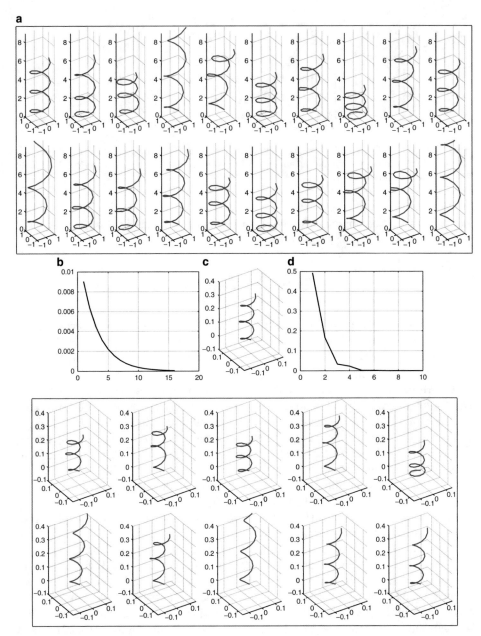

Fig. 10.9 (a) A collection of 20 spiral curves used in this experiment, (b) the decrease in the Karcher variance function during mean estimation, (c) the estimated Karcher mean and (d) the estimated singular values of the covariance matrix. (e)–(i) random samples from the estimated wrapped-normal density in $\mathscr{S}_2$

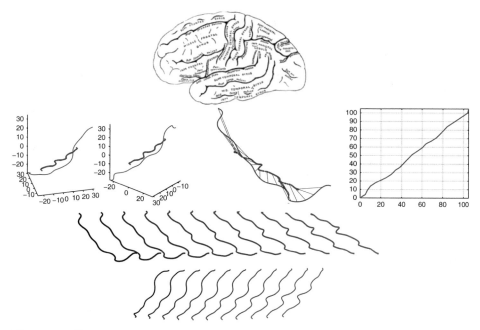

Fig. 10.10 The top cartoon shows the sulcal curves formed by wrinkled surface of a brain. The next row shows two example sulcal curves and an optimal registration between them. The bottom two rows show the geodesic path between then in $\mathscr{S}_2$, shown from two different viewpoints

10.5 Registration of Curves

In this section, we take a look at the problem of registering curves, both pairwise and group-wise, in $\mathbb{R}^n$. These curves may come, for example, as descriptions of a dynamic system evolving over time. Let $\beta(t) \in \mathbb{R}^n$ denote a feature vector describing properties of interest in a system and let $\beta = \{\beta(t)|t \in [0, T]\}$ define the full trajectory of this evolution. One goal may be to develop statistical models for analyzing observations of such systems. One can either use time series models, e.g., autoregressive models, or standard function data analysis, using the Hilbert structure of the space of such curves, to perform modeling and analysis. However, the following issue makes this problem both interesting and challenging. The temporal rate of evolution of the system is captured in the parameterization of the curve: $t \mapsto \beta(t)$. If the system goes through the same feature values, in the same order, but at different rates, then we can express the new curve as a re-parameterization of the earlier trajectory: $\tilde{\beta}(t) = \beta(\gamma(t))$. The challenge is to compare trajectories in such a way that the rate of evolution of the trajectories does not affect the inference. In other words, the trajectories β and $\tilde{\beta}$ are equivalent from the perspective of analysis. Another related problem is that of registration: which point on one curve β_1 is registered with which point on another curve β_2. In order to improve comparisons between observations of dynamic systems, it is important to find a suitable registration of time points across observations.

10.5.1 Pairwise Registration of Curves in $\mathbb{R}^n$

Pairwise registration of curves in $\mathbb{R}^n$ is essentially an extension of the problem studied in Sect. 4.4, where we discussed registration of real-valued functions. Now we are considering vector-valued functions. Also, this problem is very similar to shape analysis of contours in $\mathbb{R}^n$ except we do not perform rotational alignment. Following Definition 4.7, the pairwise registration problem can be stated as:

Definition 10.3 (Registration Problem). For any two curves β_1 and β_2, with SRVFs denoted by q_1 and q_2, the registration for β_1 and β_2 is solved using:

$$\inf_{\gamma \in \Gamma_I} \|q_1 - (q_2, \gamma)\| \,. \tag{10.20}$$

We illustrate this idea using an application in computer vision. Here one tries to identify human action using a depth camera (e.g., Microsoft's Kinect) that provides a representation of human shape in the form of a body skeleton at each observation time. The observation of an action over a time interval provides a sequence of skeletons. This sequence shows the changes in body pose and articulation during performance of that action. Figures 10.11, 10.12, 10.13 show some examples of these sequences. The first figure shows two observations of the same

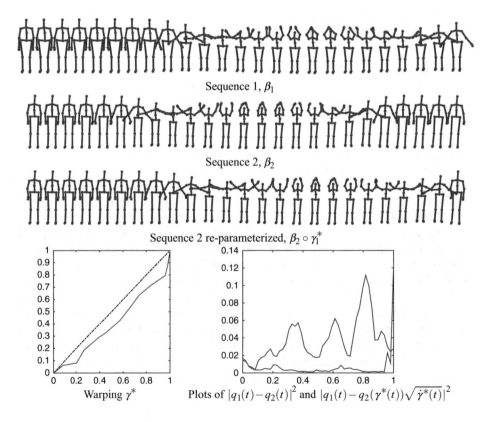

Sequence 1, β_1

Sequence 2, β_2

Sequence 2 re-parameterized, $\beta_2 \circ \gamma_1^*$

Warping γ^*

Plots of $|q_1(t) - q_2(t)|^2$ and $|q_1(t) - q_2(\gamma^*(t))\sqrt{\dot{\gamma}^*(t)}|^2$

Fig. 10.11 First two rows: Sequences of skeletons display the action "two-hand wave" by difference people. Notice the time lag in β_1 before the swing starts. Third row: Re-parameterized β_2 after temporal alignment with the second sequence

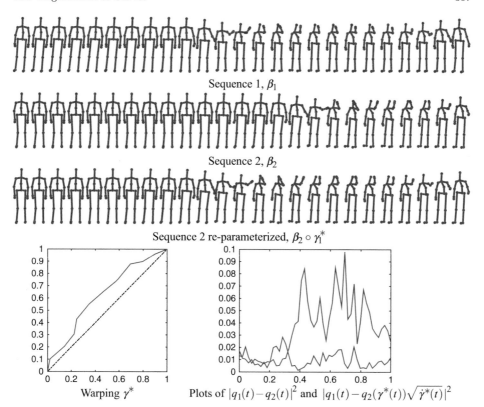

Sequence 1, β_1

Sequence 2, β_2

Sequence 2 re-parameterized, $\beta_2 \circ \gamma_1^*$

Warping γ^*

Plots of $|q_1(t) - q_2(t)|^2$ and $|q_1(t) - q_2(\gamma^*(t))\sqrt{\dot{\gamma}^*(t)}|^2$

Fig. 10.12 First two rows: Sequences of skeletons display the action "one-arm wave" by difference people. Notice the time lag in β_1 before the swing starts. Third row: Re-parameterized β_2 after temporal alignment with the second sequence

action *two-hand wave*, the second shows *one-arm wave*, and the third shows *pickup and throw*. Since a skeleton is made up of 20 joints, each having a position in $\mathbb{R}^3$, we can represent a sequence as a trajectory $\beta : [0, T] \to \mathbb{R}^{60}$. The SRVF of a trajectory is defined as usual by $q(t) = \frac{\dot{\beta}(t)}{\sqrt{|\dot{\beta}(t)|}}$, for $\dot{\beta}(t) \neq 0$, and $q(t) = 0$ if $\dot{\beta}(t) = 0$. Such an SRVF is the mapping $q : [0, T] \to \mathbb{R}^{60}$.

Given SRVFs q_1 and q_2 of two such sequences, we find the optimal alignment between them using Eq. 10.20 and the dynamic-programming algorithm. The implementation details are same as in the scalar alignment (Sect. 4.4) and are not repeated here. We present some experimental results in Figs. 10.11–10.13. In each case, we take two sequences of the same action—a two-hand wave in the first case, a one-arm wave in the second, and a pickup and throw in the last—and show their alignment. The top two rows in each figure show the given sequences β_1 and β_2, and one can see the time lags between them in starting of the actions. Then, we show the aligned sequences $\beta_2 \circ \gamma^*$ in the third row, and one can see the improvement in the level of alignment between β_1 and $\beta_2 \circ \gamma^*$. This can be also seen in the plot of $|q_1(t) - q_2(\gamma^*(t))\sqrt{\dot{\gamma}^*(t)}|^2$ versus t; this plot generally lies below the plot of $|q_1(t) - q_2(t)|^2$, quantifying the improvements made in temporal registration of two sequences. The bottom left shows the optimal warping function γ^* in each case.

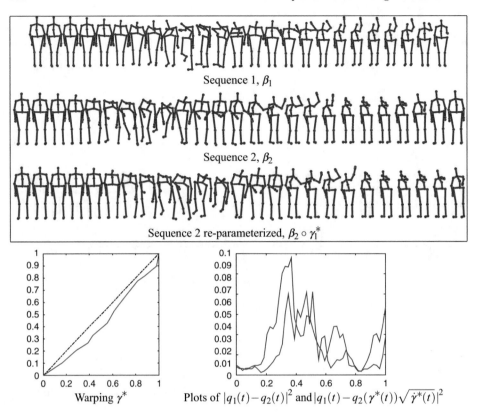

Fig. 10.13 Same as Fig. 10.11 but with the action "pickup and throw"

10.5.2 Registration of Multiple Curves

This process can be extended to the alignment of multiple curves using the framework described in Sect. 8.3. While we studied scalar functions there, the focus here is on vector-valued functions. This, however, does not make much of a difference in the underlying concepts and even the algorithms remain mostly the same. Following that section, we can define the notions of phases and amplitudes of vector-valued functions and use the same definitions of phase and amplitude distances. In view of the close similarity of the solutions in the two cases, we will not cover the problem of registration of multiple curves in detail but present an interesting example to illustrate the idea. Shown in the top panel of Fig. 10.14 are three examples of skeletal sequences involving the action "one-arm horizontal wave." Treating each sequence as a curve in $\mathbb{R}^{60}$, we use a simple extension of the phase-amplitude separation procedure presented in Algorithm 33 to align six such sequences. The middle panel shows the sequence mean before and after the alignment and the bottom panel shows the decrease in the cost function during alignment. Although the two mean sequences appear very similar, a closer look shows that the action is more clearly defined in the aligned mean, especially toward the end of the action.

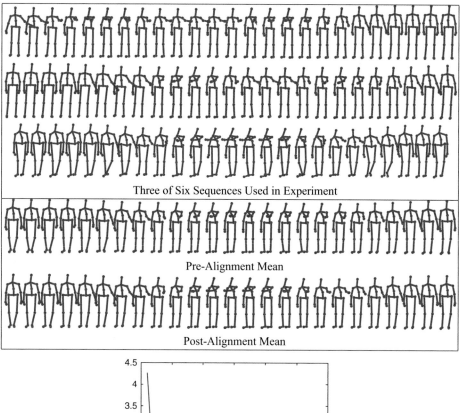

Three of Six Sequences Used in Experiment

Pre-Alignment Mean

Post-Alignment Mean

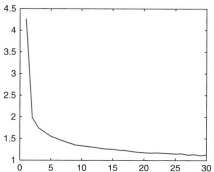

Decrease in Objective Function During Aligment

Fig. 10.14 An example of alignment six action sequences, all involving a one-arm horizontal wave, using a simple extension of Algorithm 33. Each sequence is viewed as a curve in $\mathbb{R}^{60}$

10.6 Shapes of Closed Curves in $\mathbb{R}^n$

Next we look at extending our analysis of curves in higher dimensions to *closed* curves. We will briefly describe the bending-only framework for handing closed curves and will largely focus on the elastic framework for handling this type of data.

10.6.1 Non-elastic Closed Curves

Let $\beta : [0, 1] \to \mathbb{R}^n$ be a differentiable unit-length curve, parameterized by the arc length, and let $\theta(s) = \dot{\beta}(s) \in \mathbb{S}^{n-1}$ be its direction function, as defined earlier.

Since we are interested in *closed* curves, we describe that subset next. Define a map Φ_1 by $\Phi_1(\theta) = \int_0^1 \theta(s)ds$, and let:

$$\mathscr{C}_1^c = \Phi_1^{-1}(\mathbf{0}) \equiv \{\theta : [0,1] \to \mathbb{S}^{n-1} | \Phi_1(\theta) = \mathbf{0}\} \ . \tag{10.21}$$

Here $\mathbf{0}$ denotes the origin in $\mathbb{R}^n$. Since Φ_1 is a smooth map and $\mathbf{0}$ is a regular point of $\mathbb{R}^n$, the set $\mathscr{C}_1^c$ is a submanifold. $\mathscr{C}_1^c$ is actually the set of all closed curves in $\mathbb{R}^n$, each represented by its angle function. For any $\theta \in \mathscr{C}_1^c$, the tangent space $T_\theta(\mathscr{C}_1^c)$ is given by:

$$T_\theta(\mathscr{C}_1^c) = \{v : [0,1] \to \mathbb{R}^n | v(s) \perp \theta(s), \ \forall s \in [0,1], \ \text{and} \ \int_0^1 v(s)ds = 0\} \ . \tag{10.22}$$

We will continue to use the $\mathbb{L}^2$ metric to impose a Riemannian structure on the manifold $\mathscr{C}_1^c$. By now the reader will be familiar with the non-elastic deformation resulting from the use of the $\mathbb{L}^2$ metric in the space of direction functions.

The next goal is to develop a tool for computing geodesic paths between elements of $\mathscr{C}_1^c$. Due to the closure constraint, $\mathscr{C}_1^c$ is a complicated nonlinear manifold and there is no ready expression for evaluating geodesics in it. We will use the path-straightening method, first described in Sect. 6.6, for this problem. In order to use this method, we need several basic tools relating to projections on $\mathscr{C}_1^c$ and its tangent spaces.

1. **Projection of $\theta \in \mathscr{C}_1$ into $\mathscr{C}_1^c$:** If $\theta \notin \mathscr{C}_1^c$, i.e., it denotes an open curve, we want to close it by projecting it into $\mathscr{C}_1^c$. We perform this iteratively by moving perpendicular to the level sets of the map Φ_1 in such a way that Φ_1 of the resulting point moves toward the origin in $\mathbb{R}^n$. In other words, we seek a direction g such that a perturbation of θ in that direction results in $\Phi_1(\theta + g) = \mathbf{0}$, to the first order.

   We develop some theory for implementing this idea. Consider the linear mapping $d\Phi_{1,\theta} : T_\theta(\mathbb{L}^2([0,1],\mathbb{R}^n)) \mapsto \mathbb{R}^n$ defined by $d\Phi_{1,\theta}(v) = \int_0^1 v(s)ds$. It can be shown that $d\Phi_{1,\theta}$ is surjective, as long as $\theta([0,1])$ is not contained in a one-dimensional subspace of $\mathbb{R}^n$, and therefore $\mathscr{C}_1^c$ is a codimension n submanifold of $\mathbb{L}^2([0,1],\mathbb{R}^n)$. The adjoint of $d\Phi_{1,\theta}$, $d\Phi_{1,\theta}^* : \mathbb{R}^n \to T_\theta(\mathbb{L}^2([0,1],\mathbb{R}^n))$ is the unique linear transformation with the property that for all $v \in T_\theta(\mathbb{L}^2([0,1],\mathbb{R}^n))$ and $x \in \mathbb{R}^n$, $(d\Phi_{1,\theta}(v) \cdot x) = \langle v, d\Phi_{1,\theta}^*(x) \rangle$. This adjoint is given by $d\Phi_{1,\theta}^*(x) \equiv v$ such that $v(s) = x - (x \cdot \theta(s))\theta(s)$. In other words, $d\Phi_{1,\theta}^*$ takes a vector x in $\mathbb{R}^n$ and forms a tangent vector field on f by making x perpendicular to $\theta(s)$ for all s (or by projecting x onto the tangent space $T_{\theta(s)}(\mathbb{S}^{n-1})$ for each s). This formula makes explicit the role of θ in definition of $d\Phi_{1,\theta}^*$.

   Proposition 10.1. *The range space of $d\Phi_{1,\theta}^*$ is the orthogonal complement of the null space of $d\Phi_{1,\theta}$:*

   $$\{v \in T_\theta(\mathscr{P}) | v = d\Phi_{1,\theta}^*(x) \ \text{for some} \ x \in \mathbb{R}^n\}$$
   $$= \{v \in T_\theta(\mathbb{L}^2([0,1],\mathbb{R}^n)) | d\Phi_{1,\theta}(v) = 0\}^\perp \ . \tag{10.23}$$

   Proof. If v is in range space of $d\Phi_{1,\theta}^*$, i.e., there exists a x such that $d\Phi_{1,\theta}^*(x) = v$. Then, for any g in the null space of $d\Phi_{1,\theta}$, we have $\langle v, g \rangle = \langle d\Phi_{1,\theta}^*(x), g \rangle = (x \cdot d\Phi_{1,\theta}(g)) = 0$. That is, v is perpendicular to the null space of $\Phi_{1,\theta}$. The proposition follows from the fact that the dimension of the range space of $d\Phi_{1,\theta}^*$ is equal to the codimension of the null space of $d\Phi_{1,\theta}$. $\square$

Let $e_1, e_2, \ldots, e_n$ be the canonical basis of $\mathbb{R}^n$ and define $E_{\theta,i} = d\Phi_{1,\theta}^*(e_i) \in T_\theta(\mathbb{L}^2([0,1],\mathbb{R}^n))$ for all i. The functions $E_{\theta,i}$ form a basis of the range space of $d\Phi_{1,\theta}^*$. Using Eq. 10.23, these three functions also form a basis for the orthogonal complement of null space of $d\Phi_{1,\theta}$.

Since g is perpendicular to the level set of Φ_1 at θ, it is orthogonal to the null space of $d\Phi_{1,\theta}$ and is in the range space of $d\Phi_{1,\theta}^*$. Therefore, g can be written as a linear combination of the functions $E_{i,\theta}$, $i = 1, 2, \ldots, n$. Using Taylor's approximation, we get:

$$\Phi_1(\theta + g) \approx \Phi_1(\theta) + d\Phi_{1,\theta}(g) = 0 \quad \text{if} \quad g = -A_\theta^{-1}\Phi_1(\theta) \ ,$$

where $A_\theta \in \mathbb{R}^{n \times n}$ is the Jacobian of $d\Phi_\theta$ restricted to the space orthogonal to the null space of $d\Phi_{1,\theta}$. The Jacobian A_θ can also be specified using the same basis functions. The algorithm to implement this projection is given next.

Algorithm 47 (Projection of Arbitrary θ into $\mathscr{C}_1^c$).

a. For $i = 1, 2, \ldots, n$, compute $E_{i,\theta} = d\Phi_{1,\theta}(\mathbf{e_i})$ and $A_\theta(:,i) = d\Phi_{1,\theta}(E_{i,\theta})$.
b. Compute $b = -A_\theta^{-1}\Phi_1(\theta)$.
c. Set $g = \sum_{i=1}^n b(i)E_{i,\theta}$.
d. For all s in $[0,1]$, update $\theta(s)$ using:

$$\theta(s) \mapsto \cos(|g(s)|)\theta(s) + \sin(|g(s)|)\frac{g(s)}{|g(s)|} \ .$$

e. If $|\Phi_1(\theta)| < \epsilon$, stop. Else, return to Step 1.

Shown in Fig. 10.15 are three examples of this projection. There we see three open curves in solid plane lines and their projections into $\mathscr{C}_1^c$ using marked lines.

2. **Projecting $v \in \mathbb{L}^2([0,1],\mathbb{S}^{n-1})$ into the subspace $T_\theta(\mathscr{C}_1^c)$.** For a vector field $v \in T_\theta(\mathbb{L}^2([0,1],\mathbb{S}^{n-1}))$, we have that $(v(s) \cdot \theta(s)) = 0$. However, $\int_0^1 v(s)ds$ may

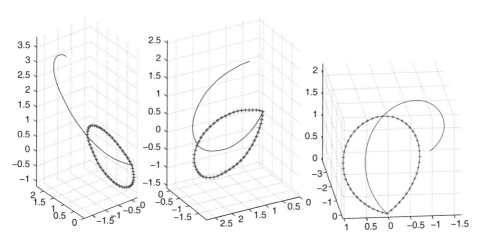

Fig. 10.15 Projection of open curves into $\mathscr{C}_1^c$ using Algorithm 47. The plots show open curves (*plane lines*) and their projections (*marked lines*) in $\mathbb{R}^3$

not be zero, and we need to ensure that for it to be an element of $T_\theta(\mathscr{C}_1^c)$. This can be achieved using a one-time projection:

Algorithm 48 (Projection of Arbitrary v into $T_\theta(\mathscr{C}_1^c)$).

 a. *For $i = 1, 2, \ldots, n$, compute $E_{i,\theta} = d\Phi_{1,\theta}^*(e_i)$ and $A_\theta(:,i) = d\Phi_{1,\theta}(E_{i,\theta})$.*
 b. *Set $b = -A_\theta^{-1}\Phi_1(\theta)$.*
 c. *Compute $g = \sum_{i=1}^n b(i)E_{i,\theta}$.*
 d. *Update: for all s in $[0,1)$, set $v(:,s) = v(s) - g(s)$.*

These two algorithms are sufficient to specify the path-straightening algorithm for computing a geodesic between any two points $\theta_1, \theta_2 \in \mathscr{C}_1^c$. This procedure is basically same as Algorithm 19 with a few modifications for the pre-shape space here.

Algorithm 49 (Path Straightening on $\mathscr{C}_1^c$).

1. *Initialize a path α between θ_1 and θ_2. One way is to form a geodesic between them in $\mathscr{C}_1$ (using Eq. 10.8) and then project each point on the path into $\mathscr{C}_1^c$. Note that $\alpha(0) = \theta_1$ and $\alpha(1) = \theta_2$. The projection into $\mathscr{C}_1^c$ is accomplished using Algorithm 47.*
2. *Compute the velocity vector $\frac{d\alpha}{dt}$ using finite differences (similar to Algorithm 12). (This and the next two steps use Algorithm 48.)*
3. *Compute the covariant integral u of $\frac{d\alpha}{dt}$ using finite sums (similar to Algorithm 13).*
4. *Compute the backward parallel transport of $u(1)$ along α (similar to Algorithm 14). Call it $\tilde{u}$.*
5. *Compute the gradient w of the path energy according to:*

$$w(\tau/k) = u(\tau/k) - (\tau/k)\tilde{u}(\tau/k) , \quad \tau = 1, 2, \ldots, k .$$

6. *Update the path α in the direction of w using the exponential map on $\mathscr{C}_1$. (For each time τ, this is same as computing $\exp_{\alpha(\tau)}(\delta w(\tau))$, where $\exp$ is the exponential map on the unit sphere $\mathbb{S}^{n-1}$.) Then, project each point on the resulting path to $\mathscr{C}_1^c$ using Algorithm 47. If $\|w\|$ is small, then stop. Else, return to Step 2.*

Figure 10.16 illustrates a geodesic computed using this algorithm. In this figure, the two end shapes (top-left two panels), evolution of the energy (top right), and two views of the final geodesic path (middle and lower panels) are displayed.

10.6.2 Elastic Closed Curves

Now we look at the task of analyzing shapes of closed curves in $\mathbb{R}^n$ using the elastic representation. Given a parameterized closed curve $\beta : [0,1] \to \mathbb{R}^n$, we choose to represent it using the square-root velocity function $q(t) = \frac{\dot{\beta}(t)}{\sqrt{|\dot{\beta}(t)|}}$, and since β is of length one, $\int_0^1 |\dot{\beta}(t)|dt = 1$. Additionally, in order to restrict to the closed curves,

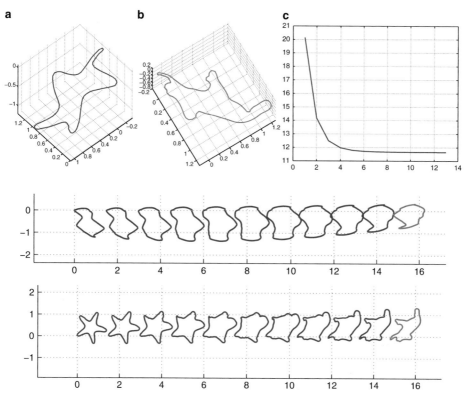

Fig. 10.16 (a) and (b) show the two curves β_1 and β_2, and (c) shows the evolution of E as Algorithm 49 proceeds. Bottom two rows show two views of the resulting geodesic path between β_1 and β_2

we impose the condition, $\int_0^1 q(t)|q(t)|dt = 0$. Thus, we have a space of unit-length closed curves represented by their square-root velocity functions:

$$\mathscr{C}_2^c = \{q \in \mathbb{L}^2([0,1], \mathbb{R}^n)| \int_0^1 |q(t)|^2 dt = 1, \int_0^1 q(t)|q(t)|dt = 0\}.$$

$\mathscr{C}_2^c$ is a submanifold of the Hilbert space $\mathbb{L}^2([0,1], \mathbb{R}^n)$. The tangent space to $\mathscr{C}_2^c$ at a point q is a subset of $\mathbb{L}^2([0,1], \mathbb{R}^n)$ with the standard $\mathbb{L}^2$ inner product. Due to the nature of $\mathscr{C}_2^c$, it is easier to specify the normal space, i.e., the space of functions in $\mathbb{L}^2([0,1], \mathbb{R}^n)$ perpendicular to $T_q(\mathscr{C}_2^c)$. The normal space is given by:

$$N_q(\mathscr{C}_2^c) = \text{span } \{q, (\frac{q_i}{||q||}q + ||q||\mathbf{e}^i), \ i = 1, \dots, n\} , \qquad (10.24)$$

where $\mathbf{e}^i$ is a unit vector in $\mathbb{R}^n$ along the ith coordinate axis. Hence:

$$T_q(\mathscr{C}_2^c) = \{w \in \mathbb{L}^2([0,1], \mathbb{R}^n)| \langle w, v \rangle = 0, \ \ \forall v \in N_q(\mathscr{C}_2^c)\} ,$$

where $\langle , \rangle$ denotes the standard inner product on $\mathbb{L}^2([0,1], \mathbb{R}^n)$.

Projection on $\mathscr{C}_2^c$ This procedure requires an important step of *closing* the curves, which we describe next. We particularize the general approach presented in Sect. 6.3 to project arbitrary elements of $\mathscr{C}_2$ into $\mathscr{C}_2^c$. We will basically repeat Algorithm 18 but this time for curves in $\mathbb{R}^n$. We define a mapping $\Psi : \mathscr{C}_2 \to \mathbb{R}^n$, where:

$$\Psi(q) = \left(\int_0^1 q^1(t)|q(t)|dt, \quad \int_0^1 q^2(t)|q(t)|dt, \quad \ldots, \quad \int_0^1 q^n(t)|q(t)|dt \right) .$$

Also, for $i = 1, 2, \ldots, n$, define:

$$b_i(t) \equiv \frac{q^i(t)}{|q(t)|}q(t) + |q(t)|\mathbf{e}^i - q(t)\int_0^1 \left\langle q(u), \frac{q^i(u)}{|q(u)|}q(u) + |q(u)|\mathbf{e}^i \right\rangle du$$

$$= \frac{q^i(t)}{|q(t)|}q(t) + |q(t)|\mathbf{e}^i - 2q(t)\int_0^1 q^i(u)|q(u)|du. \tag{10.25}$$

The next step is to define a residual vector $\mathbf{r}(q) = \Psi(q) \in \mathbb{R}^n$ and to evolve q in the direction perpendicular to the level set of Ψ so as to move its Ψ image toward the origin in $\mathbb{R}^n$. Algorithm 50 describes the procedure to project an open curve $q \in \mathscr{C}_2$ into $\mathscr{C}_2^c$. The Jacobian for this projection is an $n \times n$ matrix whose elements are given by $J_{ij} = \langle b_i(t), b_j(t) \rangle$, with b_is defined in Eq. 10.25. The algorithm for projection is as follows:

Algorithm 50 (Projection of $q \in \mathscr{C}_2$ to $\mathscr{C}_2^c$). Let $\epsilon > 0$.

1. *Compute $\mathbf{r}(q) = \Psi(q) \in \mathbb{R}^n$. If $|r(q)| < \epsilon$, stop; otherwise, continue.*
2. *Calculate the Jacobian matrix $J(q)$ given above.*
3. *Solve the equation $J(q)\beta = -\mathbf{r}(q)$ for β.*
4. *Define $dq = \sum_{i=1}^2 \beta_i b_i$, where b_is are given in Eq. 10.25.*
5. *Update using $q \mapsto \cos(\|dq\|)q + \sin(\|dq\|)\frac{dq}{\|dq\|}$.*
6. *Go to Step 1.*

Geodesic Computations The next task is to develop a computational tool that, for a given pair $q_0, q_1 \in \mathscr{C}$, computes a geodesic path $\alpha : [0,1] \to \mathscr{C}_2$ such that $\alpha(0) = q_0$ and $\alpha(1) = q_1$. For this task, we will use the path-straightening approach described in Chap. 6. In this approach, the given shapes are connected by an initial arbitrary path that is iteratively "straightened" until it becomes a geodesic. This iteration is performed using the gradient of an energy $E[\alpha]$, where:

$$E[\alpha] = \frac{1}{2}\int_0^1 \langle \dot{\alpha}(t), \dot{\alpha}(t) \rangle \; dt . \tag{10.26}$$

Let $d_c(q_0, q_1)$ be the length of the resulting geodesic between them.

Algorithm 51 (Path Straightening on $\mathscr{C}_2^c$).

1. *Initialize a path α between q_1 and q_2. One way is to form a geodesic between them in $\mathscr{C}_2$ (unit hypersphere) and then project each point on the path in $\mathscr{C}_2^c$. The geodesic on a hypersphere is given by, for $\tau = 0, 1/k, \ldots, 1$:*

$$\alpha(\tau) = \frac{1}{\sin(\theta)}\left[\sin(\theta - \tau\theta)q_1 + \sin(\tau\theta)q_2 \right] \tag{10.27}$$

where $\theta = \cos^{-1}(\langle q_1, q_2 \rangle)$. Note that $\alpha(0) = q_1$ and $\alpha(1) = q_2$. The projection into $\mathscr{C}_2^c$ is accomplished using Algorithm 50.

2. *Compute the velocity vector $\frac{d\alpha}{dt}$ (similar to Algorithm 12 but for curves in $\mathbb{R}^n$ rather than planar curves).*
3. *Compute the covariant integral u of $\frac{d\alpha}{dt}$ (similar to Algorithm 13).*
4. *Compute the backward parallel transport of $u(1)$ along α (similar to Algorithm 14).*
5. *Compute the gradient w of the path energy (similar to Algorithm 15).*
6. *Update the path α in the direction of w (similar to Algorithm 16). If $\|w\|$ is small, then stop. Else, return to Step 2.*

Now we have a numerical procedure for constructing geodesics between points in $\mathscr{C}_2^c$. An element in $\mathscr{C}_2^c$ is invariant to translation and scaling of the curve it represents but is affected by rotations and re-parameterizations of curves. This remaining variability is modeled as the actions of the groups $SO(n)$ and Γ_S on $\mathscr{C}_2^c$ as follows:

$$SO(n) \times \mathscr{C}_2^c \to \mathscr{C}_2, \quad (O, q) = Oq$$
$$\Gamma_S \times \mathscr{C}_2^c \to \mathscr{C}_2, \quad (\gamma, q) = (q \circ \gamma)\sqrt{\dot{\gamma}} .$$

It has been discussed previously in Chap. 5 that the actions of $SO(n)$ and Γ_S on $\mathscr{C}_2^c$ commute. Therefore, we can form a joint action of the product group $SO(n) \times \Gamma_S$ on $\mathscr{C}_2^c$ according to:

$$((O, \gamma), q) = O(q \circ \gamma)\sqrt{\dot{\gamma}} .$$

Furthermore, this joint action is by isometries of $\mathscr{C}_2^c$. Therefore, we can define a quotient space $\mathscr{S}_2^c = \mathscr{C}_2^c / (SO(n) \times \Gamma_S)$ with the $\mathbb{L}^2$ metric inherited from the larger space $\mathscr{C}_2^c$. Let $[q]$ denote the orbit of $q \in \mathscr{C}_2^c$ under the joint action of $SO(n) \times \Gamma_S$. The geodesic distance between any two points in $\mathscr{S}_2^c$ is given by:

$$d_s([q]_0, [q]_1) = \underset{(O, \gamma) \in SO(n) \times \Gamma_S}{\mathrm{arginf}} d_c(q_0, O(q_1 \circ \gamma)\sqrt{\dot{\gamma}}) . \tag{10.28}$$

Thus, we need to solve a joint minimization problem on $SO(n) \times \Gamma_S$ to form geodesics in $\mathscr{S}$. We will use a gradient-based search to minimize over the two groups.

The procedure for computing geodesics in the shape space $\mathscr{S}_2^c$ is very similar to Algorithm 22, except for the changes due to rotation space.

Algorithm 52 (Geodesic Computation in $\mathscr{S}_2^c$). *Let q_1 and q_2 be two elements of $\mathscr{C}_2^c$. Set $\tilde{q}_2 = q_2$, $\gamma = \gamma_{id}$, and $O = I_2$.*

1. *Compute a geodesic path from $\tilde{q}_2$ to q_1 using Algorithm 51. Let w be the initial velocity of this geodesic path. That is, $w = \exp_{\tilde{q}_2}^{-1}(q_1)$.*
2. *Compute the two gradients as follows:*

   a. *Compute the gradient of the cost function with respect to the rotation O using $x_i = \langle w, E_i \tilde{q}_2 \rangle$, where E_i forms an orthonormal basis of the space $T_{I_n}(SO(n))$. Set $dO = \exp(\epsilon_1 \sum_i x_i E_i)$.*
   b. *Compute the gradient of the cost function with respect to the re-parameterization according to:*

   $$d\gamma = \epsilon_2 \sum_{i=1}^n \langle w, d\Upsilon_{\gamma_{id}}(b_i) \rangle b_i . \tag{10.29}$$

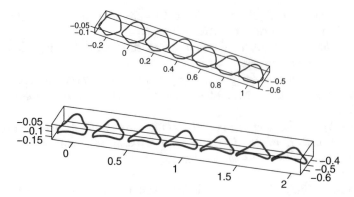

Fig. 10.17 Examples of geodesic paths between closed curves in $\mathbb{R}^3$ in the shape space $\mathscr{S}_2^c$

where $d\Upsilon_{\gamma_{id}}$ is defined in Eq. 6.46 and $\{b_i\}, i = 1, \ldots, d$ are as given in Eq. 6.45.

3. Update $\gamma \mapsto \gamma + d\gamma$ and $O \mapsto O.dO$.

4. Update β_2 according to $\tilde{\beta}_2 = O(\beta_2 \circ \gamma)$. Compute $\tilde{q}_2 = \dfrac{\dot{\tilde{\beta}}_2}{\sqrt{|\dot{\tilde{\beta}}_2|}}$.

5. If $|x|$ and $\|d\gamma\|$ are small, then stop. Otherwise, return to Step 1.

Figure 10.17 displays two examples of computing geodesic paths between closed curves in $\mathbb{R}^3$, treated as elements of the shape space $\mathscr{S}_2^c$.

10.7 Shape Analysis of Augmented Curves

So far we have looked at the shapes formed by curves in Euclidean spaces and have developed techniques for analyzing these shapes. In case there is any additional information associated with curves, other than just their shapes, it can also be used in this framework to help improve comparisons and registrations of curves. We will term this additional information as *auxiliary* information and seek to involve it in matching, deformation, and comparison of curves. Let's motivate this goal with a number of examples.

- **Shape analysis of tubes**: Consider the problems of studying shapes of generalized cylinders—3D structure represented by medial axes (a non-self-intersecting curve in $\mathbb{R}^3$) and scalar functions along axes representing thickness or radii. These objects can be viewed as flexible tubes with a continuous (and variable) radius function along the medial axis. If the thickness were constant, we could just focus on the medial axis of the tube and analyze its shape. In general cases, however, where the thickness changes along the curve, we would like to analyze shapes of tubes by taking into account both the shape of the medial axis and the varying thickness of tube around that axis. Thus, we treat the medial axis as a parameterized curve in $\mathbb{R}^3$ and the thickness as an auxiliary function along the curve.

- **Shape analysis of colored curves**: Consider color images of objects on the right side of Fig. 10.18. There are two types of features that characterize objects in these images: shapes of their boundaries and textures (or colors) along

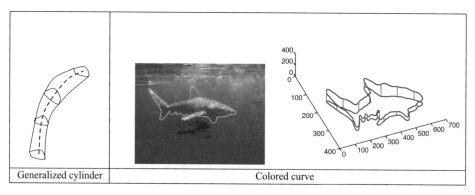

| Generalized cylinder | Colored curve |

Fig. 10.18 High-dimensional functions representing both shape and color coordinates along objects' boundaries

these boundaries. Together these two features capture the full appearance of the boundary in an image.

Thus, we are interested in studying colored contours—their shapes as well as their colors. The shape coordinates are viewed as a simple closed curve in the image plane, while the color coordinates are viewed as real- or vector-valued functions along that curve. If the image is gray scaled, then the auxiliary function is one-dimensional, but if the image is colored (with RGB scale), then this function is three-dimensional. More generally, one can form a vector-valued function, of arbitrary dimensions, from texture values in and around the boundary curves.

- **Shape analysis of curves with landmarks**: Another possible scenario for augmented curves is the curves having some predetermined landmarks associated with them. Let a curve represent the boundary of an anatomical object in an MRI or an X-ray image. Very often there are anatomical landmarks, such as corners, junctions, etc., associated with the objects that can be detected and labeled separately (using manual techniques or otherwise). When comparing shapes of similar anatomical objects, it becomes relevant or even necessary to ensure that the corresponding landmarks match each other. In this case, we can create an artificial auxiliary function, a scalar function to represent landmark information. This function peaks at given landmarks and smoothly decays to zero away from them.

- **Shape analysis of functional curves**: Yet another scenario, related to the earlier examples, is when a curve has an arbitrary function defined on it that can potentially influence shape matching and comparisons. Take the example from brain anatomy where one is interested in the shapes of neuronal fiber tracks using the diffusion MRI image data. Along the estimated fiber tracts, one can characterize certain anatomically relevant functions—diffusivity, anisotropy, etc.—that provide clues about the anatomical functionality of that region. We can formulate the given information into two components: one, the coordinates of a fiber tract in $\mathbb{R}^3$ depicting its shape and the other a scalar-valued anatomically relevant function along the curve. The latter forms the auxiliary component in this application.

In these and other related scenarios, we obtain scalar- or vector-valued auxiliary functions, along with the coordinate function of curves, and we would like the process of matching, deformation, and comparison to be based on joint shape-auxiliary information. All this while maintaining appropriate constraints and invariances.

What are the main challenges in doing so? We are going to concatenate the shape and the auxiliary coordinates to form curves in higher dimensions and will try to perform shape analysis of these resulting curves. The challenge, however, comes from the fact that shape and auxiliary coordinates feature different invariances. The shape coordinates are invariant to rotation, translation, and scale transformations, but the auxiliary coordinates may have only some or none of these invariances. For instance, one cannot rotate the larger vector as the shape and auxiliary components will get mixed with each other. Also, note that the registration of shape and auxiliary coordinates goes together. In other words, one has to be re-parameterized in the same way as the other. We cannot treat them as two independent re-parameterization (registration) problems.

10.7.1 Joint Representation of Augmented Curves

Let the shape coordinate function along a closed curves be given by $\beta_s : D \to \mathbb{R}^n$ and the auxiliary function be given by $\beta_t : D \to \mathbb{R}^k$, where k is an arbitrary dimension related to the auxiliary function. Here, $D = [0,1]$ for a general curve and $D = \mathbb{S}^1$ for a closed curve. For a planar shape, β_s simply represents the xy coordinates of the curve in the plane. We can combine these two components to form a joint shape and texture curve: $\beta(t) = \begin{bmatrix} \beta_s(t) \\ b\beta_t(t) \end{bmatrix} \in \mathbb{R}^{n+k}$. Here $b > 0$ is a parameter introduced to control the influence of the auxiliary function, relative to the shape function. The image example in Fig. 10.18 (right side) shows an example where the auxiliary function is given by the magnitude of the color vector (RGB values) along the curve. In this simple case, the auxiliary function is simply the height function with the augmented curve becoming a parameterized curve in $\mathbb{R}^3$ ($n = 2$, $k = 1$).

For the resulting augmented curve, we can define the square-root velocity function q using Eq. 10.2. Note that, since q is defined using the time derivative of β, q is invariant to the translation of β. That is, if we add a constant vector to β, its mathematical representative q will not change. While that is desirable for the shape component, it may or may not be desirable for the auxiliary component. In case we want the auxiliary component to be dependent on its translation, we simply add its mean value separately in the representation, as described later. Continuing, we can define the pre-space, this time called the *augmented pre-space*, as:

$$\mathscr{C}_2^c = \{q : D \to \mathbb{R}^{(n+k)} | \int_D \langle q(t), q(t) \rangle \, dt = 1, \ \int_D |q(t)| q(t) \, dt = 0\} . \quad (10.30)$$

The last condition is needed only if the original curves are closed. An element of $\mathscr{C}_2^c$ is a $(n + k)$-dimensional curve that is both closed and of fixed length. $\mathscr{C}_2^c$ is an infinite-dimensional nonlinear manifold and we endow it with the $\mathbb{L}^2$ Riemannian metric. For a point $q \in \mathscr{C}_2^c$, and any two tangents $g, f \in T_q(\mathscr{C}_2^c)$, we define the inner product $\langle g, f \rangle = \int_0^1 \langle g(t), f(t) \rangle \, dt$, where the inner product inside the integral is the standard Euclidean product in $\mathbb{R}^{n+k}$. With this Riemannian metric, the augmented pre-space becomes a Riemannian manifold and one can compute geodesics in it using the path-straightening approach. The algorithm for computing geodesics in this space is identical to Algorithm 51 (presented in the previous section) and is not repeated here.

10.7.2 Invariances and Equivalence Classes

Now we consider the desired invariances in the representation and use the notion of equivalence classes to achieve those invariances. We remind the reader that the shape component should be invariant to translation, rotation, scale, and re-parameterization, while the auxiliary component should, in a general situation, only be invariant to re-parameterization. If the auxiliary function has an additional invariance, we can include it in the representation accordingly. In case of images, where the auxiliary function denotes the intensity or colors along the curves, these intensity values are often centered and scaled to a unit norm, in order to make them invariant to translation and scale.

1. **Scaling**: In shape analysis of a curve, we need to remove the scaling variability but we may or may not want to rescale the auxiliary component. If we look at the augmented pre-space, the condition $\int_D \langle q(t), q(t) \rangle \, dt = 1$ ensures that the two components have together been rescaled to a standard value. Note that this condition scales the two components of β by the same amount. If one wants to preserve the original scale of the auxiliary components, this can be done using the constant b in the definition of β. This constant can also be used to scale the two components differently.

2. **Rotation**: For shape analysis, we need to remove the rotational variance and we have done that in the past by forming equivalence classes (orbits) under the action of the rotation group. In the current situation, we seldom want to remove the rotation for the augmented curves. This removal is restricted only to the shape components, leaving the auxiliary components unchanged. This can be done as follows: Define a $n(n-1)/2$ dimensional subgroup of the rotation group $SO(n+k)$ according to:

$$\mathscr{R} = \begin{bmatrix} SO(n) & 0 \\ 0 & I_k \end{bmatrix} \subset SO(n+k),$$

where I_k is the $k \times k$ identity matrix. For any $q \in \mathscr{C}_2^c$ and $O \in \mathscr{R}$, the function Oq represents an augmented function whose shape component has been rotated while the texture component remains unchanged. We define the rotational equivalence class q as the set:

$$[q]_R = \{Oq | O \in \mathscr{R}\} \, .$$

Two elements of $[q]_R$ differ only by a rotation of the shape component. For reaching the augmented shape space of augmented curves, we will remove only the subgroup $\mathscr{R}$ and not the full rotation space $SO(n+k)$.

3. **Translation**: While the shape component is invariant to translations of the object, the auxiliary component may or may not change with the translation.

As noted earlier, the definition of q, involving the time derivative of β, has already removed the translation for both the components. In case we want the auxiliary component to be dependent on its translation, we need to bring back the translation of β_0 in the representation. We do so by introducing a constant vector:

$$\bar{\beta}_0 = \frac{\int_D \beta_t(t)dt}{\int_D dt} \quad \in \mathbb{R}^k \ .$$

This is the mean value of the auxiliary function along the contour. The new representation of β is given by the pair $(q, \bar{\beta}_0)$.

4. **Re-parameterization**: We seek to use both shape and auxiliary coordinates in performing curve registrations. Setting the shape of the augmented curve to be invariant to re-parameterizations, we simply follow the same approach as earlier. Let $\tilde{\Gamma}_D$ be the re-parameterization group on a domain D. For any $q \in \mathscr{C}_2$, we define its equivalence class under re-parameterization to be:

$$[q]_{\widetilde{\Gamma_D}} = \text{closure}\{(q \circ \gamma)\sqrt{\dot{\gamma}} | \gamma \in \tilde{\Gamma}_D\} \ .$$

For reaching the shape space of augmented curves, we will remove $\tilde{\Gamma}_D$ from the pre-shape space $\mathscr{C}_2$. It should be emphasized that the two components are not being re-parameterized differently. Any point on the curve has a shape coordinate and an auxiliary coordinate, and these two coordinates remain coupled together despite a re-parameterization of this curve.

Combining different equivalence relations, we obtain the shape space $\mathscr{S}_2^c = \mathscr{C}_2^c / (\mathscr{R} \times \tilde{\Gamma}_D)$ whose elements are equivalence classes of the type:

$$[q] = \{O(q \circ \gamma)\sqrt{\dot{\gamma}} | O \in \mathscr{R}, \gamma \in \tilde{\Gamma}_D\} \ .$$

The full representation now includes $[q]$, the equivalence class of q, and $\bar{\beta}_0$, the mean of the texture function. Therefore, the total space for joint analysis of shapes and textures is given by:

$$\mathscr{T} = \mathscr{S}_2^c \times \mathbb{R}^k \ ,$$

and $b \in \mathbb{R}_+$ acts as a control parameter.

To compare any two objects, represented by $([q^1], \bar{\beta}_0^1)$ and $([q^2], \bar{\beta}_0^2)$, we use the distance function:

$$d(\beta^1, \beta^2; b) = \left(\sqrt{d_s([q^1], [q^2])^2 + |\bar{\beta}_0^1 - \bar{\beta}_0^2|^2} \right) \tag{10.31}$$

where $d_s(q_1, q_2)$ is the geodesic distance in $\mathscr{S}_2^c$ and $|\cdot|$ is the Euclidean distance in $\mathbb{R}^k$. Note that for different values of b, the resulting geodesic path and the geodesic distance will be different.

The resulting geodesic provides (i) an optimal deformation of one augmented curve to another and (ii) an optimal registration of points across these two curves. The latter needs more elaboration. The geodesic matching across the two curves is influenced by both the shape and the auxiliary coordinates; the actual level of influence depends on the control parameter b. Consider the planar shapes of two hand outlines and immerse them in an artificial texture image made of Gaussian

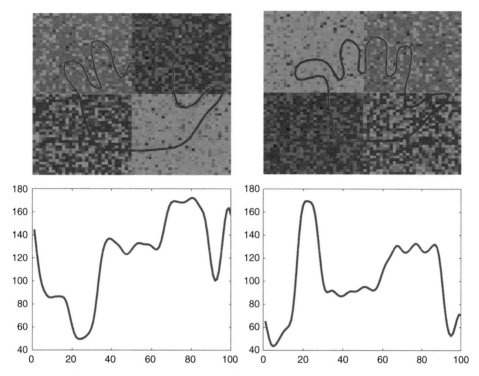

Fig. 10.19 *Top row*: two hand shapes immersed in artificial texture. *Bottom row*: the texture functions along the two curves after smoothing

random fields with different means, as shown in the top row of Fig. 10.19. For convenience, we show the corresponding texture functions (obtained after some smoothing for noise reduction) along the two curves in the bottom row. Combining the 2D coordinate function β_s and the 1D texture function β_t, we obtain augmented 3D shape-texture function β for each image.

We compute the square-root velocity function q for each object and compute geodesic paths between them in $\mathscr{S}_2^c$ using the path straightening described in Algorithm 52. Shown in Fig. 10.20 are the resulting geodesic paths for different values of the control parameter b. Rather than drawing augmented curves in $\mathbb{R}^3$, we use a color scheme to show the third (texture) coordinate. All the curves are drawn using their shape components and are colored using their texture components. For a very small value of b, the contribution from the texture function is very small and the geodesic is primarily based on the shape components of the two curves. In the top row, where b is rather small, the matching of points across curves is based mostly on the shape features. The fingers in one hand are accurately matched to the fingers in the second hand, irrespective of the texture values associated with them. As we go down the rows, the role of texture becomes increasingly prominent in matching and deformations of curves. In the bottom row, where b is quite large, the matching is completely governed by the texture function. Here, the blue part of one hand is matched to the blue part of the second hand, irrespective of their shapes. The actual matchings of points on the two curves, for the two extreme cases (b very small and b very large) are displayed in Fig. 10.21.

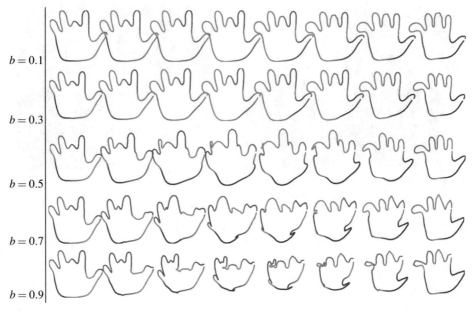

Fig. 10.20 (Colored figure) Geodesics between joint shape-texture functions in the space $\mathscr{S}_2$ for different values of b. A small value of b puts more emphasis on the shape component, while a large value of b emphasizes the texture component more. All curves are drawn using their shape components and are colored using their texture components

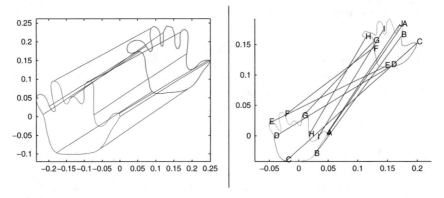

Fig. 10.21 Registration of points across curves. The matching on the left is for 2D shape analysis, while that on the right is for 3D joint shape-texture analysis when a high weight (b) is placed on the texture component

10.8 Problems

10.8.1 Theoretical Problems

1. Let $\theta(s) = \dot{\beta}(s)$ be the direction function of an arc-length parameterized curve β in $\mathbb{R}^n$. If $\theta \in \mathbb{L}^2([0,1], \mathbb{R}^n)$, then what space does the corresponding β lie in?
2. For the mapping $F : \mathbb{R}^n \to \mathbb{R}^n$ as defined in Eq. 10.1, show that F is continuous at $0 \in \mathbb{R}^n$.
3. Using the definition of SRVF, verify the formula $\beta(t) = \beta(0) + \int_0^t |q(s)|q(s)ds$.

4. Assuming that all the components of $\beta : [0, 1] \to \mathbb{R}^n$ are absolutely continuous, with the absolute continuity as defined in Eq. 4.1, show that the corresponding SRVF q is in $\mathbb{L}^2([0, 1], \mathbb{R}^n)$.

5. For a differentiable, parameterized curve $\beta : [0, 1] \to \mathbb{R}^n$, let θ and ϕ be the direction and the log-speed functions of β, respectively. Verify the following statements:

   a. If (ϕ, θ) is the representation of β, then $(\phi, O\theta)$ is the representation of the rotated curve $O\beta$ for any $O \in SO(n)$.

   b. Similarly, show that $(\phi \circ \gamma + \ln \circ \dot{\gamma}, \theta \circ \gamma)$ is the representation of the re-parameterized curve $\beta \circ \gamma$, for any $\gamma \in \Gamma_I$.

6. Show that the actions of the rotation and re-parameterizations groups are by isometries under the elastic metric. In other words, verify Eq. 10.3.

7. Derive an expression for the SRVF q of a curve β in terms of its direction and log-speed functions, θ and ϕ, respectively.

8. Show that the elastic metric, given in Eq. 10.3, reduces to the non-elastic metric, given in Eq. 10.6, when the log-speed function is forced to be a zero function.

9. Prove that Eq. 10.8 provides an expression for the shortest geodesic path between any two (non-degenerate) curves under the non-elastic metric. In other words, given any two direction functions θ_1, θ_2, that are curves themselves on $\mathbb{S}^{n-1}$, show that α satisfies the properties of a geodesic path (as defined in Sect. 3.2).

10. Verify the expression for the gradient given in Eq. 10.11.

11. Show that the action of the product group $SO(n) \times \widetilde{\Gamma_I}$ on $\mathscr{C}_2$, given by $((O, \gamma), q) = O(q \circ \gamma)\sqrt{\dot{\gamma}}$, is by isometries under the $\mathbb{L}^2$ metric.

12. Verify the expression for the tangent space $T_\theta(\mathscr{C}_1^c)$ given in Eq. 10.22.

13. Similarly, verify the expression for the normal space of $\mathscr{C}_2^c$ inside $\mathbb{L}^2([0, 1], \mathbb{R}^n)$ given by Eq. 10.24.

14. Define a curve $\beta_1 : [0, 1] \to \mathbb{R}^n$ by $\beta_1(t) = tu$, where $u \in \mathbb{R}^n$ is any constant vector. Let β_2 be an absolutely continuous curve with the property that $\left\langle u, \dot{\beta}_2(t) \right\rangle > 0$ for all $t \in [0, 1]$. Find a formula for re-parameterization $\gamma : [0, 1] \to [0, 1]$ such that $\beta_1 \circ \gamma$ is optimally matched to β_2 according to Eq. 10.20.

15. Repeat the previous problem without the assumption that $\left\langle u, \dot{\beta}_2(t) \right\rangle > 0$ for all $t \in [0, 1]$. (You must carefully consider the subintervals of $[0, 1]$ where $\left\langle u, \dot{\beta}_2(t) \right\rangle > 0$ and $\left\langle u, \dot{\beta}_2(t) \right\rangle < 0$.)

16. Show that embedding of $SO(n)$ inside $SO(n+k)$, denoted by R in this chapter, forms a subgroup of $SO(n + k)$.

17. Prove that $d(\beta^1, \beta^2; b)$, given in Eq. 10.31, forms a proper distance on the space of augmented curves, for all $0 < b < \infty$. Provide interpretations for the extreme cases: $b = 0$ and $b \to \infty$.

10.8.2 Computational Problems

1. Write a program to re-parameterize a given curve in $\mathbb{R}^n$ using the arc-length parameterization and provide a given number T of sample points along the curve. Also, add a procedure to rescale the curve to be of length one.
2. Write a program that takes an arc-length parameterized curve and computes its angle function for at T uniform points in the interval $[0, 1]$.
3. Write a program to implement Algorithm 44 and to compute the non-elastic geodesic path between any two given curves in $\mathbb{R}^n$.
4. Write a program to compute the SRVF of an arbitrary parameterized curve $\beta : [0, 1] \to \mathbb{R}^n$.
5. Write a program to compute a geodesic path between any two parameterized curves in the pre-shape space $\mathscr{C}_2$ using Eq. 10.8.
6. Implement Algorithm 45 to compute a geodesic path between any two curves, represented by their SRVFs q_1 and q_2, in the shape space $\mathscr{S}_2$.
7. Write a program to compute mean of k given parameterized curves using Algorithm 46.
8. Write a program to estimate parameters of the truncated wrapped-normal density using PCA of the shooting vectors obtained in the previous problem. Additionally, write a code that randomly samples from this TWN model and shows the corresponding random shapes generated from TWN model.
9. Write a program to perform temporal alignment of two arbitrary curves in $\mathbb{R}^n$ using Eq. 10.20. Display the function $|q_1(t) - q_2(t)|^2$ versus t before and after the alignment.
10. Write a program to temporally align k given curves in $\mathbb{R}^n$, following Algorithm 33, modified suitably to handle vector-valued functions.
11. Implement Algorithm 47 to project an arbitrary $\theta \in \mathbb{L}^2([0, 1], \mathbb{S}^{n-1})$ into $\mathscr{C}_1^c$.
12. Implement Algorithm 49 to compute a geodesic path between arbitrary arc-length parameterized, closed curves in $\mathbb{R}^n$.
13. Write a program to project an arbitrary (non-degenerate) element of $\mathscr{C}_2$ into the set $\mathscr{C}_2^c$, using Algorithm 50.
14. Implement the path-straightening Algorithm 51 for computing geodesic paths in $\mathscr{C}_2^c$, the pre-shape space of elastic, closed curves.
15. Implement Algorithm 52 for computing geodesic paths in $\mathscr{S}_2^c$, the shape space of elastic, closed curves in $\mathbb{R}^n$.

10.9 Bibliographic Notes

The framework for shape analysis of curves in Euclidean spaces was described in the paper [106]. The use of shape analysis of protein backbones in protein structure analysis was explored in a series of papers, including [70]. Shape analysis of 3D curves has also been used successfully in face recognition using range images [107, 98, 27], RNA shape analysis [60], shape-based classification of sulcal curves and DTI fiber tracts [73], and the use of skeletal trajectories for characterizing human actions [8, 13]. Shape analysis of augmented curves was introduced in [69, 68]. More recently, the idea of shape analysis of augmented curves has been used to analyze axonal trees in [82, 81].

Chapter 11
Related Topics in Shape Analysis of Curves

Previously we have developed theory and computational solutions for shape analysis associated with curves in Euclidean spaces. We end this textbook by presenting some related topics that fall outside the main framework. These miscellaneous topics do not fit an organized theme but are bunched together in this chapter for convenience. They include: (1) Investigate the use of shape in conjunction with other features, such as scale, pose, and position, to characterize curves. (2) Extend the group of shape-invariant transformations, from the similarity group to the affine group, in the case of planar closed curves. While most shape analysis works consider similarity transformations (rigid motions and global scales) as the main shape-preserving transformations, some applications, including imaging, may require us to nullify affine distortions of curves. (3) Develop techniques for analyzing trajectories on nonlinear Riemannian manifolds. While we have studied only the Euclidean curves so far, there is also a strong need for analyzing curves on other, perhaps nonlinear, domains. One may not use the word *shape* for characterizing the desired properties of these curves, but this analysis should be invariant at least to parameterizations of these trajectories.

11.1 Goals and Challenges

The goals and challenges associated with the topics covered in this section are the following:

1. **Joint analysis of shape and related features**: This book emphasizes that the removal of unwanted (nuisance) variables is the main challenge in shape analysis. One often starts with Euclidean representations and moves toward nonlinearity (and quotient spaces) as the nuisance variables are sequentially removed. So, what happens if we want to involve some of these nuisance variables in our studies? One way of doing this is to form product spaces, involving shape spaces and spaces of other variables, with the natural product metrics. For instance, to combine shape and scale, we can form the product space $\mathscr{S}_2^c \times \mathbb{R}_+$, with the elastic metric on $\mathscr{S}_2^c$ and a Euclidean metric on $\mathbb{R}_+$. While this is a perfectly valid approach, it leaves an extra degree of freedom in choosing the relative weight of the two components. Another approach, perhaps preferable

© Springer-Verlag New York 2016
A. Srivastava, E.P. Klassen, *Functional and Shape Data Analysis*,
Springer Series in Statistics, DOI 10.1007/978-1-4939-4020-2_11

in some situations, is to *not remove* the variables that are desirable, since they are no longer considered nuisance. As the removal of these variables actually complicates shape analysis, the non-removal simplifies it. Taking the second approach, we will form representation spaces that leave some variables of interest involved while still removing the remaining nuisance variables. A common theme here is that the parameterization variability of curves is always treated as nuisance, and, consequently, the analysis is always based on the elastic Riemannian framework.

2. **Affine-invariant shape analysis**: Here we want to consider the affine group $GL(2)$, and we would like to quotient out the action of this group from a chosen representation of planar closed curves. Additionally, we want to maintain the invariance to the re-parameterization group as earlier. In summary, we seek elastic shape analysis of planar curves while defining equivalence relations under the larger (affine) group. The difficulty comes from the fact that for all the metrics discussed earlier in this text, the action of the affine group is not by isometries. Hence, we cannot define shape space as the quotient space and inherit the metric from the larger representation space. This case was discussed in Sect. 3.9 where: (1) the action is not by isometries and (2) a group action admits a section that is not orthogonal. We will take the approach suggested there: identify a section S of the manifold M under the action of the group G and define a metric on S. (It turns out that the chosen section is orthogonal.) Then, use this metric structure to *define* a metric structure on the quotient space M/G.

3. **Rate-invariant analysis of trajectories on Riemannian manifolds**: We explain the goals and challenges for this item by considering a relatively simple manifold $\mathbb{S}^2$. Consider a set of smooth curves of the type $\beta_i : [0,1] \to \mathbb{S}^2$ and our goal is to develop a framework for comparisons, registration, and summarization of these trajectories. Furthermore, we want this framework to be invariant to how the trajectories are parameterized. That is, the metric between β_i and β_j has the same value as that for $\beta_i \circ \gamma_i$ and $\beta_j \circ \gamma_j$, where $\gamma_i, \gamma_j \in \Gamma_I$. In the same spirit as the elastic comparison of curves, the principal challenge here is the registration of points across curves, in a principled manner. The use of SRVFs is not feasible since derivatives of these trajectories are vector fields along the curves and cannot be directly compared due to nonlinearity of the domain. The solution comes from transporting these vector fields to a common domain where different curves can be naturally compared.

11.2 Joint Analysis of Shape and Other Features

While shape analysis is important in many applications, there has also been a need to study shape in conjunction with other features. For example, in the studies of biological growth, it is important to measure the overall size of the growing objects in addition to their changing shapes. Similarly, in certain anatomical structures, it is important to take their relative locations and orientations into account while deciding on their normality or abnormality. If we consider a continuous curve, then all its physical characteristics can be summarized using *shape, scale, location, and orientation*. While it is convenient to work with parameterized curves, a parameterization is merely for the convenience of analysis and is not an intrinsic property of a curve, like the previous four properties. Similar to the Chaps. 5 and 6, we seek a framework for analyzing curves that can incorporate any arbitrary subset of these properties, depending upon the needs of an application. This allows for generating meaningful comparisons between subjects and populations as well as for performing classification. Furthermore, it provides tools for computation of statistics such as the mean and covariance. To demonstrate these ideas in a concrete setting, we shall develop tools for comparing, clustering, and classifying white matter fibers in the human brain, obtained from diffusion-tensor MRI (DT-MRI) data as parameterized curves in $\mathbb{R}^3$.

Similar to the framework laid out in Chap. 10, consider the set of all absolutely continuous parameterized curves in $\mathbb{R}^n$. The set of all rigid rotations of these curves is $SO(n)$, the set of all re-parameterizations of curves is $\widetilde{\Gamma_I}$, the set of scales is $\mathbb{R}_+$, and for translation is $\mathbb{R}^n$. Let $\beta : [0,1] \to \mathbb{R}^n$ be a parameterized curve.

11.2.1 Geodesics and Geodesic Distances on Feature Spaces

In this section we describe different feature spaces, their Riemannian structures and the corresponding geodesic equations.

1. **Shape, Scale, and Orientation**: In case we are interested in comparing curves using their shapes, scales and orientations, we can do so directly using their SRVFs in $\mathbb{L}^2$ with the Riemannian structure described in Chap. 10. The only variability we have to remove here is the re-parameterization. To remove the re-parameterization variability, we utilize the algebraic operation of forming a quotient space $\mathcal{S}_{sso} \equiv \mathbb{L}^2([0,1], \mathbb{R}^n)/\widetilde{\Gamma_I}$. Recall that the action of $\widetilde{\Gamma_I}$ on $\mathbb{L}^2([0,1], \mathbb{R}^n)$ is given by: $(q, \gamma)(t) \equiv q(\gamma(t))\sqrt{\dot{\gamma}(t)}$. Hence, an orbit under $\tilde{\Gamma}_I$ is given by $[q] = \{q \circ \gamma)\sqrt{\dot{\gamma}} | \gamma \in \tilde{\Gamma}_I\} = \text{closure}\{(q \circ \gamma)\sqrt{\dot{\gamma}} | \gamma \in \Gamma_I\}$, and $\mathcal{S}_{sso} = \{[q] | q \in \mathbb{L}^2\}$. The space $\mathcal{S}_{sso}$ inherits a Riemannian structure from $\mathbb{L}^2$ and one can define geodesics and distances in $\mathcal{S}_{sso}$ as follows. For any q_1, $q_2 \in \mathbb{L}^2$, define an approximation to the optimal re-parameterization according to: $\gamma^* = \text{arginf}_{\gamma \in \Gamma_I} \|q_1 - (q_2 \circ \gamma)\sqrt{\dot{\gamma}}\|$. Let $q_2^* = (q_2 \circ \gamma^*)\sqrt{\dot{\gamma}^*}$ be the optimal representation of the second curve. Then, the geodesic distance between the orbits of q_1 and q_2 is $d_{sso}([q_1], [q_2]) \doteq \|q_1 - q_2^*\|$ and the geodesic path between them in $\mathcal{S}_a$ is the straight line $\alpha(\tau)(t) = (1-\tau)q_1(t) + \tau q_2^*(t)$. Figure 11.1a displays an example of a geodesic path and correspondence between two curves with different shapes, scales, and orientations.

2. **Shape and Scale**: In this case we want to include only the shape, and the size of the curves in the analysis and the orientations are no longer relevant. We can modify the previous setup to reflect this change as follows. The quotient space of equivalent curves is now defined by $\mathcal{S}_{ss} \doteq \mathcal{S}_{sso}/SO(n) = \mathbb{L}^2([0,1], \mathbb{R}^n)/(\tilde{\Gamma}_I \times SO(n))$. The orbits are now given by $[q] = \{O(q \circ \gamma)\sqrt{\dot{\gamma}} | \gamma \in \tilde{\Gamma}_I, O \in SO(n)\}$, and $\mathcal{S}_{ss} = \{[q] | q \in \mathbb{L}^2([0,1], \mathbb{R}^n)\}$. For any q_1, $q_2 \in \mathbb{L}^2$, define the optimal re-parameterization and rotation of the second curve according to: $(\gamma^*, O^*) = \text{arginf}_{\gamma \in \Gamma_I, O \in SO(n)} \|q_1 - O(q_2 \circ \gamma)\sqrt{\dot{\gamma}}\|$. The optimization over Γ_I is performed using the DP algorithm and over $SO(n)$ is performed using the SVD (Procrustes alignment). Let $q_2^* = O^*(q_2 \circ \gamma^*)\sqrt{\dot{\gamma}^*}$ be the optimal representation of the

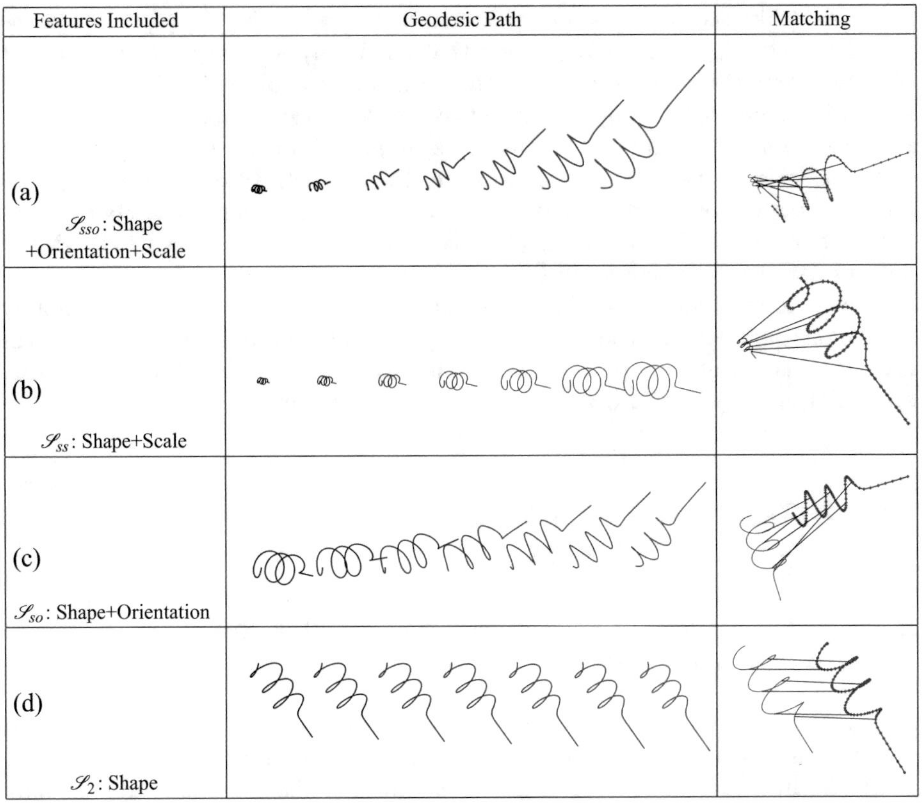

Features Included	Geodesic Path	Matching
(a) $\mathscr{S}_{sso}$: Shape +Orientation+Scale		
(b) $\mathscr{S}_{ss}$: Shape+Scale		
(c) $\mathscr{S}_{so}$: Shape+Orientation		
(d) $\mathscr{S}_2$: Shape		

Fig. 11.1 An example of geodesic paths in different feature spaces

second curve. Then, the geodesic distance between the orbits of q_1 and q_2 is $d_{ss}([q_1], [q_2]) \doteq \|q_1 - q_2^*\|$. Similarly, the geodesic path between them in $\mathscr{S}_b$ is simply the straight line. Figure 11.1b displays an example of a geodesic path and optimal correspondence between two curves with different shapes and scales.

3. **Shape and Orientation**: If we are interested in comparing curves according to their shapes and orientations, but not their scales and positions, we can simply rescale all the curves to a fixed length, say one, and then analyze the SRVFs. As discussed earlier, the set of all SRVFs representing curves of length one is $\mathscr{C}_2 = \mathbb{S}_\infty$. While the scale variability has been removed, the parameterization variability still remains. To take care of that, we define a quotient space $\mathscr{S}_{so} \doteq \mathscr{C}_2/\tilde{\Gamma}_I$. The elements of this quotient space are the orbits given by $[q] = \{(q \circ \gamma)\sqrt{\dot{\gamma}} | q \in \mathscr{C}_2, \gamma \in \widetilde{\Gamma}_I\}$. The geodesic distance in $\mathscr{S}_c$ is calculated by solving the following minimization problem: $\gamma^* = \arginf_{\gamma \in \Gamma_I} \cos^{-1}(\langle q_1, \sqrt{\dot{\gamma}}(q_2 \circ \gamma) \rangle) = \arginf_{\gamma \in \Gamma_I} \|q_1 - \sqrt{\dot{\gamma}}(q_2 \circ \gamma)\|$. The last equality comes from the fact that minimizing the arc length between points on a sphere is the same as minimizing the chord length between them. This latter minimization can be solved using the DP algorithm. Let $q_2^* = \sqrt{\dot{\gamma}^*}(q_2 \circ \gamma^*)$. Then, the geodesic path between $[q_1]$ and $[q_2]$ in $\mathscr{S}_c$ is given by the great circle connecting them in $\mathscr{C}_2$, and the geodesic distance between them is $d_{so}([q_1], [q_2]) \doteq \cos^{-1}(\langle q_1, \sqrt{\dot{\gamma}^*}(q_2 \circ \gamma^*) \rangle)$. Figure 11.1c displays an example of

a geodesic path and correspondence between two curves with different shapes and orientations.

4. **Incorporating Curve Positions in Analysis**: In some applications there is a need for including the position in the analysis of curves. For example, in the problem of classifying and labeling sulcal curves in a human brain, the relative locations of these curves can play an important role. There are two solutions to this issue. The first solution is to use the SRVFs directly to compute distances between curves and then add a weighted portion of a translational distance. This allows for some flexibility, in that the weight of translation in the final distance is decided by the user. On the other hand, we may be inclined to use a different representation for curves that automatically includes their relative positions in the metric and still remains invariant to the group action of $\widetilde{\Gamma_I}$.

In order to study shape and other features of curves, we will represent parameterized curves using either their SRVFs or a new representation introduced next.

Definition 11.1 (Square-Root Function (SRF)). Define the SRF to be a function $h : [0,1] \to \mathbb{R}^3$ given by $h(t) = \sqrt{|\dot{\beta}(t)|}\beta(t)$. If the curve β is reparameterized by a $\gamma \in \Gamma_I$, its SRF changes to $h(t) \mapsto (h,\gamma)(t) \equiv h(\gamma(t))\sqrt{\dot{\gamma}(t)}$.

This representation has the advantage (from our current perspective) that it includes the position information about the curve. The drawback is that it is not straightforward to reconstruct a curve from its SRF. While this limits our ability to draw a geodesic path between any two curves, or to compute sample statistics of curves, we can still obtain a distance with all the desired properties. The group actions of $SO(n)$ and Γ_I on SRFs remain the same as they were for SRVFs. The shape space is defined as $\mathscr{S}_{all} \doteq \mathbb{L}^2([0,1],\mathbb{R}^n)/\widetilde{\Gamma_I}$. The distance between two SRFs in $\mathbb{L}^2$ is defined as $\|h_1 - h_2\|$, while the distance in $\mathscr{S}_{all}$ is: $d_{all}([h_1],[h_2]) \doteq \operatorname{arginf}_{\gamma \in \Gamma_I} \|h_1 - \sqrt{\dot{\gamma}}(h_2 \circ \gamma)\|$. Once again, the optimization over γ is performed using the dynamic programming algorithm.

11.2.2 Feature-Based Clustering

We illustrate these definitions by clustering a set of curves several times, each time using a different metric defined above. We present clustering results using an artificial dataset first and then DT-MRI brain fibers.

1. **Clustering of Artificial Data**: It is easier to evaluate the clustering performance for an artificial dataset since the ground truth is known. First, we will study clustering of artificial curves based on their shapes (d_s); shapes and orientations (d_{so}); shapes and scales (d_{ss}); shape, scales, and orientations (d_{sso}); and shapes, scales, orientations, and positions (d_{all}), using the Ward clustering technique (in MATLAB) with a complete linkage function.

   Figure 11.2 (top row) displays some examples the data used here that consists of curves with combinations of three distinct shapes, two scales, two rotations, and two different positions. The total data consists of 96 different curves such that each combination of shape, scale, rotation, and translation is included in the data three times. The bottom row shows the (96 × 96) geodesic distance

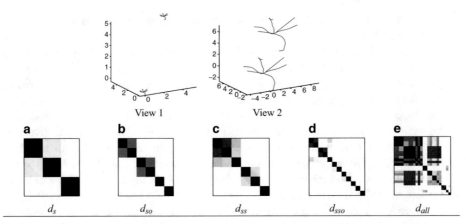

Fig. 11.2 *Top*: types of 3D curves used in the clustering experiment. *Bottom*: matrices of distances using features: (**a**) shape, (**b**) shape+orientation, (**c**) shape+scale, (**d**) shape+orientation+scale, and (**e**) shape+orientation+scale+translation

matrices as gray scale images for each combination of features for this artificial data. We see a very nice separation between curves with different features based on our distances. This validates our claim that SRVF and SRF representations, and the associated geodesic distances are successful in clustering curves according to different feature combinations.

2. **Clustering of DT-MRI Brain Fibers**: Now we consider fiber tract data from speech regions of a human brain. The fiber clustering here partitions brain fibers into groups that provide physical evidence of the complexity and intricacy of the connecting patterns between the Broca's and Wernicke's areas. It should be pointed out that this clustering is solely data driven—whether or not it represents the true physiological picture remains unknown. One way to validate this is to correlate the clustering results with fMRI data, which can be used to depict functional connections between the language areas.

    The datasets considered here consist of 388 fibers from four subjects: subject 1 had 176 total fibers, subject 2 had 68 total fibers, subject 3 had 48 total fibers, and subject 4 had 88 total fibers. Since we consider the locations of fibers as an important feature, we use SRFs (d_{all}) to compute pairwise distances between them. The results are shown in Fig. 11.3. We see a clear separation of clusters based on geodesic distances, which take into account shape, translation, scale, and orientation. In order to enhance the display of the clusters, we have used a multidimensional scaling (MDS) method to obtain 2D coordinates of each fiber and displayed them in a scatter plot. The distance matrix and the MDS plot for subject 1 indicate that there should be two clear clusters. This is in fact the case, and we see that translation plays a very important role in this particular case. The distance matrix and the MDS plot for subject 2 indicate that there should be two clear clusters here as well. But, looking at the plotted fibers, this separation is not as clear. We note that the green cluster should be split up into two separate clusters, based on the different shapes, translations, and orientations of the fibers. In the case of subject 3, there are three clusters. The blue and the green clusters differ most in shape, while the green and the red clusters differ a lot in their position in $\mathbb{R}^3$. There are two clusters present for subject 4, but it appears that some of the fibers in the red cluster have a

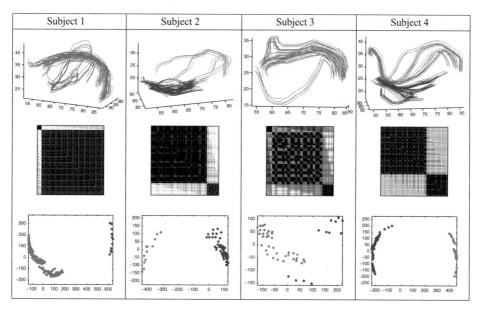

Fig. 11.3 *Top*: clustered fibers. *Middle*: distance matrix between clusters. *Bottom*: MDS plot (different colors denote different fiber clusters)

significantly different orientation and could possibly be split into two different clusters. The shapes in the green cluster are fairly similar.

Similar to Chaps. 9 and 10, one can develop tools for computing mean, covariances, principal subspaces, and generating statistical models for curves under different combinations of features.

11.3 Affine-Invariant Shape Analysis of Planar Curves

As mentioned at the start of this chapter, the past chapters have focused on making the framework invariant to similarity transformations (translation, rotation, and global scaling). However, in a variety of practical situations, especially those arising in imaging contexts, the observations are affected by transformations that are more complicated than the similarity group alone. These types of transformations occur, for example, when the image plane of a camera is not parallel to the plane containing the defining part of the shape or when a camera images the same scene from different viewing angles. Here shapes become transformed by perspective effects, and one often needs to go beyond the similarity group to define shape equivalences. The affine and projective groups are both larger than (and contain) the similarity group and are commonly used to model such shape deformations brought about by perspective skew. In this context, the focus is now on developing shape analysis methods that are invariant to affine and projective transformations.

1. **Affine Group**: The affine group for a plane is the semi-direct product $\mathscr{G}_A \equiv GL(2) \ltimes \mathbb{R}^2$ with the action given by: $\mathscr{G}_A \times \mathbb{R}^2 \to \mathbb{R}^2$

$$((A, b), x) = Ax + b \ .$$

In case the matrix component A is restricted to matrices with determinant $+1$, the resulting group is called the *special affine* group. In general, A has all four degrees of freedom, as opposed to the similarity shape analysis where the A matrix has only two degrees of freedom (one for rotation and one for scaling). Along with the translation, $\mathscr{G}_A$ has six degrees of freedom.

2. **Projective Group**: The projective transformation of a point in $\mathbb{R}^2$ is based on the action of the *projective general linear* (PGL) group as outlined in one of the exercises at the end of this chapter. It is expressed as the quotient group $PGL(3) = GL(3)/\Omega$ where Ω is a one-dimensional group of 3×3 diagonal matrices with all diagonal entries being the same. The action of $PGL(3)$ on $\mathbb{R}^2$ is defined by embedding $\mathbb{R}^2$ in $\mathbb{R}^3$, applying $GL(3)$, and then projecting back to $\mathbb{R}^2$. One way to embed $\mathbb{R}^2$ in $\mathbb{R}^3$ is to add a fixed value, say 1, as the last component, i.e., $x \equiv (x_1, x_2) \in \mathbb{R}^2 \mapsto (x_1, x_2, 1) \in \mathbb{R}^3$. With this choice of embedding, the projective transformation $PGL(3) \times \mathbb{R}^2 \to \mathbb{R}^2$ is given by:

$$(B, x) = \tilde{x}, \quad \text{where} \begin{pmatrix} \tilde{x} \\ 1 \end{pmatrix} = \frac{B \begin{pmatrix} x \\ 1 \end{pmatrix}}{\left(B \begin{pmatrix} x \\ 1 \end{pmatrix} \right)_3} .$$

The number of degrees of freedom in choosing a projective transformation is $9 - 1 = 8$, 9 for $GL(3)$ and 1 for Ω.

These transformations and their degrees of freedom are summarized in Table 11.1.

Figure 11.4 shows examples of similarity, affine, and projective transformations applied to a square, to help explain their differences. The affine and projective

Table 11.1 Summaries of the different group actions on $\mathbb{R}^2$

Groups	Their actions ($\mathbb{R}^2 \to \mathbb{R}^2$)	Space	Degrees of freedom
Similarity	$x \mapsto aOx + y$	$O \in SO(2)$, $y \in \mathbb{R}^2$, $a \in \mathbb{R}$	Four
Affine	$x \mapsto Ax + b$	$A \in GL(2)$, $b \in \mathbb{R}^2$	Six
Projective	$x \mapsto \frac{B[x,1]^T}{(B[x,1]^T)_3}$	$B \in PGL(3)$	Eight

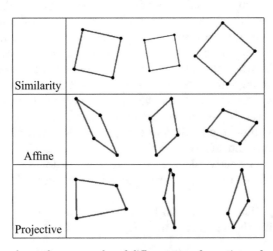

Fig. 11.4 Each row shows three examples of different transformations of a unit square

transformations can better capture the distortion introduced in imaged contours when objects are imaged from different angles. Although the affine group allows for skew effects, parallel lines still remain parallel after affine transformations. The extra degrees of freedom present in the projective group can model perspective distortions where parallel lines are skewed to converge to vanishing points located at infinity. The affine group can be used to approximate these perspective effects seen in some situations, such as when the camera is far enough away from the imaging plane or when there is only a slight change in viewing angle. In this chapter, we will focus only on affine-invariant shape analysis of curves, especially using elastic Riemannian metrics. The corresponding analysis for projective-invariant shape analysis remains to be developed.

11.3.1 Global Section Under the Affine Action

Can we directly apply the previously developed framework (Chap. 6) for affine-invariant shape analysis of planar closed curves? The answer is no, and the reason is that the action of the affine group on $\mathbb{R}^2$ is not by isometries under the Euclidean norm (this is simple to show and is left as an exercise for the reader). Suppose we choose the SRVF representation of curves (and the $\mathbb{L}^2$ norm) for analyzing shapes of curves, then the curves are viewed as elements of the set $\mathbb{L}^2([0,1], \mathbb{R}^2)$. In order to form an affine-invariant shape space, we will need to construct the quotient space $\mathbb{L}^2([0,1], \mathbb{R}^2)/(\mathscr{G}_A \times \tilde{\Gamma}_S)$ and seek to inherit the $\mathbb{L}^2$ norm from the larger space $\mathbb{L}^2([0,1], \mathbb{R}^2)$ to this quotient space. Ignoring $\tilde{\Gamma}_S$ momentarily, we focus only on the action of $\mathscr{G}_A$. As mentioned above, $\mathscr{G}_A$ does not act by isometries on the set $\mathbb{L}^2([0,1], \mathbb{R}^2)$ under the $\mathbb{L}^2$ norm. So, we cannot inherit the $\mathbb{L}^2$ metric on to the quotient space $\mathbb{L}^2([0,1], \mathbb{R}^2)/\mathscr{G}_A$. Thus, we will take the approach called **Method (3)** in Table 3.1 (Sect. 3.9). This method involves deriving a global section of the larger space under the action of the group and choosing a metric structure on this section. (Recall from Definition 3.18 that the global section of a set M, under the action of a group G, is the subset of M such that it intersects each orbit of G in one and only one point.) In the current case, where the larger space is $\mathbb{L}^2([0,1], \mathbb{R}^2)$ and the group is $\mathscr{G}_A$, we will reach a global section in two steps. We will first specify a set $\mathscr{F}_a^c$ of *standardized curves* such that each orbit of $\mathscr{G}_A$ in $\mathbb{L}^2([0,1], \mathbb{R}^2)$ is associated with a standardized curve but under an arbitrary rotation. In other words, $\mathscr{F}_a^c$ is a section of $\mathbb{L}^2([0,1], \mathbb{R}^2)$, under the action of $\mathscr{G}_A$, only up to the action of $SO(2)$. By itself, it is not quite a global section, but its quotient space $\mathscr{F}_a^c/SO(2)$ will act as the desired global section. Since this quotient space is bijective to the quotient space $\mathbb{L}^2([0,1], \mathbb{R}^2)/\mathscr{G}_A$, we can simply induce a metric on the latter using a metric on the former. As described next, it is natural to use the elastic shape metric on that section. We remark that the section identified here is a global section but not an *orthogonal* section of $\mathbb{L}^2([0,1], \mathbb{R}^2)$ under the action of $\mathscr{G}_A$.

Sectioning of Curve Space Next we derive a global section of $\mathbb{L}^2([0,1], \mathbb{R}^2)$ under $\mathscr{G}_A$. We already know (see Sect. 5.4) that the SRVF representation provides a bijection between $\mathscr{F}_2$, the set of unit-length absolutely continuous curves that start at the origin (defined first in Sect. 5.4), and $\mathscr{C}_2$. Similarly, $\mathscr{F}_2^c \subset \mathscr{F}_2$, the subset of closed curves, has a bijection with $\mathscr{C}_2^c$. Therefore, instead of defining a section of $\mathscr{C}_2^c$, we can equivalently define a section of $\mathscr{F}_2^c$, especially since it is an easier task. In fact, this task of finding a canonical element in the affine orbit of any given curve has been termed *affine standardization* in the literature. Let $\beta : \mathbb{S}^1 \to \mathbb{R}^2$ be an element of $\mathscr{F}_2^c$, and the $\mathscr{G}_A$-orbit of β is the set

$$[\beta]_{\mathscr{G}_A} = \left\{ A\beta + b \mid A \in GL(2), b \in \mathbb{R}^2 \right\}.$$

Recall that $L_\beta = \int_{\mathbb{S}^1} |\dot{\beta}(t)| dt$ is the length of the curve β. Additionally, define $C_\beta = \frac{1}{L_\beta} \int_{\mathbb{S}^1} \beta(t) |\dot{\beta}(t)| dt \in \mathbb{R}^2$ as the *centroid* and

$$\Sigma_\beta = \frac{1}{L_\beta} \int_{s^1} (\beta(t) - C_\beta)(\beta(t) - C_\beta)^T |\dot{\beta}(t)| dt \in \mathbb{R}^{2 \times 2}$$

as the *covariance* of β. Note that for our purposes below, a closed curve is called degenerate if it is completely contained in a line.

Theorem 11.1. *For any non-degenerate β there exists a standardized element $\beta^* \in [\beta]_{\mathscr{G}_A}$, the $\mathscr{G}_A$-orbit of β, that satisfies the following three conditions: (1) $L_{\beta^*} = 1$, (2) $C_{\beta^*} = 0$, and (3) $\Sigma_{\beta^*} \propto I$. Furthermore, for any two curves $\beta_1, \beta_2 \in [\beta]_{\mathscr{G}_A}$, the corresponding standardized elements, β_1^* and β_2^*, are related by a rotation and re-parameterization, $\beta_2^* = O(\beta_1^* \circ \gamma)$, where $O \in SO(2)$ and $\gamma \in \Gamma_S$.*

For a proof of this result, we direct the reader to the online supplementary material associated with the paper [21]. We define the set of *affine-standardized curves* as

$$\mathscr{F}_a^c = \{\beta \in \mathscr{F}_2 \mid L_\beta = 1, C_\beta = 0, \Sigma_\beta = c_\beta I_2, \text{ and } \beta \text{ is closed}\} \subset \mathscr{F}_2^c, \quad (11.1)$$

where c_β is a real scalar. Note that due to the unit-length condition, we may not achieve $\Sigma_\beta = I_2$ exactly but only within a constant. This section $\mathscr{F}_a^c$ implicitly defines the *curve-affine pre-shape space*. Furthermore, the quotient space $\mathscr{F}_a^c / SO(2)$ forms a bijection with the quotient space $\mathbb{L}^2([0,1], \mathbb{R}^2) / \mathscr{G}_A$.

Affine Standardization of Curves This approach requires that given an arbitrary element $\beta \in \mathscr{F}_2^c$ we should be able to project it in $\mathscr{F}_a^c$ and find its affine-standardized representative. Since an analytical expression for standardization is not known, we develop an iterative algorithm to standardize a given curve β; call that standardized curve β^*. It is trivial to find the transformations that satisfy the first two conditions, i.e., scaling and centering, and therefore we focus on the more complicated task of satisfying the third condition, i.e., the covariance condition, and assume we are given β with unit length and centroid located at the origin. Thus, we focus on finding a group element $A^* \in GL(2)$ such that $\Sigma_{A^*\beta} \propto I$. In fact, it can be shown that it is sufficient to search only over the subspace $S(2) \cap GL(2) \subset GL(2)$, where $S(2)$ is the space of all 2×2 symmetric matrices. $S(2)$ is a three-dimensional vector space $(T_I(S(2)) = S(2))$ and let B_1, B_2, B_3 form an orthonormal basis of $S(2)$. For example,

$$B_1 = \begin{bmatrix} 1 & 0 \\ 0 & 0 \end{bmatrix}, B_2 = \begin{bmatrix} 0 & 0 \\ 0 & 1 \end{bmatrix}, \text{ and } B_3 = \begin{bmatrix} 0 & \frac{1}{\sqrt{2}} \\ \frac{1}{\sqrt{2}} & 0 \end{bmatrix}.$$

Any matrix in $S(2)$ can be expressed as an element of $\mathbb{R}^3$ consisting of coefficients from a linear combination of these basis vectors.

Define a function $F : S(2) \times \mathscr{F}_2^c \to S(2)$ as $F(A; \beta) = L_{A\beta} \Sigma_{A\beta}$. We will use a Newton-Raphson procedure to find $A^* \in S(2)$ such that $F(A^*; \beta) - I = 0$. Then, we can simply scale A^* accordingly to satisfy the unit-length condition on β^*. The Newton-Raphson algorithm updates the estimate of A^* and β^* incrementally at each iteration. Let the $(m+1)$-th estimate of A^* be $A^{(m+1)} = A_{m+1}A_m \cdots A_1$, and let the $(m+1)$-th estimate of β^* be $\beta_{m+1} = A_{m+1}\beta_m - C_{A_{m+1}\beta_m}$. At each iteration we calculate A_{m+1} as a perturbation from identity, and thus we must define the directional derivative of $F(I; \beta_m)$. For any $B \in T_I(S(2))$, and let $a = -C_{\beta_m}$. The directional derivative of F at identity in the direction of B is denoted by $dF_I(B; \beta_m)$, which turns out to be:

$$\int_0^1 (B\beta_m\beta_m^T + \beta_m\beta_m^T B + B\beta_m a^T + a\beta_m^T B + aa^T)|\dot\beta_m|dt$$

$$+ \int_0^1 \frac{1}{|\dot\beta_m|}(\beta_m\beta_m^T + \beta_m a^T + a\beta_m^T + aa^T)\dot\beta_m{}^T B\dot\beta_m dt. \tag{11.2}$$

We can express dF_I as a 3×3 matrix J under the chosen basis B_1, B_2, B_3. That is $J_{ij} = \langle dF_I(B_i, \beta_m), B_j \rangle$.

Algorithm 53 (Curve-Affine Standardization). *Given β_0 with $L_{\beta_0} = 1$ and $C_{\beta_0} = 0$, initialize $A^{(0)} = A_0 = I$ and select a step size δ. Let $m = 0$.*

1. *Calculate the residual $r_m = F(I; \beta_m) - I$. Rewrite r_m as a vector of coefficients with respect to the basis vectors B_1, B_2, B_3. If $|r_m| < \epsilon$ go to (7), else*
2. *Calculate $dF_I(B_i; \beta_m)$ for $i = 1, 2, 3$ using Eq. 11.2, and form the Jacobian J.*
3. *Compute $x = J^{-1}r_m$ and let $D = \sum_{i=1}^3 x_i B_i$.*
4. *Let $A_{m+1} = I - \delta D$.*
5. *Update $\beta_{m+1} = A_{m+1}\beta_m - C_{A_{m+1}\beta_m}$ and $A^{(m+1)} = A_{m+1}A^{(m)}$.*
6. *Return to (1) and let $m = m + 1$.*
7. *Let $A^* = A^{(m+1)}/L_{\beta_{m+1}}$, and let $\beta^* = \beta_{m+1}/L_{\beta_{m+1}}$.*

Figure 11.5 shows the result of applying this subroutine to curves from four different affine orbits; each orbit is shown as a different row. The left panel shows curves before and the right panel shows the corresponding curves after the standardization. Note that a standardization differs only in rotation and re-parameterization within each row. This standardization provides us with a technique to take any arbitrary closed curve in $\mathscr{F}_2^c$ (unit length, absolutely continuous) and project it uniquely into an element of $\mathscr{F}_a^c$.

Fig. 11.5 Affine standardization of curves. The original curves are shown in the left, and their standardizations are shown in the right

11.3.2 Geodesics Using Path-Straightening Algorithm

After standardization our problem reduces to comparing elements of $\mathscr{F}_a^c$, the manifold of affine-standardized, unit-length closed curves modulo rotation and re-parameterization. This task is a repeat of the ideas presented in Chap. 6 for similarity shape analysis of closed curves. Since each curve is represented by the standard element of its affine orbit, we need to further restrict the analysis to SRVFs of affine-standardized curves by imposing a few added conditions beyond that of the closure condition.

For a parameterized curve $\beta : \mathbb{S}^1 \to \mathbb{R}^2$, let $q : [0,1] \to \mathbb{R}^2$ denote its SRVF. Let $x(q;t) = \int_0^t q(u)|q(u)|du$ be our original curve $\beta(t)$ but with the starting point located at the origin, i.e., $\beta(0) = 0$. As noted in Chap. 5, $|q(t)|^2 = |\dot{\beta}(t)|$ and the set of all unit-length curves is the unit hypersphere $\mathbb{S}_\infty$. The centroid and covariance of a curve $x(q;t)$ can be stated in terms of the q-function as follows. The centroid in $\mathbb{R}^2$ is given by $C_q = \int_0^1 x(q;t)|q(t)|^2 dt$, and the covariance in $\mathbb{R}^{2\times 2}$ is given by $\Sigma_q = \int_0^1 (a + x(q;t))(a + x(q;t))^T |q(t)|^2 dt$, where $a = -C_q$. In order to impose the condition that the curve β be closed, we set $x(q;1) = 0$, i.e., the endpoint of the curve $x(q;t)$ is equal to the initial point, the origin. Define a mapping $\Psi : \mathbb{S}^\infty \to \mathbb{R}^4$ as:

$$
\begin{aligned}
\Psi_1(q) &= \int_{\mathbb{S}^1} \left((a_1 + x_1(q;t))^2 - (a_2 + x_2(q;t))^2 \right) |q(t)|^2 dt \\
\Psi_2(q) &= \int_{\mathbb{S}^1} (a_1 + x_1(q;t))(a_2 + x_2(q;t)) |q(t)|^2 dt \\
\Psi_3(q) &= \int_{\mathbb{S}^1} q_1(t)|q(t)|dt \\
\Psi_4(q) &= \int_{\mathbb{S}^1} q_2(t)|q(t)|dt,
\end{aligned}
\tag{11.3}
$$

where a subscript indicates the ith coordinate in Euclidean space, that is, $\Psi_i(q) \in \mathbb{R}$ for each i. $\Psi_1(q) = 0$ implies that the difference of diagonal entries in the covariance matrix is 0. $\Psi_2(q) = 0$ implies that the off-diagonal entry in the covariance matrix is 0. Together, they imply that $\Sigma_\beta = c_\beta I$. The constraint $\Psi_3(q) = \Psi_4(q) = 0$ implies the closure condition. Since SRVFs are translation invariant, we don't need any explicit condition on the centroid. The space of all affine-standardized, unit length, closed curves is therefore the level set $\mathscr{C}_a^c = \Psi^{-1}((0,0,0,0)) \subset \mathbb{S}^\infty$. The section $\mathscr{C}_a^c$ has a one-to-one correspondence with $\mathscr{F}_a^c$ and is therefore also called *curve-affine pre-shape space*. We shall use the notation $\mathscr{S}_a^c$ to denote *curve-affine shape space* $\mathscr{C}_a^c/(SO(2) \times \tilde{\Gamma}_S)$. In order to compute geodesics on $\mathscr{C}_a^c$ and its quotient space, we will use the path-straightening algorithm introduced earlier in Sect. 6.6 of Chap. 6.

Defining the Normal Space of $\mathscr{C}_a^c$ An integral step in the subroutines necessary for path-straightening on $\mathscr{C}_a^c$ is obtaining a basis for the normal space $N_q(\mathscr{C}_a^c)$, the set of elements that are normal to $\mathscr{C}_a^c$ inside $\mathbb{L}^2(\mathbb{S}^1, \mathbb{R}^2)$. The normal space is a four-dimensional space, and here we provide analytical expressions for the functions $\{h^j(t), j = 1, \ldots, 4\}$ that serve as a basis for this space. These functions arise from the calculation of the directional derivative of Ψ in Eq. 11.3, and we leave the full derivation as an exercise. In order to define the normal space, we must first find the directional derivative of Ψ (see Eq. 11.3) at any point $q \in \mathscr{C}_a^c$. Let $w \in \mathbb{L}^2(\mathbb{S}^1, \mathbb{R}^2)$. We write the derivative $d\Psi_{j,q}(w) = \frac{d}{ds}\Psi_j(q(t) + sw(t))|_{s=0}$, for $j = 1, 2, 3, 4$. By expressing the directional derivative as an inner product as follows: $d\Psi_{j,q}(w) = \langle w(t), h^j(t) \rangle$ for some functions $h^1, \ldots, h^4$, where $\langle \cdot, \cdot \rangle$ is the standard $\mathbb{L}^2$ inner product, these functions will serve as a basis for the normal space. The restriction of these functions inside $T_q(\mathscr{C}_a^c)$ is obtained by removing the projection along the function q: $b^j(t) = h^j(t) - q(t)\langle q(t), h^j(t) \rangle$ and $\{b^j\}$ span the normal space $N_q(\mathscr{C}_a^c)$ inside $T_q(\mathscr{B})$. Let $f^i(t) = |q(t)|\mathbf{e}_i + \frac{q_i(t)}{|q(t)|}q(t)$ for $i = 1, 2$. Recall that $\frac{q_i(t)}{|q(t)|}q(t) = 0$ in case $|q(t)| = 0$ at some $t \in \mathbb{S}^1$. Let $Q(t) = \int_0^t |q(\tau)|^2 d\tau$ and let $G_i(t) = \int_0^t |q(\tau)|^2 x_i(q;\tau)d\tau$ for $i = 1, 2$. In this notation,

$$h^1(t) = 4q(t)a_1x_1(q;t) - 4q(t)a_2x_2(q;t) + 2a_1f^1(t)(1 - Q(t)) - 2a_2f^2(t)(1 - Q(t))$$
$$+ 2q(t)x_1(q;t)^2 - 2q(t)x_2(q;t)^2 + 2f^1(t)(a_1 - G_1(t)) - 2f^2(t)(a_2 - G_2(t)) ,$$
$$(11.4)$$

and

$$h^2(t) = 2a_2q(t)x_1(q;t) + 2a_1q(t)x_2(q;t) + a_2f^1(t)(1 - Q(t)) + a_1f^2(t)(1 - Q(t))$$
$$+ 2q(t)x_1(q;t)x_2(q;t) + f^1(t)(a_2 - G_2(t)) + f^2(t)(a_1 - G_1(t)). \quad (11.5)$$

The remaining two are given by $h^3(t) = f^1(t)$, and $h^4(t) = f^2(t)$. The restriction of these functions inside $T_q(\mathbb{S}_\infty)$ is obtained by removing the projection along the function q: $b^j(t) = h^j(t) - q(t)\langle q(t), h^j(t)\rangle$ and $\{b^j\}$ span the normal space $N_q(\mathscr{C}_a^c)$ inside $T_q(\mathbb{S}_\infty)$. Next we describe three additional subroutines needed to implement the path-straightening algorithm on $\mathscr{C}_a^c$.

For any point $q \in \mathbb{S}_\infty$, we need a tool to project q to the nearest point in $\mathscr{C}_a^c$. The procedure presented below is different from Algorithm 53 in that, although they accomplish a similar task, standardization is restricted to the same orbit, while the projection is not. To clarify further, Algorithm 53 is used on coordinate functions for curve standardization, and Algorithm 54 is used on SRVFs within the path-straightening algorithm. These algorithms are very similar to their counterparts used in Chap. 6.

Algorithm 54 (Projection onto $\mathscr{C}_a^c$). *Let $\epsilon > 0$.*

1. *Compute the residual vector $r = \Psi(q) \in \mathbb{R}^4$, with Ψ as given in Eq. 11.3. If $|r| < \epsilon$, stop. Otherwise continue to step 2.*
2. *Calculate the basis vectors $\{b^j, j = 1, \ldots, 4\}$, for $N_q(\mathscr{C}_a^c)$, and form the Jacobian matrix J via $J_{ij} = \langle b^i(t), b^j(t)\rangle$.*
3. *Solve $Jx = -r$ for x.*
4. *Define $dq = \sum_{j=1}^4 x_j b^j$. Update q using $q \mapsto \cos(\|dq\|)q + \sin(\|dq\|)\frac{dq}{\|dq\|}$. Go to step 1.*

Algorithm 55 (Projection onto $T_q(\mathscr{C}_a^c)$). *Given any function $w \in \mathbb{L}^2(\mathbb{S}^1, \mathbb{R}^2)$,*

1. *If $w \notin T_q(\mathbb{S}_\infty)$, then project w to $T_q(\mathbb{S}_\infty)$ via $w \mapsto w - \langle w, q\rangle q$.*
2. *Compute a basis $\{b^j\}$, $j = 1, 2, 3, 4$, for $N_q(\mathscr{C}_a^c)$ as earlier, and then obtain a Gram-Schmidt orthonormalization, $\{b_o^j\}$.*
3. *Project w into $T_q(\mathscr{C}_a^c)$ via $w \mapsto w - \sum_{j=1}^4 \langle w, b_o^j\rangle b_o^j$.*

Note that we can skip the first step even if $w \notin T_q(\mathbb{S}_\infty)$, but it will compromise numerical stability.

In case two points $q_1, q_2 \in \mathscr{C}_a^c$ are close to each other, we can approximately parallel translate a tangent vector from q_1 to q_2 using a projection.

Algorithm 56 (Parallel Translation). *Given two points* $q_1, q_2 \in \mathscr{C}_a^c$ *and* $w \in T_{q_1}(\mathscr{C}_a^c)$.

1. Compute the analytic expression for parallel translation on $\mathbb{S}^\infty$ *as*

$$w \mapsto \tilde{w} \equiv \frac{2\langle w, q_2 \rangle}{\|q_1 + q_2\|^2}(q_1 + q_2).$$

2. Let $l = |\tilde{w}|$. *Project* $\tilde{w}$ *onto* $T_{q_2}(\mathscr{C}_a^c)$ *using Algorithm 55 to obtain* $\bar{w}$.
3. Rescale $\bar{w}$ *via* $\bar{w} \mapsto \frac{\bar{w}l}{\|\bar{w}\|}$.

We now have all the tools necessary to compute geodesics via path-straightening on $\mathscr{S}_a^c$, and the general procedure of affine-invariant, elastic shape analysis is as follows, assuming we are given two curves β_1 and β_2. First, use Algorithm 53 to obtain β_1^* and β_2^*. Then, convert to SRVF representation to obtain $q_1^*, q_2^* \in \mathscr{C}_a^c$ and hence the unique orbits $[q_1^*]$ and $[q_2^*]$ under the group action of $SO(2) \times \tilde{\Gamma}_S$. Since this group action is by isometries on $\mathscr{C}$, we define the distance on $\mathscr{S}_a^c$ as

$$d_{\mathscr{S}_a^c}([q_1^*], [q_2^*]) = \inf_{O \in SO(2), \gamma \in \Gamma_S} d_{\mathscr{C}_a^c}(q_1^*, O(q_2^*, \gamma)). \qquad (11.6)$$

The optimal group elements that achieve this distance minimization, say O^* and γ^*, are solved, respectively, by Procrustes rigid body alignment and by either dynamic programming or gradient-based optimization (see Chap. 5). The distance $d_{\mathscr{C}_a^c}(q_1^*, O^*(q_2^*, \gamma^*))$ is thus computed with path straightening.

In order to highlight the deformations resulting along geodesic paths between shapes, we use an additional step of de-standardization in our displays. Note that our algorithms start by standardization of given curves β_1 and β_2 into standardized curves $\beta_1^* = A_1^* \beta_1$ and $\beta_2^* = A_2^* \beta_2$. Here A_1^*, A_2^* are symmetric, non-singular matrices found using Algorithm 53. In order to revert to the original curve, one can use their inverses, A_1^{*-1} and A_2^{*-1}, on the corresponding curves. In order to de-standardize shapes along the geodesic path between the standardized shapes, we use a geodesic between A_1^{*-1} and A_2^{*-1} in $S(2)$ and apply the transformation at time τ to the corresponding shape at time τ along the geodesic.

Various examples of geodesic paths in $\mathscr{S}_a^c$ can be seen in Figs. 11.6, 11.7, and 11.8. Figure 11.6 shows an example of path-straightening iterations that lead to a geodesic path in $\mathscr{S}_a^c$. Likewise, Fig. 11.7 shows four examples of final geodesic paths in affine shape space along with their corresponding de-standardized paths.

Figure 11.8 provides a visual illustration of the benefit of affine invariance as well as elasticity in a shape analysis framework. Here we compute the geodesic path between two shapes in various shape spaces, whereby the two shapes differ in placement of bumps and an affine transformation (shown in top and bottom of the last column in Fig. 11.8). We compute the geodesics in the following shape spaces: (a) closed curve shape space with a non-elastic, bending only metric in $\mathscr{S}_1^c$ (b) closed curve, similarity-invariant shape space with the elastic metric in $\mathscr{S}_2^c$, (c) closed curve, affine-invariant shape space with the elastic metric in $\mathscr{S}_a^c$, and for display purposes (d) the de-standardization of (c). We can see that the deformation in geodesic path (b) is more natural and smaller than that of path (a) due to the addition of elasticity to appropriately stretch and match features. Furthermore, path (c) shows a smaller deformation than that of path (b) due to the addition of affine invariance. Note that any affine transformation of either the

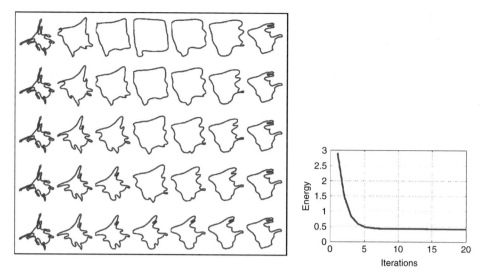

Fig. 11.6 Path straightening on $\mathscr{S}_a^c$. The left side shows iterations of the path-straightening algorithm from top (initial path) to bottom (final path). The right panel shows the corresponding evolution of the path energy

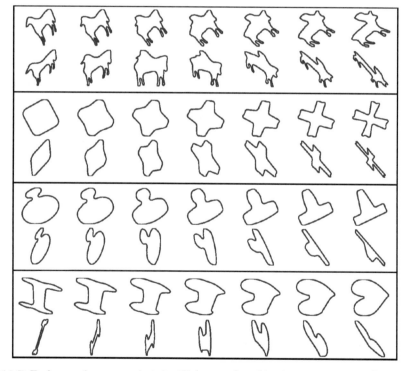

Fig. 11.7 Each case shows a geodesic in $\mathscr{S}_a^c$ (*top row*) and its de-standardization (*bottom row*)

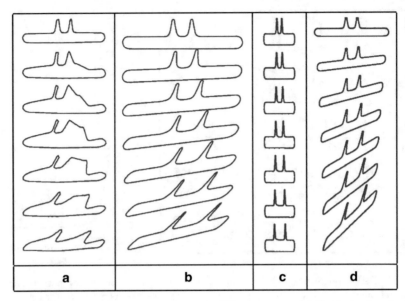

Fig. 11.8 Geodesic paths between two shapes in different spaces. (**a**) Similarity invariant with bending-only metric, (**b**) similarity invariant with elastic metric, (**c**) $\mathscr{S}_a^c$ (**d**) de-standardized version of path (**c**)

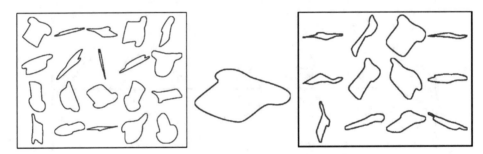

Fig. 11.9 *From left to right*: original data, the mean in the similarity shape space, and random samples

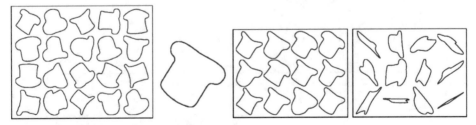

Fig. 11.10 *From left to right*: standardized versions of shapes in Fig. 11.9, the Karcher mean in $\mathscr{S}_a^c$, random samples, and their de-standardizations

beginning or ending shape will not change the geodesic path shown in (c) due to our affine-invariant shape analysis framework.

Figures 11.9 and 11.10 demonstrate the advantage of using statistical models in $\mathscr{S}_a^c$ over the similarity shape space. Here, we use the MPEG-7 shape class "hat" under random affine transformations. In Fig. 11.10 we see that we obtain gains in

Table 11.2 Leave-one-out classification rates for human activity sequence dataset under various simulated camera angles

	Broadside level	Narrow-side level	Broadside elevated	Narrow-side elevated
Similarity (%)	98	97	51	87
Affine (%)	98	98	98	98

modeling these affine shapes by separating the variability into the spaces $\mathscr{S}_a^c$ and $GL(2)$.

Pose-Invariant Activity Classification An important application of affine-invariant shape analysis is in the field of human activity or human motion analysis where a major need here is to be invariant under differing pose or camera angles. Note that if the pose changes so much that certain body parts are occluded and new parts become visible, then the shape changes are too complicated to be modeled as a simple transformation. However, in case of moderate changes ($< 45°$), one can model these changes using affine or projective transformations. In that situation the proposed algorithms can be used for a pose-invariant activity classification.

In this experiment we use the UMD activity dataset, which consists of 100 sequences of 80 shapes each, where each sequence represents a frame-by-frame outline of a person performing a task. The dataset is divided into 10 classes, or "activities," of 10 sequences each. Our goal is to classify a test sequence under an arbitrary viewing angle; we simulate different viewing angles by applying appropriate affine transformations on a given sequence. Since these data were captured using broadside imaging, we simulate test sequences for different views: original (camera level and broadside), narrow side (camera at level height and slightly facing the subject), top view (camera at an elevated position and broadside), and top left (camera elevated and slightly facing the subject). To generate a test sequence, we simulate a stochastic process on $GL(2)$ with an appropriate mean and apply each point of the process to the corresponding shape in the sequence. Figure 11.11 shows an example of an activity sequence under these simulated views. Then, we classify this test sequence using the nearest-neighbor classifier under different metrics. (The distance for classifying a sequence of shapes is the sum of the distances for individual shapes.)

Note that the classification rate of the affine-standardized test sequences will match that of the original, un-transformed sequences. Since classification rate decreases under different camera angles without standardization, we conclude that an affine-invariant metric helps in classifying sequences of human activity under various camera angles (Table 11.2). From the drastic improvement in classification rate from 51 % to 98 % in the top camera angle, we can see that standardization would be especially useful when applied to sequences from mounted surveillance cameras.

11.4 Registration of Trajectories on Nonlinear Manifolds

In the third and final part of this chapter, we will study techniques for comparing objects that are observations of stochastic processes on nonlinear Riemannian manifolds. In other words, we are interested in objects of the type $\beta : [0, 1] \rightarrow M$,

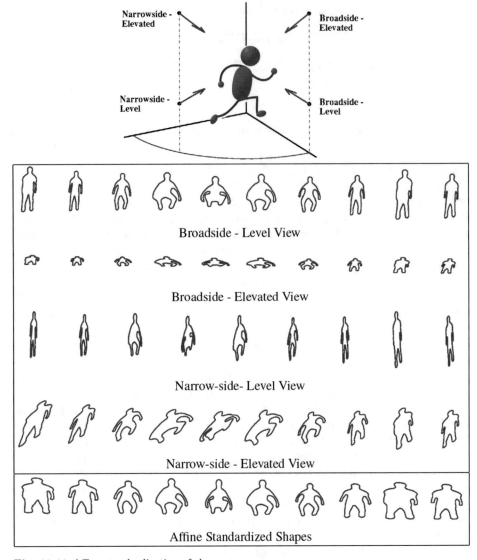

Fig. 11.11 Affine standardization of shape sequences

where M is a nonlinear Riemannian manifold. Although they are essentially curves, we will call them trajectories to distinguish them from curves in Euclidean spaces studied in previous chapters.

The need to summarize and model trajectories arises in many situations, especially in pattern recognition of complex systems. An important challenge in handling real data is that trajectories are often not observed at standard times; in fact, they are often observed at arbitrary times. If this temporal variability is not accounted for in the analysis, then the resulting statistical summaries will not be precise. The mean trajectory may not be a representative of individual trajectories and the cross-sectional variance will be artificially inflated. This, in turn, will greatly reduce the effectiveness of any subsequent modeling or analysis based on estimated mean and covariance. As a simple example consider the trajectory on

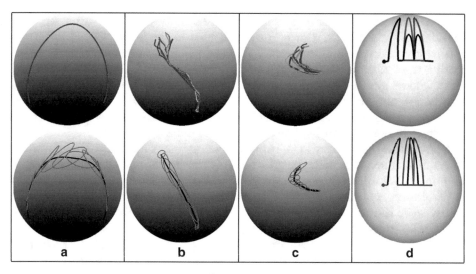

Fig. 11.12 Summary of trajectories on $\mathbb{S}^2$: (**a**) a simulated example, (**b**) bird migration paths, (**c**) hurricane tracks, and (**d**) cross-sectional mean of two trajectories without (*top*) and with (*bottom*) registration (colored picture)

$\mathbb{S}^2$ shown in the top panel of Fig. 11.12a. We simulate a set of random, discrete observation times and generate observations of this trajectory at these random times. These simulated trajectories are identical in terms of the points traversed but their evolutions, or parameterizations are quite different. If we compute cross-sectional mean and variance, the results are shown in the bottom panel. We draw the sample mean trajectory in black and the sample variance at discrete times using tangential ellipses. Not only is the mean fairly different from the original curve, the variance is purely due to randomness in observation times and is somewhat artificial. If we had observed the trajectory at fixed, synchronized times, then this problem will not exist.

To motivate this issue further, consider the phenomenon of bird migration that is the regular seasonal journey undertaken by many species of birds. There are variabilities in migration trajectories, even within the same species, including the variability in their rates of travels. In other words, either birds can travel along different paths or, even if they travel the same path, different birds (or subgroups) can fly at different speed patterns along that path. This results in variability in observation times of migration paths for different birds and artificially inflates the cross-sectional variance in the data. Another issue is that such trajectories are naturally studied as paths on a unit sphere, which is a nonlinear manifold. We will study the migration data for Swainson's Hawk, with some example paths shown in the top of Fig. 11.12b. Swainson's Hawk inhabits North America mainly in the spring and summer and winters in South America. The bottom panel in Fig. 11.12b shows cross-sectional sample mean and variance of the trajectories. Another motivating application comes from hurricane tracking, where one is interested in studying the shapes of hurricane tracks in certain geographical regions. As in the previous application, the hurricane tracks are also naturally treated as trajectories on a unit sphere. The top panel of Fig. 11.12c shows a set of hurricane tracks originating from the Atlantic region. The sample mean and the variance of these trajectories are adversely affected by this phase variability present in data, as shown in the bottom row of Fig. 11.12c. As the last motivating example, consider two trajectories, drawn in red and blue in the top of Fig. 11.12d These two trajectories have the same shape, i.e., two bumps each, and a curve representing their mean is also expected to have two bumps. A simple cross-sectional mean, shown by the black trajectory in the same picture, has three bumps! If we incorporate optimal temporal alignment in our analysis, as we did for function data and curves previously, then such inconsistencies are avoided and the black trajectory in bottom panel shows the resulting mean.

Let's look at the problem in more mathematical terms: Let $\alpha : [0, 1] \to M$, where M is a Riemannian manifold, be an absolutely continuous map; it denotes a trajectory on M. We will study such trajectories as elements of an appropriate subset of $M^{[0,1]}$, the set of all maps from $[0, 1]$ to M. Rather than observing a trajectory α directly, say in the form of time observations $\alpha(t_1), \alpha(t_2), \ldots$, we instead observe the (arbitrarily) time-warped trajectory $\alpha(\gamma(t_1)), \alpha(\gamma(t_2)), \ldots$, where $\gamma \in \Gamma_I$ governs the rate of evolution. The mean and variance of $\{\alpha_1(t), \alpha_2(t), \ldots, \alpha_n(t)\}$ for any t, where n is the number of observed trajectories, are termed the *cross-sectional mean* and *variance* at that t. If we use the observed samples $\{\alpha_i(\gamma_i(t)), i = 1, 2, \ldots, n\}$ for analysis, then the cross-sectional variance maybe inflated due to random γ_is. Our hypothesis is that this problem can be mitigated by temporally registering the trajectories. Thus, we are interested in the following four tasks:

1. **Temporal Registration**: This is a process of establishing a one-to-one correspondence between points along multiple trajectories. That is, given any n trajectories, say $\alpha_1, \alpha_2, \ldots, \alpha_n$, we are interested in finding functions $\gamma_1, \gamma_2, \ldots, \gamma_n$ such that the points $\alpha_i(\gamma_i(t))$ are matched optimally for all t.
2. **Metric Comparison**: We want to develop a metric that is invariant to different evolution rates of trajectories. Specifically, we want to define a distance $d_\alpha(\cdot, \cdot)$ such that for arbitrary evolution functions γ_1, γ_2 and arbitrary trajectories α_1 and α_2, we have $d_\alpha(\alpha_1, \alpha_2) = d_\alpha(\alpha_1 \circ \gamma_1, \alpha_2 \circ \gamma_2)$.
3. **Statistical Summaries**: The main use of this metric will be in defining and computing a (Karcher) mean trajectory $\mu(t)$ and a cross-sectional variance function $\rho(t)$, associated with any given set of trajectories. The main reason for performing registration is to reduce the cross-sectional variance that is artificially introduced in the data due to random observation times. The reduction in variance is quantified using ρ.
4. **Statistical Modeling and Evaluation**: We will use the estimated mean and covariance of registered trajectories to define a "Gaussian-type" model on random trajectories. This model will then be used to evaluate p-values associated with new trajectories. Here p-value implies the proportion of trajectories with density under the model smaller than the current trajectory.

For performing comparison and summarization of trajectories, we need a metric and, at first, we consider a more conventional solution. Since M is a Riemannian manifold, we have a natural distance d_m, i.e., the geodesic distance under the given Riemannian metric, between points on M. Using d_m, one can compare any two trajectories: $\alpha_1, \alpha_2 : [0, 1] \to M$, as

$$d_x(\alpha_1, \alpha_2) = \int_0^1 d_m(\alpha_1(t), \alpha_2(t))dt \ . \tag{11.7}$$

See Sect. 3.3 for a proof that d_x is a distance between trajectories. Although this quantity represents a natural extension of d_m from M to $M^{[0,1]}$, it suffers from the problem that $d_x(\alpha_1, \alpha_2) \neq d_x(\alpha_1 \circ \gamma, \alpha_2 \circ \gamma)$ generally. If this equality held, for all $\gamma \in \Gamma_I$, then one could develop a fully invariant distance and use it to properly register trajectories. So, the failure to have this equality is in fact a key issue that forces us to look for other solutions in situations where trajectories are observed at arbitrary temporal evolutions. When a trajectory α is observed as $\alpha \circ \gamma$, for an arbitrary temporal re-parameterization γ, we call this perturbation *compositional noise*. In these terms, d_x is not useful in comparing trajectories observed under compositional noise.

11.4.1 Transported SRVF for Trajectories

Continue to use α to denote an absolutely continuous trajectory on a Riemannian manifold of interest M, where M is endowed with a Riemannian metric $\langle \cdot, \cdot \rangle_p$. Let $\mathcal{M}$ denote the set of all such trajectories: $\mathcal{M} = \{\alpha : [0,1] \to M | \alpha$ is absolutely continuous$\}$. If α is a trajectory on M, then $\alpha \circ \gamma$, for any $\gamma \in \tilde{\Gamma}_I$, is a trajectory that follows the same sequence of points as α but at the evolution rate governed by γ. More technically, the group $\tilde{\Gamma}_I$ acts on $\mathcal{M}$, $\mathcal{M} \times \tilde{\Gamma}_I \to \mathcal{M}$, according to $(\alpha, \gamma) = \alpha \circ \gamma$.

Given two smooth trajectories $\alpha_1, \alpha_2 \in \mathcal{M}$, we want to register points along the trajectories and compute a time-warping invariant distance between them. A seemingly natural choice for this purpose would be:

$$\inf_{\gamma \in \tilde{\Gamma}_I} \left(\int_0^1 d_m(\alpha_1(t), \alpha_2(\gamma(t))) dt \right),$$

but it fails for the same reasons as those highlighted in Sects. 4.4 and 10.5, including the fact that it is not even symmetric. Fundamentally speaking, this and other quantities used in previous literature are not appropriate for solving the registration problem because they are not measuring registration in the first place. To highlight this issue, take the registration of points between the pair (α_1, α_2) and the pair $(\alpha_1 \circ \gamma, \alpha_2 \circ \gamma)$, for any $\gamma \in \Gamma_I$. It can be seen that the pairs (α_1, α_2) and $(\alpha_1 \circ \gamma, \alpha_2 \circ \gamma)$ have exactly the same registration of points. In fact, any identical time warping of two trajectories does not change the registration of points between them. But the quantity given in Eq. 11.7 and the equation written above, both provide different values for these pairs, despite them having the same registration. Hence, they are not good measures of registration. We emphasize that the invariance under identical time warping is a key property that is needed in the desired framework.

We introduce a new representation of trajectories that will be used to compare and register them. We will assume that for any two points $p, q \in M$, we have an expression for parallel transporting any vector $v \in T_p(M)$ along the shortest geodesic from p to q, denoted by $(v)_{p \to q}$. As long as p and q do not fall in the cut loci of each other, the shortest geodesic between them is unique and the parallel transport is well defined. The measure of the set of cut locus on the manifolds of our interest is typically zero. So, the practical implications of this limitation are negligible. Let c be a point in M that we will designate as a reference point.

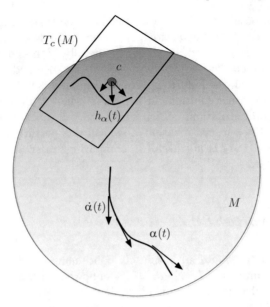

Fig. 11.13 Parallel translation of scaled $\dot{\alpha}(t)$ from $\alpha(t)$ to c to form the TSRVF $h_\alpha(t)$

We will assume that none of the observed trajectories pass through the cut locus of c to avoid the problem mentioned above.

Definition 11.2 (Transported Square-Root Vector Field). For any absolutely continuous trajectory $\alpha \in \mathscr{M}$, define its transported square-root vector field (TSRVF) to be a parallel transport of a scaled velocity vector field of α to a reference point $c \in M$ according to: $h_\alpha(t) = \frac{\dot{\alpha}(t)_{\alpha(t)\to c}}{\sqrt{|\dot{\alpha}(t)|_{\alpha(t)}}} \in T_c(M)$, where $|\cdot|$ denotes the norm related to the Riemannian metric on M.

This definition is illustrated in Fig. 11.13 using a spherical manifold M. Since α is absolutely continuous, the vector field h_α is square integrable. Let $\mathscr{H} \subset T_c(M)^{[0,1]}$ be the set of square-integrable curves in $T_c(M)$ obtained as TSRVFs of trajectories in M, $\mathscr{H} = \{h_\alpha | \alpha \in \mathscr{M}\}$. If $M = \mathbb{R}^n$ with the Euclidean metric, then h reduces to the SRVF defined in previous chapters.

The choice of reference point c used in Definition 11.2 is important in this framework and can potentially affect the results. The choice of c would typically depend on the application, the data and the manifold M under study. In case all the trajectories pass through a point or pass close to a point, then that point is a natural candidate for c. This would be true, for example, in the case of hurricane tracks, if we are focused on all hurricanes starting from the same region in the Atlantic Ocean. While the choice of c can, in principle, affect distances, some experiments suggest that the results of registration, distance-based clustering and classification are quite stable with respect to this choice. An example is presented later in Fig. 11.14.

We will mathematically represent a trajectory $\alpha \in \mathscr{M}$ by the pair $(\alpha(0), h_\alpha) \in (M \times \mathscr{H})$. Given this representation, we can reconstruct the path, an element of $\mathscr{M}$, as follows. For any time t, let V_t be a time-varying tangent vector field on M obtained by parallel transporting $h_\alpha(t)$ (along respective geodesics) over the whole M (except the cut locus of $\alpha(t)$), i.e., for any $p \in M$, $V_p(t) = |h_\alpha(t)|(h_\alpha(t))_{c\to p}$.

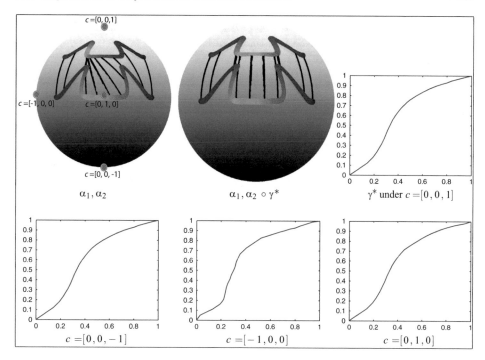

Fig. 11.14 Registration of trajectories on $\mathbb{S}^2$

Then, define an integral curve β such that $\dot{\beta}(t) = V_{\beta(t)}(t)$ with the starting point $\beta(0) = \alpha(0) \in M$. This resulting curve β will exactly be same as the original curve α.

The starting points of different curves can be compared using the Riemannian distance d_m on M. However, these points do not play an important role in the alignment of trajectories since they are already matched to each other. Therefore, the main focus of analysis, in terms of alignment, is on TSRVFs. Since a TSRVF is a path in $T_c(M)$, one can use the $\mathbb{L}^2$ norm to compare such paths.

Definition 11.3 (TSRVF Distance). Let α_1 and α_2 be two smooth trajectories on M and let h_{α_1} and h_{α_2} be the corresponding TSRVFs. The distance between them is:

$$d_h(h_{\alpha_1}, h_{\alpha_2}) \equiv \left(\int_0^1 |h_{\alpha_1}(t) - h_{\alpha_2}(t)|^2 dt \right)^{\frac{1}{2}}.$$

The distance d_h, being the standard $\mathbb{L}^2$ norm, satisfies symmetry, positive definiteness, and triangle inequality. Also, note that due to invertibility of mapping from $\mathscr{M}$ to $(M \times \mathscr{H})$, one can use d_h (along with d_m) to define a distance on $\mathscr{M}$. The main motivation of this setup—TSRVF representation and $\mathbb{L}^2$ norm—comes from the following fact. If a trajectory α is warped by γ, to result in $\alpha \circ \gamma$, the TSRVF of $\alpha \circ \gamma$ is given by:

$$h_{\alpha \circ \gamma}(t) = \frac{(\dot{\alpha}(\gamma(t))\dot{\gamma}(t))_{\alpha(\gamma(t)) \to c}}{\sqrt{|\dot{\alpha}(\gamma(t))\dot{\gamma}(t)|}} = \frac{(\dot{\alpha}(\gamma(t)))_{\alpha(\gamma(t)) \to c}\sqrt{\dot{\gamma}(t)}}{\sqrt{|\dot{\alpha}(\gamma(t))|}}$$
$$= h_{\alpha}(\gamma(t))\sqrt{\dot{\gamma}(t)}.$$

We will often use (h_α, γ) to denote $h_{\alpha \circ \gamma}$. As stated earlier, we need a distance for registration that is invariant to identical time warping of trajectories. Next, we show that d_h satisfies this property.

Theorem 11.1. *For any $\alpha_1, \alpha_2 \in \mathcal{M}$ and $\gamma \in \tilde{\Gamma}_I$, the distance d_h satisfies $d_h(h_{\alpha_1 \circ \gamma}, h_{\alpha_2 \circ \gamma}) = d_h(h_{\alpha_1}, h_{\alpha_2})$. In other words, the action of $\tilde{\Gamma}_I$ on $\mathcal{H}$ under the $\mathbb{L}^2$ metric is by isometries.*

Proof. Starting with the left side,

$$
d_h(h_{\alpha_1 \circ \gamma}, h_{\alpha_2 \circ \gamma}) = \left(\int_0^1 |h_{\alpha_1}(\gamma(t))\sqrt{\dot\gamma(t)} - h_{\alpha_2}(\gamma(t))\sqrt{\dot\gamma(t)}|^2 dt \right)^{\frac{1}{2}}
$$

$$
= \left(\int_0^1 |h_{\alpha_1}(s) - h_{\alpha_2}(s)|^2 ds \right)^{\frac{1}{2}} = d_h(h_{\alpha_1}, h_{\alpha_2}), \quad \text{where } s = \gamma(t). \qquad \square
$$

Next we define a quantity that can be used as a distance between trajectories while being invariant to their temporal variability. To set up this definition, we first introduce an equivalence relation between trajectories. For any two trajectories α_1 and α_2, we define them to be equivalent, $\alpha_1 \sim \alpha_2$, when:

1. $\alpha_1(0) = \alpha_2(0)$, and
2. there exists a sequence $\{\gamma_k\} \in \Gamma_I$ such that $\lim_{k \to \infty} h_{(\alpha_1 \circ \gamma_k)} = h_{\alpha_2}$; this convergence is measured under the $\mathbb{L}^2$ metric.

In other words, any two trajectories are equivalent if they have the same starting point and the TSRVF of one can be time-warped into the TSRVF of the other using a sequence of warpings. It can be easily checked that $\sim$ forms an equivalence relation on $\mathcal{H}$ (and correspondingly $\mathcal{M}$).

Since we want our distance to be invariant to time warping of trajectories, we wish to compare trajectories by comparing their equivalence classes. Thus, our next step is to inherit the distance d_h to the set of such equivalence classes. Toward this goal, we use the action of the monoid $\tilde{\Gamma}_I$. For a TSRVF $h_\alpha \in \mathcal{H}$, its equivalence class, or *orbit* under $\tilde{\Gamma}_I$, is given by $[h_\alpha] = \{(h_\alpha, \gamma) | h_\alpha \in \mathcal{H}, \ \gamma \in \tilde{\Gamma}_I\}$ It can be shown that the orbits under $\tilde{\Gamma}_I$ are exactly same as the closures of the orbits of Γ_I, defined as $[h_\alpha]_0 = \{(h_\alpha, \gamma) | \gamma \in \Gamma_I\}$, as long as α has nonvanishing derivatives. (The last condition is not restrictive since we can always re-parameterize α by the arc length.) The closure is with respect to the $\mathbb{L}^2$ metric on $\mathcal{H}$.

Now we are ready to define the quantity that will serve as both the cost function for registration and the distance for comparison. This quantity is essentially d_h measured between not the individual trajectories but their equivalence classes.

Definition 11.4 (Trajectory Shape Distance). Define a distance d_s on $\mathcal{H}/\sim$ by computing the shortest d_h distance between equivalence classes in $\mathcal{H}$:

$$
d_s([h_{\alpha_1}], [h_{\alpha_2}]) \equiv \inf_{\gamma_1, \gamma_2 \in \tilde{\Gamma}_I} d_h((h_{\alpha_1}, \gamma_1), (h_{\alpha_2}, \gamma_2))
$$

$$
= \inf_{\gamma_1, \gamma_2 \in \tilde{\Gamma}_I} \left(\int_0^1 |h_{\alpha_1}(\gamma_1(t))\sqrt{\dot\gamma_1(t)} - h_{\alpha_2}(\gamma_2(t))\sqrt{\dot\gamma_2(t)}|^2 dt \right)^{\frac{1}{2}}.
$$

$$(11.8)$$

Theorem 11.2. *The distance d_s is a proper distance on $\mathscr{H}/\sim$.*

Proof. The symmetry of d_s comes directly from the symmetry of d_h. For positive definiteness, we need to show that $d_s([h_{\alpha_1}], [h_{\alpha_2}]) = 0 \Rightarrow [h_{\alpha_1}] = [h_{\alpha_2}]$. Suppose that $d_s([h_{\alpha_1}], [h_{\alpha_2}]) = 0$. By definition, it follows immediately that for all $\epsilon > 0$, there exists a $\gamma \in \Gamma_I$ such that $d_h(h_{\alpha_1}, (h_{\alpha_2}, \gamma)) < \epsilon$. From this, it follows that h_{α_1} is in the orbit h_{α_2}. Since we are assuming that orbits are closed, it follows that $h_{\alpha_1} \in [h_{\alpha_2}]$, so $[h_{\alpha_1}] = [h_{\alpha_2}]$.

To establish the triangle inequality, we need to prove $d_s([h_{\alpha_1}], [h_{\alpha_3}]) \leq d_s([h_{\alpha_1}], [h_{\alpha_2}]) + d_s([h_{\alpha_2}], [h_{\alpha_3}])$, for any $h_{\alpha_1}, h_{\alpha_2}, h_{\alpha_3} \in \mathscr{H}$. Seeking contradiction, suppose that $d_s([h_{\alpha_1}], [h_{\alpha_3}]) > d_s([h_{\alpha_1}], [h_{\alpha_2}]) + \tilde{d}([h_{\alpha_2}], [h_{\alpha_3}])$. Let $\epsilon = \frac{1}{3}(d_s([h_{\alpha_1}], [h_{\alpha_3}]) - d_s([h_{\alpha_1}], [h_{\alpha_2}]) - d_s([h_{\alpha_2}], [h_{\alpha_3}]))$; by our supposition, $\epsilon > 0$. From the definition of ϵ, it follows that $d_s([h_{\alpha_1}], [h_{\alpha_3}]) = d_s([h_{\alpha_1}], [h_{\alpha_2}]) + d_s([h_{\alpha_2}], [h_{\alpha_3}]) + 3\epsilon$. By the definition of d_s, we can choose $\gamma_1, \gamma_2 \in \tilde{\Gamma}_I$, such that $d_h((h_{\alpha_1}, \gamma_1), h_{\alpha_2}) \leq d_s([h_{\alpha_1}], [h_{\alpha_2}]) + \epsilon$ and $d_h(h_{\alpha_2}, (h_{\alpha_3}, \gamma_2)) \leq d_s([h_{\alpha_2}], [h_{\alpha_3}]) + \epsilon$. Now by the triangle inequality for d_h, we know that $d_h((h_{\alpha_1}, \gamma_1), (h_{\alpha_3}, \gamma_2)) \leq d_h((h_{\alpha_1}, \gamma_1), h_{\alpha_2}) + d_h(h_{\alpha_2}, (h_{\alpha_3}, \gamma_2)) \leq d_s([h_{\alpha_1}], [h_{\alpha_2}]) + d_s([h_{\alpha_2}], [h_{\alpha_3}]) + 2\epsilon$. It follows that $d_s([h_{\alpha_1}], [h_{\alpha_3}]) \leq d_s([h_{\alpha_1}], [h_{\alpha_2}]) + d_s([h_{\alpha_2}], [h_{\alpha_3}]) + 2\epsilon$. But this contradicts that fact that $d_s([h_{\alpha_1}], [h_{\alpha_3}]) = d_s([h_{\alpha_1}], [h_{\alpha_1}]) + d_s([h_{\alpha_2}], [h_{\alpha_3}]) + 3\epsilon$. Hence our supposition that $d_s([h_{\alpha_1}], [h_{\alpha_3}]) > d_s([h_{\alpha_1}], [h_{\alpha_2}]) + d_s([h_{\alpha_2}], [h_{\alpha_3}])$ must be false. The triangle inequality follows. $\square$

Now, since Γ_I is dense in $\tilde{\Gamma}_I$, for any $\delta > 0$, there exists a $\gamma^* \in \Gamma_I$ such that:

$$|d_h(h_{\alpha_1}, h_{\alpha_2 \circ \gamma^*}) - d_s([h_{\alpha_1}], [h_{\alpha_2}])| < \delta . \tag{11.9}$$

This γ^* may not be unique but any such γ^* is sufficient for our purpose. Furthermore, since $\gamma^* \in \Gamma_I$, it has an inverse that can be used in further analysis. The minimization over Γ_I in Eq. 11.9 is performed in using the dynamic programming (DP) algorithm presented in Appendix B.

Our goal of warping-invariant comparisons of trajectories is achieved using d_s. For any $\gamma_1, \gamma_2 \in \Gamma_I$ and $\alpha_1, \alpha_2 \in \mathscr{M}$, we have

$$[h_{\alpha_1 \circ \gamma_1}] = [h_{\alpha_1}], \quad [h_{\alpha_2 \circ \gamma_2}] = [h_{\alpha_2}],$$

and, therefore, we get $d_s([h_{\alpha_1 \circ \gamma_1}], [h_{\alpha_2 \circ \gamma_2}]) = d_s([h_{\alpha_1}], [h_{\alpha_2}])$. The next goal is to perform registration of points along trajectories. Let our approximation to the optimal warping be as defined in Eq. 11.9. This solves for the registration between α_1 and α_2. It says that the point $\alpha_1(t)$ on the first trajectory is optimally matched to the point $\alpha_2(\gamma^*(t))$ on the second trajectory.

Equation 11.8 also solves for the registration between trajectories. Therefore, we are able to achieve our goal of joint registration and comparison (via a proper metric) in a unified fashion. In general registration methods, the cost function has two separate terms, one for matching and one for regularization, with an arbitrary weight parameter that is to be chosen by the user. However, in Eq. 11.8, the two terms have been merged into a single natural form. Recall that the change in TSRVF h due to the time warping of α by γ is given by $(h, \gamma) = (h \circ \gamma)\sqrt{\dot{\gamma}}$, and the distance d_s is based on these warped TSRVFs. It turns out that the term $\sqrt{\dot{\gamma}}$ provides an intrinsic regularization on γ in the matching process. This term provides an elastic penalty against excessive warping since $\dot{\gamma}$ becomes large at those places. Lastly, the optimal registration in Eq. 11.8 remains the same if we change the order of the input functions. That is, the registration process is inverse consistent!

11.4.1.1 Summarization and Registration of Multiple Trajectories

An additional advantage of this framework is that one can compute an average of several trajectories and use it as a *template* for future classification. Furthermore, this template can, in turn, be used for registering multiple trajectories. We will use the notion of the Karcher mean to define and compute average trajectories. Given a set of sample trajectories $\alpha_1, \ldots, \alpha_n$ on M, we represent them using the corresponding pairs $(\alpha_1(0), h_{\alpha_1}), (\alpha_2(0), h_{\alpha_2}), \ldots, (\alpha_n(0), h_{\alpha_n})$. We will compute the Karcher means of each component in its respective space: (1) the Karcher mean of $\alpha_i(0)$s are computed with respect to d_m in M, and (2) the Karcher mean of h_{α_i}s are with respect to d_s in $\mathscr{H}/\sim$. The latter Karcher mean is defined by: $[h_\mu] = \text{argmin}_{[h_\alpha] \in \mathscr{H}/\sim} \sum_{i=1}^{n} d_s([h_\alpha], [h_{\alpha_i}])^2$. Note that $[h_\mu]$ is actually an equivalence class of trajectories and one can select any element of this mean class to help in alignment of multiple trajectories. The standard algorithm to compute the Karcher mean, adapted to this problem, is as follows:

Algorithm 57 (Karcher Mean of Multiple Trajectories). *Compute the Karcher Mean of $\{\alpha_i(0)\}$s and set it to be $\mu(0)$.*

1. *Initialization step: Select μ to be one of the original trajectories and compute its TSRVF h_μ.*
2. *Align each h_{α_i}, $i = 1$, ,n, to h_μ according to Eq. 11.9. That is, solve for γ_i^* using the DP algorithm and set $\tilde{\alpha}_i = \alpha_i \circ \gamma_i^*$.*
3. *Compute TSRVFs of the warped trajectories, $h_{\tilde{\alpha}_i}$, $i = 1, 2, \ldots, n$, and update h_μ as a curve in $T_c(M)$ according to: $h_\mu(t) = \frac{1}{n} \sum_{i=1}^{n} h_{\tilde{\alpha}_i}(t)$.*
4. *Define μ to be the integral curve associated with a time-varying vector field on M generated using h_μ, i.e., $\frac{d\mu(t)}{dt} = (h_\mu)(t)_{c \to \mu(t)}$, and the initial condition $\mu(0)$.*
5. *Compute $E = \sum_{i=1}^{n} d_s([h_\mu], [h_{\tilde{\alpha}_i}])^2 = \sum_{i=1}^{n} d_h(h_\mu, h_{\tilde{\alpha}_i})^2$ and check it for convergence. If the change in E from the previous iteration is larger than the stopping criterion, then return to step 2.*

For computing and analyzing the second and higher moments of a sample trajectory, the tangent space $T_{\mu(t)}(M)$, for $t \in [0, 1]$, is used. This is convenient because it is a vector space and one can apply more traditional methods here. First, for each aligned trajectory $\tilde{\alpha}_i(t)$ at time t, the vector $v_i(t) \in T_{\mu(t)}(M)$ is computed such that a geodesic that goes from $\mu(t)$ to $\tilde{\alpha}_i(t)$ in unit time has the initial velocity $v_i(t)$. This is also called the *shooting vector* from $\mu(t)$ to $\tilde{\alpha}_i(t)$. Let $\hat{K}(t)$ be the sample covariance matrix of all the shooting vectors from $\mu(t)$ to $\tilde{\alpha}_i(t)$. The sample Karcher covariance at time t is given by

$$\hat{K}(t) = \frac{1}{n-1} \sum_{i=1}^{n} v_i(t) v_i(t)^T, \quad \text{with} \quad \hat{\rho}(t) = \text{trace}(\hat{K}(t)) . \qquad (11.10)$$

This $\hat{\rho}(t)$ represents a quantification of the cross-sectional variance, as a function of t, and can be used to study the level of alignment of trajectories. Also, for capturing the essential variability in the data, one can perform principal component analysis (PCA) of the shooting vectors. The basic idea is to compute the singular value decomposition (SVD) $\hat{K}(t) = U(t)\Sigma(t)U^T(t)$, where $U(t)$ is an orthogonal matrix and $\Sigma(t)$ is the diagonal matrix of singular values. Assuming that the entries along the diagonal in $\Sigma(t)$ are organized in a non-increasing order, $U_1(t), U_2(t)$, etc. represent the dominant directions of variability in the data.

11.4.2 Analysis of Trajectories on $\mathbb{S}^2$

To illustrate this framework, in a relatively simple setting, we take the case of $M = \mathbb{S}^2$, with the standard Euclidean Riemannian metric. For any two points p, $q \in \mathbb{S}^2$ ($p \neq -q$) and a tangent vector $v \in T_p(\mathbb{S}^2)$, the parallel transport $(v)_{p \to q}$ along the shortest geodesic (i.e., great circle) from p to q is given by $v - \frac{2\langle v, q \rangle}{|p+q|^2}(p+q)$.

Registration of Trajectories As mentioned earlier, for any two trajectories on $\mathbb{S}^2$, we can use their TSRVFs and the DP algorithm in Eq. 11.9 to find the optimal registration between them. In Fig. 11.14 we show one example of registering such trajectories. The parameterization of trajectories is displayed using colors. In the top row, the left column shows the given trajectories α_1 and α_2, the middle column shows α_1 and $\alpha_2 \circ \gamma^*$, and the right column shows γ^* using $c = [0, 0, 1]$. The correspondences between two trajectories are depicted by black lines connecting points along them. Due to optimization over γ in Eq. 11.9, the d_h value between them reduces from 1.67 to 0.36, and the correspondences appear quite reasonable after the alignment. We also try different choices of c's ($c = [0, 0, -1], [-1, 0, 0], [0, 1, 0]$). The registration results are very close despite different c's as shown in the bottom row.

To emphasize practical utility of this framework, we will apply it to two specific applications: bird migration data and hurricane tracks, and show how the cross-sectional variance of mean trajectories is reduced by registration. We use the mean of starting points of trajectories as the reference point c in Definition 11.2 for both applications.

Analysis of Bird Migration Data This dataset has 35 migration trajectories of Swainson's Hawk, measured from 1995 to 1997, each having geographic coordinates measured at some random times. Several sample paths are shown in the top row in Fig. 11.15a. In the bottom panel of Fig. 11.15a, we show the optimal warping functions $\{\gamma_i^*\}$ used in aligning them, and this clearly underscores the significant temporal variability present in the data. In Fig. 11.15b and c, we show the Karcher mean μ and the cross-sectional variance $\hat{\rho}$ without and with registration, respectively. For the definition of cress-sectional variance $\hat{\rho}$, please refer to Eq. 11.10. In the top row, $\hat{\rho}$ is displayed using colors, where red areas correspond to higher variability in the given data. In the bottom row, the principal modes of variation are displayed using ellipses on tangent spaces. We use the first and second principal tangential directions as the major and minor axes of ellipses and the corresponding singular values as their lengths. We observe that: (1) the mean after registration better preserves the shapes of trajectories and (2) the variance ellipses before registration have major axis along the trajectory, while the ellipses after registration exhibit the actual variability in the data. Most of the variability

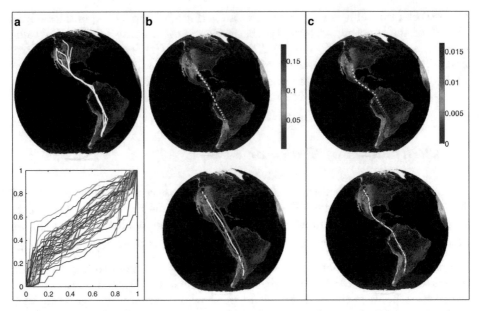

Fig. 11.15 Swainson's Hawk migration: (**a**) $\{\alpha_i\}$ (*top*) and $\{\gamma_i^*\}$ (*bottom*), (**b**) μ and $\hat{\rho}$ without registration, (**c**) μ and $\hat{\rho}$ with registration

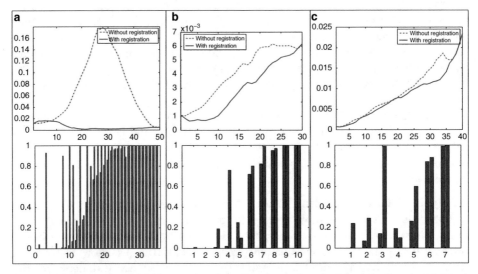

Fig. 11.16 Comparison of $\hat{\rho}$ (*first row*) and p-values (*second row*) without (*red*) and with (*blue*) registration. (**a**) Bird migration. (**b**) Hurricane tracks subset 1. (**c**) Hurricane tracks subset 2

after registration is limited to the top end where the original trajectories indeed had differences. The top row of Fig. 11.16a shows a decrease in the function $\hat{\rho}$ due to the registration.

Next we construct a "Gaussian-type" model for these trajectories using estimated summaries for the two cases (with and without temporal registration), as described previously, and compute p-values of individual trajectories using Monte Carlo simulation. The results are shown in the bottom of Fig. 11.16a, where we note a general increase in the p-values for the original trajectories after the alignment.

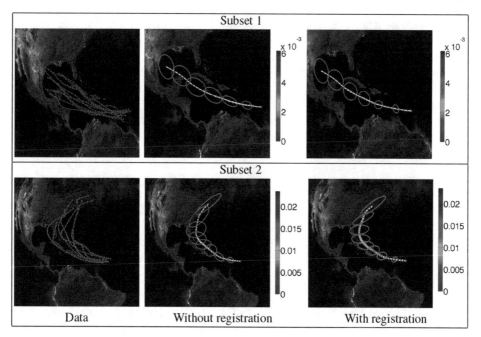

Fig. 11.17 Summary of hurricane tracks without and with temporal registration

This is attributed to a reduced variance in the model due to temporal alignment and the resulting movement of individual samples closer to the mean values.

Analysis of Hurricane Tracks We choose two subsets of Atlantic Tracks File 1851–2011, available on the National Hurricane Center website.[1] The first subset has ten tracks and another has seven tracks, with observations at 6-hour separation. We show the data, their Karcher mean, and variance without and with registration in Fig. 11.17 for each subset. The decrease in the value of $\hat{\rho}$ is shown in the top of Fig. 11.16b and c. Although the decrease here is not as large as the previous example, we observe about 20 % reduction in $\hat{\rho}$ in average due to registration. In the bottom plots of Fig. 11.16b and c, it is also shown that there is a general increase of p-values after registration although it decreases in a few cases.

11.5 Problems

11.5.1 Theoretical Problems

1. The pre-shape space for a joint analysis of curves using their orientation, scale and shape is the full $\mathbb{L}^2([0,1], \mathbb{R}^n)$. Show that a geodesic path in this space under the standard $\mathbb{L}^2$ metric is a straight line.
2. In the case where both shape and scale are included in the analysis of curves, the pre-shape space (before removal of $\tilde{\Gamma}_I$) is given by $\mathbb{L}^2([0,1], \mathbb{R}^n)/SO(n)$. Derive an expression for a geodesic path in this space.

---

[1] http://www.nhc.noaa.gov/pastall.shtml.

3. In the case where both shape and orientation are included in the analysis, the pre-shape space (before removal of $\tilde{\Gamma}_I$) is given by $\mathscr{C}_2$. Derive an expression for geodesic path in this space.

4. For the square-root function given in Definition 11.1, show that the action of the re-parameterization group $\tilde{\Gamma}_I$ is given by $(h, \gamma) = (h \circ \gamma)\sqrt{\dot{\gamma}(t)}$.

5. Show that the mapping $(GL(2) \ltimes \mathbb{R}^2) \times \mathbb{L}^2([0,1], \mathbb{R}^2) \to \mathbb{L}^2([0,1], \mathbb{R}^2)$, given by $((A,x), q)(t) \equiv (Aq(t) + x)$ forms an action of $GL(2) \ltimes \mathbb{R}^2$ on $\mathbb{L}^2([0,1], \mathbb{R}^2)$. Furthermore, verify that this action is not by isometries under the $\mathbb{L}^2$ metric. That is, in general, we have:
$$\langle q_1, q_2 \rangle \neq \langle ((A,x), q_1), ((A,x), q_2) \rangle, \quad (A,x) \in GL(2) \ltimes \mathbb{R}^2 .$$

6. Define $\Omega = \{\omega I_3 | \omega \in \mathbb{R}_+\}$ to be the set of 3×3 positive, scalar matrices and show that it is a subgroup of $GL(3)$. (In fact, it is a normal subgroup.) Define the natural action of Ω on $GL(3)$ according to the map $\Omega \times GL(3) \to GL(3)$ given by $(\omega, A) = \omega A$. Under this action, the orbit of any $A \in GL(3)$ is given by:
$$[A] = \{AB | B \in \Omega\} = \{\omega A | \omega \in \mathbb{R}_+\} .$$

Form the quotient space $PGL(3) \equiv GL(3)/\Omega = \{[A] | A \in GL(3)\}$ and show that PGL forms a group under matrix multiplication. (This set is called the *projective linear group* or *projective general linear group*.)

7. Show that the mapping $PGL(3) \times \mathbb{L}^2([0,1], \mathbb{R}^2) \to \mathbb{L}^2([0,1], \mathbb{R}^2)$ given by $([A], q)(t) \equiv \tilde{q}(t)$, where $\tilde{q}$ is defined using
$$\left(B, \begin{bmatrix} q(t) \\ 1 \end{bmatrix} \right) \mapsto \frac{B \begin{bmatrix} q(t) \\ 1 \end{bmatrix}}{\left(B \begin{bmatrix} q(t) \\ 1 \end{bmatrix} \right)_3} = \begin{bmatrix} \tilde{q}(t) \\ 1 \end{bmatrix}, \quad \text{for any} \quad B \in [A] .$$

forms an action of $PGL(3)$ on $\mathbb{L}^2([0,1], \mathbb{R}^2)$. (Here, $(\cdot)_3$ denotes the third element of a 3-vector. Note that this action is not defined for a q if, for some t, $\left(B \begin{bmatrix} q(t) \\ 1 \end{bmatrix} \right)_3 = 0$.) Furthermore, verify that this action is not by isometries under the $\mathbb{L}^2$ metric.

8. Show that the section $\mathscr{C}_a^c$ is not a section of the set $\mathscr{C}_2^c$.

9. Derive analytical expressions for the basis elements h_1, h_2, h_3, and h_4 given in Eqs. 11.4–11.5 and afterward.

10. Consider the problem of finding a group element $A^* \in GL(2)$ such that $\Sigma_{A^* \beta} \propto I_2$, for a given $\beta \in \mathscr{F}_2$. Show that it is sufficient to search for A^* only over the subspace $S(2) \cap GL(2)$, where $S(2)$ is the set of all 2×2 symmetric matrices.

11. Using the definition given in Sect. 11.4.1, show that $\tilde{\Gamma}_I$ acts on $\mathscr{M}$ according to $(\alpha, \gamma) = \alpha \circ \gamma$.

12. For a Riemannian manifold M, let d_m denote the geodesic distance between points and let $d_x(\alpha_1, \alpha_2) = \int_0^1 d_m(\alpha_1(t), \alpha_2(t))dt$ be the distance between trajectories on M. Show that, in general,
$$d_x(\alpha_1, \alpha_2) \neq d_x(\alpha_1 \circ \gamma, \alpha_2 \circ \gamma), \quad \text{for} \quad \gamma \in \Gamma_I .$$

13. Verify that the following is an equivalence relation. For any two trajectories α_1 and α_2, define $\alpha_1 \sim \alpha_2$, when:
    a. $\alpha_1(0) = \alpha_2(0)$, and
    b. there exists a sequence $\{\gamma_k\} \in \Gamma_I$ such that $\lim_{k \to \infty} h_{(\alpha_1 \circ \gamma_k)} = h_{\alpha_2}$; this convergence is measured under the $\mathbb{L}^2$ metric.

11.5.2 Computational Problems

1. Write a program to compute distances between any two curves in $\mathbb{R}^n$ using the following combination of features: (1) shape, orientation, and scale, (2) shape and orientation, (3) shape and scale, (4) shape.
2. Write a program to compare any two Euclidean curves based on all their features (shape, position, scale, and orientation) using the square-root representation.
3. Write a program to cluster a set of given Euclidean curves using distances derived in the previous two programs.
4. Write a program to affine-standardize any planar closed curve using Algorithm 53.
5. Write a program to implement the path-straightening algorithm for computing geodesic paths between planar curves as elements of $\mathscr{C}_a^c$.
6. Develop and implement an expression for geodesic path in $GL(n)$ for de-standardizing geodesics obtained in the last problem.
7. Write a program to compute TSRVF h_α of a given trajectory α on a Riemannian manifold M, using finite differences and parallel transport.
8. Implement the optimization suggested in Eqn. 11.9 for alignment of TSRVFs of any two trajectories on a Riemannian manifold M.
9. Implement Algorithm 57 to compute Karcher mean of multiple trajectories.
10. Write a program to compute the cross-sectional scalar variance function $\hat{\rho}(t)$ for a set of trajectories. Evaluate this function on a dataset before and after registration of trajectories.

11.6 Bibliographic Notes

The material on analyzing curves using shape and other features is primarily taken from the paper [57]. An application of this framework for clustering white matter fiber tracts is presented in [73].

Some of the early work on projective shape analysis of landmark configurations has been described in [51, 74, 52]. The material on affine-invariant elastic shape analysis of planar curves was first presented in [21]. This paper also studied projective shape analysis but only for landmark representations, not curves. The UMD dataset used here is take from [118].

An interesting problem of estimating smooth trajectories (splines) on nonlinear manifolds, from discrete-time and noisy observations, is discussed in [109]. The representation of trajectories using vector fields have been discussed by many papers. For instance, forming a curve in a Euclidean space using translation of vector field $\dot{\alpha}(t)$ to the starting point of a trajectory is described in [44]. The TSRVF representation was introduced by Su et al. in [111]. The use of bird migration data here is motivated by the work of Owen and Moore [86] who studied migration patterns for Swainson's Hawk birds and their effects on their immune system. This framework has more recently been applied for human action recognition using depth sensors in [8]. Also, different mathematical representations, involving SRVFs in the tangent spaces of starting points of curves, have been introduced for a more intrinsic analysis of trajectories [128].

Appendix A
Background Material

A.1 Basic Differential Geometry

In this section, we will introduce the concepts of differentiable manifolds, tangent spaces, exponential maps, and integral flows. We start with the question: What is a manifold? Loosely speaking, a manifold is a space that can be flattened locally, although this may not be possible globally. We will start by defining the finite-dimensional manifolds, taking a few steps to develop a formal definition.

Definition A.1. A **topology** on a set X is a collection $\mathscr{T}$ of subsets of X, called *open sets*, satisfying

1. X and the empty set $\emptyset$ are in $\mathscr{T}$.
2. The union of an arbitrary collection of open sets in $\mathscr{T}$ is in $\mathscr{T}$.
3. The intersection of a finite collection of open sets in $\mathscr{T}$ is in $\mathscr{T}$.

The pair $(X, \mathscr{T})$ is referred to as a topological space. Given a topological space $(X, \mathscr{T})$, a collection $\mathscr{U}$ of subsets of X forms a **basis** for the topology $\mathscr{T}$, if

1. Every set in $\mathscr{U}$ is open (i.e., $\mathscr{U} \subset \mathscr{T}$).
2. For every open set $V \in \mathscr{T}$ and for every $x \in V$, there exists a set $U \in \mathscr{U}$ such that $x \in U \subset V$.

One way to obtain a topological space is to start with a metric space, which is a set with a distance function defined on it.

Definition A.2. A **metric space** is a set X equipped with a **distance function** or a **metric** $d : X \times X \to \mathbb{R}$ such that for all $x, y, z \in X$:

1. $d(x, y) > 0$ for $x \neq y$ and $d(x, x) = 0$.
2. $d(x, y) = d(y, x)$.
3. $d(x, z) \leq d(x, y) + d(y, z)$.

Given a metric space X with a distance function d, and given $x \in X$ and $\epsilon > 0$, we define the open ϵ-ball centered at x by

$$B_\epsilon(x) = \{y \in X : d(x, y) < \epsilon\}.$$

© Springer-Verlag New York 2016
A. Srivastava, E.P. Klassen, *Functional and Shape Data Analysis*,
Springer Series in Statistics, DOI 10.1007/978-1-4939-4020-2

A metric on a space X gives rise to a very natural topology, called the **metric topology**, in which a set $U \subset X$ to defined to be open if for every $x \in U$, there exists an $\epsilon > 0$ such that $B_\epsilon(x) \subset U$. It is easy to verify that this defines a topology on X and that the set of open ϵ-balls in X provides a basis for this topology.

Definition A.3. A topological space is said to be **Hausdorff** if for every pair of distinct points $p, q \in X$, there exist disjoint open subsets $U \subset X$ containing p and $V \subset X$ containing q.

It is an easy exercise to show that every metric space is Hausdorff when endowed with the metric topology. Let X and Y be topological spaces. A map $f : X \to Y$ is said to be **continuous** if for every open set $U \subset Y$, the inverse image $f^{-1}(U)$ is open in X. If it is both one-to-one (injective) and onto (surjective), then it is called a bijective map. A continuous, bijective map $f : X \to Y$ with continuous inverse is called a **homeomorphism**. X and Y are homeomorphic to each other when there exists a homeomorphism between them. A **manifold** is a topological space that is locally homeomorphic to $\mathbb{R}^n$. To be more precise, we have the following definition. In this definition, for a point $p \in M$, any open set U containing p is termed a neighborhood of p.

Definition A.4 (Manifold). A topological space M is called a **manifold** of dimension n if:

1. It is Hausdorff,
2. It has a countable basis, and
3. For each point $p \in M$, there is a neighborhood U of p that is homeomorphic to an open subset of $\mathbb{R}^n$.

The last property is termed the **locally Euclidean** property of M. According to this property, for each $p \in M$, there exists an open neighborhood U of p and a mapping $\phi : U \to \mathbb{R}^n$ such that $\phi(U)$ is open in $\mathbb{R}^n$ and $\phi : U \to \phi(U)$ is a homeomorphism. The pair (U, ϕ) is called a *coordinate chart* for the points that fall in U; for any point $y \in U$, one can view the Euclidean coordinates $\phi(y) = (\phi_1(y), \phi_2(y), \ldots, \phi_n(y))$ as the coordinates of y. The dimension of the manifold M is n. This is a way of flattening the manifold locally. Using ϕ and ϕ^{-1}, one can move between the sets U and $\phi(U)$ and perform calculations in the more convenient Euclidean space. (See Fig. A.1.)

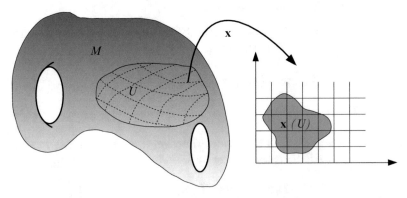

Fig. A.1 A coordinate chart $(U, \mathbf{x})$ on a manifold (The mapping should be ϕ not x)

In our applications of geometry, having just homeomorphic charts is not sufficient. We would like to be able to evaluate first and higher derivatives of functions
on our manifold. Two charts (U, ϕ) and (V, ψ) are called **smoothly compatible**
if the functions $\phi \circ \psi^{-1} : \psi(U \cap V) \to \phi(U \cap V)$ and $\psi \circ \phi^{-1} : \phi(U \cap V) \to \psi(U \cap V)$
(which are homeomorphisms between open sets in $\mathbb{R}^n$) are smooth, i.e., they have
continuous partial derivatives of all orders. A **smooth atlas** $\mathscr{A}$ is a collection
of charts which are pairwise smoothly compatible and whose domains cover M,
i.e. $M = \cup_{U \in \mathscr{A}} U$. For a given manifold, one can imagine having many different
atlases. A smooth atlas $\mathscr{A}$ on M is said to be **maximal** if every chart on M
which is smoothly compatible with all the charts in $\mathscr{A}$ is already in $\mathscr{A}$. A **smooth
structure** on a manifold M is a maximal smooth atlas. A manifold with such a
smooth structure is termed a **differentiable manifold** . The nice thing about a
differentiable manifold M is that we can calculate derivatives of functions on M in
terms of any coordinate chart we wish, and if we change coordinate charts, these
derivatives will be related by the chain rule.

Example A.1. 1. The Euclidean space $\mathbb{R}^n$ is an n-dimensional differentiable manifold which can be covered by the single chart $(\mathbb{R}^n, \phi)$, $\phi(x) = x$. Similarly, any
   open subset of $\mathbb{R}^n$ is also an n-dimensional differentiable manifold which can,
   again, be covered by a single chart.
2. Any open subset of a differentiable manifold is itself a differentiable manifold.
   (Obtain charts on the open set by intersecting it with charts on the original
   manifold.) A well-known example of this idea comes from linear algebra. Let
   $M(n)$ be the set of all $n \times n$ matrices; $M(n)$ can be identified with the set $\mathbb{R}^{n \times n}$
   and is, therefore, a differentiable manifold. Define the subset $GL(n)$ as the set
   of non-singular matrices, i.e.

$$GL(n) = \{A \in M(n) | \det(A) \neq 0\} \ ,$$

   where $\det(\cdot)$ denotes the determinant of a matrix. Since $GL(n)$ is an open
   subset of $M(n)$, it is also a differentiable manifold. It is, in fact, an important
   manifold and some of its subsets will play a crucial role in formulating invariance
   of shapes.
3. If M and N are two differentiable manifolds of dimensions m and n, respectively, then their Cartesian product $M \times N$ is also a differentiable manifold, of
   dimension $m + n$. For example, the Cartesian product $GL(n) \times \mathbb{R}^n$ is a manifold
   of dimension $n^2 + n$.
4. An important example of a differentiable manifold, both for our treatment of
   shapes and in general, is the unit sphere in $\mathbb{R}^{n+1}$:

$$\mathbb{S}^n = \{p \in \mathbb{R}^{n+1} | \sum_{i=1}^{n+1} p_i^2 = 1\}$$

To show that it is a differentiable manifold, we consider the case $n = 2$. The
local charts on $\mathbb{S}^2$ can be obtained using stereographic projection onto the plane.
Stereographic projection maps points on $\mathbb{S}^2$ to the xy plane in Euclidean space
$\mathbb{R}^3$. This projection is obtained by taking the intersection of the line connecting
the north pole $N = (0, 0, 1)$ and the point to be projected, with the xy plane,
and can be written mathematically as: $\phi^{-1} : \mathbb{R}^2 \to \mathbb{S}^2$ as

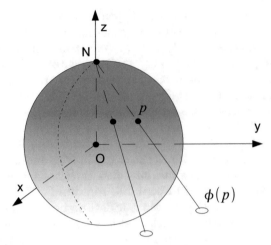

Fig. A.2 Illustration of the stereographic projection of $\mathbb{S}^2$

$$\phi^{-1}(u, v) \mapsto \left(\frac{2u}{u^2 + v^2 + 1}, \frac{2v}{u^2 + v^2 + 1}, \frac{u^2 + v^2 - 1}{u^2 + v^2 + 1} \right) .$$

Under ϕ, the lower hemisphere of $\mathbb{S}^2$ maps to the region inside the unit circle in the xy plane, the equator maps to the unit circle, and the upper hemisphere (except for N) maps to the region outside the unit circle. Figure A.2 shows the stereographic projection of a point p on the xy plane. This chart covers all of $\mathbb{S}^2$ except the north pole N. A similar chart can be constructed by projecting from the south pole $(0, 0, -1)$. The domain of this second chart is the complement of the south pole. These two charts are smoothly compatible and cover $\mathbb{S}^2$, so together they form a smooth atlas on $\mathbb{S}^2$.

In this book we will deal only with the differentiable manifolds and from now on we will refer to them simply as manifolds.

Shifting attention to mappings between manifolds, how does one define their smoothness? For Euclidean spaces, the smoothness of maps is defined using partial derivatives. A mapping $f : \mathbb{R}^m \to \mathbb{R}^n$ is smooth if all its partial derivatives exist and are continuous. Now in a more general case, let $f : M \to N$ be a mapping between two manifolds M and N. For any point $p \in M$, consider local charts (U, ϕ) with $p \in U$ and (V, ψ) with $f(p) \in V$. If the mapping: $\psi \circ f \circ \phi^{-1} : \phi(U \cap f^{-1}(V)) \to \psi(V)$ is smooth at the point $\phi(p)$, in the sense that all its partial derivatives exist and are continuous, then f is called **smooth** at the point p. Figure A.3 shows a pictorial illustration of this idea.

Because of the smooth compatibility of charts, this determination will not depend on which charts are used. If f is smooth for all points on M, then it is called a *smooth mapping* from M to N. Additionally, if f is a bijective smooth mapping with an inverse that is also smooth, then it is called a **diffeomorphism**. We will discuss these mappings on simple domains such as $[0, 1]$, $\mathbb{S}^1$, and $[0, 1]^2$ in detail later in this chapter. As an example, let a point on a unit circle be represented by $(\cos(\theta), \sin(\theta))$. We can define a mapping $f : \mathbb{S}^1 \to \mathbb{S}^1$ by $f(\cos(\theta), \sin(\theta)) = (\cos(\theta + 1), \sin(\theta + 1))$. This f is in fact a diffeomorphism from the unit circle to itself. (f is simply a rotation by one radian.) Later in Chap. 6, in the context of parameterized closed curves, the diffeomorphisms of this kind will be called translational re-parameterizations.

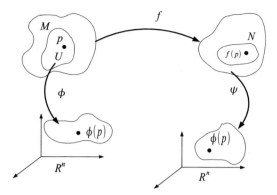

Fig. A.3 Illustration of compositions of smooth mappings

A.1.1 Tangent Spaces on a Manifold

In order to perform differential calculus, i.e. to compute gradients, directional derivatives, critical points, etc., of functions on manifolds, one needs to understand the tangent structure of those manifolds. The central idea here is to define tangent spaces at different points on the manifold and to relate tangent vectors with differential operators. Although there are several ways to define tangent spaces, one intuitive approach is to consider differentiable (C^1) curves on the manifold passing through the point of interest and to study the velocity vectors of these curves at that point. Different velocity vectors, corresponding to all possible curves, will be tangential to the manifold at that point and form a vector space that is termed as the tangent space at that point (see Fig. A.5). In simple words, the tangent space at a point is the set of all directions one can be traveling in while traversing the manifold at that point.

More formally, let M be an n-dimensional manifold and, for a point $p \in M$, consider a differentiable curve $\gamma : (-\epsilon, \epsilon) \to M$ such that $\gamma(0) = p$. Let (U, ϕ) be a coordinate chart that includes p. Since γ is differentiable and ϕ is smooth, the composition $\mu \equiv (\phi \circ \gamma) : (-\epsilon, \epsilon) \mapsto \mathbb{R}^n$ is also differentiable. The derivative $\dot{\mu}(0) \equiv \frac{d}{dt}\mu(t)|_{t=0}$ denotes the velocity of γ at p in local coordinates. This vector has the same dimension as the manifold M itself. For any two differentiable curves γ_1 and γ_2 passing through p at $t = 0$, define an equivalence relation by $\gamma_1 \sim \gamma_2$ if

$$\frac{d}{dt}\mu_1(0) = \frac{d}{dt}\mu_2(0) \ , \quad \text{where } \mu_i = \phi \circ \gamma_i, \ i = 1, 2 \ .$$

This defines an equivalence relation on the space of differentiable curves on M passing through p. Each equivalence class takes the form

$$[\gamma] = \{\beta : (-\epsilon, \epsilon) \to M | \ \beta(0) = p \text{ and } \frac{d}{dt}(\phi \circ \beta)(t)|_{t=0} = \frac{d}{dt}(\phi \circ \gamma)(t)|_{t=0}\} \ .$$

A **tangent vector** to M at p is defined to be one of these equivalence classes. The set of all such tangent vectors (equivalence classes) is called the **tangent space** to M at p, or $T_p(M)$. There is a bijective map $T_p(M) \to \mathbb{R}^n$ defined by $[\gamma] \mapsto \frac{d}{dt}\phi \circ \gamma|_{t=0}$. Using this map, one transfers the operations of addition and scalar multiplication from $\mathbb{R}^n$ to $T_p(M)$, making $T_p(M)$ a real vector space of

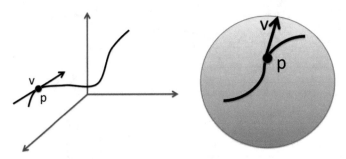

Fig. A.4 Tangent vector v defined as the instantaneous velocity of a C^1 curve passing through p for $\mathbb{R}^3$ and $\mathbb{S}^2$

dimension n. (Using the chain rule, it is easy to see that the structure of $T_p(M)$ as a vector space does not depend on which chart ϕ we used.) This is a very important fact that we will rely on heavily in statistical analysis of shapes. Even though the manifold M may be nonlinear, the tangent space $T_p(M)$ is always linear and one can impose probability models on it using more traditional approaches.

Example A.2. 1. In the case of the Euclidean space $\mathbb{R}^n$, the tangent space $T_p(\mathbb{R}^n) = \mathbb{R}^n$ for all $p \in \mathbb{R}^n$. To see this, consider a curve $\gamma : (-\epsilon, \epsilon) \to \mathbb{R}^n$ such that $\gamma(0) = p$; for this curve $\dot{\gamma}(0)$ will be a vector in $\mathbb{R}^n$; see the left panel of Fig. A.4. (Recall that for $\mathbb{R}^n$ the coordinate chart is simply the identity map $\phi(x) = x$.) In fact, for any vector $v \in \mathbb{R}^n$, we can form a curve $\gamma(t) = v\,t + p$, such that $\gamma(0) = p$ and $\dot{\gamma}(0) = v$. It follows that the tangent space to $\mathbb{R}^n$ at p, $T_p(\mathbb{R}^n)$, will be all of $\mathbb{R}^n$.

2. For $GL(n)$, the space of non-singular matrices and for an $A \in GL(n)$, let $\gamma(t)$ be a path in $GL(n)$ passing through $A \in GL(n)$ at $t = 0$. Its velocity vector at A, $\dot{\gamma}(0)$, is an element of $M(n)$, the set of all $n \times n$ matrices. As in the previous example, given any element $V \in M(n)$, we can construct a differentiable path $\gamma(t) = Vt + A$; hence, V is in the tangent space at that point. (One has to be careful to keep the domain of γ small enough to ensure that its image lies in $GL(n)$.) Thus, the tangent space $T_A(GL(n))$ is all of $M(n)$ and this holds for any $A \in GL(n)$.

3. For the unit circle, $\mathbb{S}^1 \subset \mathbb{R}^2$, let $\gamma : (-\epsilon, \epsilon) \to \mathbb{S}^1$ be a differentiable curve such that $\gamma(0) = p \equiv (p_1, p_2)$. Since $\mathbb{S}^1 \subset \mathbb{R}^2$, we can write $\gamma(t) = [\gamma_1(t) \; \gamma_2(t)]$, such that $\gamma_1(t)^2 + \gamma_2(t)^2 = 1$. Taking derivatives with respect to t, we get:

$$\gamma_1(t)\dot{\gamma}_1(t) + \gamma_2(t)\dot{\gamma}_2(t) = 0, \quad \text{or} \quad \langle \gamma(t), \dot{\gamma}(t) \rangle = 0 ,$$

where $\langle \cdot, \cdot \rangle$ denotes the Euclidean inner product. Setting $t = 0$, we see that the tangent vectors to $\mathbb{S}^1$ at p are orthogonal to p under the Euclidean inner product. Since all the orthogonal vectors take a specific form, we can write the tangent space as $T_p(\mathbb{S}^1) = \{\alpha(-p_2, p_1)|\alpha \in \mathbb{R}\}$. This can also be seen by taking the velocity vector of a curve, e.g. $\gamma(t) = (p_1 \cos(t\alpha) - p_2 \sin(t\alpha), p_1 \sin(t\alpha) + p_2 \cos(t\alpha))$. γ is a differentiable curve such that $\gamma(0) = (p_1, p_2)$ and $\dot{\gamma}(0) = \alpha(-p_2, p_1)$.

4. Similar to the circle, the tangent space of $\mathbb{S}^n$ at a point p is the set of all velocity vectors associated with the C^1 curves passing through p (see the right panel of Fig. A.4 for an example). The resulting space is an n-dimensional hyperplane

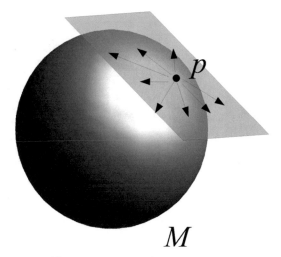

Fig. A.5 Tangent space to M at p

in $\mathbb{R}^{n+1}$ orthogonal to the vector p under the Euclidean metric:

$$T_p(\mathbb{S}^n) = \{v \in \mathbb{R}^{n+1} | \langle v, p \rangle = 0\}$$

Among other things, the tangent vectors play an important role in computing derivatives of functions where they are called *derivations*. In fact, one can completely develop tangent vectors and their spaces from this point of view. Let $C^\infty(p)$ be the set of all functions from $\mathbb{R}^n \mapsto \mathbb{R}$ that are smooth at $p \in \mathbb{R}^n$. For a point $p \in \mathbb{R}^n$, a linear map $X : C^\infty(p) \to \mathbb{R}$ is called a **derivation** at p if it satisfies the Leibnitz rule:

$$X(fg) = X(f)g(p) + X(g)f(p), \text{ for all } f, g \in C^\infty(p)$$

For instance, the partial derivative operators $\frac{\partial}{\partial x_i}$ are examples of derivations. It turns out that the space of all such derivations can be identified with the tangent space $T_p(M)$, and there is a one-to-one correspondence between tangent vectors and derivations. Since $T_p(\mathbb{R}^n) = \mathbb{R}^n$, any tangent vector v is simply an element of $\mathbb{R}^n$. To identify it with a derivation, define the mapping from $\mathbb{R}^n$ to the space of derivations as $v \mapsto X_v \equiv \sum_{i=1}^n v_i \frac{\partial}{\partial x_i}$. This mapping is a bijection and can be used to interchange between the two interpretations of tangent vectors. Associated with any tangent vector v is a unique derivation X_v and vice versa. This association underlines the usage of tangent vectors for computing directional derivatives. As mentioned earlier, a tangent vector provides a valid direction of traversal on the manifold, while a derivation provides derivatives of functions. Their identification naturally leads to the notion of **directional derivatives**. For example, for a function $f \in C^\infty(p)$, the notation $vf(p)$ stands for the directional derivative of f in the direction of v at p and is defined to be $vf(p) \equiv X_v f(p) = \sum_{i=1}^n v_i \frac{\partial f}{\partial x_i}(p)$. This can also be done by computing the directional derivative of the composition $f \circ \gamma(t)$, where γ is a differentiable curve in $\mathbb{R}^n$ such that $\gamma(0) = p$ and $\dot{\gamma}(0) = v$, and then evaluating its derivative at $t = 0$, i.e. $vf(p) = \frac{d(f \circ \gamma)}{dt}|_{t=0}$.

This identification between tangent vectors and derivations extends more generally to arbitrary differentiable manifolds (see Fig. A.6). For a manifold M and a

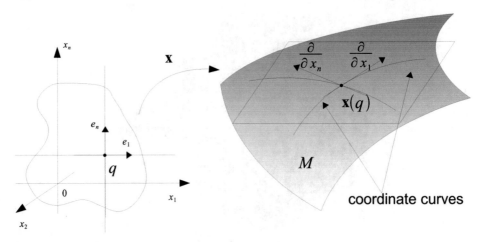

Fig. A.6 Coordinate curves w.r.t. the parametrization $\mathbf{x}$

point $p \in M$, let $C^{\infty}(p)$ be the set of all smooth functions $M \mapsto \mathbb{R}$ whose domain
includes p. As in the case of $\mathbb{R}^n$, define a linear transformation $X : C^{\infty}(p) \to \mathbb{R}$ to
be a **derivation** at p if it satisfies the Leibnitz rule

$$X(fg) = X(f)g(p) + X(g)f(p).$$

To see how each tangent vector in $T_p(M)$ corresponds to a derivation at p, suppose
$\gamma : (-\epsilon, \epsilon) \to M$ is a differentiable curve such that $\gamma(0) = p$, and let $v = [\gamma]$, the
equivalence class of C^1 curves with the velocity v at that point. Then, define

$$X_v(f) = \frac{d}{dt}(f \circ \gamma(t))|_{t=0} .$$

It is not difficult to verify that X_v is a derivation at p and that the correspondence
$v \mapsto X_v$ provides a bijection between $T_p(M)$ and the set of derivations on M at p.
So, one can interchangeably consider a tangent vector as a derivative operator for
functions at p or velocity vector of a curve passing through p.

Example A.3. 1. For any vector $v \in \mathbb{R}^n$ and a smooth function $f : \mathbb{R}^n \mapsto \mathbb{R}$, the directional derivative of f in the direction of v is given by $vf(p) = \sum_{i=1}^{n} v_i \frac{\partial f}{\partial x_i}(p)$.
2. Similarly, in $GL(n)$, the directional derivative of a smooth function f :
   $GL(n) \mapsto \mathbb{R}$ at $A \in GL(n)$, in the direction $V \in M(n)$ is given by:
   $Vf(A) = \sum_{i,j=1}^{n} V_{ij} \frac{\partial f}{\partial x_{ij}}(A)$.
3. On a unit circle, let $f : \mathbb{S}^1 \to \mathbb{R}$ be a smooth function; we will now compute the
   directional derivative of f in the direction of $\alpha(-p_2, p_1) \in T_P(\mathbb{S}^1)$, at $p \in \mathbb{S}^1$.
   Define a curve $\gamma(t) = (p_1 \cos(t\alpha) - p_2 \sin(t\alpha), p_1 \sin(t\alpha) + p_2 \cos(t\alpha))$ passing
   through p at $t = 0$. Note that $\dot{\gamma}(0) = \alpha(-p_2, p_1)$. Thus, the directional derivative
   of f at p is:
   $$\frac{d(f \circ \gamma)}{dt}|_{t=0} = (-p_2 \frac{\partial f}{\partial p_1} + p_1 \frac{\partial f}{\partial p_2})(p) .$$

(To make sense of this formula, we assume that f is not only defined on $\mathbb{S}^1$,
but it has been extended to a small neighborhood of $\mathbb{S}^1$ in $\mathbb{R}^2$. Otherwise, the
derivatives $\frac{\partial f}{\partial p_1}$ and $\frac{\partial f}{\partial p_2}$ would not be defined.) On more complicated manifolds,
it is not straightforward to come up with curves, associated with given tangent

vectors, that can be used to evaluate directional derivatives. Later on we will introduce the notion of integral curves that can be used to define and compute directional derivatives of functions. We postpone additional examples for later.

A.1.2 Submanifolds

Although many sets of interest can be studied intrinsically as manifolds, some others are better seen as submanifolds of larger manifolds. This will be especially useful in the case of infinite-dimensional manifolds where most of the sets we will encounter will be treated as submanifolds of larger better-known spaces.

Definition A.5 (Submanifold). Let M be a smooth m-dimensional manifold. A subset $N \subset M$ is an n-dimensional **submanifold** if for every $p \in N$, there is a coordinate chart (U, ϕ) for M, with $p \in U$, such that $N \cap U = \phi^{-1}(\mathbb{R}^n \times \{\mathbf{0}\}^{(m-n)})$.

This means that although under the chart ϕ, the points of $U \subset M$ correspond to points of $\mathbb{R}^m$, and the points in the subset $N \cap U$ correspond to points of $\mathbb{R}^m$ for which the last $m - n$ coordinates are 0.

Next we introduce the notion of a differential which is important in defining the submanifolds of interest to us.

Definition A.6 (Differential of a Mapping). The differential of a smooth mapping $f : M \to N$ at $p \in M$, denoted by df_p, is a linear map $df_p : T_p(M) \to T_{f(p)}(N)$ specified as follows. Let $g : N \to \mathbb{R}$ be a smooth function. Then, for any $v \in T_p(M)$, define

$$(df_p(v))(g) = v(g \circ f)(p) .$$

Here, $df_p(v) \in T_{f(p)}(N)$ is treated as a derivation which when applied to g results in the directional derivative of g. According to the right side, this derivative is defined by forming a new function $f \circ g : M \to \mathbb{R}$ and computing its directional derivative at p using v. In other words, the directional derivative of g on N is defined as the directional derivative of the composition $f \circ g$ on M.

Example A.4. 1. In case $M = \mathbb{R}^m$ and $N = \mathbb{R}^n$, and for any $f : M \to N$, the differential $df_p : \mathbb{R}^m \to \mathbb{R}^n$ is represented by the matrix of first partial derivatives of the components of f; $df_p = \{\frac{\partial f_i}{\partial x_j}\}$.

2. Let $M = GL(n)$ and $N = \mathbb{R}$ and define $f : M \to N$ by $f(A) = \frac{1}{2}\text{trace}(A^T A)$. Given a tangent vector $V \in T_A(GL(n)) \equiv M(n)$, we compute the directional derivative:

$$Vf(A) = \frac{d}{dt}\left(\frac{1}{2}\text{trace}((A + tV)^T(A + tV))\right)\Big|_{t=0} = \text{trace}(V^T A) .$$

The differential of f, $df_A : M(n) \to \mathbb{R}$, is given by the mapping $df_A(V) = \text{trace}(V^T A)$.

If we think of tangent vectors at p as equivalence classes of curves passing through p that share the same derivative at p, we can define the differential by $df_p([\gamma]) = [f \circ \gamma]$. The right side is an equivalence class of differentiable curves passing through an $f(p)$ in N. This definition is equivalent to the one just given.

As we will see, the rank of the differential of f is an important descriptor of the behavior of f near p.

We now describe a useful way of obtaining submanifolds. A point $p \in M$ is defined to be a **critical point** of f if the differential df_p is not onto and is called a **regular point** if df_p is onto. The image of a critical point, $f(p)$, is called a **critical value** of f. Any point $q \in N$ that is not a critical value is said to be a **regular value** of f, i.e., a point $q \in N$ is a regular value if every point in $f^{-1}(q)$ is a regular point.

Example A.5. 1. For $f : \mathbb{R} \mapsto \mathbb{R}^2$ given by $t \mapsto (\sin(t), \cos(t))$, the differential is $df_t(x) = (\cos(t)x, -\sin(t)x)$. Since this df_t is not onto (no linear map $\mathbb{R} \to \mathbb{R}^2$ is onto!), all the points in $\mathbb{R}$ are critical points of f.

2. If we define $f : \mathbb{R}^2 \mapsto \mathbb{R}$ by $f(x, y) = x^2 + y^2$, then we get a differential $df_{(x,y)}(u, v) = 2xu + 2yv$. This $df_{(x,y)}$ is onto at every point (x, y) except the origin $(0, 0)$. Hence, $(0, 0)$ is a critical point of f, while every other point of $\mathbb{R}^2$ is a regular point.

Theorem A.1. *Suppose M and N are manifolds of dimensions m and n, respectively, and let $f : M \to N$ be a smooth map, with a regular value $y \in N$. Then $f^{-1}(y)$ is a submanifold of M of dimension $m - n$. Furthermore, the tangent space of $f^{-1}(y)$ at a point p is given by the kernel of df_p.*

(Note that the kernel of a linear transform, denoted by $\ker(\cdot)$, is the set of vectors in its domain whose image under that mapping is zero. For example, the kernel of df_p is given by $ker(df_p) = \{x \in T_p(M) | df_p(x) = 0\}$.) An elementary proof of this theorem can be found in [78] pp.11.

Example A.6. 1. **Unit Sphere**: Using this theorem, let us check if $\mathbb{S}^n$ is indeed a submanifold of $\mathbb{R}^{n+1}$. Let $f : \mathbb{R}^{n+1} \to \mathbb{R}$ be a map given by $f(p) = \sum_{i=1}^{n+1} p_i^2$, where $p = (p_1, \ldots, p_{n+1})$. The differential of f is given $df_p(u) = 2\langle p, u \rangle$, which is clearly onto for all $p \in f^{-1}(1)$. Thus, 1 is a regular value of f and the set $f^{-1}(1)$ given by $\mathbb{S}^n$ is an n-dimensional submanifold of $\mathbb{R}^{n+1}$. Also, the tangent space $T_p\mathbb{S}^n = \ker(df_p)$, which is just the orthogonal complement of $p \in \mathbb{R}^{n+1}$, confirming a result we first derived in Example A.2.

2. **Orthogonal Matrices**: We now consider the set $O(n)$ of orthogonal matrices, which is a subset of the manifold $GL(n)$. We define $O(n)$ to be the set of all $n \times n$ invertible matrices A that satisfy $AA^T = I$. We will apply Theorem A.1 to prove that $O(n)$ is a submanifold of $GL(n)$. To do this, define $S(n)$ to be the set of $n \times n$ symmetric matrices, and then define $f : GL(n) \to S(n)$ by $f(A) = AA^T$. It can easily be shown that I is a regular value of f, and, hence, $f^{-1}(I) = O(n)$ is a submanifold of $GL(n)$. Note that $O(n)$ is not connected but has two components: those orthogonal matrices with determinant $+1$ and those with determinant -1. The set of orthogonal matrices with determinant 1 is called the **special orthogonal group** and denoted by $SO(n)$. The dimension of $O(n)$ can be determined by the above theorem; it is $n^2 - n(n+1)/2 = n(n-1)/2$.

Since $O(n)$ is a very important example which will be of use to us later, we examine it in a little more detail. First, note that if A is an element of $O(n)$, so is A^{-1}; also, if A and B are elements of $O(n)$, so is AB. By definition, this makes $O(n)$ a group (see Sect. A.2). Let us calculate $T_I O(n)$, the tangent space of $O(n)$ at the identity matrix. We compute

$$df_I(X) = \frac{d}{dt}(I + tX)(I + tX)^T|_{t=0} = X + X^T.$$

It follows that $T_I O(n) = \ker(df_I) = \{X : X + X^T = 0\}$, which is the set of skew-symmetric matrices. We now determine the tangent space of $O(n)$ at an arbitrary matrix $A \in O(n)$. Since left multiplication by A is a diffeomorphism $O(n) \to O(n)$ (its inverse is left multiplication by A^{-1}), it follows that $T_A O(n) = \{AX | X \text{ is skew symmetric}\}$.

A.2 Basic Algebra

An important property in our notion of shape is its invariance under operations like rotations, translations, and rescalings of objects. To take such invariance into account in a mathematical framework, the main mathematical tool is group theory. Here one considers the set of all transformations, for example rotations, as a group, where the group operation is concatenation (or composition) of two transformations. For instance, if an object is transformed by two rotations, one after another, what is the cumulative effect of these two transformations? It will be a rotation that is composed of the original two rotations combined through a group operation. The same holds for the translations, scalings, and others. In this section, we introduce the definitions and concepts associated with group theoretic representations of transformations of shapes.

We start with the definition of a group.

Definition A.7 (Group). A group G is a set having an associative binary operation, denoted by $\cdot$, such that:

1. there is an element e in G such that $e \cdot g = g \cdot e = g$ for all $g \in G$, and
2. for every $g \in G$, there exists a unique h such that $g \cdot h = h \cdot g = e$.

e is called the identity element of G and h is called the inverse of g, denoted by g^{-1}.

Consider some simple examples of groups:

Example A.7. • The set of all real numbers is a group with addition being the binary operation, and the identity element being 0. Similarly, $\mathbb{R}^n$ is a group with vector addition as the group operation and the zero vector as the identity element. This group is called the **translation group** for the following reason. If we take an object defined with respect to some coordinate system and add a vector $v \in \mathbb{R}^n$ to all points of this object, the result will be the same object located at a new point in that coordinate system, i.e. the object will be translated by the vector v. Two translations v_1 and v_2, one after another, will result in a total translation of $v_1 + v_2$ in the translation group. Addition by a zero vector does not change the location of the object.

• The set of positive real numbers is a group with binary operation being multiplication and the identity element being one. We will denote this group by $\mathbb{R}_+^\times$. It is called the **scaling group** because for any $a \in \mathbb{R}_+^\times$, a vector av, for a $v \in \mathbb{R}^n$ is simply a scaled version of v. Two scalings, a_1 and a_2, applied to the same vector result in a single scaling by $a_1 a_2$, and a scaling by $a = 1$ does not change v.

• Consider the set of $n \times n$ non-singular matrices, $GL(n)$, studied earlier. $GL(n)$ is a group with the group operation being matrix multiplication and the identity element being the $n \times n$ identity matrix I. Due to this group structure, $GL(n)$ is also called the general linear group. This is the group of all non-singular **linear transformations** on $\mathbb{R}^n$—any $A \in GL(n)$ is a linear transformation on $\mathbb{R}^n$. In other words, for any $a_1, a_2 \in \mathbb{R}$ and $v_1, v_2 \in \mathbb{R}^n$, we have $A(a_1 v_1 + a_2 v_2) = a_1(Av_1) + a_2(Av_2)$.

- The direct product of any two groups is also a group. If G_1 and G_2 are two groups with their respective group operations, then $G = G_1 \times G_2$ is a group with the group operation $(g_1, h_1) \cdot (g_2, h_2) = (g_1 \cdot g_2, h_1 \cdot h_2)$. For example, $GL(n) \times \mathbb{R}^n$ is a group with the group operation $(A_1, v_1) \cdot (A_2, v_2) = (A_1 A_2, v_1 + v_2)$.

Some additional properties of a group can be useful in the analysis of shapes. A group is called **abelian** if it satisfies the commutative law, that is $g \cdot h = h \cdot g$ for all $g, h \in G$. Otherwise, it is called a **non-abelian group**. As an example, the real line with addition operation is an abelian group, while $GL(n)$ is a non-abelian group for $n > 1$.

$GL(n)$ is a rather large group containing many types of linear transformations that can be applied to elements of $\mathbb{R}^n$. To focus on more specific transformations, we introduce the notion of subgroups. A subset S of a group G is said to be **closed** under multiplication if for every pair $g, h \in S$, the product is also in S, i.e. $g \cdot h \in S$.

Definition A.8 (Subgroup). A subset H of a group G is said to be a subgroup if it is nonempty, closed under the group operation, and for each $g \in H$, the inverse g^{-1} is also in H.

Example A.8. • The set $\mathbb{Z}$ of all integers is a subgroup of $\mathbb{R}$ under addition.
- There are several interesting subgroups of $GL(n)$ that will be central to shape analysis. For instance, consider the set of all $n \times n$ matrices that have determinant $+1$:

$$SL(n) = \{A \in GL(n) \big| \det(A) = +1\}$$

$SL(n)$ is called the **special linear group** and is a subgroup of $GL(n)$. It is a subgroup because: (i) it is closed, i.e. for $A_1, A_2 \in SL(n)$, we have $\det(A_1 A_2) = \det(A_1) \det(A_2) = 1$ and (ii) the inverse of $A \in SL(n)$ is in $SL(n)$, i.e. $\det(A^{-1}) = (\det(A))^{-1} = 1$. $SL(n)$ is the set of all volume-preserving, linear transformations of $\mathbb{R}^n$. The volume enclosed by a parallelepiped formed by vectors $v_1, v_2, \ldots, v_n \in \mathbb{R}^n$ is given by the determinant of matrix V with columns given by v_i. So $\det(AV) = \det(A) \det(V) = \det(V)$ if $A \in SL(n)$.
- **Rotation Group**: Consider the orthogonal group, $O(n) = \{A \in GL(n) | A^T A = I\}$, and the special orthogonal group, $SO(n) = \{A \in GL(n) | det(A) = +1 | A^T A = I_n\}$, which we have introduced as examples of submanifolds of $GL(n)$ in Example A.6. $SO(n)$ is a subgroup of $GL(n)$, $O(n)$ and $SL(n)$. $SO(n)$ is important in shape analysis because it is the set of all rotations of $\mathbb{R}^n$. If $x \in \mathbb{R}^n, O \in SO(n)$, then $O \cdot x$ is simply a rotation of x; it does not change the length of x, i.e. $\|Ox\|^2 = x^T O^T O x = x^T x = \|x\|^2$. If $O_1, O_2 \in SO(n)$ denote two rotations of an object, then $O_1 \cdot O_2$ denotes the cumulative rotation applied to that object. Similarly, for $O \in SO(n)$, the inverse $O^{-1} = O^T$ provides the rotation to undo the effect of O on an object. For $n = 2$, elements of $SO(2)$ take the form: $\begin{pmatrix} \cos(\theta) & -\sin(\theta) \\ \sin(\theta) & \cos(\theta) \end{pmatrix}$. Recall that $SO(n)$ is an $n(n-1)/2$-dimensional manifold; thus, $SO(2)$ is one dimensional and $SO(3)$ is three dimensional.

Next, we look at the mappings from one group to another.

Definition A.9 (Homomorphism). For two groups G_1 and G_2, a mapping $f : G_1 \rightarrow G_2$ is a **homomorphism** if $f(g \cdot h) = f(g) \cdot f(h), \forall g, h \in G_1$.

The operation on the left is performed in G_1 while the operation on the right is performed in G_2. A homomorphism implies that the group operation can be performed either before or after the mapping f is applied, without changing the result. The evaluation of determinant of a matrix, as a mapping from $GL(n)$ to $\mathbb{R}^{\times}$, is in fact, a homomorphism under the respective group operations, since $\det(A_1 A_2) = \det(A_1) \det(A_2)$. As a counterexample, the exponential of a matrix: $\exp : M(n) \mapsto GL(n)$, defined earlier, is not a homomorphism since $\exp(A + B) \neq \exp(A) \exp(B)$ in general.

Definition A.10 (Isomorphism). An **isomorphism** between two groups is a homomorphism $f : G_1 \rightarrow G_2$ that is one-to-one and onto. G_1 is said to be isomorphic to G_2, and vice versa, if there exists an isomorphism from G_1 to G_2.

If $f : G_1 \rightarrow G_2$ and $g : G_2 \rightarrow G_3$ are two isomorphisms, then $(g \circ f) : G_1 \rightarrow G_3$ is also a isomorphism. If f is an isomorphism, then f^{-1} is also an isomorphism.

As mentioned earlier in Sect. 3.1, the notion of equivalence relations is important in introducing invariance of shapes. An equivalence relation $\sim$ on a set X leads to a quotient set $X/\sim$ whose elements are equivalence classes under that relation; elements of an equivalence class are deemed equivalent under that relation. In group theory, an equivalence relation occurs naturally due to the group operations. This involves subgroups and cosets of subgroups that are defined next.

Definition A.11 (Coset). Let H be a subgroup of G. For any element $g \in G$, define a left **coset** of H in G by $gH = \{g \cdot h | h \in H\}$.

In general, the cosets are not subgroups and the only coset that is a subgroup of G is H itself (eH). For different elements g_1 and g_2, the cosets $g_1 H$ and $g_2 H$ will either be identical or disjoint. They will be identical when $g_2^{-1} g_1$ is an element of H; otherwise they will be disjoint. This is similar to an equivalence relation that partitions a set into disjoint equivalence classes. In fact, one can define an equivalence relation using membership of these cosets: we define $g_1 \sim g_2$ if $g_1 \in g_2 H$, i.e. $g_1 = g_2 h$ for some $h \in H$. In the notation of equivalence classes, we have $[g] = gH$. The quotient space $G/\sim$, also denoted by G/H to emphasize the role of H in defining $\sim$, is the set of all left cosets of H in G. In general, the quotient space is not a group; however, an exception results when H is a *normal* subgroup. This case is presented later in this chapter. The quotient space G/H is also called the space G *modulo* H, or the space that results when H *is removed from* G.

Similar to the left cosets, one can define the right cosets, $Hg = \{h \cdot g | h \in H\}$, and an equivalence relation based on right cosets. If G is an abelian group ($g_1 \cdot g_2 = g_2 \cdot g_1$), then the left and the right cosets coincide. The rotation group $SO(2)$ is abelian, since for any $O_1, O_2 \in SO(2)$, $O_1 O_2 = O_2 O_1$. In this case the left and the right cosets are same. On the other hand $SO(3)$ is non-abelian, i.e. $O_1, O_2 \in SO(3), O_1 O_2 \neq O_2 O_1$ and, consequently, the left and the right cosets are different. Here are some examples of the left cosets:

Example A.9. 1. We know that the rotation group $H = SO(n)$ forms a subgroup of the generalized linear group $GL(n)$. The left cosets of H are given by: $A \, SO(n) = \{AO | O \in SO(n)\}$, for $A \in GL(n)$. The quotient space is the set of such cosets: $G/H = \{A\cdot SO(n) | A \in GL(n)\}$. A physical interpretation of this operation is the following. Let n rows of a matrix $A \in GL(n)$ denote n points in $\mathbb{R}^n$. This set can be used to denote a rigid object represented by n landmarks. For example, for $n = 3$ the three rows of A represent the three vertices of a triangle in $\mathbb{R}^3$. Then, the coset $A \cdot SO(3)$ consists of all sets obtained by rigidly rotating this triangle about the origin. By rigid rotation we mean that all the points in A are rotated by the same rotation.

2. An interesting quotient space results when a set of lower dimensional rotations is removed from a set of higher dimensional rotations. Let G be the rotation group $SO(n)$, and H be the rotation group $SO(n-d)$ for $d < n$. H is a subgroup of G with the embedding:

$$SO(n-d) \longrightarrow \left\{ \begin{bmatrix} I_d & | & \\ -- & - & - \\ & | & O \end{bmatrix} | O \in SO(n-d) \right\}$$

The left cosets H in G are

$$V\ SO(n-d) = \left\{ V \begin{bmatrix} I_d & | & \\ -- & - & - \\ & | & O \end{bmatrix} | O \in SO(n-d) \right\},$$

for $V \in SO(n)$, with the quotient space G/H being the set of left cosets $V\ SO(n-d)$ for all $V \in SO(n)$. How can we interpret this quotient space? Take an $n \times n$ rotation matrix V and consider the set $V\ SO(n-d)$ as defined above. Each element of this set is denoted an $n \times n$ matrix whose first d columns are the same as those of V and the remaining $n-d$ columns have been rotated by a $(n-d) \times (n-d)$ rotation matrix. Setting all these elements of $V\ SO(n-d)$ equivalent, by putting them in a left coset, results in an equivalence class of matrices whose first d columns are same and the remaining columns include all possible ways of completing the first d columns to a positively oriented orthonormal basis. In other words, every $W \in V\ SO(n-d)$ is of the type $W = [V_1\ W_2]$, where $W_2 = V_2 O$ for some $O \in SO(n-d)$. Hence, each coset represents a unique $n \times d$ matrix whose columns are orthonormal, and the quotient space $SO(n)/SO(n-d)$ is nothing but the set of all such matrices. This set is also called a **Stiefel manifold** . Summarizing this discussion, the quotient set of $SO(n)$ modulo $SO(n-d)$ is a Stiefel manifold. A particular case of this situation is when $d = n-1$ and the resulting quotient set is a unit sphere $\mathbb{S}^{n-1}$ in $\mathbb{R}^n$.

A further extension of this idea is the quotient set $O(n)/(O(d) \times O(n-d))$ which is called a **Grassmann manifold** . Here, the subgroup $O(d) \times O(n-d)$ is embedded in $O(n)$ according to:

$$\left\{ \begin{bmatrix} O_1 & | & \\ -- & - & - \\ & | & O_2 \end{bmatrix} | O_1 \in O(d), O_2 \in O(n-d) \right\}$$

and the quotient set is computed as earlier. Each element of the quotient set represents a d-dimensional subspace in $\mathbb{R}^n$ and $O(n)/(O(d) \times O(n-d))$ is the set of all d-dimensional subspaces of $\mathbb{R}^n$.

A.3 Basic Geometry of Function Spaces

We start by introducing some basic notation. A set V is called a **vector space** if for all $v_1, v_2 \in V$ and $\alpha_1, \alpha_2 \in \mathbb{R}$, we have $\alpha_1 v_1 + \alpha_2 v_2 \in V$. A vector space V is called **normed** if there exists a function $\|\cdot\| : V \to \mathbb{R}$ such that: (i) $\|v\| \geq 0$ for all $v \in V$, (ii) $\|\alpha v\| = |\alpha| \|v\|$ for all $\alpha \in \mathbb{R}$, $v \in V$, and (iii) $\|v_1 + v_2\| \leq \|v_1\| + \|v_2\|$ for all $v_1, v_2 \in V$. Given a normed vector space, we can make it into a metric space by defining the distance function $d(v, w) = \|v - w\|$. A normed vector space V is defined to be **complete** if every Cauchy sequence in V converges to a limit in V.

Definition A.12 (Banach Space). A **Banach space** is a complete, normed vector space.

Example A.10. 1. $\mathbb{L}^p$ **Spaces**: Let $\mathscr{F}$ be the set of all measurable functions defined on the interval $[0, 1]$. For an $f \in \mathscr{F}$, define the $\mathbb{L}^p$-norm as:

$$\|f\|_p = \left(\int_0^1 |f(x)|^p dx \right)^{1/p} . \tag{A.1}$$

With the $\mathbb{L}^p$ norm, we can define the $\mathbb{L}^p([0, 1], \mathbb{R})$ space:

$$\mathbb{L}^p([0, 1], \mathbb{R}) = \{f : [0, 1] \mapsto \mathbb{R} \| \|f\|_p < \infty\} .$$

$\mathbb{L}^p([0, 1], \mathbb{R})$ is a Banach space for all $p \geq 1$ (see for example [72]).

2. **Products of Banach Spaces**: If V and W are two Banach spaces with the norms $\|\cdot\|_1$ and $\|\cdot\|_2$, then the product space $V \times W$ is also a Banach space with the norm $\|(v, w)\| \equiv \|v\|_1 + \|w\|_2$. (There are other ways of defining norm on the product space, although we have listed one as an example.)

If V and W are Banach spaces and $T : V \to W$ is a linear transformation, then T is defined to be **bounded** if there exists a real number B such that $\|T(v)\| \leq B\|v\|$ for all $v \in V$. We denote by $L(V, W)$ the set of all bounded linear transformations from V to W. (It's easy to prove that a linear transformation $V \to W$ is bounded if and only if it is continuous.) We now observe that $L(V, W)$ is itself a Banach space. The norm is defined by $\|T\| = B$, where B is the smallest real number with the property that $\|T(v)\| \leq B\|v\|$ for all $v \in V$.

Suppose V and W are Banach spaces, $U \subset V$ is an open subset, and $f : U \to W$ is any function. We will define the **derivative** of f at a point $v \in U$, when it exists, to be the bounded linear transformation $A : V \to W$ which best approximates f near v. More precisely:

Definition A.13. Suppose V and W are Banach spaces, $U \subset V$ is an open subset, $f : U \to W$ is a function, and $v \in U$. If there is a bounded linear transformation $A : V \to W$ satisfying the equation

$$\lim_{x \to 0} \frac{f(v + x) - f(v) - A(x)}{\|x\|} = 0,$$

then we say that f is **differentiable** at v, and define A to be the **derivative** of f at v. In this case, we write $df_v = A$.

We say that f is **differentiable** on the entire open subset $U \subset V$ if it is differentiable at every $v \in U$. Note that if f is differentiable on U, then the assignment $v \mapsto df_v$ defines a function $f' : U \to L(V, W)$. Since $L(V, W)$ is a Banach space, we can ask if the function f' is differentiable at a point $v \in U$. If it is, we call this derivative the **second derivative** of f at v. Note that the second derivative of f at v is a bounded linear transformation $V \to L(V, W)$. We can continue this process in the obvious way: If f' is differentiable on all of U, then we obtain a function $f'' : U \to L(V, L(V, W))$. If this process can be continued indefinitely, i.e., if f has an n-th derivative $f^{(n)}$ for all n, then we say that f is **smooth**. If $U \subset V$ and $X \subset W$ are both open sets, a function $f : U \to X$ is defined to be a **smooth isomorphism** if f is smooth and bijective and f^{-1} is also smooth.

A.3.1 Hilbert Manifolds and Submanifolds

Definition A.14 (Hilbert Space). A **Hilbert space** is a Banach space in which the norm is defined in terms of an inner product, $\| \cdot \| = \sqrt{\langle \cdot, \cdot \rangle}$.

An interesting instance of this is the space $\mathbb{L}^2([0, 1], \mathbb{R})$. This space has the following inner product: for $f_1, f_2 \in \mathbb{L}^2([0, 1], \mathbb{R})$,

$$\langle f_1, f_2 \rangle = \int_0^1 f_1(x) f_2(x) dx . \tag{A.2}$$

$\mathbb{L}^2([0, 1], \mathbb{R})$ is a complete space under this norm, i.e. it is a Hilbert space. Shape analysis of continuous curves and surfaces will invariably involve representing them using functions on different domains. Although the eventual spaces representing shapes will become more restricted and specific, as shape-related constraints are applied, the starting point in many discussions will be the $\mathbb{L}^2$ spaces. These are infinite-dimensional vector spaces with several choices of complete orthonormal bases. For example, one can use the Fourier series to decompose any $f \in \mathbb{L}^2([0, 1], \mathbb{R})$ into components according to:

$$f(x) = a_0 + \sum_{i=1}^{\infty}(a_i \cos(2\pi ix) + b_i \sin(2\pi ix)) , \quad x \in [0, 1] ,$$

where $a_i = \langle f, \cos(2\pi i \cdot) \rangle$ and $b_i = \langle f, \sin(2\pi i \cdot) \rangle$, with the inner product given in Eq. A.2. The convergence in the equation above is with respect to the $\mathbb{L}^2$ metric. If we restrict to the first m basis functions of each kind, sines and cosines, we obtain a subspace V_m of $\mathbb{L}^2$:

$$V_m = 1, \operatorname{span}\{(\cos(2\pi ix), \sin(2\pi ix)) | i = 1, 2, \ldots, m - 1\} .$$

V_m is a finite-dimensional space that allows a more traditional multivariate calculus and statistics for analysis, as opposed to $\mathbb{L}^2([0, 1], \mathbb{R})$ whose infinite dimensionality is a major obstacle. In the later chapters, involving statistical analysis of shapes, we will regularly restrict to such finite-dimensional subspaces of $\mathbb{L}^2([0, 1], \mathbb{R})$ for the purposes of statistical analysis. We can write the larger Hilbert space as the direct sum: $\mathbb{L}^2([0, 1], \mathbb{R}) = V_m \oplus V_m^{\perp}$, and the projection from $\mathbb{L}^2([0, 1], \mathbb{R})$ to V_m will be used for approximating functions. This example used the Fourier basis of $\mathbb{L}^2$ but other bases may perform better in different applications. In particular, the use of principal component analysis (PCA) to find empirical orthogonal functions is quite common in functional statistics.

We will need to take the derivatives of functionals on $\mathbb{L}^2$ spaces. Let $E : \mathbb{L}^2([0,1], \mathbb{R}) \to \mathbb{R}$ be a real-valued functional defined on square-integrable functions on $[0,1]$. At a point $f \in \mathbb{L}^2([0,1], \mathbb{R})$, the directional derivative of E, in the direction $g \in \mathbb{L}^2([0,1], \mathbb{R})$, is given by:

$$\nabla E[g] = \lim_{t \to 0} \frac{1}{t}(E[f + tg] - E[f]), \quad t \in \mathbb{R} . \tag{A.3}$$

The derivative of a function $E : \mathbb{L}^2([0,1], \mathbb{R}) \to \mathbb{R}$ is the same derivative defined above in Definition A.14 for functions between Banach spaces, since $\mathbb{L}^2([0,1], \mathbb{R})$ and $\mathbb{R}$ are both Banach spaces.

In this textbook, we are greatly interested in manifolds and submanifolds formed by constraining functions and forming submanifolds of $\mathbb{L}^2$. These are infinite-dimensional manifolds and require a slightly different introduction than the finite-dimensional case treated in Appendix A. We first define the notion of a smooth atlas and then use that to define a smooth manifold modeled on an infinite-dimensional space. Most of the material in this introduction is taken from Lang [63] and we encourage the reader to consult that book for more details.

Definition A.15 (Smooth Atlas). Let X be a topological space. A **smooth atlas** on X is a collection of pairs (U_i, ϕ_i) satisfying the following conditions:

1. Each U_i is an open subset of X and the U_i's cover X.
2. Each ϕ_i is a homeomorphism of U_i onto the open set $\phi_i(U_i)$ of some Banach space E_i, and for any i, j, the set $\phi_i(U_i \cap U_j)$ is open in E_i.
3. The map:
$$\phi_j \circ \phi_i^{-1} : \phi_i(U_i \cap U_j) \to \phi_j(U_i \cap U_j)$$

is a smooth isomorphism for each pair i, j.

Definition A.16. A new chart (U, ϕ) is called **compatible** with a given smooth atlas $\{(U_i, \phi_i)\}$ if the map:

$$\phi_i \circ \phi^{-1} : \phi(U \cap U_i) \to \phi_i(U \cap U_i) ,$$

is a smooth isomorphism for all i.

Two smooth atlases are called **compatible** if every chart in one is compatible with the other atlas. This notion of compatibility can be used to define an equivalence relation between smooth atlases. Any two atlases are defined to be equivalent if they are compatible. An equivalence class of smooth atlases defines a structure on the space X that makes it a smooth manifold.

Definition A.17 (Smooth Manifold). A topological space X with a choice of an equivalence class of smooth atlases is called a **smooth manifold**.

If E used in the definition of the atlas is a Banach space, as we have assumed, then X is called a **smooth manifold modeled on a Banach space** or a **Banach manifold**. Similarly, if E is a Hilbert space, then X is called a **Hilbert manifold**. While comparing to the similar definition for the finite-dimensional case (Definition A.4), the readers may notice that there is basically only one difference. In the previous cases E was $\mathbb{R}^n$ but now E is an infinite-dimensional Banach or Hilbert space.

Definition A.18 (Submanifold). Let X be a smooth manifold. A subset $N \subset X$ is a **submanifold** if for every $p \in N$, there is a coordinate chart (U, ϕ) for X, with $p \in U$, such that ϕ is a smooth isomorphism from U to $V_1 \times V_2$, where V_1 and V_2 are open subsets of Banach spaces E_1 and E_2 (respectively) and where $\phi(N \cap U) = V_1 \times \{a_2\}$ for some point $a_2 \in V_2$.

This definition is similar to Definition A.1.2 for the finite-dimensional manifolds. Although this provides a way of defining a submanifold, we realize submanifolds in practice using the inverse images of smooth maps; recall Theorem A.1 for the finite-dimensional case. We will state a similar result for the general submanifolds.

Let X and Y be two smooth Banach manifolds and let $f : X \to Y$ be a smooth mapping. For a point $y \in Y$, consider the pullback set $f^{-1}(y) \subset X$. The map f is called **transversal** over y if the differential of f at every point of $f^{-1}(y)$ is onto. (For finite-dimensional cases, we used the term **regular value** for such a g.) This leads to a more useful characterization of a submanifold of X.

Theorem A.2. *If f is transversal over y, then $f^{-1}(y)$ is a submanifold of X.*

Let us take some examples of submanifolds of $\mathbb{L}^2$.

Example A.11. **Infinite-Dimensional Sphere**: $\mathbb{L}^2([0,1])$ and $\mathbb{R}$ are Hilbert spaces, and a function $f : \mathbb{L}^2([0,1]) \to \mathbb{R}$ defined by $f(g) = \langle g, g \rangle = \|g\|$ is a smooth map. Its derivative is given by $f'(g)h = 2\langle g, h \rangle$ and for any $g \neq 0$, this derivative is onto. Therefore, f is a transversal function over $1 \in \mathbb{R}$ and the pullback set of $\{1\}$,

$$\mathbb{S}_\infty \equiv \{g \in \mathbb{L}^2([0,1]) | \|g\| = 1\} \tag{A.4}$$

is a submanifold of $\mathbb{L}^2$. We will call this set an **infinite-dimensional sphere** or a **hypersphere**. Geometrically, it is rather similar to its finite-dimensional counterpart $\mathbb{S}^n$ that we have seen earlier. The center of this sphere is given by the zero function. For any $g \in \mathbb{S}_\infty$, the tangent space $T_g(\mathbb{S}_\infty)$ is given by:

$$T_g(\mathbb{S}_\infty) = \{h \in \mathbb{L}^2([0,1]) | \langle g, h \rangle = 0\} .$$

(See Example A.2 for a derivation.) It is a Riemannian manifold with the $\mathbb{L}^2$ inner product on the tangent spaces.

Appendix B
The Dynamic Programming Algorithm

As described in earlier chapters, diffeomorphisms of a certain type are synonymous with re-parameterizations of functions and curves. In order to make shape analysis invariant to re-parameterizations, an optimization problem on the space of diffeomorphisms Γ_I needs to be solved. With that motivation, we consider situations where there is a need for finding optimal diffeomorphisms of $[0, 1]$, with the optimality defined using a certain type of cost function. A key requirement for the ensuing solution to be applicable is that the objective function is additive over $[0, 1]$. This is true for most of the scenarios considered in this textbook. Then, there exists an efficient numerical procedure for approximating the solution to such problems. This solution is based on the idea of dynamic programming (DP), a class of algorithms described in great detail in [14], and is summarized here next.

B.1 Theoretical Setup

First we set up a typical optimization problem that we will face. Let $f, g : [0, 1] \to \mathbb{R}$ be two given functions and we want to solve for:

$$\hat{\gamma} = \underset{\gamma \in \Gamma_I}{\text{argmin}} \int_0^1 |f(t) - g(\gamma(t))|^2 dt . \tag{B.1}$$

In the integrand, γ is a function that matches the point $g(\gamma(t))$ with the point $f(t)$, and $\hat{\gamma}$ is the optimal matching function. We can solve a discrete approximation of this problem using DP. As mentioned above, a necessary condition for applying DP to such problems is that the cost function is additive in time t. We will conveniently view γ as a graph from $(0, 0)$ to $(1, 1)$ in $\mathbb{R}^2$ such that the slope of this graph is always strictly between 0 and 90 degrees. We can verify that the cost function in Eqn. B.1 is indeed additive over the graph. To decompose the large problem into several subproblems, define a partial cost function:

$$E(s, t; \gamma) = \int_s^t |f(\tau) - g(\gamma(\tau))|^2 d\tau \tag{B.2}$$

so that our original cost function is simply $E(0, 1; \gamma)$. Our goal is to find an optimal path from $(0, 0)$ to $(1, 1)$ in $\mathbb{R}^2$, corresponding to $(t, \gamma(t))$, that minimizes this cost function.

© Springer-Verlag New York 2016
A. Srivastava, E.P. Klassen, *Functional and Shape Data Analysis*,
Springer Series in Statistics, DOI 10.1007/978-1-4939-4020-2

B.2 Computer Implementation

In order to use a numerical approach, we will replace the domain $[0,1] \times [0,1]$ with a finite grid and restrict our search to that grid. Although it is not necessary, we will simplify the setup by using the uniform partition $G_n = \{0, 1/n, 2/n, \ldots, (n-1)/n, 1\}$ of $[0,1]$ and form a grid $G_n \times G_n$. We will search over the set of all restrictions of γ to this grid; any such restriction is now a piecewise linear path which is never vertical or horizontal (see the left panel of Fig. B.1). The total cost associated with the path is now a sum of the costs associated with its linear segments. On an $n \times n$ grid, there are only a finite number of paths, more so when we impose any additional constraint. However, this number grows exponentially with n, and we cannot possibly search over all possible paths in an exhaustive enumerative fashion. The DP algorithm, however, finds the optimal path in $O(n^2)$ time!

Denote a point on the grid $(i/n, j/n)$ by (i, j). We will impose an additional constraint by bounding the slope of the path at any point. As a result, there are only certain nodes that are allowed to go to (i, j); denote by N_{ij} be the set of nodes that are allowed to go to (i, j). For instance:

$$N_{ij} = \{(k/n, l/n) | 0 \leq k < i, 0 \leq l < j\} .$$

is a valid set. In practice, one often restricts to a smaller subset to seek a computational speed up. The net effect is that the number of possible values for the slope along the path are further restricted (see the middle panel of Fig. B.1). Let $L(k, l; i, j)$ denote a straight line joining the nodes $(k/n, l/n)$ and $(i/n, j/n)$; for $(k/n, l/n) \in N_{ij}$ this is a line with slope strictly between 0 and 90 degrees. This sets up the iterative optimization problem:

$$(\hat{k}/n, \hat{l}/n) = \operatorname*{argmin}_{(k/n, l/n) \in N_{ij}} [E(k/n, l/n; L(k, l; i, j)) + H(k/n, l/n)] , \qquad (B.3)$$

with E as defined in Eq. B.2. Define the minimum energy of reaching the point (i, j), in an iterative fashion as:

$$H(i/n, j/n) = E(\hat{k}/n, \hat{l}/n; L(\hat{k}, \hat{l}; i, j)) + H(\hat{k}/n, \hat{l}/n) , \quad \text{with} \quad H(0,0) = 0 .$$

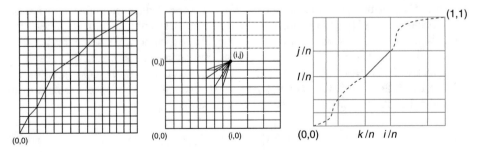

Fig. B.1 *Left*: an example of a γ function restricted to a finite graph. *Right*: an illustration of some nodes that are allowed to go to the $(i/n, j/n)$ point on the graph

This subproblem is solved sequentially for each node (i, j), starting from $(1/n, 1/n)$ and increasing i, j until one reaches the node $(1, 1)$. Tracing the path that results in the optimal energy $H(1, 1)$ provides a discrete version of the optimal γ.

Algorithm 58 (Dynamic Programming Algorithm).

```
E = zeros(n,n); E(1,:) = 1; E(:,1) = 1; E(1,1) = 0;
        for i = 2 : n
            for j = 2 : n
                for Num = 1:size(𝒩,1)
                    k = i - 𝒩(Num,1);
                    l = j - 𝒩(Num,2);
                    if (k> 0 & l > 0)
                        Hc(Num) = H(k,l) + FunctionE(f,g,k,i,l,j);
                    else
                        Hc(Num) = ∞;
                    end
                    H(i,j) = min(Hc);
                end
            end
        end
```

Here $\mathcal{N}$ is a list of sites used to define N_{ij}. Typically, $\mathcal{N}$ is a two-column matrix of type $\{(1, 1); (1, 2); (2, 1); (1, 3); (2, 3); (3, 1); (3, 2); \dots, \}$ depending on the number of preceding neighbors included in the implementation. FunctionE is a subroutine that computes $E(k/n, l/n; L(k, l; i, j))$, the partial cost function specified in Eq. B.2:

Algorithm 59 (FunctionE).

Input: f, g, k, i, l, j, n
Output: E

```
        m = size(g,2);
        x = [k:1:i];
        y = (x-k)*m + l;
        idx = round(y*m/n);
        v = q2(:,idx);
        E = norm(q1(:,x) - v,'fro')²/n;
```

Algorithm 58 stores the values of $(\hat{k}_{(i,j)}, \hat{l}_{(i,j)})$, the optimal incoming node, for each (i, j) and uses it to reconstruct the optimal $\hat{\gamma}$. One starts from (n, n) and goes to $(\hat{k}_{(n,n)}, \hat{l}_{(n,n)})$, the optimal incoming nodes at (n, n). Then, the next step is go to optimal nodes coming into $(\hat{k}_{(n,n)}, \hat{l}_{(n,n)})$ and so on. This piecewise-linear path is $\hat{\gamma}$.

Example B.1. To illustrate the dynamic programming algorithm, we consider the problem of matching two functions:

$$f(x) = e^{\frac{-(x-\mu_1)^2}{2\sigma^2}} \quad \text{and} \quad g(x) = e^{\frac{-(x-\mu_2)^2}{2\sigma^2}},$$

for a certain fixed σ. In Fig. B.2, we show results of using the dynamic programming algorithm for finding the optimal γ that minimizes the cost function given in Eq. B.1. In the first row, we have $\mu_1 = \mu_2$ but the separation is increasing in the

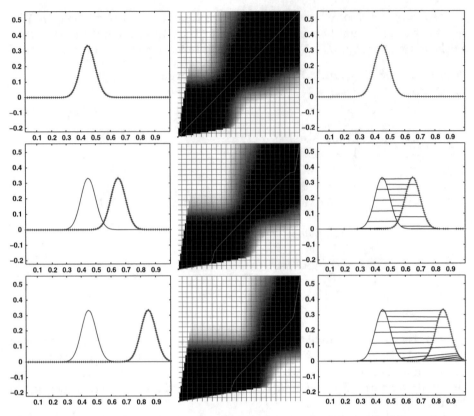

Fig. B.2 Matching of functions using dynamic programming. In each row the left panel shows two function f and g. The middle row shows the optimal $\hat{\gamma}$ that minimizes the cost function in Eq. B.1, drawn over the partial cost function H. The right panel shows the functions f and $g(\hat{\gamma})$ with the resulting correspondences

lower rows. The two original functions for each case are shown in the left panel. The middle panel of each row shows the estimated optimal $\hat{\gamma}$ drawn over the partial cost function H. The darker intensity in the image denotes lower values of H. Notice that the optimal diffeomorphism stays close to the darker regions in the image. Finally, in the right panel of each row, we see the functions $f(t)$ and $g(\hat{\gamma}(t))$ and some line correspondences between the two functions. In case of $\mu_1 = \mu_2$, the optimal γ is naturally identity, but when the two means are pulled further apart, the optimal matching $\hat{\gamma}$ also moves further away from identity to account for this mismatch.

References

1. M.F. Abdelkader, W. Abd-Almageed, A. Srivastava, R. Chellappa, Gesture and action recognition via modeling trajectories on shape manifolds. Comput. Vis. Image Underst. J. **115**(3), 439–455 (2011)
2. A. Abou-Elailah, F. Dufaux, J. Farah, M. Cagnazzo, A. Srivastava, B. Pesquet-Popescu, Fusion of global and local motion estimation using foreground objects for distributed video coding. IEEE Trans. Circ. Syst. Video Technol. **25**(6), 973–987 (2015)
3. M. Adams, T. Ratiu, R. Schmid, The lie group structure of diffeomorphism groups and invertible fourier integral operators, with applications. In: *Infinite-Dimensional Groups with Applications*, ed. by V. Kac (Springer, New York, 1985)
4. S. Amari, *Differential Geometric Methods in Statistics*. Lecture Notes in Statistics, Vol. 28 (Springer, New York, 1985)
5. S. Amari, H. Nagaoka, *Methods of Information Geometry, Mathematical Monographs Volume 191* (Oxford University Press, Oxford, 2000)
6. S.-I. Amari, O.E. Barndorff-Nielsen, R.E. Kass, S.L. Lauritzen, C.R. Rao, *Differential Geometry in Statistical Inference*, Monograph Series (Institute of Mathematical Statistics, Hayward, 1987)
7. Y. Amit, U. Grenander, M. Piccioni, Structural image restoration through deformable templates. J. Am. Stat. Assoc. **86**(414), 376–387 (1991)
8. B. Ben Amor, J. Su, A. Srivastava, Action recognition using rate-invariant analysis of skeletal shape trajectories. IEEE Trans. Pattern Anal. Mach. Intell. **38**(1), 1–13 (2016)
9. K.S. Arun, T.S. Huang, S.D. Blostein, Least-squares fitting of two 3-d point sets. IEEE Trans. Pattern Anal. Mach. Intell. **PAMI-9**(5), 698–700 (1987)
10. M. Bauer, M. Bruveris, P. Harms, J. Mller-Andersen, Second order elastic metrics on the shape space of curves. arXiv preprint, arXiv:1507.08816 (2015)
11. M. Bauer, M. Bruveris, S. Marsland, P.W. Michor, Constructing reparameterization invariant metrics on spaces of plane curves. Differ. Geom. Appl. **34**, 139–165 (2014)
12. M. Bauer, M. Bruveris, P.W. Michor, R-transforms for sobolev h^2-metrics on spaces of plane curves. Geom. Imaging Comput. **1**(1), 1–56 (2014)
13. M. Bauer, M. Eslitzbichler, M. Grasmair, Landmark-gauided elastic shape analysis of human character motions. arXiv, arXiv:1502.07666 (2015)
14. D.P. Bertsekas, *Dynamic Programming and Optimal Control* (Athena Scientific, Belmont, 1995)
15. P.J. Besl, N.D. McKay, A method for registration of 3-D shapes. IEEE TPAMI **14**(2), 239–256 (1992)
16. A. Bhattacharya, On a measure of divergence between two statistical populations defined by their probability distributions. Bull. Calcutta Math. Soc. **35**, 99–109 (1943)
17. A. Bhattacharya, R. Bhattacharya, *Nonparametric Inference on Manifolds: with Applications to Shape Spaces* (Cambridge University Press, Cambridge, 2012)
18. F.L. Bookstein, *Morphometric Tools for Landmark Data: Geometry and Biology* (Cambridge University Press, Cambridge, 1991)
19. W.M. Boothby, *An Introduction to Differential Manifolds and Riemannian Geometry* (Academic Press, New York, 1986)

© Springer-Verlag New York 2016

A. Srivastava, E.P. Klassen, *Functional and Shape Data Analysis*, Springer Series in Statistics, DOI 10.1007/978-1-4939-4020-2

20. M. Bruveris, Optimal reparameterizations in square root velocity framework. arXiv, arXiv:1507.02728 (2015)

21. D. Bryner, E. Klassen, H. Le, A. Srivastava, 2d affine and projective shape analysis. IEEE Trans. Pattern Anal. Mach. Intell. **36**(5), 998–1011 (2014)

22. D. Bryner, A. Srivastava, Q. Huynh, Elastic shape models for improving segmentation of object boundaries in synthetic aperture sonar images. Comput. Vis. Image Underst. **117**(12), 1695–1710 (2013)

23. M.P. Do Carmo, *Differential Geometry of Curves and Surfaces* (Prentice-Hall, Englewood Cliffs, 1976)

24. N.N. Čencov, *Statistical Decision Rules and Optimal Inferences*, volume 53 of *Translations of Mathematical Monographs* (AMS, Providence, 1982)

25. Y. Chen, G. Medioni, Object modeling by registration of multiple range images. Image Vis. Comput. **10**(3), 145–155 (1992)

26. T.F. Cootes, C.J. Taylor, D.H. Cooper, J. Graham, Active shape models: Their training and application. Comput. Vis. Image Underst. **61**(1), 38–59 (1995)

27. H. Drira, B. Ben Amor, A. Srivastava, M. Daoudi, R. Slama, 3d face recognition under expressions, occlusions, and pose variations. IEEE Trans. Pattern Anal. Mach. Intell. **35**(9), 2270–2283 (2013)

28. I.L. Dryden, K.V. Mardia, *Statistical Shape Analysis* (Wiley, London, 1998)

29. D.G. Ebin, J. Marsden, Groups of diffeomorphisms and the motion of an incompressible fluid. Ann. Math. Second Ser. **92**(1), 102–163 (1970)

30. B. Efron, Defining the curvature of a statistical problem (with applications to second order efficiency). Ann. Stat. **3**, 1189–1242 (1975)

31. J.K. Ghosh, R.V. Ramamoorthi, *Bayesian Nonparametrics*. Springer Series in Statistics (Springer, New York, 2003)

32. U. Grenander, *General Pattern Theory* (Oxford University Press, Oxford, 1993)

33. U. Grenander, M.I. Miller, Computational anatomy: An emerging discipline. Q. Appl. Math. **LVI**(4), 617–694 (1998)

34. U. Grenander, M.I. Miller, *Pattern Theory: From Representation to Inference* (Oxford University Press, Oxford, 2007)

35. U. Grenander, M.I. Miller, A. Srivastava, Hilbert-schmidt bounds on matrix lie groups for atr. IEEE Trans. Pattern Anal. Mach. Intell. **20**(8), 790–800 (1998)

36. S. Helgason, *Differential Geometry, Lie Groups and Symmetric Spaces* (Academic Press, New York, 1978)

37. T. Hofmann, J.M. Buhmann, Pairwise data clustering by deterministic annealing. IEEE Trans. Pattern Anal. Mach. Intell. **19**(1), 1–14 (1997)

38. L. Horvath, P. Kkozska, *Inference for Functional Data with Applications*. Springer Series in Statistics (Springer, New York, 2012)

39. W. Huang, K.A. Gallivan, A. Srivastava, P.-A. Absil, Riemannian optimization for registration of curves in elastic shape analysis. J. Math. Imag. Vis. **54**(3), 320–343 (2016)

40. A.K. Jain, R.C. Dubes, *Algorithms for Clustering Data* (Prentice-Hall, Englewood Cliffs, 1988)

41. I.T. Jolliffe, *Principal Component Analysis* Springer Series in Statistics (Springer, New York, 2002)

42. S.H. Joshi, E. Klassen, A. Srivastava, I.H. Jermyn, A novel representation for efficient computation of geodesics between n-dimensional curves. In: *IEEE CVPR*, 2007

43. S.H. Joshi, E. Klassen, A. Srivastava, I.H. Jermyn, Removing shape-preserving transformations in square-root elastic (SRE) framework for shape analysis of curves. In: *EMMCVPR, LNCS 4679*, ed. by A. Yuille et al., pp. 387–398 (2007)

44. P.E. Jupp, J.T. Kent, Fitting smooth paths to spherical data. J. R. Stat. Soc. Ser. C (Appl. Stat.) **36**(1), 34–46 (1987)

45. V.G. Kac, *Infinite-Dimensional Lie Algebras*, 3rd edn. (Cambridge University Press, Cambridge, 1990)

46. H. Karcher, Riemann center of mass and mollifier smoothing. Commun. Pure Appl. Math. **30**, 509–541 (1977)

47. R.E. Kass, P.W. Vos, *Geometric Foundations of Asymptotic Inference* (Wiley, London, 1997)

48. D. Kaziska, A. Srivastava, The karcher mean of a class of symmetric distributions on a unit circle. Stat. Probab. Lett. **78**, 1314–1316 (2008)

49. D.G. Kendall, D. Barden, T.K. Carne, H. Le, *Shape and Shape Theory* (Wiley, London, 1999)

50. D.G. Kendall, Shape manifolds, procrustean metrics and complex projective spaces. Bull. Lond. Math. Soc. **16**, 81–121 (1984)

51. J. Kent, K. Mardia, Procrustes methods for projective shape. Syst. Biol. Stat. Bioinf. 37–40 (2007)

52. J. Kent, K. Mardia, A geometric approach to projective shape and the cross ratio. Biometrika **99**(4), 833–849 (2012)

53. E. Klassen, A. Srivastava, Geodesics between 3d closed curves using path-straightening. In: *European Conference on Computer Vision, LNCS 3951*, ed. by A. Leonardia, H. Bischof, A. Pinz (2006)

54. E. Klassen, A. Srivastava, W. Mio, S. Joshi, Analysis of planar shapes using geodesic paths on shape spaces. IEEE Pattern Anal. Mach. Intell. **26**(3), 372–383 (2004)

55. A. Kneip, T. Gasser, Statistical tools to analyze data representing a sample of curves. Ann. Stat. **20**, 1266–1305 (1992)

56. A. Kume, I.L. Dryden, H. Le, Shape-space smoothing splines for planar landmark data. Biometrika **94**, 513–528 (2007)

57. S. Kurtek, A. Srivastava, E. Klassen, Z. Ding, Statistical modeling of curves using shapes and related features. J. Am. Stat. Assoc. **107**(499), 1152–1165 (2012)

58. S. Kurtek, J. Su, C. Grimm, M. Vaughan, R.T. Sowell, A. Srivastava, Statistical analysis of manual segmentations of structures in medical images. Comput. Vis. Image Underst. **117**(9), 1036–1050 (2013)

59. S. Kurtek, Q. Xie, A. Srivastava, Analysis of juggling data: Alignment, extraction, and modeling of juggling cycles. Electron. J. Stat. **8**, 1865–1873 (2014)

60. J. Laborde, D. Robinson, A. Srivastava, E. Klassen, J. Zhang, Rna alignment in the joint sequence-structure space using elastic shape analysis. J. Nucleic Acids Res. **41**(11, e114) (2013)

61. H. Laga, S. Kurtek, A. Srivastava, S.J. Miklavcic, Landmark-free statistical analysis of the shape of plant leaves. J. Theor. Biol. **363**, 41–52 (2014)

62. S. Lahiri, D. Robinson, E. Klassen, Precise matching of PL curves in R^N in square root velocity framework. Geom. Imaging Comput. **2**(3), 133–186 (2015)

63. S. Lang, *Differential and Riemannian Manifolds, Third Edition* (Springer: Graduate Texts in Mathematics, New York, 1995)

64. S. Lang, *Fundamentals of Differential Geometry* (Springer, New York, 1999)

65. S. Lang, *Algebra* (Springer, New York, 2002)

66. H. Le, D.G. Kendall, The Riemannian structure of euclidean shape spaces: A novel environment for statistics. Ann. Stat. **21**(3), 1225–1271 (1993)

67. X. Leng, H.G. Mueller, Time ordering of gene coexpression. Biostatistics **7**(4), 569–584 (2006)

68. W. Liu, A Riemannian framework for annotated curve analysis. PhD thesis, Florida State University, August 2011

69. W. Liu, A. Srivastava, E. Klassen, Joint shape and texture analysis of objects boundaries in images using a Riemannian approach. In: *Asilomar Conference on Signals, Systems, and Computers*, October, 2008

70. W. Liu, A. Srivastava, J. Zheng, A mathematical framework for protein structure comparison. PLOS Comput. Biol. **7**(2), 1–10 (2011)

71. X. Liu, H.G. Mueller, Functional convex averaging and synchronization for time-warped random curves. J. Am. Stat. Assoc. **99**, 687–699 (2004)

72. D.G. Luenberger, *Optimization by Vector Space Methods* (Wiley, New York, 1969)

73. M. Mani, S. Kurtek, A. Srivastava, C. Barillot, A comprehensive riemannian framework for analysis of white matter fiber tracts. In: *Proc. of International Symposium on Biomedical Imaging (ISBI)*, 2010

74. K. Mardia, J. Kent, A new representation for projective shape. In: *Proceedings in Interdisciplinary Statistics and Bioinformatics*, pp. 75–78 (2006)

75. K.V. Mardia, P. Jupp, *Directional Statistics (2nd edition)* (Wiley, London, 2000)

76. J.S. Marron, J.O. Ramsay, L.M. Sangalli, A. Srivastava, Statistics of time warpings and phase variations. Electron. J. Stat. **8**(2), 1697–1702 (2014)

77. P.W. Michor, D. Mumford, Riemannian geometries on spaces of plane curves. J. Eur. Math. Soc. **8**, 1–48 (2006)

78. J.W. Milnor, *Topology from the Differentiable Viewpoint* (Princeton University Press, Princeton, 1997)

79. W. Mio, A. Srivastava, S. Joshi, On shape of plane elastic curves. Int. J. Comput. Vis. **73**(3), 307–324 (2007)

80. F. Mokhtarian, S. Abbasi, J. Kittler, Efficient and robust shape retrieval by shape content through curvature scale space. In: *Proceedings of First International Conference on Image Database and MultiSearch*, 1996

81. A. Mottini, Axon Morphology Analysis: From Image Processing to Modelling. PhD thesis, University of Nice Sophia Antipolis, 2014

82. A. Mottini, X. Descombes, F. Besse, From curves to trees: A tree-like shapes distance using the elastic shape analysis framework. Neuroinformatics **13**(2), 175–191 (2015)

83. J. Munkres, *Topology* (Prentice-Hall, Englewood Cliffs, 2000)

84. S. Osher, R. Fedkiw, *Level Set Methods and Dynamic Implicit Surfaces* (Springer, New York, 2003)

85. S.J. Osher, R.P. Fedkiw, *Level Set Methods and Dynamic Implicit Surfaces* (Springer, New York, 2000)

86. J.C. Owen, F.R. Moore, Swainson's thrushes in migratory disposition exhibit reduced immune function. J. Ethol. **26**, 383–388 (2008)

87. R.S. Palais, Morse theory on Hilbert manifolds. Topology **2**, 299–340 (1963)

88. V. Patrangenaru, R. Bhattacharya, Large sample theory of intrinsic and extrinsic sample means on manifolds. Ann. Stat. **31**(1), 1–29 (2003)

89. B. Pelletier, Non-parametric regression estimation on closed riemannian manifolds. Nonparametric Stat. **18**(1), 57–67 (2006)

90. J.O. Ramsay, X. Li, Curve registration. J. R. Stat. Soc. Ser. B **60**, 351–363 (1998)

91. J.O. Ramsay, B.W. Silverman, *Functional Data Analysis, Second Edition*. Springer Series in Statistics (Springer, New York, 2005)

92. C.R. Rao, Information and accuracy attainable in the estimation of statistical parameters. Bull. Calcutta Math. Soc. **37**, 81–91 (1945)

93. C.P. Robert, G. Casella, *Monte Carlo Statistical Methods*. Springer Text in Statistics (Springer, New York, 1999)

94. D. Robinson, Functional Analysis and Partial Matching in the Square Root Velocity Framework. PhD thesis, Florida State University, August 2012

95. K. Rose, Deterministic annealing for clustering, compression, classification, regression, and related optimization problems. Proc. IEEE **86**(11), 2210–2239 (1998)

96. W. Rudin, *Functional Analysis, 2nd Edition* (McGraw-Hill Higher Education, New York, 1991)

97. C. Samir, A. Srivastava, M. Daoudi, Three-dimensional face recognition using shapes of facial curves. IEEE Trans. Pattern Anal. Mach. Intell. **28**(11), 1858–1863 (2006)

98. C. Samir, A. Srivastava, M. Daoudi, E. Klassen, An intrinsic framework for analysis of facial surfaces. Int. J. Comput. Vis. **82**(1), 80–95 (2009)

99. C. Samir, A. Srivastava, M. Daoudi, S. Kurtek, On analyzing symmetry of objects using elastic deformations. In: *4th International Conference on Computer Vision Theory and Applications*, pp. 194–200, 2009

100. T.B. Sebastian, P.N. Klein, B.B. Kimia, On aligning curves. IEEE Trans. Pattern Anal. Mach. Intell. **25**(1), 116–125 (2003)

101. B.W. Silverman, *Density Estimation for Statistics and Data Analysis* (Chapman and Hall, London, 1985)

102. C.G. Small, *The Statistical Theory of Shape* (Springer, New York, 1996)

103. A. Srivastava, A bayesian approach to geometric subspace estimation. IEEE Trans. Signal Process. **48**(5), 1390–1400 (2000)

104. A. Srivastava, I.H. Jermyn, Looking for shapes in two-dimensional, cluttered point cloud. IEEE Trans. Pattern Anal. Mach. Intell. **31**(9), 1616–1629 (2009)

105. A. Srivastava, S. Joshi, W. Mio, X. Liu, Statistical shape analysis: Clustering, learning, and testing. IEEE Trans. Pattern Anal. Mach. Intell. **27**(4), 590–602 (2005)

106. A. Srivastava, E. Klassen, S.H. Joshi, I.H. Jermyn, Shape analysis of elastic curves in Euclidean spaces. IEEE Trans. PAMI **33**, 1415–1428 (2011)

107. A. Srivastava, C. Samir, S.H. Joshi, M. Daoudi, Elastic shape models for face analysis using curvilinear coordinates. J. Math. Imaging Vis. **33**(2), 253–265 (2009)

108. A. Srivastava, W. Wu, S. Kurtek, E. Klassen, J.S. Marron, Registration of functional data using fisher-rao metric. arXiv, arXiv:1103.3817 (2011)

109. J. Su, I.L. Dryden, E. Klassen, H. Le, A. Srivastava, Fitting optimal curves to time-indexed, noisy observations on nonlinear manifolds. J. Image Vis. Comput. **30**(6–7), 428–442 (2012)

110. J. Su, F. Huffer, A. Srivastava, Detection, classification and estimation of shapes in 2d and 3d point clouds. Comput. Stat. Data Anal. **58**, 227–241 (2013)

111. J. Su, S. Kurtek, E. Klassen, A. Srivastava, Statistical analysis of trajectories on riemannian manifolds: Bird migration, hurricane tracking, and video surveillance. Ann. Appl. Stat. **8**(1), 530–552 (2014)

112. G. Sundaramoorthi, A. Mennucci, S. Soatto, A.J. Yezzi, A new geometric metric in the space of curves, and applications to tracking deforming objects by prediction and filtering. SIAM J. Imaging Sci. **4**(1), 109–145 (2011)

113. R. Tang, H.G. Mueller, Pairwise curve synchronization for functional data. Biometrika **95**(4), 875–889 (2008)

114. D.W. Thompson, *On Growth and Form: The Complete Revised Edition* (Dover, Cambridge, 1992)

115. A. Trouve, Diffemorphisms groups and pattern matching in image analysis. Int. J. Comput. Vis. **28**(3), 213–221 (1998)

116. D. Tucker, W. Wu, A. Srivastava, Analysis of proteomics data: Phase amplitude separation using an extended fisher-rao metric. Electron. J. Stat. **8**, 1724–1733 (2014)

117. J.D. Tucker, W. Wu, A. Srivastava, Generative models for functional data using phase and amplitude separation. Comput. Stat. Data Anal. **61**, 50–66 (2013)

118. A. Veeraraghavan, A. Srivastava, A.K. Roy-Chowdhury, R. Chellappa, Rate-invariant recognition of humans and their activities. IEEE Trans. Image Process. **8**(6), 1326–1339 (2009)

119. P.W. Vosm, R.E. Kass, *Geometrical Foundations of Asymptotic Inference* (Wiley-Interscience, New York, 1997)

120. F.W. Warner, *Foundations of Differentiable Manifolds and Lie Groups* (Springer, New York, 1994)

121. L.C. White, Shape Analysis of Curves in Higher Dimensions. PhD thesis, Florida State University, April 2013

122. W. Wu, A. Srivastava, Analysis of spike train data: Alignment and comparisons using extended the fisher-rao metric. Electron. J. Stat. **8**(2), 1786–1792 (2014)

123. Q. Xie, S. Kurtek, A. Srivastava, Analysis of aneurisk65 data: Elastic shape registration of curves. Electron. J. Stat. **8**, 1920–1929 (2014)

124. L. Younes, Computable elastic distance between shapes. SIAM J. Appl. Math. **58**, 565–586 (1998)

125. L. Younes, Optimal matching between shapes via elastic deformations. J. Image Vis. Comput. **17**(5/6), 381–389 (1999)

126. L. Younes, P.W. Michor, J. Shah, D. Mumford, R. Lincei, A metric on shape space with explicit geodesics. Matematica E Applicazioni **19**(1), 25–57 (2008)

127. Z. Zhang, D. Pati, A. Srivastava, Bayesian clustering of shapes. J. Stat. Plann. Inference **166**, 171–186 (2015)

128. Z. Zhang, J. Su, E. Klassen, H. Le, A. Srivastava, Video-based action recognition using rate-invariant analysis of covariancetrajectories. arXiv, arXiv:1503.06699 (2015)

Index

© Springer-Verlag New York 2016
A. Srivastava, E.P. Klassen, *Functional and Shape Data Analysis*,
Springer Series in Statistics, DOI 10.1007/978-1-4939-4020-2

Printed in the United States
By Bookmasters